W0232282

Springer Series in
CLUSTER PHYSICS

Springer-Verlag Berlin Heidelberg GmbH

Springer Series in
CLUSTER PHYSICS

Series Editors:
A. W. Castleman, Jr.   R. S. Berry   H. Haberland   J. Jortner   T. Kondow

The intent of the Springer Series in Cluster Physics is to provide systematic information on developments in this rapidly expanding field of physics. In comprehensive books prepared by leading scholars, the current state-of-the-art in theory and experiment in cluster physics is presented.

Mesoscopic Materials and Clusters
Their Physical and Chemical Properties
Editors: T. Arai, K. Mihama, K. Yamamoto and S. Sugano

Cluster Beam Synthesis of Nanostructured Materials
By P. Milani and S. Iannotta

Theory of Atomic and Molecular Clusters
Editor: J. Jellinek

Dynamics of Clusters and Thin Films on Crystal Surfaces
By G. Rosenfeld

T. Arai K. Mihama K. Yamamoto
S. Sugano (Eds.)

Mesoscopic Materials and Clusters

Their Physical and Chemical Properties

With 394 Figures and 30 Tables

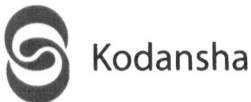

Kodansha

Springer

Toshihiro Arai
University of Tsukuba, Tsukuba-shi
Ibaraki 305-0006, Japan

Kazuhiro Mihama
Department of Applied Electronics
Daido Institute of Technology
Minami-ku, Nagoya 457-0811, Japan

Keiichi Yamamoto
Department of Electrical
and Electronics Engineering
Faculty of Engineering, Kobe University
Nada-ku, Kobe 657-0013, Japan

Satoru Sugano
2471-2849 Ageya, Nagano 380-0888, Japan

Series Editors:

Professor A. W. Castleman, Jr.
(*Editor-in-Chief*)
Department of Chemistry
The Pennsylvania State University
152 Davey Laboratory
University Park, PA 16802, USA

Professor R. Stephen Berry
Department of Chemistry
The University of Chicago
5735 South Ellis Avenue
Chicago, IL 60637, USA

Professor Dr. Hellmut Haberland
Albert-Ludwigs-Universität Freiburg
Fakultät für Physik
Hermann-Herder-Strasse 3
D-79104 Freiburg, Germany

Professor Dr. Joshua Jortner
School of Chemistry, Tel Aviv University
Raymond and Beverly Sackler
Faculty of Sciences
Ramat Aviv, Tel Aviv 69978, Israel

Dr. Tamotsu Kondow
Toyota Technological Institute
Cluster Research Laboratory
East Tokyo Laboratory
Genesis Research Institute Inc.
Futamata 717-86
Ichikawa, Chiba 272-0001, Japan

ISSN 1437-0395
ISBN 978-3-662-08676-6

Library of Congress Cataloging-in-Publication Data applied for
Die Deutsche Bibliothek – CIP Einheitsaufnahme
Mesoscopic materials and clusters: their physical and chemical properties; with 30 tables / T. Arai ... (ed.).
(Springer series in cluster physics)
ISBN 978-3-662-08676-6   ISBN 978-3-662-08674-2 (eBook)
DOI 10.1007/978-3-662-08674-2

The use of general descriptive names, registered names, trademarks, etc. in this publication does not imply, even in the absence of a specific statement, that such names are exempt from the relevant protective laws and regulations and therefore free for general use.

Cover concept: eStudio Calamar Steinen
Cover production: *design & production* GmbH, Heidelberg
SPIN: 10690831   57/3144 - 5 4 3 2 1 0

List of Contributors

Numbers in parentheses refer to the chapters.

Achiba, Y. (38) Department of Chemistry, Tokyo Metropolitan University, Hachioji-shi, Tokyo 192-0397, Japan

Akagi, K. (36) Department of Physics, Graduate School of Science, University of Tokyo, Bunkyo-ku, Tokyo 113-0033, Japan

Amemiya, K. (17) Department of Physics, Faculty of Science and Technology, Keio University, Kohoku-ku, Yokohama 223-8522, Japan

Aoki, S. (41) Electron Optics Division, JEOL Ltd., Akishima-shi, Tokyo 196-0021, Japan

Arai, T. (3) University of Tsukuba, Tsukuba-shi, Ibaraki 305-0006, Japan / present address: Department of Physical Engineering, Ishinomaki Senshu University, Ishinomaki-shi, Miyagi 986-0031, Japan

Arita, M. (27) Department of Electronics and Information Engineering, Faculty of Engineering, Hokkaido University, Kita-ku, Sapporo 060-0813, Japan

Edamatsu, K. (4) Department of Applied Physics, Tohoku University, Aoba-ku, Sendai 980-8579, Japan

Egami, A. (21) Institute of Physics, University of Tsukuba, Tsukuba-shi, Ibaraki 305-0006, Japan

Ekimov, A. (4) A. F. Ioffe Physico-Technical Institute, 194021 St. Petersburg, Russia

Endo, H. (19) Faculty of Engineering, Fukui Institute of Technology, 3-6-1 Gakuen, Fukui 910-0028, Japan

Eto, M. (17) Department of Physics, Faculty of Science and Technology, Keio University, Kohoku-ku, Yokohama 223-8522, Japan

Ezaki, H. (2) Department of Applied Physics, University of Tokyo, Bunkyo-ku, Tokyo 113-0033, Japan / present address: Faculty of Engineering, Tokyo Institute of Polytechnics, Atsugi-shi, Kanagawa 243-0213, Japan

Fujima, N. (24) Faculty of Engineering, Shizuoka University, Hamamatsu-shi, Shizuoka 432-8561, Japan

Fukushima, M. (32) (33) Department of Basic Science, Ishinomaki Senshu University, Ishinomaki-shi, Miyagi 986-0031, Japan

Fukutani, H. (21) Institute of Physics, University of Tsukuba, Tsukuba-shi, Ibaraki 305-0006, Japan

Furumiya, M. (4) Department of Physics, Tohoku University, Aoba-ku, Sendai 980-8578, Japan

Goto, H. (11) Department of Physics, Faculty of Science, University of Tokyo, Bunkyo-ku, Tokyo 113-0033, Japan

Goto, T. (6) Department of Physics, Graduate School of Science, Tohoku University, Aoba-ku, Sendai 980-8578, Japan

Gourdon, C. (4) Groupe de Physique des Solides, Université Paris 6&7-CNRS, 2 place Jussieu, 75251 Paris, France

Gunji, S. (37) Institute of Physics, University of Tsukuba, Tsukuba-shi, Ibaraki 305-0006, Japan

Hanamura, E. (2) Department of Applied Physics, University of Tokyo, Bunkyo-ku, Tokyo 113-0033, Japan

Haus, Joseph W. (18) The Department of Physics, Rensselaer Polytechnic Institute, Troy, NY 12180-3590, USA

Hayashi, S. (7) Department of Electrical and Electronics Engineering, Faculty of Engineering, Kobe University, Nada-ku, Kobe 657-0013, Japan

Honma, I. (18) Electrotechnical Laboratory (ETL), Tsukuba-shi, Ibaraki 305-0045, Japan

Horie, C. (33) Department of Basic Science, Ishinomaki Senshu University, Ishinomaki-shi, Miyagi 986-0031, Japan

Hoshino, K. (23) Faculty of Integrated Arts and Science, Hiroshima University, Higashihiroshima-shi, Hiroshima 739-0046, Japan

Hosoi, S. (21) Institute of Physics, University of Tsukuba, Tsukuba-shi, Ibaraki 305-0006, Japan

Ichihara, T. (25) Department of Earth and Space Science, Graduate School of Science, Osaka University, Toyonaka-shi, Osaka 560-0043, Japan

Ikehara, T. (4) Department of Physics, Tohoku University, Aoba-ku, Sendai 980-8578, Japan

Ishikawa, Y. (41) Mechanical Engineering Research Laboratory, Hitachi Ltd., 503 Kandatsu, Tsuchiura-shi, Ibaraki 300-0013, Japan

Ito, H. (25) Department of Physics, Faculty of Science, Osaka University, Toyonaka-shi, Osaka 560-0043, Japan

Itoh, T. (4) Department of Applied Physics, Tohoku University, Aoba-ku, Sendai 980-8579, Japan

Iwabuchi, Y. (4) Department of Physics, Tohoku University, Aoba-ku, Sendai 980-8578, Japan

Iwai, S. (4) Department of Applied Physics, Tohoku University, Aoba-ku, Sendai 980-8579, Japan

Iwama, S. (12) (26) Department of Applied Electronics, Daido Institute of Technology, Minami-ku, Nagoya 457-0811, Japan

Iwatsuki, M. (41) Electron Optics Division, JEOL Ltd., Akishima-shi, Tokyo 196-0021, Japan

Kamimura, H. (37) Institute of Physics, Graduate School of Science, Science University of Tokyo, Shinjuku-Ku, Tokyo 162-0825, Japan

Kasuya, A. (32) (33) Institute for Materials Research, Tohoku University, Aoba-ku, Sendai 980-8577, Japan

Katagiri, N. (4) Department of Physics, Tohoku University, Aoba-ku, Sendai 980-8578, Japan

Katakuse, I. (25) Department of Earth and Space Science, Graduate School of Science, Osaka University, Toyonaka-shi, Osaka 560-0043, Japan

Katsumoto, S. (11) Department of Physics, Faculty of Science, University of Tokyo, Bunkyo-ku, Tokyo 113-0033, Japan

Kawabata, A. (1) Department of Physics, Gakushuin University, Toshima-ku, Tokyo 171-0031, Japan

Kawamura, K. (17) Department of Physics, Faculty of Science and Technology, Keio University, Kohoku-ku, Yokohama 223-8522, Japan

Kaya, K. (29) Department of Chemistry, Keio University, Kohoku-ku, Yokohama 223-8522, Japan / Institute of Physical and Chemical Research (RIKEN), Wako-shi, Saitama 351-0106, Japan

Kimoto, M. (41) Department of Materials Science and Engineering, Kanazawa Institute of Technology, Nonoichi-cho, Ishikawa 921-8501, Japan

Kimura, K. (16) Department of Material Science, Faculty of Science, Himeji Institute of Technology, Ako-gun, Hyogo 678-1231, Japan

Kizuka, T. (43) Department of Applied Physics, School of Engineering, Nagoya University, Chikusa-ku, Nagoya 464-8603, Japan / Research Center for Advanced Waste and Emission Management, Nagoya University, Chikusa-ku, Nagoya 464-8603, Japan

Kobayashi, N. (30) Department of Physics, Tokyo Metropolitan University, Hachioji-shi, Tokyo 192-0364, Japan

Kobayashi, S. (11) Department of Physics, Faculty of Science, University of Tokyo, Bunkyo-ku, Tokyo 113-0033, Japan

Komiyama, H. (18) Department of Chemical System Engineering, Faculty of Engineering, University of Tokyo, Bunkyo-ku, Tokyo 113-0033, Japan

Kondow, T. (28) Department of Chemistry, School of Science, University of Tokyo, Bunkyo-ku, Tokyo 113-0033, Japan / present address: Cluster Research Laboratory, Toyota Technological Institute: in East Tokyo Laboratory, Genesis Research Laboratory Inc., Ichikawa-shi, Chiba 272-0001, Japan

Maeda, T. (33) Department of Basic Science, Ishinomaki Senshu University, Ishinomaki-shi, Miyagi 986-8577, Japan

Makino, T. (3) University of Tsukuba, Tsukuba-shi, Ibaraki 305-0006, Japan /

Present adress: Opto-electro Mechanics Reserch Laboratory, Matsushita Research Institute Tokyo, Inc., Kawasaki-shi, Kanagawa 214-0033, Japan

Matsuishi, K. (3) University of Tsukuba, Tsukuba-shi, Ibaraki 305-0006, Japan

Matsuo, T. (25) Department of Physics, Faculty of Science, Osaka University, Toyonaka-shi, Osaka 560-0043, Japan

Matsushita, M. (15) Department of Physics, Chuo University, Bunkyo-ku, Tokyo 112-0003, Japan

Mihama, K. (12) (13) (31) (43) Department of Applied Electronics, Daido Institute of Technology, Minami-ku, Nagoya 457-0811, Japan

Mitsugashira, T. (32) Institute for Materials Research, Tohoku University, Aoba-ku, Sendai 980-8577, Japan

Miura, N. (6) Institute for Solid State Physics, University of Tokyo, Minato-ku, Tokyo 106-0032, Japan

Miyashita, A. (34) Takasaki Research Establishment, Japan Atomic Energy Research Institute, Takasaki-shi, Gunma 370-1207, Japan

Miyauchi, H. (21) Institute of Physics, University of Tsukuba, Tsukuba-shi, Ibaraki 305-0006, Japan

Mizuno, T. (17) Department of Physics, Faculty of Science and Technology, Keio University, Kohoku-ku, Yokohama 223-8522, Japan

Mochizuki, S. (35) Department of Physics, College of Humanities and Sciences, Nihon University, Setagaya-ku, Tokyo 156-0045, Japan

Mori, H. (14) Research Center for Ultra-High Voltage Electron Microscopy, Osaka University, Suita-shi, Osaka 565-0871, Japan

Motokawa, M. (9) Institute for Materials Research, Tohoku University, Aoba-ku, Sendai 980-8577, Japan

Murakami, K. (34) Institute of Materials Science, University of Tsukuba, Tsukuba-shi, Ibaraki 305-0006, Japan

Nagata, T. (28) Department of Chemistry, School of Science, University of Tokyo, Bunkyo-ku, Tokyo 113-0033, Japan

Nakahara, A. (15) Department of Physics, Chuo University, Bunkyo-ku, Tokyo 112-0003, Japan

Nakajima, A. (29) Department of Chemistry, Keio University, Kohoku-ku, Yokohama 223-8522, Japan / Institute of Physical and Chemical Research (RIKEN), Wako-shi, Saitama 351-0106, Japan

Nakamura, A. (5) Department of Applied Physics, Graduate School of Engineering, Nagoya University, Chikusa-ku, Nagoya 464-0814, Japan

Nakata, H. (35) Department of Physics, College of Humanities and Sciences, Nihon University, Setagaya-ku, Tokyo 156-0045, Japan

Nakayama, T. (15) Department of Physics, Chuo University, Bunkyo-ku, Tokyo 112-0003, Japan

Nanba, T. (9) Department of Physics, Faculty of Science, Kobe University, Nada-ku, Kobe 657-8501, Japan

Nishida, I. (27) Deceased, Former Professer, Faculty of Science, Nagoya University, Chikusa-ku, Nagoya 464-0814, Japan

Nishikawa, O. (41) Department of Materials Science and Engineering, Kanazawa Institute of Technology, Nonoichi-cho, Ishikawa 921-8501, Japan

Nishina, Y. (32) (33) Institute for Materials Research, Tohoku University, Aoba-ku, Sendai 980-8577, Japan

Noda, T. (31) (39) Toyohashi Junior College, Ushikawa-cho, Toyohashi-shi, Aichi 440-0016, Japan

Nonose, S. (28) Department of Chemistry, School of Science, University of Tokyo, Bunkyo-ku, Tokyo 113-0033, Japan

Oka, Y. (10) Research Institute for Scientific Measurements, Tohoku University, Aoba-ku, Sendai 980-8577, Japan

Okamoto, H. (10) Research Institute for Scientific Measurements, Tohoku University, Aoba-ku, Sendai 980-8577, Japan

Onari, S. (3) University of Tsukuba, Tsukuba-shi, Ibaraki 305-0006, Japan

Orii, T. (3) University of Tsukuba, Tsukuba-shi, Ibaraki 305-0006, Japan / present address: Applied Laser Chemistry Laboratory, Institute of Physical and Chemical Research (RIKEN), Wako-shi, Saitama 351-0106, Japan

Saito, S. (6) Department of Physics, Graduate School of Science, Tohoku University, Aoba-ku, Sendai 980-8578, Japan

Saito, Y. (31) (32) (33) (39) Department of Electrical and Electronic Engineering, Mie University, 1515 Kamihama-cho, Tsu-shi, Mie 514-8507, Japan

Sakurai, T. (25) Department of Physics, Faculty of Science, Osaka University, Toyonaka-shi, Osaka 560-0043, Japan

Sasaki, S. (6) Institute for Solid State Physics, University of Tokyo, Minato-ku, Tokyo 106-0032, Japan

Sasaki, Y. (33) Department of Basic Science, Ishinomaki Senshu University, Ishinomaki-shi, Miyagi 986-0031, Japan

Sato, T. (22) Department of Applied Physics and Physico-Informatics, Faculty of Science and Technology, Keio University, Kohoku-ku, Yokohama 223-8522, Japan

Satoh, I. (32) Institute for Materials Research, Tohoku University, Aoba-ku, Sendai 980-8577, Japan

Shibayama, T. (32) Institute for Materials Research, Tohoku University, Aoba-ku, Sendai 980-8577, Japan

Shimojo, F. (23) Faculty of Integrated Arts and Science, Hiroshima University, Higashihiroshima-shi, Hiroshima 739-0046, Japan

Shinohara, H. (40) Department of Chemistry, Nagoya University, Chikusa-ku,

Nagoya 464-0814, Japan

Shiokawa, Y. (32) Institute for Materials Research, Tohoku University, Aoba-ku, Sendai 980-8577, Japan

Shiromaru, H. (38) Department of Chemistry, Tokyo Metropolitan University, Hachioji-shi, Tokyo 192-0397, Japan

Sonoda, K. (23) Venture Business Laboratory, Hiroshima University, Higashihiroshima-shi, Hiroshima 739-0046, Japan

Suezawa, M. (8) Institute for Materials Research, Tohoku University, Aoba-ku, Sendai 980-8577, Japan

Suto, S. (44) Department of Physics, Graduate School of Science, Tohoku University, Aoba-ku, Sendai 980-8578, Japan

Suzuki, N. (27) Kawasaki Steel Co., Kurashiki-shi, Okayama 712-8511, Japan

Suzuki, S. (38) Department of Chemistry, Tokyo Metropolitan University, Hachioji-shi, Tokyo 192-0397, Japan

Takahashi, H. (32) Institute for Materials Research, Tohoku University, Aoba-ku, Sendai 980-8577, Japan

Takahashi, M. (10) Research Institute for Scientific Measurements, Tohoku University, Aoba-ku, Sendai 980-8577, Japan

Taketomi, K. (21) Institute of Physics, University of Tsukuba, Tsukuba-shi, Ibaraki 305-0006, Japan

Tamura, R. (36) Department of Physics, Graduate School of Science, University of Tokyo, Bunkyo-ku, Tokyo 113-0033, Japan

Tanaka, H. (6) Department of Physics, Graduate School of Science, Tohoku University, Aoba-ku, Sendai 980-8578, Japan

Tanaka, M. (6) Research Institute for Scientific Measurements, Tohoku University, Aoba-ku, Sendai 980-8577, Japan

Tanaka, N. (12) (13) (43) Department of Applied Physics, School of Engineering, Nagoya University, Chikusa-ku, Nagoya 464-8603, Japan

Taniyama, T. (22) Department of Materials Science, Faculty of Science and Technology, Keio University, Kohoku-ku, Yokohama 223-8522, Japan / National Research Institute for Metals, Tsukuba-shi, Ibaraki 305-0047, Japan

Terasaki, A. (28) Department of Chemistry, School of Science, University of Tokyo, Bunkyo-ku, Tokyo 113-0033, Japan / present address: Cluster Research Laboratory, Toyota Technological Institute: in East Tokyo Laboratory, Genesis Research Laboratory Inc., Ichikawa-shi, Chiba 272-0001, Japan

Tokihiro, T. (2) Department of Applied Physics, University of Tokyo, Bunkyo-ku, Tokyo 113-0033, Japan / present address: Graduate School of Mathematical Science, University of Tokyo, Meguro-ku, Tokyo 153-0041, Japan

Tokizaki, T. (5) Department of Applied Physics, Graduate School of Engineering, Nagoya University, Chikusa-ku, Nagoya 464-0814, Japan

Tomitori, M. (42) School of Materials Science, Japan Advanced Institute of Science and Technology, Hokuriku, Nomi-gun, Ishikawa 923-1212, Japan

Tsukada, M. (36) Department of Physics, Graduate School of Science, University of Tokyo, Bunkyo-ku, Tokyo 113-0033, Japan

Uchida, H. (20) Laboratory of Electrochemical Energy Conversion, Yamanashi University, 4-3 Takeda, Kofu-shi, Yamanashi 400-0511, Japan

Urata, N. (17) Department of Physics, Faculty of Science and Technology, Keio University, Kohoku-ku, Yokohama 223-8522, Japan

Wakabayashi, T. (38) Department of Chemistry, Tokyo Metropolitan University, Hachioji-shi, Tokyo 192-0397, Japan / present address: Department of Chemistry, Kyoto University, Sakyo-ku, Kyoto 606-8224, Japan

Watabe, M. (23) Faculty of Integrated Arts and Science, Hiroshima University, Higashihiroshima-shi, Hiroshima 739-0046, Japan

Yamaguchi, F. (17) Department of Physics, Faculty of Science and Technology, Keio University, Kohoku-ku, Yokohama 223-8522, Japan

Yamaguchi, T. (24) Faculty of Engineering, Shizuoka University, Shizuoka-shi, Shizuoka 422-8529, Japan

Yamamoto, I. (19) Faculty of Education, Hirosaki University, Hirosaki-shi, Aomori 036-8224, Japan

Yamamoto, K. (7) Department of Electrical and Electronics Engineering, Faculty of Engineering, Kobe University, Nada-ku, Kobe 657-0013, Japan

Yanata, K. (10) Research Institute for Scientific Measurements, Tohoku University, Aoba-ku, Sendai 980-8577, Japan

Yano, S. (4) Department of Physics, Tohoku University, Aoba-ku, Sendai 980-8578, Japan

Yao, M. (19) Department of Physics, Graduate School of Science, Kyoto University, Sakyo-ku, Kyoto 606-8224, Japan

Yasuda, H. (14) Research Center for Ultra-High Voltage Electron Microscopy, Osaka University, Suita-shi, Osaka 565-0871, Japan

Yoda, O. (34) Takasaki Research Establishment, Japan Atomic Energy Research Institute, Takasaki-shi, Gunma 370-1207, Japan

Zhou, H.-S. (18) Frontier Research Program, Institute of Physical and Chemical Research (RIKEN), Wako-shi, Saitama 351-0106, Japan

Preface

The present volume is a collection of research works for the project, "Physics of Mesoscopic Phase between Solids and Molecules," which was supported by a Grant-in-Aid for Scientific Research under the auspices of the Ministry of Education, Science, Sports and Culture of Japan for the fiscal years 1992 through 1995. The research group for this project was organized in view of the rapidly expanding activities in mesoscopic sciences, involving the physical as well as chemical properties of various materials on a nanometer scale. Considerable effort was applied towards prompt publication, so the topics of the works presented here cover the most recent developments in Japan of the last decade in materials research of mesoscopic particles and clusters of atoms and molecules.

It is clear that the properties of mesoscopic materials and clusters depend critically on their size and physical properties as well as chemical environment. Consequently, the contents of this volume are classified into two parts. The first deals with mesoscopic materials and the second with clusters. Each of these parts consists of two sections:

Part I Mesoscopic Materials
Sec. I-A Mesoscopic Particles Embedded in Matrix and Some Porous Materials
Sec. I-B Reaction and Coated Materials
Part II Clusters
Sec. II-A Properties in Vacuum
Sec. II-B Properties Held on Surfaces

Possible applications of quantum size effects observed in mesoscopic systems to technological fields may be indicated in quantum dots, wires and catalyzers, and nonlinear optical elements which are being studied intensively with rapid progress. All of these items are important subjects of interest in the papers presented here, and we hope this publication will contribute to further technological developments in the near future.

We gratefully acknowledge the Ministry of Education, Science, Sports and Culture of Japan for its financial support of this project and the publication of the results. We also thank to Mr. Ippei Ohta for his cooperation in editing this book.

Toshihiro Arai
January 1999

Contents

I Mesoscopic Materials

II Clusters

II -A Properties in Vacuum

II -B Properties Held on Surfaces

1

Effects of Manybody Interaction on Properties of Mesoscopic Particles

Arisato Kawabata

Department of Physics, Gakushuin University
Toshima-ku, Tokyo 171-0031, Japan

Purpose of the present study

How manybody interaction between electrons affects the electronic properties of mesoscopic particles is clarified. The investigation foucuses mainly on the problem of charge neutrality of particles, which is one of the crucial assumption in Kubo's theory of metallic fine particles. The main issues are: 1) under what conditions the charge neutrality is maintained and 2) how the properties of the mesoscopic particle are modified if the charge neutrality is not maintained.

Contents of the present study

1. Historical survey

1.1 Introduction

Studies on metallic fine particles was initiated by Kubo in 1962.[a] This paper was the first pioneering work on mesoscopic physics, in the sense that Kubo suggested the possibilities of observing various kinds of quantum effects in a system the size of which is between macroscopic and microscopic, although the problems treated there are different from those which became fashionable in the 1980's, *e.g.* Aharonov-Bohm effect, etc.

Anyone who has studied elementary quantum mechanics will note that in a small system the electronic levels are discrete and that its thermodynamic properties are different from those of bulk systems. What is important is, however, that Kubo predicted the possibility of observing the mesoscopic effects under realistic conditions with the following assumptions:

1. Charge neutrality of metallic fine particles.
2. Discreteness of the electronic levels.
3. Statistical properties of the electronic levels.

These assumptions are closely related to the manybody interaction effects, *i.e.* the subject of the present studies, and among them, the charge neutrality of the particle has the closest relation to the interaction effects.

In the following we will clarify: 1) under what conditions the charge neutrality is maintained, 2) how we can know whether the charge neutrality is maintained or not, and 3) if the charge neutrality is broken, how it is reflected in the measurable quantities.

1.2 Role of manybody interaction in Kubo's theory

In the paper by Kubo, he does not mention "manybody effects," but the charge neutrality of the particles is due to the coulomb interaction between electrons, and in this sense the manybody interaction plays an important role in Kubo's theory. According to Kubo, the excess charge of $\pm e$ on a particle of radius a gives the charging energy

$$E_c = \frac{e^2}{2a},\qquad(1.1)$$

and it amounts to 800 K (times k_B) for $a = 100$ Å. Therefore, at room temperatures the probability of the charging of a particle by thermal excitation is very small.

This argument is, however, too simple to be applied to real experimental situations, and needs some corrections in order to obtain quantitative results. For example, in the system of particles embedded in an insulator with dielectric constant ε (or an aggregate of particles coated with insulator), the charging energy must be corrected because of the dielectrics around the particles, and it might be reduced by a factor 10 (Fig. 1). In addition, the energy required to transfer an electron from a particle to another is given by

$$\Delta E = \frac{e^2}{\varepsilon a} - \frac{e^2}{\varepsilon r},\qquad(1.2)$$

within the same argument as in deriving eq. (1.1), where r is the distance between the centers of the particles and we have assumed that the particles are of the same size. The second term of the right-hand side of the above equation is the interaction energy between two charged particles and it reduces the charging energy. These problems have been discussed by many investigators, but the correspondence between these arguments and the experimental results are not yet clear.

Moreover, if the insulator between the particles is not thick enough, quantum mechanical charge fluctuation due to electron tunneling between the particles may take place. This problem has already been discussed by Abeles and Sheng,[b)]

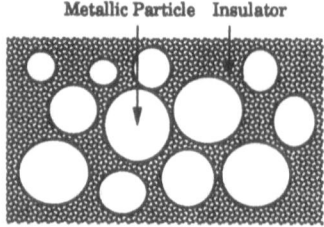

Fig. 1 Fine particles embedded in an insulator.

and by Strongin [c] phenomenologically, and it will be discussed in section 3 below.

2. Breakdown of the charge neutrality (Thermodynamic effects)

2.1 Theory of charging energy in metallic fine particles

The arguments in the previous section are based on eq. (1.1). There have been a great deal of discussions on this expression. One of the important issues is that the expression is based on classical electromagnetism and the quantum effects are neglected. Here one of the quantum mechanical effects is the self-interaction of electrons: From eq. (1.1) we must remove the energy of the interaction of an electron with itself. Then the charging energy is given by

$$E_S = \frac{e^2}{2a} N(N-1).\tag{2.1}$$

In fact eq. (1.1) is obtained if we calculate the charging energy by Hartree approximation, and eq. (2.1) is obtained if we take into account the Fock term, which corresponds to the interaction of an electron with itself. It might look reasonable from the point of view of the self-interaction. In this approximation, however, only a part of the Fock term is taken into account, and eq. (2.1) is still inadequate to give the correct charging energy.

In addition to the above mentioned effects, one-body energies like work function should be taken into account in calculating the energy required to remove an electron from a particle. Therefore, generally, the charging energy is of the form

$$E_g = \frac{e^2}{2a} N(N-\alpha),\tag{2.2}$$

where $\alpha = W/(e^2/2a)$, W being the one-electron energy required to remove an electron from a particle.

From this equation one may conclude that if $|\alpha| > 1$, a charged state has the lowest energy, but the total energy of the system is dependent on where the electron removed from a particle is placed. For example, for the system shown in Fig. 1, the energy required to transfer an electron the particle 1 to particle 2 is given by

$$\Delta E = \frac{e^2}{2a}\left(1 - \frac{\alpha_2 - \alpha_1}{2}\right),\tag{2.3}$$

where α_1 and α_2 are the parameter α of the particle 1 and 2, respectively. We find that eq. (2.3) gives $\Delta E < 0$ for $\alpha_2 - \alpha_1 > 2$; hence what matters is not the average value of α's but their fluctuation. Generally, the charge neutrality of the particlescould be broken if the mean deviation of α's from the average is of order 1. As regards the origin of the fluctuation of α, Pollak and Adkins [d] have proposed a mechanism due to that of the work function.

Generally, the value of the work function depends on the crystallographic direction of the surface. For example, the work functions of tungsten are 4.56 eV and 4.39 eV for (001) and (111), respectively. It is well known that the charge transfer between two metal blocks takes place when contact between them is made. The situation is the same for two metallic particles of the same metal with different crystallographic directions of the surfaces. Moreover, the values of the work function are sensitive to the method of measurement, with a fluctuation of about 0.5 eV. From those facts we can guess that the fluctuation of the work function is at least about 0.1 eV. Then we find that the mean deviation of α is larger than 1 if

$$a \gtrsim 70 \text{ Å} \qquad\qquad (2.4)$$

and hence that there is a possibility of broken charge neutrality for particles in most experiments.

In the next section we will discuss how the Kubo theory must be modified if the charge neutrality is broken. So far we have been discussing the thermodynamic charge fluctuations, and they must be distinguished from those caused by the quantum mechanical mixing of the charged states in the ground state.

2.2 Charge neutrality and the properties of metallic fine particles

Even when the charge neutrality is broken, all the results of Kubo's theory are not invalidated by it. In his paper of 1962, Kubo discussed mainly electronic specific heat and spin susceptibility, and the fact that their characteristics are determined by the discreteness of the excitation energies and their statistical properties. Even if a particle is charged, these conditions on the excitation energies are fulfilled provided the amount of the charge does not vary in time. Therefore, the fundamental characteristics of those quantities are not so much modified by the charging of the particles as they are determined by the excitations within a particle.

It should be noted, however, that a particle composed of an element with an even number of valence electrons may have an odd number of electrons. As was pointed out by Kubo, it gives rise to a qualitative difference in the behaviors of the spin susceptibility whether the total number of the electrons in a particle is even or odd. Thus the effects of the broken charge neutrality can be observed in this way.

On the other hand, the charging of the particles gives important effects on the phenomena associated with the charge transfer between particles. This problem is discussed in the next section.

3. Breakdown of charge neutrality in metallic fine particles (Quantum mechanical effects)

3.1 Charge neutrality of metallic fine particles and electrical conduction

Here we will discuss the charge fluctuations of particles due to the tunneling between them and its effects on electrical conduction.

Here again we consider the system shown in Fig. 1. According to refs. b and c, the condition for the suppression of the electron transfer between the particles due to the charging energy (coulomb blockade) is given by

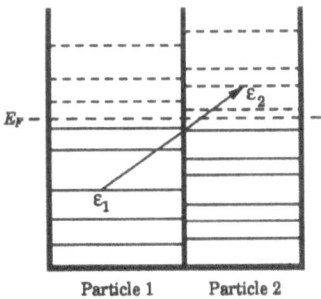

Particle 1   Particle 2

Fig. 2 Electron transfer between the fine particles.

$$R \gg \frac{h}{e^2},$$ (3.1)

where R is the two-dimensional resistance per area at high temperatures where the charging effects do not show up. The arguments in these papers are semiclassical, and the quantum mechanical investigations are given in ref. 1. Consider two particles 1 and 2. Let t be the average of the tunneling matrix elements between the states in particles 1 and 2. Then the probability that one electron is transferred from particle 1 to 2 is given by

$$A = 2D_2 \int_{E_F}^{\infty} d\varepsilon_2 D_1 \int_0^{E_F} d\varepsilon_1 \frac{t^2}{(\Delta E + \varepsilon_2 - \varepsilon_1)^2},$$ (3.2)

where D_i and ε_i are the density of states and the one-electron energy levels, respectively, and the subscripts 1 and 2 indicate particles 1 and 2. Here E_F is the Fermi energy and ΔE is the energy required for the charge transfer (see Fig. 2).

Carrying out the integral we obtain

$$A = 2D_1 D_2 t^2 \log(E_F/\Delta E).$$ (3.3)

If $A \ll 1$, the quantum mechanical charge fluctuation of a particle is very small and the coulomb blockade is effective. It should be noted that A does not depend on ΔE very much. In fact, the value of ΔE is not involved in the condition (3.1) and the condition

$$A \ll 1,$$ (3.4)

is nearly independent of ΔE.

Next we consider the relation between this condition and eq. (3.1). First, we will investigate the electrical resistance R' between two particles at high temperature regions where the effects of the charging energy do not show up. Suppose the electric potential difference V is given between particles 1 and 2. Among the electrons in particle 1, those above the Fermi level of particle 2 can tunnel to particle 2, and the number of such electrons is given by

$$N_V = \frac{2eV}{\delta_1},$$ (3.5)

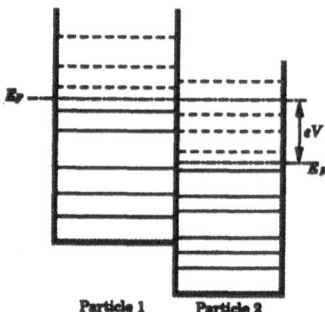

Fig. 3 Electronic states in two fine particles with a potential difference.

where $\delta_1 \equiv 1/D_1$ is the average level spacing in the particle 1 (see Fig. 3).

On the other hand, the probability of the tunneling of an electron per unit time is given by

$$\frac{1}{\tau} = \frac{2\pi}{\hbar} D_2 t^2. \tag{3.6}$$

The current between the particles is given by

$$I = \frac{eN_V}{\tau}, \tag{3.7}$$

hence the high temperature resistance is given by

$$R' = \frac{\hbar}{4\pi e^2 D_1 D_2 t^2}. \tag{3.8}$$

From this equation, we find that the condition (3.4) can be written in the form

$$R' \gg \frac{h}{e^2}, \tag{3.9}$$

provided that $\log(E_F/\Delta E) \sim 1$ in eq. (3.3).

A model for the system shown in Fig. 1 is a network of the resistors of resistance R'. For such a system, the 2-dimentional resistance per area is R' too, and we find that eq. (3.9) is equivalent to eq. (3.1).

Thus we have confirmed that eq. (3.1) is a correct condition for the coulomb blockade by purely quantum mechanical investigation. We expect the conductivity to be of the activation type at low temperatures because of the coulomb blockade. It may seem unreasonable that the charging energy does not enter eq. (3.1). But in order that the summations over the energy can be replaced by integrals in eq. (3.1), the condition

$$\Delta E \gg \delta_1, \delta_2, \tag{3.10}$$

is required. This condition is mostly fulfilled for metallic particles but is not always fulfilled for quantum dots.

4. Summary

In this article we have investigated the effects of the manybody interaction on the properties of metallic fine particles. We discussed mainly the breakdown of the charge neutralit and the fluctuations of the charge. As for the charge fluctuation, we must take into account several factors discussed in ref. a. Even if the charge neutrality is broken, however, it gives little effect on the quantities such as magnetic susceptibility and specific heat, in which only intra particle electron excitations are involved.

The breakdown of the charge neutrality gives essential effects on the transport properties of the aggregates of the particles. A detailed investigation on this problem is reported in ref. d. Here we have investigated the quantum mechanical charge fluctuations due to the electron tunneling between the particles and have derived a condition for the coulomb blockade, which is almost the same as those derived by semiclassical theories.[b,c] In addition, we have found that eq. (3.10) must be the satisfied in order for eq. (3.1) to be correct condition for the coulomb blockade. This condition cannot be obtained by a semiclassical theory.

There are many problems in which the coulomb blockade is involved. One of the important problems is the Kondo effect in a quantum dot.[2-4] In principle this phenomenon can be observed in a metallic particle and further investigations, both theoretical and experimental, are required.

PUBLICATIONS

1. A. Kawabata, *Physica A*, **204**, 359–366 (1994).
2. A. Kawabata, *J.Phys. Soc. Jpn.*, **60**, 3222–3225 (1991).
3. A. Kawabata, *Proc. 4th Int. Symp. on Foundations of Quantum Mechanics in the Light of New Technology* (1992, Tokyo), Japanese J. Appl. Phys, 1993, pp.34–36.
4. A. Kawabata, *Proc. Int. Symp. on the Science and Technology of Mesoscopic Structures* (1991, Nara), Springer-Verlag, 1992, pp. 268–278.

REFERENCES

a. R. Kubo: *J.Phys. Soc. Jpn.,* **17**, 975–986 (1962).
b. B. Abeles and P. Sheng, *Electrical Transport and Optical Properties of Inhomogeneous Media*, J.G. Garland and D.B. Tanner, eds., AIP Conf. Proc. No. 40, AIP, New York (1978).
c. Y. Imry and M. Strongin, *Phys. Rev. B*, **24**, 6353 (1981).
d. M. Pollak and J.C. Adkins, *Phil. Mag. B*, **65**, 855–860 (1992).

2

Nonlinear Optical Responses of Frenkel Excitons

Eiichi Hanamura, Hiromi Ezaki[a] and Tetsuji Tokihiro[b]

Department of Applied Physics, University of Tokyo
Bunkyo-ku, Tokyo 113-0033, Japan

present address:
[a] Faculty of Engineering, Tokyo Institute of Polytechnics
   Atsugi-shi, Kanagawa 243-0213, Japan
[b] Graduate School of Mathematical Science, University of Tokyo
   Meguro-ku, Tokyo 153-0041, Japan

Purpose of the present study

In contrast to the highly excited system of Wannier excitons, we know the exact solution for any degree of excitation of Frenkel excitons in a one-dimensional mesoscopic system. The present work proposes a new family of elementary excitations called an excitonic n-string in terms of this solution, and derives new features of superradiance from highly excited Frenkel excitons. Both of these have important roles in the nonlinear optical responses of a mesoscopic system. Experimental evidence of excitonic 2- and 3-strings is also shown.

Based on these results, we discuss important roles of highly excited excitons in mutual and quantum-mechanical control of radiation fields and material systems in microcavities.

Contents of the present study

1. Introduction

Many celebrated concepts for highly excited semiconductors have been proposed and experimentally confirmed, regarding strong excitation effects on optical properties of semiconductors.[a] For example, electron-hole liquid droplets have been observed in strongly pumped semiconductors Si and Ge.[b] Excitonic molecule [c] was confirmed first in CuCl crystal and subsequently in II-VI and III-V semiconductors. Elementary excitations in these semiconductors are well described in terms of Wannier excitons. On the other hand, Frenkel exciton is a well-defined elementary excitation for the linear response of many organic crystals. However, nonlinear optical responses due to these Frenkel excitons have been poorly studied. Sometimes these electronic structures can be well described as quasi-one-dimensional systems, e.g., in J- and H-aggregate crystal of dye molecules and donor-acceptor complex of molecules such as Anthracene-PMDA (pyromellitic acid dianhydride). From the theoretical point of view, we have exact solutions for these electronic structures at any degree of excitation so that we can introduce a new family of elementary excitations, i.e., excitonic n-string,[d] which is missing in the system of Wannier excitons in semiconductors. This will be discussed in section 2. We will be able to derive optical response functions under any degree of excitation for the 1D system of Frenkel excitons. The peculiar features of superradiance from this system are also discussed by comparing those with Dicke's superradiance of two-level atoms. This is done in section 3. Both concepts are more clearly observed in the mesoscopic system.

Recently we have established high-level technology of material manipulations. Thus we can make a microcavity in which electronic excitation and the radiation mode are both well quantized and couple very strongly. Here the spontaneous emission is controlled so that the thresholdless semiconductor laser with 100% quantum efficiency becomes possible. The last section 4 is a brief discussion and we look at future problems, where it is shown that the efficient squeezing of light becomes possible for the Frenkel excitons in the microcavity.

2. Excitonic n-strings

An excitonic molecule of two Wannier excitons has been observed in several inorganic crystals.[c] The exchange interaction between the electrons of two excitons is responsible for the formation of the excitonic molecule.[e] In crystals with nondegenerate conduction and valence bands, the exchange interaction does not support bound states of more than two Wannier excitons. We will show in this section that a bound state of several Frenkel excitons is possible in a linear chain of molecules when the exciton is accompanied by a static dipole moment larger than the transition dipole moment along the chain direction.

We consider a system of linear chains of organic molecules in which a Frenkel exciton can propagate to the neighboring unit cell with transfer matrix element $-J$ and is associated with a static dipole moment μ . Here two Frenkel excitons attract each other with the static dipole interaction $-2J_z$ when they are on the nearest unit cells. Then the Hamiltonian of this system is written in terms of spin half operator $\hat{s}_j$ at the j -th unit cell as follows:

$$
\begin{aligned}
H &= \omega_0 \sum_j \hat{s}_j^z - J \sum_j (\hat{s}_j^+ \hat{s}_{j+1}^- + \hat{s}_j^- \hat{s}_{j+1}^+) - 2J_z \sum_j \left(\hat{s}_j^z + \frac{1}{2} \right) \left(\hat{s}_{j+1}^z + \frac{1}{2} \right) \\
&= -\frac{NJ_z}{2} + (\omega_0 - 2J_z) \sum_j \hat{s}_j^z - 2\{ J \sum_j \hat{s}_j \cdot \hat{s}_{j+1} + (J_z - J) \sum_j \hat{s}_j^z \cdot \hat{s}_{j+1}^z \}.
\end{aligned}
\tag{1}
$$

The operators $s_j^+ \equiv \hat{s}_j^x + i\hat{s}_j^y$ and $\hat{s}_j^- \equiv \hat{s}_j^x - i\hat{s}_j^y$ indicate excitation and deexcitation between two levels with an excitation energy ω_o. The unit $\hbar = 1$ is used here and hereafter. Only the nearest neighbor interaction is kept in Eq. (1) because the dipolar interaction in the chain is of rather short range. The exciton propagation J is associated with the transition dipole moment d between two levels. The expressions of J and J_z are given by

$$
J = \frac{-d^2}{4\pi\varepsilon_0 a^3} \left\{ 1 - \frac{3(\varepsilon_d \cdot a)^2}{a^2} \right\},
\tag{2a}
$$

$$
J_z = \frac{-\mu^2}{4\pi\varepsilon_0 a^3} \left\{ 1 - \frac{3(\varepsilon_\mu \cdot a)^2}{a^2} \right\},
\tag{2b}
$$

where a is a vector connecting two nearest-neighbor unit-cells and $a = |a|$. When

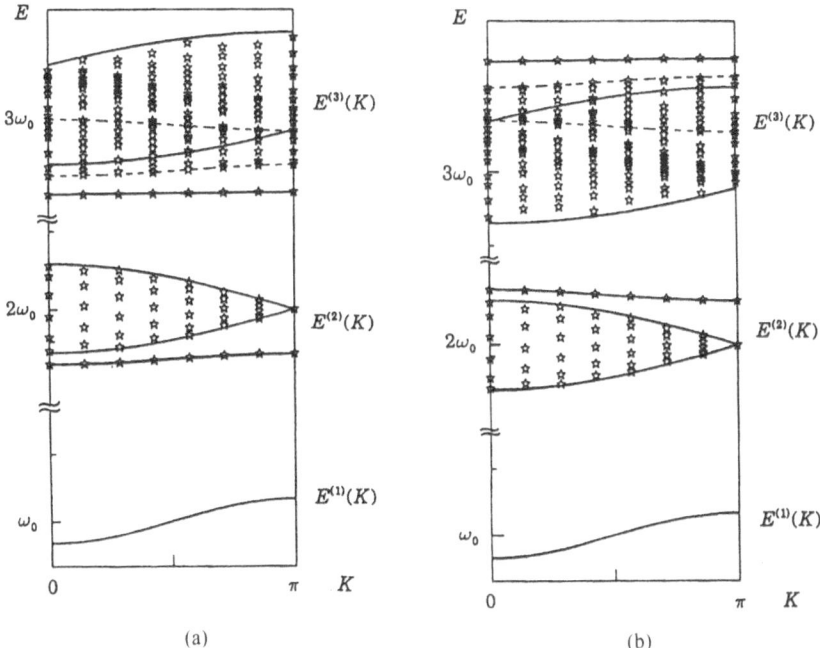

Fig. 1 Dispersion relations of one, two, and three excitations against the wave number for the center-of-mass motion. The system size $N = 14$ and (a) $\Delta = 2.0$ (b) $\Delta = -2.0$.

the transition and static dipole moment vectors $d\varepsilon_d$ and $\mu\varepsilon_\mu$ (ε_d and ε_μ are unit vectors) are parallel to the chain axis a, both J and J_z become positive.

The eigenenergies and eigenstates of the Hamiltonian (1) are calculated numerically with the use of Ising basis functions: $|j> \equiv \hat{s}_j^+ |g>$ for the one-exciton state, $|j, j + m> \equiv \hat{s}_j^+ s_{j+m}^+ |g>$ for the two-exciton state, $|j, j + l, j + m > = \hat{s}_j^+ \hat{s}_{j+l}^+ \hat{s}_{j+m}^+ |g>$ for the three-exciton state, ... , where $|g>$ is the N direct products of a molecular ground state. The energy dispersion relations for the system of infinite size are obtained analytically by the quantum inverse scattering method. For the excitonic n-string, $i.e.$, a bound state of n excitons, the eigenenergy $E_s^{(n)}$ as a function of wave number K for the center-of-mass motion is obtained in the following form:

$$E_s^{(n)} = n(\bar{\omega}_0 - 2\Delta) - \frac{2\sinh \lambda \left[(-1)^{n-1} \cos K + \cosh n\lambda \right]}{\sinh n\lambda} \qquad (n = 2, 3, 4, ...), \quad (3)$$

where $\bar{\omega}_0 = \omega_0/J$, $\Delta \equiv J_z/J$, and $-\cosh \lambda = \Delta$ with J being the energy unit. Calculated eigenenergies of up to two or three excitons are drawn in Figs. 1(a) and 1(b), respectively, for the attractive case $J_z > 0$ and the repulsive one $J_z < 0$. These figures show that bound states exist for both repulsive and attractive interactions at a total wave number $K = 0$ if $|\Delta| > 1$. When $0 \le \Delta \le 1$, we have no excitonic n-string state for $K = 0$ which is directly connected by the optical transitions from the ground state.

The difference in effective mass is an additional factor which gives rise to the qualitative difference between the excitonic n-string and the excitonic molecule of Wannier excitons. Next we discuss the effective masses for center-of-mass motion of the excitonic n-strings. The free exciton mass $M_f^{(1)}$ is defined at small

K by $E^{(1)}(K) \equiv \omega_0 - 2J + K^2/2M_f^{(1)}$. This is obtained to be $M_f^{(1)} = 1/(2J)$ by expanding $E^{(1)}(K) = \omega_0 - 2J \cos K$ in K. The effective masses $M_s^{(2)}$ and $M_s^{(3)}$ of two- and three-strings are derived from Eq. (3) by the substitutions $n = 2$ and $n = 3$:

$$M_s^{(2)} = J_z/J^2 = M_f^{(1)} \times 2\Delta > 2M_f^{(1)} \quad (\Delta > 1), \tag{4}$$

$$M_s^{(3)} = (4J_z^2 - J^2)/2J^3 = M_f^{(1)} \times (4\Delta^2 - 1) > 3M_f^{(1)} \quad (\Delta > 1). \tag{5}$$

In the case of Wannier excitons, a mass of the excitonic molecule is just twice as large as that of a free Wannier exciton. In the case of Frenkel excitons, however, the mass of the excitonic two-string depends on Δ as Eq. (4) shows, and is always heavier than twice the free exciton mass for the case in which the bound state is expected at $K = 0$, i.e., $\Delta > 1$. The three-string turns out to have a mass that is much heavier than three times $M_f^{(1)}$. These results come from the complicated propagation mechanism of n-strings; for the n-string, n steps are necessary in order to propagate as a whole by a single site. Here the n-string must be dissolved virtually and successively into (n-m)- and m-strings in the intermediate states ($1 \leq m \leq n-1$), in each of the n steps. As a result, the effective mass of the excitonic n-string increases rapidly with n; the wave function of the n-string in real space tends to be strongly localized with increasing n.

The differential transmission (DT) spectrum $\Delta\alpha(\Omega_2)$ is one of the most sensitive methods to detect excitonic n-strings. In this technique, the relative change of probe light (Ω_2) transmission due to the presence of pump light (Ω_1) is detected. The DT spectrum is well described in terms of the nonlinear optical polarization $P^{nl}(\Omega_2)$ as follows:

$$DT(\Omega_2) = -\Delta\alpha(\Omega_2)l$$

$$= -\frac{l\Omega_2}{cn(\Omega_2)} \, \text{Im} \, \frac{P^{nl}(\Omega_2)}{E(\Omega_2)}. \tag{6}$$

Here $n(\Omega_2)$ is a refractive index function and l a sample thickness. We evaluate $P^{nl}(\Omega_2)$ to any even orders in $E(\Omega_1)$ but to the first order in $E(\Omega_2)$ as the pump field $E(\Omega_1)$ is much stronger than the probe field $E(\Omega_2)$. We need $P^{(3)}(\Omega_2)$ ($\propto |E(\Omega_1)|^2 E(\Omega_2)$) and $P^{(5)}(\Omega_2)$ ($\propto |E(\Omega_1)|^4 E(\Omega_2)$) to discuss the excitonic 2- and 3-strings. We confine ourselves to the case relevant to the experiments on Anthracene-PMDA.[f] First, the probe field is applied before the excitations suffer from cross relaxation among the sublevels. Second, the pulse width T_p is shorter than the longitudinal and transverse relaxation times T_1 and T_2 of the exciton and the DT spectrum is measured at such a short time as $t \ll T_1$. The third-order $P^{(3)}(\Omega_2)$ and fifth-order $P^{(5)}(\Omega_2)$ polarization are evaluated for the conditions described above. Choosing the effective size $N = 14$ as discussed at the end of this section and several pump powers $\rho \equiv 2\pi\bar{\rho}T_p d^2/\hbar^2$ with $\bar{\rho}$ the energy density of the pump field per unit frequency, the DT spectrum is evaluated using Eq. (6) and is drawn in Fig. 2(a).

Some characteristic features of the DT spectrum are listed: (1) absorption saturation at $\Omega_2 = -2J$, i.e., the single-exciton frequency, (2) induced absorption at $\Omega_2 = \omega_0 - 2J_z - 2J^2/J_z + 2J$, and (3) under stronger pumping, another induced absorption at $\Omega_2 = \omega_0 - 2J_z - 2J^2/(2J_z - J)$. The induced absorption (2) indicates the transition from a single exciton into the excitonic 2-string, denoted by α in Fig. 2(a). The third process (3) is due to the transition from 2- to 3-srtings with

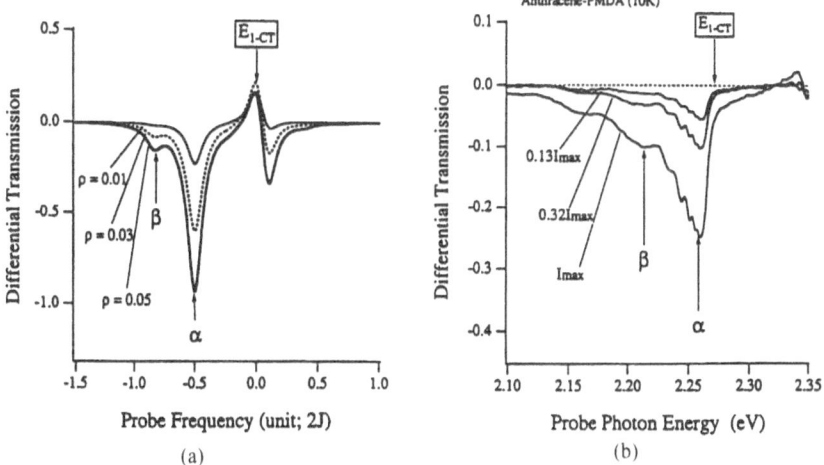

Fig. 2 (a) Calculated differential transmission spectrum for several pumping intensities. Parameters are the same as in Fig. 1. (b) Differential transmission spectrum of Anthracene-PMDA at 10 K.

large oscillator strength as marked by β in Fig. 2(a). A similar DT spectrum was observed as shown in Fig. 2(b) for Anthracene-PMDA crystal measured at 10 K using 200 fs laser pulses.[f] The calculated results of Fig. 2(a) agree with the observed ones not only qualitatively but also quantitatively when we choose $2J = 10$–12 meV and $2J_z = 24$–36 meV, and the system size $N = 14$ unit cells. This is a size in which an exciton can propagate coherently along the chain, and which is estimated from the band width $2J$ and the spectrum width $(T_2)^{-1} = 0.15J$. The absolute value of the DT is also reasonable when we use the transition dipole moment per unit cell $d = e \times 0.37$ Å, the sample thickness $l \equiv 2$ μm and $\Delta \equiv J_z/J = 2.0$. The donor (Anthracene) and acceptor (PMDA) pair constitutes the unit cell and the charge-transfer state is considered to be a Frenkel exciton because the excitation-propagation is brought about by the dipolar interaction J between the neighboring unit cells. Thus we have confirmed the new elementary excitations of the excitonic 2- and 3-strings.

The excitonic 2-string can also be observed by two-photon absorption spectrum [g] with the giant transition dipole moment [h] for the loosely bound 2-string, i.e., in the case of $\Delta \approx 1$.

3. Superradiance of frenkel excitons

We have found new characteristics of superradiance from a highly excited system of Frenkel excitons.[i–k] These are missing in Dicke's superradiance from atomic and molecular systems. Although we may expect the sharp superradiant pulse with a strong peak intensity from the crystal because of high density oscillator strength, there arises a question whether the exciton effect, i.e., the excitation propagation from molecule to molecule, prevents superradiance or not. We will answer this question in this section. The superradiance of a single Wannier exciton in microcrystallites was proposed theoretically [l] and confirmed experimentally.[m] Now we are discussing the superradiance of Frenkel excitons under any degree of excitations.

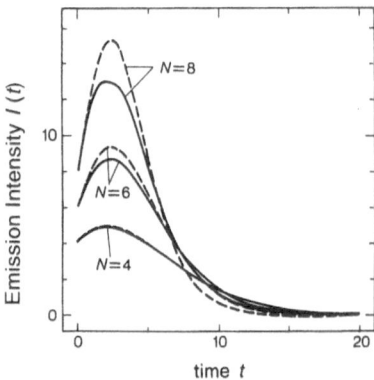

Fig. 3   The superradiant pulse profile from the fully population-inverted system of Frenkel excitons
(solid line) and that of Dicke's system (dashed line), *i.e.*, the dilute gas of atoms. The number
N denotes the system size N =4, 6, or 8,. and the decay rate of the single atom Γ =0.1. The
effect of excitation transfer is measured by the deviation of the solid lines from the dotted
lines, and this increases as N does.

We evaluate the emission pulse profile and the time-resolved emission
spectrum using the same eigenenergies and eigenstates of Eq. (1) for Frenkel
excitons as in section 2. In the case of two-level atoms, the peak intensity of the
superradiant pulse is $I = \frac{1}{2}N\left(\frac{1}{2}N+1\right)\Gamma$ and the pulse width in time is inversely
proportional to the number N of involving atoms, where Γ is a spontaneous
emission rate of a single atom.

First, the superradiant pulse profile from the fully population-inverted system
of Frenkel excitons with vanishing static dipole moment $\mu = 0$, *i.e.*, $J_z = 0$ (solid
line) is drawn in Fig. 3 in comparison to that of Dicke's system (dashed line), *i.e.*,
the dilute gas of atoms. When the system size N increases, the peak intensity
increases but more weakly than in the case of Dicke's. It increases as $N^{1.5}$ in
comparison to N^2 for Dicke's case.[n] This is because the cooperation number is
kept to the maximum for the two-level atoms while it decreases in the emission
process due to the effect of excitation transfer. However, the peak intensity
increases superlinearly against N. The coherent spontaneous emission clearly
prevails so we conclude the superradiance of Frenkel excitons.

Second, the time-resolved emission spectrum from the Frenkel excitons is
shown in Fig. 4. For the case of $J > 0$ and $J_z = 0$, the emission peak frequency is
at $\omega = \omega_0 + 2J$ at the first half of the superradiant pulse, while that at the second
half is at $\omega = \omega_0 - 2J$. Thus we can except strong frequency chirping within the
superradiant pulse. This frequency shift and the superradiant pulse profile depend
on the sign and relative magnitude of J and J_z. For example, we have the same
superradiant behavior for the case of $J = J_z$ as Dicke's because the cooperation
number is kept in this case. When $J_z > J > 0$, the sign of the frequency chirping
becomes reversed to the case of $J_z = 0$.

We have also studied the superradiance of Frenkel excitons under any degree
of excitation prepared by a short-pulse laser.[k] Depending on the pumping degree
as the initial condition, we have some characteristic features as shown in Fig. 5.
Under full population-inversion (a), the emission intensity peak appears with a
time delay, while under the half-population inversion (d) the emission intensity

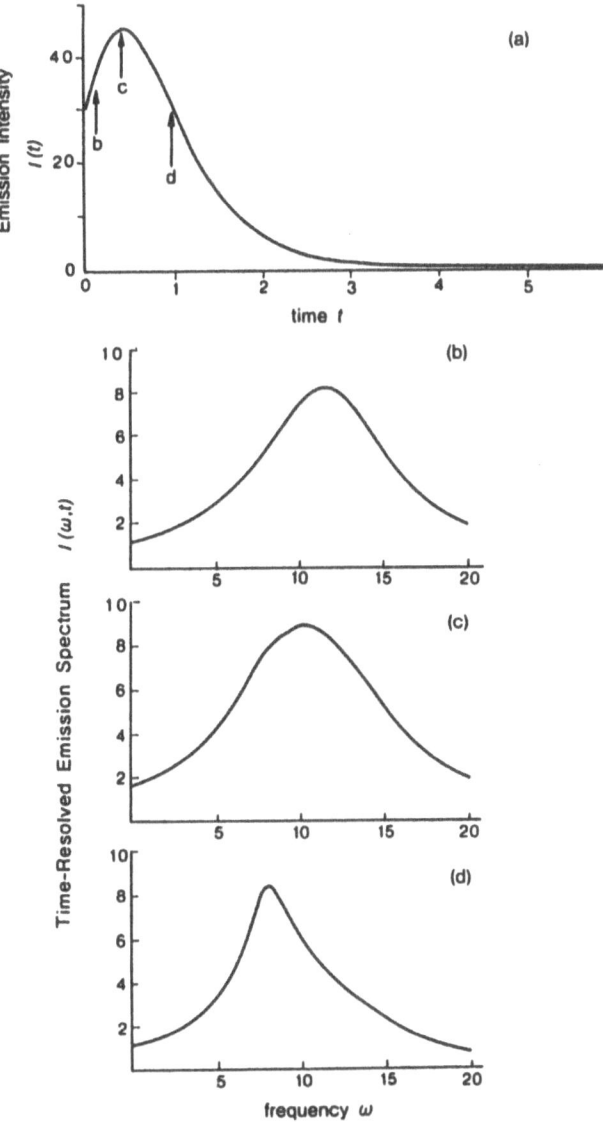

Fig. 4 (a) The superradiant pulse profile from the linear chain with $N = 6$, and the time-resolved emission spectra at $t = 0.15$ (b), $t = 0.44$ (at peak) (c), and $t = 1.0$ (d). Material constants $\omega_0 = 10.0$, $T = 1.0$, and $\Gamma = 0.5$.

decays very rapidly at the initial stage, then decays with oscillation. This comes from the interference between a few channels with different cooperation numbers for the case of $J > 0$ and $J_z > 0$. The decay behavior depends also on the relative values of J_z and J, as shown in Fig. 6. For the case of $J = J_z$, the emission intensity decays monotonically in time even under the $\pi/2$ pulse pumping as the initial state. We have a great number of kinds for the J- and H-aggregates of dyes so we can choose chain systems with arbitrary values of J_z/J and any sign for J_z and J. Therefore we will be able to observe the superradiant emission profile and time-resolved emission spectrum from these systems under any degree of initial

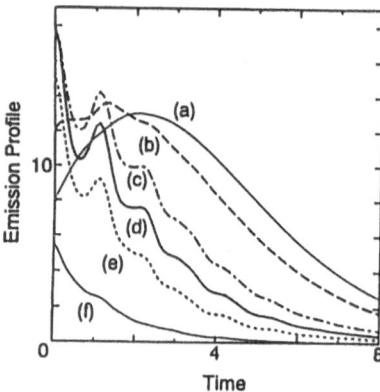

Fig. 5   Emission profile from partially excited system of the case of $J_z = 0$. $\Theta(t_f) = $ (a)π, (b)0.8π, (c)0.6π, (d)0.5π, (e)0.4π, and (f)0.2π, $N = 8$, $\gamma = 0.1$, $J_x = 2.0$, and $J_z = 0$. Here $\Theta(t_f)$ is the Bloch angle due to the pump pulse.

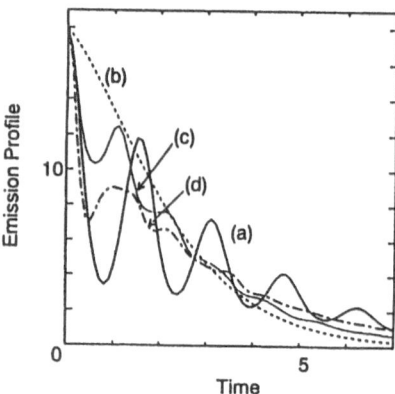

Fig. 6   Emission profile from a partially excited system for $\Theta(t_f) = 0.5$. (a) Ising case, (b) Heisenberg case, (c) XX0 case, and (d) $J_x = -J_z$.

pumping.

4. Discussion and Future Problems

A new family of elementary excitations and superradiance of Frenkel excitons are concepts of the linear chain systems put in the free field of radiation. It is interesting to study the mesoscopic system of excitons embedded in the microcavity in which both the electronic system and radiation field are well quantized. Here the coherent elementary excitations of excitons interact strongly and resonantly with the single mode in the cavity. First, the large vacuum Rabi splitting and Rabi oscillation have been observed for Wannier excitons in semiconductor quantum wells. Second, the collapse and revival of the Rabi oscillation were observed for the atomic system in a microcavity. Third, we obtain stationary squeezed light under resonant and stationary pumping of the Frenkel excitons in the microcavity and strong antibunching light under the coherent light pulse excitation of excitons. We hope that more exciting concepts of non-classical light will become available under quantum-mechanical and mutual control of electronic and radiation systems. These are fascinating future problems of excitons at high density.

PUBLICATIONS

1. H. Ezaki, T. Tokihiro, M. Kuwata-Gonokami, R. Shimano, K. Ema, E. Hanamura, B. Fluegel, K. Meissner, S. Mazumdar and N. Peyghambarian, *Solid State Commun,* **88**, 211 (1993).
2. M. Kuwata-Gonokami, N. Peyghambarian, K. Meissner, B. Fluegel, Y. Sato, K. Ema, R. Shimano, S. Mazumdar, F. Guo, T. Tokihiro, H. Ezaki and E. Hanamura, *Nature* **367**, 47 (1994).
3. H. Ezaki, T. Tokihiro and E. Hanamura, *Phys. Rev.*, **B50**, 10506 (1994).
4. T. Tokihiro, Y. Manabe and E. Hanamura, *Phys. Rev.*, **B47**, 2019 (1993).
5. Y. Manabe, T. Tokihiro and E. Hanamura, *Phys. Rev.*, **B48**, 2773 (1993).
6. T. Tokihiro, Y. Manabe and E. Hanamura, *Phys. Rev.*, **B51**, 7655 (1995).
7. H. Suzuura, T. Tokihiro and Y. Ohta, *Phys. Rev.*, **B49**, 4344 (1994).
8. H. Ezaki, S. Miyashita and E. Hanamura, *Phys. Lett.*, **A203**, 403 (1995).

REFERENCES

a. M. Voos and C. Benoit a la Guillaume, *Optical Properties of Solids, New Developments*, (B. O. Seraphin ed.), pp.143 and references therein, North-Holland, Amsterdam (1976).
b. W. F. Brinkman, T. M. Rice, P. W. Anderson and S. T. Chui, *Phys. Rev. Lett.*, **28**, 961 (1972).
c. M. Ueda, H. Kanzaki, K. Kobayashi, Y. Toyozawa and E. Hanamura, in *Excitonic Processes in Solids*, chapts. 2 and 3, and references therein, Springer-Verlag, Heidelberg (1986).
d. H. Ezaki, T. Tokihiro, M. Kuwata-Gonokami, R. Simano, K. Ema, E. Hanamura, B. Fluegel, K. Meissner, S. Mazumdar and N. Peyghambarian, *Solid State Commun.*, **88**, 211 (1993).
e. O. Akimoto and E. Hanamura, *J. Phys. Soc. Jpn.*, **33**, 1537 (1972).
f. M. Kuwata-Gonokami, N. Peyghambarian, K. Meissner, B. Fluegel, Y. Sato, K. Ema, R. Shimano, S. Mazumdar, F. Guo, T. Tokihiro, H. Ezaki and E. Hanamura, *Nature,* **367**, 47 (1994).
g. H. Ezaki, T. Tokihiro and E. Hanamura, *Phys. Rev.*, **B50**, 10506 (1994).
h. E. Hanamura, Solid State Commun., **12**, 951 (1973).
i. T. Tokihiro, Y. Manabe and E. Hanamura, *Phys. Rev.*, **B47**, 2019 (1993).
j. Y. Manabe, T. Tokihiro and E. Hanamura, *Phys. Rev.*, **B48**, 2773 (1993).
k. T. Tokihiro, Y. Manabe and E. Hanamura, *Phys. Rev.*, **B51**, 7655 (1995).
l. E. Hanamura: *Phys. Rev.*, **B39**, 1228 (1988).
m. T. Itoh, M. Furumiya and T. Ikehara, *Solid State Commun.*, **73**, 271 (1990).
n. H. Suzuura, T. Tokihiro and Y. Ohta, *Phys. Rev.*, **B49**, 4344 (1994).
o. H. Ezaki, S. Miyashita and E. Hanamura, *Phys. Lett.*, **A203**, 403 (1995).

3

Optical Properties and Structural Changes in Semiconductor Fine Particles

Tosihiro Arai[a], Seinosuke Onari, Kiyoto Matsuishi, Toshiharu Makino[b] and Takaaki Orii[c]

Institute of Applied Physics, University of Tsukuba,
Tsukuba-shi, Ibaraki 305-0006, Japan

present address:
[a] Department of Physical Engineering, Ishinomaki Senshu University
   Ishinomaki-shi, Miyagi 986-0031, Japan
[b] Opto-electro Mechanics Research Laboratory, Matsushita Research Institute Tokyo, Inc.
   Kawasaki-shi, Kanagawa 214-0033, Japan
[c] Applied Laser Chemistry Laboratory, Institute of Physical and Chemical Research (RIKEN)
   Wako-shi, Saitama 351-0106, Japan

Purpose of the present study

In mesoscopic systems quantum size effects become important. The physical properties depend not only on the size but also on the shape and state of the surfaces. Increase in surface affects the cohesive energy as well as bonding force and structural change. It is very important to study the optical properties (absorption, photoluminescence, Raman spectroscopy, etc.) and X-ray diffraction on the mesoscopic particles and the beam of mesoscopic-free particles. The effect of size on pressure induced phase transition on CdS fine particles is one of the main objects of our research. Photoluminescence and the relaxation processes of fine particles incorporated into zeolite, embedded in organic films and in glasses are studied in relation to quantum size effects. Spectroscopic studies on porous Si and C_{60} are also reported.

Contents of the present study

1. Dependence of LO phonon Raman scattering of CdSe, CdS fine particles on particle size [1,2]

Dependence of LO phonon Raman scattering intensity $I(\omega)$ of CdSe, CdS fine particles on the particle size are studied, and relaxation of the selection rules of q-vector by the phonon confinement is observed.

Dependence of the optical phonon Raman scattering intensity $I(\omega)$ of the fine particles on the particle size are expressed essentially as follows [a]

$$I(\omega) \cong \int d^3q \frac{|c(0, q)|^2}{\{\omega - \omega(q)\}^2 + \left(\dfrac{\Gamma_0}{2}\right)^2} .$$

Here, $\omega(q)$ is phonon dispersion, Γ_0 is the natural width of the Raman lines, $c(0,$

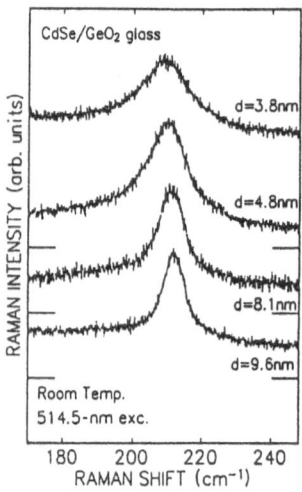

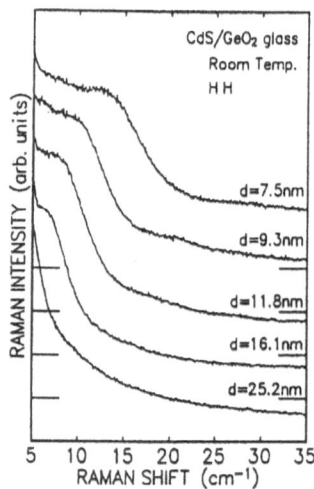

Fig. 1 Raman scattering of CdSe fine particles.

Fig. 2 Low frequency Raman scattering of CdS fine particles.

q) is the Fourier coefficient of the phonon confinement function. If we consider the confinement function

$$W(r,L) = \exp\left(-8\pi^2 r^2/L^2\right).$$

Fourier coefficient of the phonon confinement function becomes

$$c(0,q) = \exp\left(-q^2 L^2/32\pi^2\right).$$

As shown in Fig. 1, peak energy shift and the broadening are observed for LO phonon of CdSe due to the contribution of $\omega(q)$ near $q = 0$.

The range of the contributing q is therefore inversely proportional to the size L.

2. Low frequency Raman scattering of CdS fine particles [1,3,5]

Low frequency Raman scattering of CdS fine particles is studied and Raman band frequencies are observed to follow the dependence proportional to the inverse particle size.[b] They are interpreted by the analysis of the whole vibration modes of the particles. In the particle, spheroidal mode and torsional modes are considered, and have been studied well by H. Lamb. From this theory, the frequency is proportional to the inverse of the diameter d. For $l = 0$, $v_0^s = 2.07$ v_t/dc, $v_0^s = 4.52$ v_t/dc are obtained. $v_2^s = 0.85$ v_t/dc is obtained for $l = 2$.

3. Size effect of the phase transition pressure of CdS microcrystals [4,6]

The effects of finite size on the structural phase of nanometer-scale semiconductor microcrystals are of great interest. In this study, the structural phase transition of CdS microcrystals of various sizes has been studied by high pressure Raman spectroscopy in order to explore the finite size effects on the

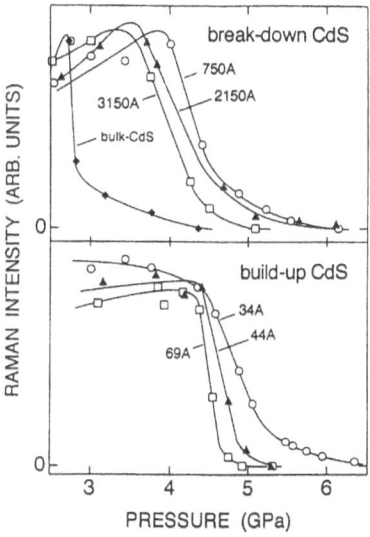

Fig. 3 Pressure dependence of 1-LO Raman intensity of CdS microcrystals of various sizes. The result of bulk CdS is also shown as a reference.

transition.

The samples used in this study were prepared in two ways. One is a break down method. Bulk CdS materials were ground into pieces and filtered into ethanol to divide into groups of different sizes (break-down CdS). The other is a build-up method. CdS microcrystals produced by evaporation in Ar gas were embedded in SiO films on cover glasses (build-up CdS). The 457.9 nm Ar^+ laser line was employed for Raman excitation. Pressure was generated up to 6.5 GPa using a Bassett-type gasketed diamond anvil cell with 4:1 methanol-ethanol mixture as the pressure medium. Pressure was determined by the ruby fluorescence technique.

The pressure dependence of the Raman intensity of 1-LO mode is shown in Fig. 3 for break-down CdS and build-up CdS. The 1-LO phonon mode increased gradually in intensity and then diminished rapidly with an increase in pressure, indicating the occurrence of the structural phase transition from a wurtzite to a rock salt phase. The pressure at which the phase transition occurred was found to be elevated gradually from 2.7 GPa (the phase transition pressure of bulk) as the particle size decreases. However, the elevation of the phase transition pressure for break-down CdS was observed at much larger size than that for build-up CdS.

As one of the reasons for this, we suspected that the crystallinity of break-down CdS was degraded during the preparation. The analysis of X-ray diffraction patterns before applying pressure shows that the inhomogeneous strain in the particle of break-down CdS increased as the particle size decreased. This suggests that the deformation which leads to an increase in the inhomogeneous strain is responsible, in part, for the elevation of the phase transition pressure.

Build-up CdS is expected not to have a significant inhomogeneous strain. The elevation of transition pressure observed in build-up CdS could be interpreted mainly by the effects of finite size and surface energy. There must also be other factors which elevate the transition pressure of CdS microcrystals, e.g., the matrix

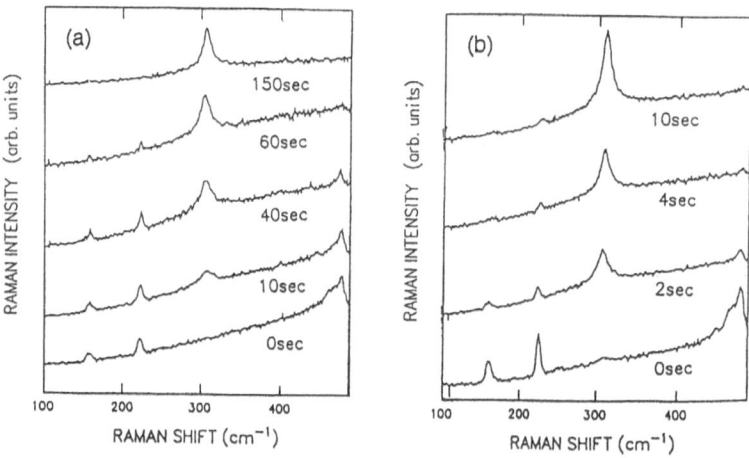

Fig. 4   Change in Raman spectra corresponds to the photochemical change of S to CdS in zeolite (Cd-X).
(a) 514.5 nm laser irradiation, (b) 476.5 nm laser irradiation.

effect surrounding the microcrystals, the difference in bulk modulus between microcrystal and matrix, the mismatch in lattice constant on the interface, the strain and dislocation inside the microcrystals, etc.

4. Photochemical processes of S and Se microcrystals in zeolite (Cd-X) [7,9]

We produced CdS in zeolite supercage by using the special properties of ion exchange and the adsorptivity. Na^+ ion of zeolite is first exchanged by the Cd^{2+} ion and then sulfur is adsorbed to produce CdS in the two kinds of zeolites 4A and 13X. We can clearly observe the process of the formation of CdS microcrystals by the Raman measurement after the laser light irradiation. However, no change to CdSe was observed for Se.

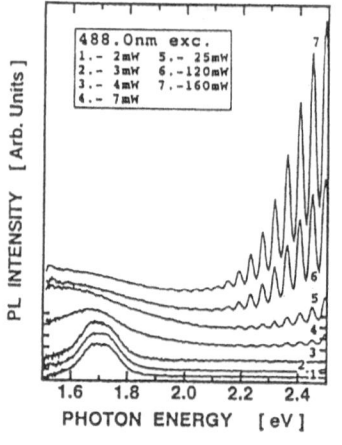

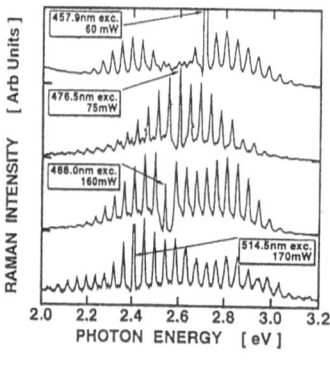

Fig. 5 The excitation power dependence of the photoluminescence for CdSe cluster beam.

Fig. 6 Resonance Raman spectra for CdSe cluster beam excited by several Ar ion laser lines.

5. Photoluminescence measurements for CdSe cluster beam[10,11)]

Since the smaller nanocrystals have a larger surface-to-volume ratio, the surface conditions of nanocrystals play an important role in their properties. Since the surface of nanocrystals is active and interacts strongly with the surrounding matter, most studies on nanocrystals have been done for nanocrystals embedded in various matrices or kept in atmosphere because of restrictions in the preparation methods. However, it is undeniable that nanocrystals are influenced by matrices or adsorbed matter. Thus, in order to investigate the optical properties of surface states in themselves, one must perform optical measurements with nanocrystals suspended in vacuum.

We have developed an apparatus for the purpose of optical measurements of nanocrystals suspended in vacuum. CdSe clusters are generated by the gas evaporation method. CdSe chunk materials were heated to 850°C in a Knudsen cell surrounded with argon gas at 100 Pa. Generated CdSe clusters form a cluster beam with argon gas by differential pumping through a skimmer. And then for the CdSe cluster beam, photoluminescence spectra were measured in the energy range between 1.5 eV to 2.5 eV. Simultaneously to estimate the average size of cluster, clusters were deposited directly onto electron microscopic grids. The average size of clusters was estimated to be about 20 nm by TEM images.

Figure 5 shows the excitation power dependence of the photoluminescence for CdSe cluster beam. At low excitation power below 4 mW, a broad peak is observed at 1.7 eV. This broad peak can be separated into two components. The one at 1.75 eV, which corresponds to the bulk energy gap, is regarded as the CdSe edge emission. The other at 1.65 eV may be attributed to the emission from an Se vacancy level. With increase of excitation power, the broad peak merges with the tail of a emission band which seems to exist in the energy region lower than 1.5 eV and many Raman series are observed around the excitation energy. The same Raman series were observed from measurements on Se cluster beam at high excitation power. Both Raman series were coincident with each other. We assigned these Raman spectra as the vibrational states of Se dimer because the Raman shifts whose intervals are about 345 cm^{-1} correspond to the phonon energy of Se dimer reported with theoretical calculations and some experiments. Many Raman series from Se dimer are observed because the excitation laser gives rise to resonance with the Se dimers. Resonance Raman spectra obtained for CdSe cluster beam excited by several laser lines are shown in Fig. 6.

Since bulk CdSe and CdSe nanocrystals deposited on substrate do not exhibit such phenomena, we explain these phenomena as properties characteristic of clusters which are suspended in vacuum and have a clean and free surface. At low excitation power, CdSe clusters absorb excitation light and the created electron-hole pairs then recombine with emission from band edge and Se vacancy. With increase of excitation power, Se atoms dissociate from free surface of CdSe clusters and generate Se dimers. We attribute the dissociation induced by low excitation power to softening of surface layer resulting in a decrease in melting point.

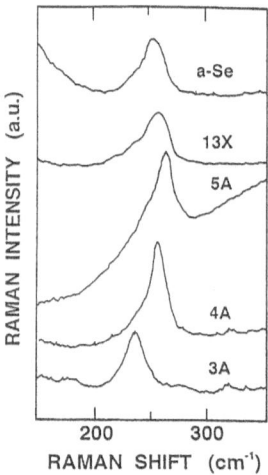

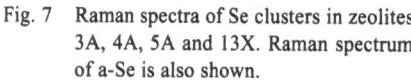

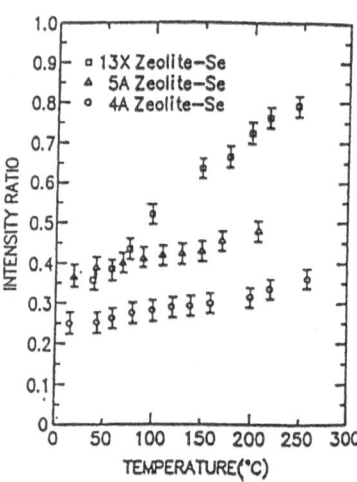

Fig. 7   Raman spectra of Se clusters in zeolites
3A, 4A, 5A and 13X. Raman spectrum
of a-Se is also shown.

Fig. 8   Temperature dependence of relative
Raman intensity of 235 cm^{-1} mode
(chain-like component) to 255 cm^{-1}
mode (ring-like component).

6. Structural phase stability and structural unit of Se clusters in zeolites [9,12]

We have produced Se clusters in zeolites of different pore sizes (3A, 4A, 5A and 13X) by a vapor method, and investigated their structural phase stability and structural units by Raman scattering measurements in the temperature range of 300–530 K.

Figure. 7 shows the Raman spectra of Se clusters in zeolites 3A, 4A, 5A and 13X at room temperature, together with that of amorphous Se (a-Se). Se clusters in 3A zeolite exhibit one Raman mode at 235 cm^{-1}, which originates from the vibration of helical chain or chain-like segment of Se, suggesting that the ring-like fragment of Se in vapor may find it geometrically difficult to go through the windows of pores in 3A zeolite to form Se_8 rings, since the diameter of window in 3A zeolite is much smaller than that of the Se_8 ring. On the other hand, the 255 cm^{-1} mode, which originates from the vibration of S_8 ring or ring-like segment of Se, is predominant for Se clusters in 4A zeolite. Those two modes are seen in 5A zeolite, though they are shifted to higher frequencies by *ca.* 10 cm^{-1} owing to a change in the electric field due to Ca^{2+} cations in the framework of zeolite. It should be noted that the 235 cm^{-1} mode in 5A zeolite is not as weak as that in 4A, suggesting that Se clusters in 5A somehow possess chain segments in the majority of ring-like structures, possibly interconnected through the windows in the zeolite. The Raman spectrum of Se clusters in 13X zeolite appears to be very similar to that of a-Se. It has been found that none of Se clusters in the zeolites show a drastic change in the Raman spectra with temperature, while a-Se exhibits the appearance of a sharp peak at 235 cm^{-1} at 380 K, indicating crystallization to the trigonal phase, and a broad feature around 250 cm^{-1} above 520 K due to melting. Even So clusters in 13X zeolite, which exhibit Raman features similar to those of a-Se at room temperature, do not show evidence of any phase changes up to 530

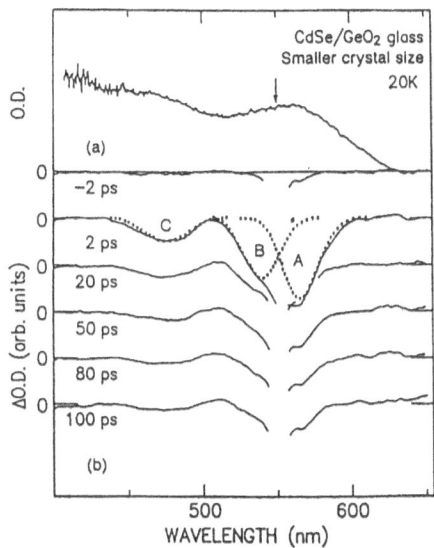

Fig. 9 (a) The steady-state absorption spectrum (O.D.) measured for CdSe microcrystals embedded in a germanium dioxide glass matrix in the strong confinement regime at 20 K. The arrow represents the energy position of the pump pulse ($\lambda_P = 550$ nm). (b) The transient differential absorption spectra (Δ O.D.) at various delay times between pump and probe pulses at 20 K. The dotted lines A, B and C show the decomposition of $1S_A$-$1S$, $1S_B$-$1S$ and $1S_C$-$1S$ transitions, respectively.

K, but show broadening of the Raman modes and a drastic increase in the intensity of 235 cm^{-1} mode relative to that of 255 cm^{-1} mode with increasing temperature, as shown in Fig. 8. Se clusters in 4A zeolite show little broadening of the Raman band with increasing temperature, giving evidence that the clusters are more likely isolated-Se$_8$ ring molecules. The results indicate that Se clusters in zeolites no longer have a definite thermodynamical phase transition but structural relaxation due to thermal excitation.

In conclusion, we have found that the structure of Se is strongly affected by confinement in zeolites, depending strongly on the pore size. The Se clusters in the zeolites show no clear structural phase transformation over a temperature range of 300–530 K. These results indicate that the structural phase stability and the structural unit of Se clusters depend strongly on the geometrical constraints in zeolites. These findings have also been confirmed by differential scanning calorimetry measurements.

7. Femtosecond pump-probe studies of CdSe microcrystals [8]

We have investigated the carrier dynamics of CdSe microcrystals embedded in a germanium dioxide glass matrix by means of a femtosecond pump-probe method.

Figure 9 shows the absorption spectrum and the temporal changes in the transient absorption spectra obtained for CdSe microcrystals embedded in a GeO$_2$ glass matrix in a strong confinement regime (the microcrystal size d is smaller than $2a_B$; a_B is the effective Bohr radius). Figure 9 (a) shows the steady-state absorption spectra measured at 20 K. Figure 9 (b) shows the differential transient

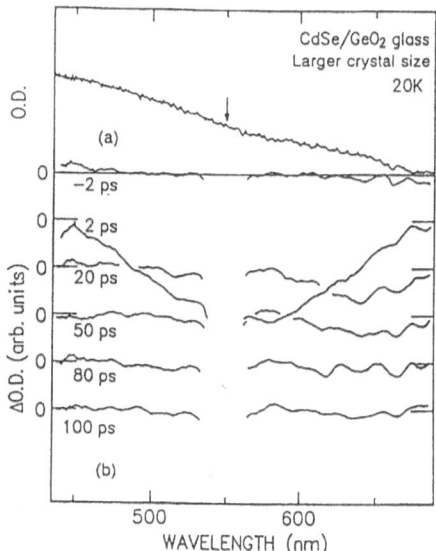

Fig. 10 (a) The steady-state absorption spectrum (O.D.) measured for CdSe microcrystals embedded
in a germanium dioxide glass matrix in the weak confine regime at 20 K. The arrow represents
the energy position of the pump pulse (λ_P = 550 nm). (b) The transient differential absorption
spectra (Δ O.D.) at various delay times between pump and probe pulses at 20 K.

absorption spectra for various delay times between the pump and probe pulses at
20 K. As shown in Fig. 9 (b), the transient absorption spectra of the sample in the
strong confinement regime show the bleaching peaks corresponding to the
transitions between quasi-zero-dimensional confined states (**A**: the transition
between the lowest hole state corresponding to the heavy hole valence band and
the lowest electron state ($1S_A$-$1S$ transition), **B**: the transition between the lowest
hole state corresponding to the light hole valence band and the lowest electron
state ($1S_B$-$1S$ transition), **C**: the transition between the lowest hole state
corresponding to the spin-orbit split-off valence band and the lowest electron
state ($1S_C$-$1S$ transition)). We have found that the absorption bleachings
corresponding to the three quantized transitions recover with the two exponential
components with decay times of about 7 ps and 200 ps, and the three bleaching
peaks recover with almost the same decay times. The fast recovery time could be
attributed to the time constant of the trapping process of excited carriers, and
the slow recovery time to the time constant of the electron-hole recombination
process. On the other hand, the transient absorption spectra of the sample in the
weak confinement regime ($d \geq 3 \sim 4a_B$) exhibit a broad bleaching band due to hot
carriers in a three-dimensional electronic state like a bulk crystal, as shown in Fig.
10.

In conclusion, we have found some important differences in the dynamical
properties of electron-hole systems of the microcrystals between the strong and
weak confinement regimes. In the strong confinement regime, the carrier
dynamics of quasi-zero-dimensional quantum confined transition was observed.
In the weak confinement regime, the dynamical properties of the hot carriers
indicating the three-dimensional electronic states were observed.

8. Effect of oxygen on the structural phase transition of solid C₆₀ and photo-chemical processes of oxygen-diffused C₆₀ films by laser irradiation [13–18]

We measured the Raman spectra of C_{60} films with and without oxygen exposure in the temperature range 18–300 K in order to investigate the effect of oxygen on the structural phase transition at a temperature near 260 K. The splitting and activation of some modes have been observed with decreasing temperature below the structural phase transition temperature, as an indication of the importance of the solid state effect in understanding the vibrational modes of solid C_{60}. While the film without oxygen exposure exhibits a drastic change in the Raman spectrum at the phase transition, the change is much less pronounced for the film with oxygen exposure. In contrast with the high temperature phase, the effect of oxygen on the Raman spectra in the low temperature phase was not seen. These observations indicate the importance of the oxygen effect as well as the effect of condensation of C_{60} molecules into the solid forms in understanding the vibrational modes in solid C_{60}. The fact that the A_g mode at 1469 cm^{-1} is strongly enhanced with oxygen exposure is the most significant feature in the influence of oxygen on the Raman spectrum of high temperature phase of solid C_{60}. The intensity of 1469 cm^{-1} mode becomes nearly 10 times stronger than that of 1461 cm^{-1} mode in one day with oxygen exposure. The appearance of the 1469 cm^{-1} mode is also seen without oxygen exposure when the temperature is

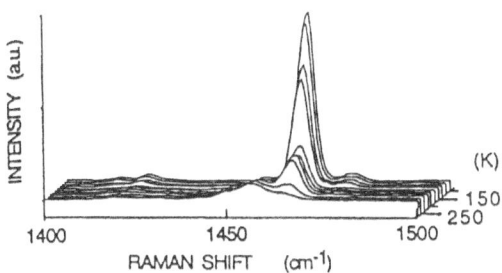

Fig. 11 Temperature dependence of Raman spectrum of C_{60} film without oxygen exposure in the frequency range of 1400–1500 cm^{-1}.

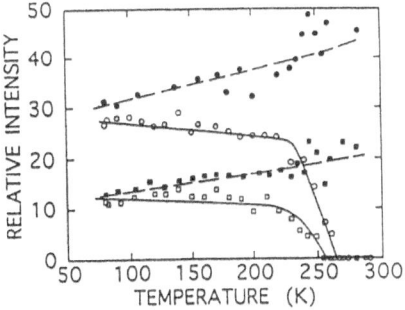

Fig. 12 Temperature dependence of relative Raman intensity of the 1469 cm^{-1} A_g mode to the 1425 cm^{-1} H_g mode in C_{60} films with (●: 514.5 nm excitation, ■: 488.0 nm excitation) and without (○: 514.5 nm excitation, □: 488.0 nm excitation) oxygen exposure.

decreased below the structural phase transition, as shown in Fig. 11. The relative intensity of the A_g mode at 1469 cm^{-1} to the H_g mode at 1425 cm^{-1} is plotted as a function of temperature in Fig. 12. Open and solid circles represent the C_{60} films with and without oxygen exposure, respectively. The oxygen diffusion into the fcc lattice of the high temperature phase could make a change in the electronic state somehow through the slowing of molecular rotation and/or the modification of crystal field due to a symmetry change. The change in the electronic state could occur also through the structural phase transition. We attribute the appearance of the enhanced A_g Raman mode observed in low temperature phase and oxygen-diffused high temperature phase (see Fig. 12) to resonance with the electronic states of the solid forms.

We have also studied in detail the formation of photo-induced chemical products in oxygen-diffused C_{60} films and their properties by Raman scattering, optical absorption and photoluminescence (PL) measurements. The laser irradiation (514. 5 or 488.0 nm) causes the A_g Raman modes at 497 and 1469 cm^{-1} to disappear and the laser irradiated spot to be discolored. In addition, some other vibrational modes are shifted in frequency by laser irradiation. The changes in the Raman spectra are irreversible at room temperature, suggesting that photo-induced chemical products are formed in oxygen-diffused C_{60} films by laser irradiation. The thermal activation energy for the photochemical processes was estimated to be about 0.24 eV using the Arrhenius plots from the dependence of the intensity of 1469 cm^{-1} Raman mode on laser irradiation time at different temperatures. Change in the optical absorption spectrum with laser irradiation is shown in Fig. 13. The photochemical change blurs the optical absorption bands at 2.7 and 3.6 eV while the optical gap at 1.74 eV does not change appreciably. The 2.7 eV band, which corresponds to an optical transition characteristic of the solid forms of C_{60}, disappears completely in 10 hours, whereas the 3.6 band, which corresponds to the lowest HOMO-LUMO transition of C_{60} molecule, remains observable. This result suggests that the form of solid state of C_{60} is modified remarkably in addition to a decrease in the molecularity of C_{60} by the photochemical processes. It has been found from PL measurements that the PL band at 1.80 eV diminishes with laser irradiation, whereas the intensities of 1.50 and 1.68 eV bands increase (the increment of the intensity of 1.50 eV band is larger than that of 1.68 eV band).

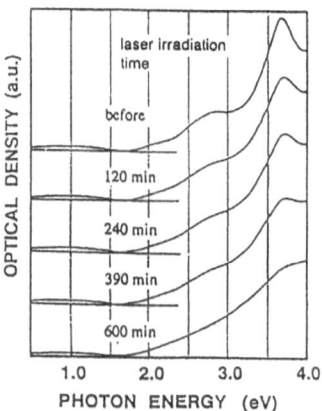

Fig. 13 Time evolution of the optical absorption spectrum of oxygen-diffused C_{60} film under laser irradiation with 488.0 nm line at room temperature.

In conclusion, the molecularity of C_{60} seems to be preserved in the vibrational properties of photo-induced products. The change in the electronic state of C_{60} with laser irradiation, however, indicates the destruction of C_{60} molecules and a remarkable change in the solid form. These observations, therefore, suggest that the electronic excitation with *ca.* 2.5 eV, which gives rise to the resonance Raman scattering of the A_g modes, initiates the photochemical processes of oxygen-diffused C_{60} film and produces a sort of fullerene network as a new type of solid forms of carbon.

9. Optical properties of porous Si

In order to see the origin of the photoluminescence from the porous Si, dependence of oxygen and hydrogen on the infrared spectra and photoluminescence has been studied. The main object was to study whether the origin of photoluminescence is a true quantum effect of Si nanostructure, or is due to the oxygen and hydrogen. The photoluminescence spectra are controlled to some extent by light irradiation during anodization. Large change in the micro structure of porous Si under irradiation is observed by microscope.

PUBLICATIONS

1. Raman scattering from CdSe microcrystals embedded in a germanate glass matrix, A. Tanaka, S. Onari and T. Arai, *Phys. Rev.*, **B45**, 6587–6592 (1992).
2. One Phonon Raman Scattering of CdS Microcrystals Embedded in a Germanium Dioxide Glass Matrix, Akinori Tanaka, Seinosuke Onari and Toshihiro Arai, *J. Phys. Soc. Jpn.*, **61**, 4222–4228 (1992).
3. Low-frequency Raman scattering from CdS microcrystals embedded in a germanium dioxide glass matrix, A. Tanaka, S. Onari and T. Arai, *Phys. Rev.*, **B47**, 1237–1243 (1993).
4. Pressure effects on CdS microcrystals embedded in germanate glasses, T. Arai, T. Inokuma and S. Onari, *Jpn. J. Appl. Phys.*, **32**, suppl.32-1, 297–299 (1993).
5. Raman scattering and photoluminescence of CdS-chalcogenide microcrystals, A. Tanaka, S. Onari and T. Arai, *Supplement to Z. Phys.*, **D26**, 222–224 (1993).
6. Pressure-induced structural phase transition of CdS microcrystals studied by Raman scattering, T. Arai, T. Makino, M. Arai, K. Matsuishi and S. Onari, *J. Phys. Chem. Solids.*, **56**, 491–494 (1995).
7. Observation of the photochemical reactions of Cd and S in zeolite by means of Raman scattering, S. Onari, M. Murai, K. Matsuishi and T. Arai, *The Physics of Semiconductors,* (D. J. Lockwood, ed.), 2039–2042, World Scientific (1995), Proc. of the 22nd Int. Conf. on the Physics of Semiconductors, Vancouver, (Aug. 1994).
8. Femtosecond pump-probe studies of CdSe microcrystals embedded in a germanium dioxide glass matrix, T. Arai, A. Tanaka, K. Matsuishi, S. Onari, Y. Maruyama and M. Ishikawa, *SPIE*, **2362**, 304–311 (1994).
9. Optical properties of CdS and Se clusters in zeolites, T. Arai, M. Murai, J. Ohmori, K. Matsuishi and S. Onari, *Surface Review and Letters*, 3, 707–710 (1996).
10. Study of photodissociation of CdSe nanocrystal beam by means of photoluminescence and Raman scattering, T. Orii, S. Kaito, K. Matsuishi, S. Onari and T. Arai, *J. Phys.: Condensed Matter*, **21**, 4483–4494 (1997).
11. Photoluminescence of granular CdSe films, T. Arai, T. Orii, H. Ichikawa, S. Onari and K. Matsuishi, *Materials Science and Engineering A*, **217/218**, 159–163 (1996).
12. Structural phase stability of Se clusters in zeolites, K. Matsuishi, K. Nogi, J. Ohmori, S. Onari and T. Arai, *Z. Phys. D*, **40**, 530–533(1997).
13. Far Infrared Spectrum of C_{60}, S. Onari, K. Tada and T. Arai, *J. Phys. Soc. Jpn.*, **60**, 4392–4392 (1991).
14. Photochemical processes of oxygen-diffused C_{60} films by laser irradiation, K. Matsuishi, K. Tada, S. Onari and T. Arai, *The Physics of Semiconductors*, (D. J. Lockwood ed.), pp2089–2092, World Scientific (1995), Proc. of the 22nd Int. Conf. on the Physics of Semiconductors, Vancouver, (Aug. 1994).
15. Enhanced Raman scattering and photodissociation of C_{60}-oxygen complexes in C_{60} films and single crystals, K. Matsuishi, R. L. Meng, Y. T. Ren, P. H. Hor and C. W. Chu, *HTS Materials*,

Bulk Processing and Bulk Applications, (C. W. Chu *et al.*,eds.) pp205–210, World Scientific, (1992).

16. Effect of condensation of C_{60} molecules on Raman spectra of solid C_{60}: Study of structural phase transition, K. Matsuishi, K. Tada, S. Onari, T. Arai, *Trans. Mat. Res. Soc. Jpn.*, **14B**, 440–443 (1994).

17. Effect of oxygen on the structural phase transition of solid C_{60} studied by Raman scattering spectroscopy, K. Matsuishi, K. Tada, S. Onari, T. Arai, R. L. Meng and C. W. Chu, *Philosophical Magazine*, **B70**, 795–807 (1994).

18. Effects of laser irradiation on photoluminescence of C_{60} single crystal with/without air exposure, T. Ohno, K. Matsuishi and S. Onari, *Solid State Communications*, **101**, 785–789 (1997).

REFERENCES

a. I. H. Campbell and P. M. Fauchet, *Solid State Commun.*, **58**, 739 (1986).

b. H. Lamb, *Proc. Lond. Math. Soc.*, **13**, 189 (1882).

4

Fundamental and Nonlinear Optical Properties of Semiconductor Mesoscopic Particles

Tadashi Itoh[a], Yasuo Iwabuchi[b], Tsuyoshi Ikehara[b], Masayuki Furumiya[b], Nobuo Katagiri[b], Satoshi Yano[b], Sadayuki Iwai[a], Keiichi Edamatsu[a], Catherine Gourdon[c] and Alexei Ekimov[d]

[a] Department of Applied Physics, Tohoku University
Aoba-ku, Sendai 980-8579, Japan
[b] Department of Physics, Tohoku University
Aoba-ku, Sendai 980-8578, Japan
[c] Groupe de Physique des Solides, Université Paris 6&7-CNRS,
2 place Jussieu, 75251 Paris, France
[d] A.F. Ioffe Physico-Technical Institute, 194021 St. Petersburg, Russia

Purpose of the present study

Semiconductor mesoscopic particles of nm size are considered to be the intermediate phase which bridges between bulk crystals and molecular clusters and are expected to show various kinds of new properties, such as quantum size effect of electronic excited states. In order to clarify these properties, it is essential to study optical phenomena which depend on the size of nanoparticles or on the number of constituent molecules, physical phenomena being characteristic of surface, interface and surrounding matrices, and furthermore, optical nonlinear phenomena enhanced by the carrier confinement.

In the present study we focus on the following optical properties of CuCl nanoparticles which pronouncedly exhibit the confinement of excitons: (1) Exciton coherence plays an important role in various size-dependent properties of mesoscopic nanoparticles. We study the exciton-phonon interaction which disturbs the exciton coherence and how it affects the excitonic optical properties with decrease in size and with change in surrounding matrices. (2) Photofatigue effect of exciton luminescence is caused by laser irradiation. We study its dependencies on nanoparticle size, matrix, temperature and laser intensity. We discuss the formation process of nonradiative center of excitons at the interface. (3) Optical nonlinearities of semiconductor nanoparticles are observed under the simultaneous creation of multi-excitons. We study their correlation with the high density excitation effect of excitons and exciton dynamics by means of ultrafast laser excitation.

The final goal of this study is to summarize and interpret the physical meaning of the above results in order to understand various properties characteristic of nanoparticles of semiconductor zero-dimensional system.

Contents of the present study

1. Introduction

1.1 Characteristic nature of CuCl nanoparticles

If we consider an infinitesimally small particle with a mass M confined inside a spherical quantum well with a radius a, the wavefunction of the particle is expressed by a spherical Bessel function and the eigen energies become discrete, showing so-called quantum size effect. The particle energy measured from the bottom of the well, that is, from the energy of the particle at rest without confinement, is given as

$$\Delta E(a) = \frac{\hbar^2}{2M} \left(\frac{\phi}{a} \right)^2.$$ (1)

Here, $\phi/\pi = 1$ (1S state), $1.43..$(1 P), $1.83..$(1 D), 2(2 S),.... Therefore, when we assume that the particle is an exciton, the mass M is the exciton translational mass M_{ex}, and the quantum well is the exciton potential well defined by the boundary of the spherical semiconductor nanoparticle, then the lowest state of the exciton which is confined inside the nanoparticle shows the high energy shift of ΔE (a, $\phi = \pi$) compared with that of the exciton in the bulk crystal.[a,b] In CuCl, the exciton binding energy (Rydberg constant) is very large, about 200 meV,[1] and the bulk-like exciton is still alive under the perturbation of confinement. In other words, on account of very small exciton Bohr radius a_{ex} of 0.7 nm,[1] the internal motion of the electron and the hole inside the exciton hardly changes and only the exciton translational motion is affected by the confinement. In this case, the exciton wavefunction becomes the standing wave, which means the full coherence of the exciton wave all over the nanoparticle.

On the contrary, in ordinary direct-gap semiconductors, such as CdS, CdSe and GaAs, the exciton internal motion is strongly modified in the nanoparticle and the individual confinement of an electron and a hole is realized because of the large exciton Bohr radius of 3, 4 and 14 nm, respectively. The mass in eq. (1) should be replaced by the exciton reduced mass μ_{ex}, and therefore, the energy shift due to the confinement amounts to the order of eV.[a,b,c] In these materials, we cannot discuss the quantum size effect along with the basis of the bulk exciton, and the concept of exciton coherence no longer exists. These differences are the main reasons why the peculiar exciton confinement effect is expected in CuCl nanoparticles.

In a real situation, the exciton center of mass cannot approach the interface because of the finite extent of the exciton internal motion represented by the exciton effective Bohr radius a_{ex}, and therefore, the peak energy shifts of the exciton spectra can be quantitatively explained by eq. (1) by replacing a to $a^* = a - a_{ex}/2$, which is empirically defined as the effective nanoparticle size.[1] Moreover, since the peak energy of the exciton absorption and luminescence critically depends on the nanoparticle size, the exciton absorption and the luminescence bands are inhomogeneously broadened due to the size distribution.[2]

Idealized exciton state is expressed by the linear superposition of the excited states of constituent molecules all over the nanoparticle. If the dipole-allowed components are mainly concentrated on the lowest exciton state due to the *in-phase* superposition, the exciton radiative decay rate is expected to be proportional to the nanoparticle volume as a result of the summation of all the oscillator strengths of the constituent molecules.[d] However, in real crystals of macroscopic size, the size-dependent lifetime of excitons has not been observed on account of the disturbance caused by the impurity or phonon scattering and by the exciton polariton effect. In contrast, when the particle size is reduced to nm size, which is smaller than the exciton coherence length in the bulk, the exciton lifetime is really size-dependent.[3),e)] According to these experimental results, the exciton full coherence is maintained in nanoparticles with radius $a*$ of less than 5–6 nm at 2 K and the upper limit of the size depends on temperature.

As for the disturbance of the exciton coherence, scattering by interface and phonon are considered. How does the existence of interface, surface or matrix holding the nanoparticles affect on the exciton confinement? Also, how does the exciton-phonon interaction change? Answers to these questions would be useful when considering the confinement effects of excitons in CuCl nanoparticles.

Furthermore, optical nonlinearities of CuCl nanoparticles obtained from absorption saturation[f] and four wave mixing[g] show pronounced size dependence, by indicating a correlation with the exciton coherence. However, in these measurements a laser with a pulse width longer than the exciton lifetime was used and a detailed analysis of the size dependence and the relation with the exciton dynamics has not yet been conducted.

1.2 Outlines of this research report

This report consists of 1 Introduction about the research background and the research concept of semiconductor nanoparticles, 2 Experimental procedures, and 3 Experimental results and discussion about the following subjects together with their mutual relation; 3.1 Exciton linewidth measured by selective excitation method, 3.2 Hole burning spectroscopy, and 3.3 exciton-lattice interaction, 3.4 Photofatigue effect, 3.5 Dependence of the optical spectra on matrices, and 3.6 Optical nonlinearities, including the size dependence and the relation between the temporal response of the nonlinear change in the exciton absorption and the exciton and/or biexciton dynamics. Finally in 4 Conclusion, we summarize the characteristic nature of exciton quantum size effects on semiconductor zero-dimensional system and discuss briefly the future prospects of the newly-found peculiar properties of this nanoparticle.

2. Experimental procedures

2.1 Sample preparation

We have investigated three different samples: CuCl nanoparticles in a crystalline matrix of NaCl and in an amorphous matrix of alumino-boro-silicate glass with molar concentration of 0.1–1 %, and CuCl smoke of 'bare' nanoparticles. Sample preparation methods for CuCl nanoparticles in a NaCl matrix and in a glass matrix are described in Refs. 1 and b, respectively. Annealing temperatures are below 400°C for a NaCl matrix and above 600°C for a glass matrix, indicating crystal growth in solid phase and liquid phase, respectively.

CuCl smoke was prepared in a cryostat by gas-evaporation method[h] using low-pressure He gas atmosphere and collected on a quartz substrate kept at 90 K in order to avoid the coalescence of CuCl nanoparticles and to make *in situ* measurements.

2.2 Experimental methods

2.2.1 Size-selective excitation

A tunable UV light is generated by doubling the frequency of a pyridine 2 dye laser pumped by the second harmonic light of a CW-ML YAG laser and used as a 5 ps tunable excitation light source. The maximum average power density at the sample surface is 300 mW/cm^2 (pulse energy density of 4 nJ/cm^2 at 76 MHz repetition rate). The spectral width is 0.5 meV.

2.2.2 High intensity excitation

For ns resonant excitation, a dye laser (B-PBD) pumped by a nitrogen laser was used. Its pulse width, spectral width and the maximum pump power at the sample surface were 7 ns, 0.5 meV and 1 MW/cm^2. The light transmitted through the sample was fed to a monochromator and monitored for intensity only at the excitation photon energy to avoid the contribution from luminescence. For the hole-burning experiment, an ASE light from the B-PBD dye cell pumped by the same nitrogen laser was used as a probe light and the absorption change was recorded by a combination of a monochromator, a high speed biplanar phototube and a boxcar integrator with a 1 ns sampling head.

For sub-ps time resolved spectroscopy, we used a UV pump-and-probe system to measure the nonlinear absorption change, as shown in Fig. 1. The ps pulse produced by a CW-ML YLF laser was compressed and converted to SH light, which pumped a Rh6G dye laser to produce 608 nm light pulse of 500 fs. The dye laser light pulse was compressed down to 150 fs and amplified in a sulforhodamine dye solution, which was pumped by the frequency-doubled output from a YLF regenerative amplifier of 1 kHz repetition rate seeded by the above CW-ML YLF laser. The output light from the amplifier had the following character: wavelength of 608 nm, repetition rate of 1 kHz, pulse energy of 30 μJ, pulse width of 300 fs. The light was converted into a UV pump light of 304 nm with 1 μJ/pulse by a BBO crystal and the unconverted part of the red light was again focused onto a water cell to produce a probe white light continuum down to a UV wavelength of 340 nm. The pump beam was fed to delay line 2 and then softly focused onto the sample surface with a pump power of 1 mJ/cm^2. The transmitted light beam of the probe beam as well as the reference beam divided from the probe beam were analyzed simultaneously by a pair combination of a spectrometer and a photodiode array to monitor the absorption change at the excited area of the sample. The overall time resolution was estimated to be better than 500 fs. Correction of the group velocity dispersion effect on the white probe pulse was properly taken into account. For the measurement of the time response of exciton luminescence spectra, a streak camera with low repetition rate was used under a time resolution of 20 ps.

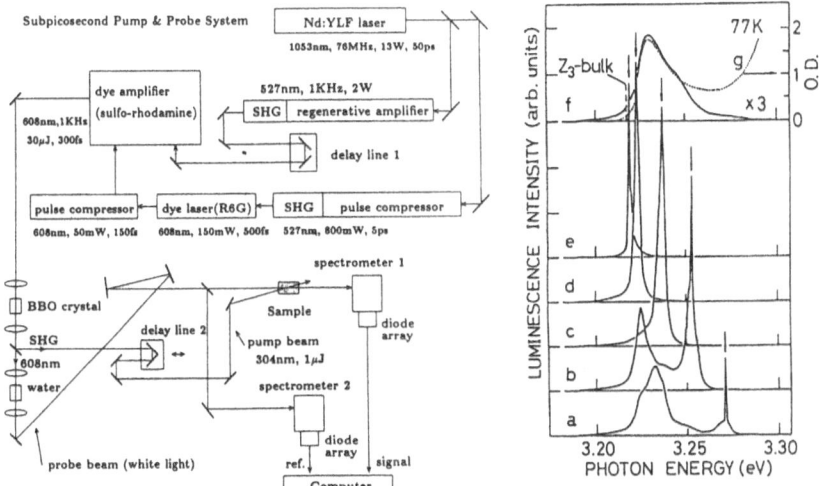

Fig. 1 Block diagram of subpicosecond UV pump-and-probe system developed for the present work. Pump and probe beams are separated by a dichroic mirror put after the BBO crystal for SHG. For details, see text.(Pub.13)
(From T. Itoh, *Mater. Sci. Eng. A.*, MSA **217 & 218**, 168 (1996)).

Fig. 2 Exciton luminescence spectra for $a^* = 3.8$ nm CuCl nanoparticles sample in NaCl at 77 K under size-selective resonant excitation, the energy of which is shown by downward arrows. The uppermost curves (f) (solid line) and (g) (dotted line) are luminescence under nonresonant excitation and absorption spectra, respectively. (Pub.4)
(From T. Itoh, *J. Lumin.*, **48 & 49**, 705 (1991)).

3. Experimental results and discussion

3.1 Exciton homogeneous linewidth

Temperature and size dependencies of homogeneous linewidth Γ_h of the lowest exciton absorption are expected to reveal the exciton-phonon interaction characteristic of confined exciton systems. Here, we study these dependencies derived from the linear optical spectra of exciton resonant luminescence under size-selective excitation. Figure 2 shows the exciton luminescence spectra of CuCl nanoparticles of 3.8 nm in average radius a^* at 77 K with different excitation energies. The excitation energy is indicated by a downward arrow in each spectrum from (a) to (e). For comparison, both the luminescence spectrum under band-to-band excitation and the absorption spectrum are also shown on the top by curves (f) and (g), respectively. Here it is noted that there exists no Stokes shift between the absorption and luminescence peaks, and the luminescence quantum yield is a few tens of percent. The exciton bands in curves (f) and (g) are found to be inhomogeneously broadened due to a wide size distribution of CuCl nanoparticles. Under size-selective excitation among the exciton bands, an intense and narrow luminescence band appears superimposed on the Rayleigh line of the laser light.[4] This narrow band is the resonant exciton luminescence from the nanoparticles of a specific radius a^*. From (a) to (e), the excitation energy is successively decreased towards the Z_3 bulk exciton energy, as indicated by a vertical dashed line in (f). This sequence corresponds to the gradual increase of the effective radius a^*. Here, increase in the linewidth of the resonant luminescence with decreasing crystal size is clearly observed. The linewidth is, in principle, twice as large as the homogeneous width, provided that the homogeneous width is sufficiently smaller than the inhomogeneous width of the exciton absorption band. However, due to successive reabsorption and reemission processes,[5] the relation between the homogeneous width and the resonant luminescence linewidth should be somewhat modified.[4]

Figure 3 shows the homogeneous width versus temperature for different sizes a^* of nanoparticles.[6] Closed circles represent the experimental data. As shown by the dotted lines, the temperature dependence of the width for $a^* = 8.1$ nm resembles that obtained for the longitudinal exciton of CuCl bulk crystal by means of two-photon absorption.[i] In the bulk crystal the temperature-dependent part of the homogeneous width is known to be caused by the intraband scattering of Z_3 exciton through the absorption and emission of an LA phonon at low temperatures, while at high temperatures it is mainly caused by the exciton-LO phonon scattering. We assume that a similar interpretation is applicable for the nanoparticles, as shown by solid lines for theoretical fitting.[4] The strength of the exciton-LA phonon coupling increases with decreasing nanoparticle size a^*. Since the interaction through the deformation potential is a short range interaction,

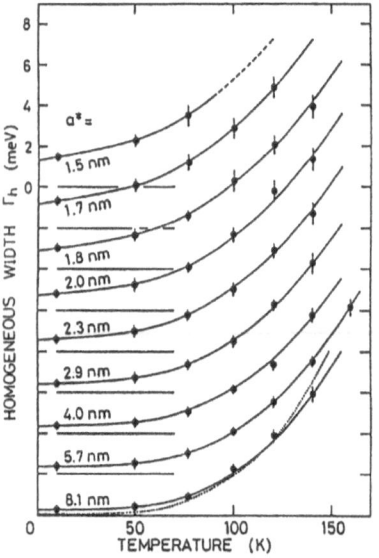

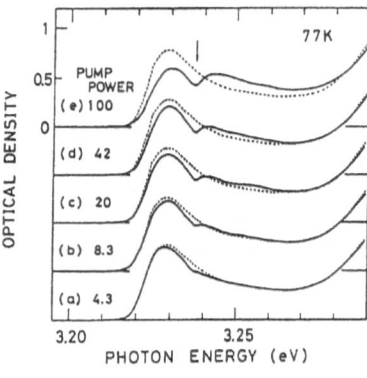

Fig. 3   Temperature dependencies of the homogeneous width for different sizes a^*. Each curve is shifted by 2 meV to avoid overlapping. Closed circles and solid lines represent the experimental data and the calculation described in the text. The dotted line is the bulk width reproduced from Ref. i. (Pub.6)

Fig. 4   Exciton absorption spectra for $a^* = 3.6$ nm sample in NaCl matrix under resonant intense laser excitation with different pump powers at the energy indicated by a downward arrow. The maximum intensity is 150 kW/cm². Dotted curves indicate the spectrum without pumping. (Pub.7)
(From T. Itoh, *Optical Properties of Solids*, (K. C. Lee, P. M. Hui and T. Kushida ed.), p.106, World Scientific (1991)).

it becomes more important for smaller size due to the localization of the vibrational motion of nanoparticles. On the other hand, the strength of the exciton-LO phonon coupling does not change much due to the long-range Froehlich interaction so long as the optical phonon is not confined in a nanoparticle.

It is noted that the condition, in which the homogeneous width is smaller than the confinement energy ΔE, holds except for nanoparticle size larger than 10 nm at 77 K. This result guarantees the observation of the size-dependent rapid radiative decay of confined, coherent excitons observed in this material at low temperatures.[3]

3.2 Hole-burning spectroscopy

Similar to the phenomenon of resonant luminescence line narrowing under size-selective excitation, one may observe the hole-burning effect on the inhomogeneously broadened exciton absorption band under high intensity resonant excitation with a narrow, intense laser, if the nonlinear change of the exciton absorption occurs due to high intensity excitation, such as saturation, shift or broadening. Figure 4 shows such a change in the exciton absorption for CuCl nanoparticles of $a^* = 3.6$ nm at 77 K.[7] Broken and solid lines represent the absorption spectra without and with intense resonant excitation at the energy indicated by a downward arrow. From (a) to (e) the excitation intensity increases as indicated by relative pump powers where the maximum intensity corresponds to 150 kW/cm². Not only the hole burning around the excitation energy but also the absorption decrease on the lower energy side and the increase on the high energy side are clearly observed as the intensity increases. The width of the hole at the lowest pump power is 3 meV, which consistently coincides with that obtained from the resonant luminescence. The absorption change on both sides of the excitation energy is explained as follows: besides the size-selective resonant effect on the lowest exciton state, the excitation energy coincides also with the higher excited states of the excitons in larger nanoparticles, causing the absorption change on the lower energy side of the pumping energy. On the other hand, the increase on the high energy side is mainly ascribed to the induced absorption of the nanoparticles size-selectively excited by the resonant excitation, since the particle already having more than one exciton may show the blue-shifted absorption band for antibonding states of two excitons or for three exciton states. At the excitation power for (e), there appears a biexciton luminescence band with an intensity similar to the exciton one, indicating the existence of more than two excitons in a nanoparticle. Since the response time of the absorption change is nearly or less than ns, its origin is not heating effect or carrier trapping with long lifetime.

3.3 Exciton-lattice interaction

Here, we will note peculiar structures observed at 2 K in CuCl nanoparticles embedded in oxide glass matrix.[8] Samples with mean radii a^* of 1.5, 2.1, and 2.3 nm were grown. The photoluminescence at 2 K is selectively excited by the tunable ps laser light as shown in Fig. 5 for $a^* = 1.5$ nm, where the structures of R, TO, LO and P are observed. The R line appears just on the lower energy side of the excitation energy and has a Stokes shift which increases with the decrease

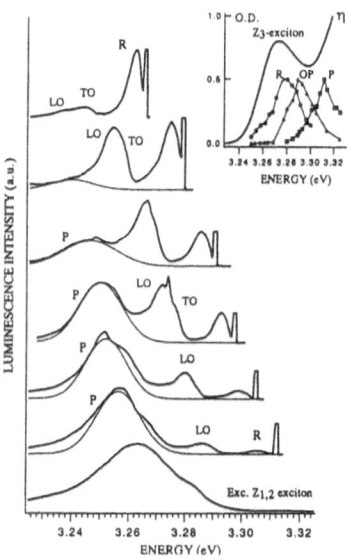

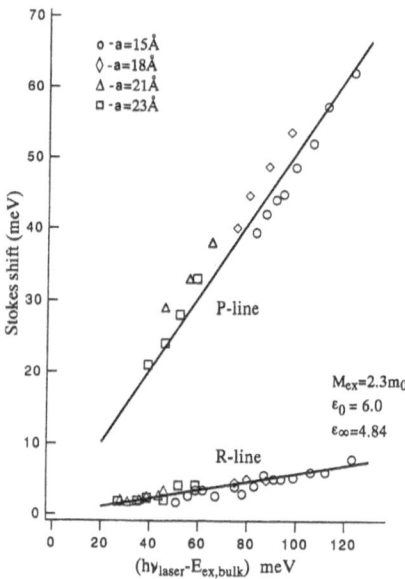

Fig. 5   Luminescence spectra for $a^* = 1.5$ nm sample in glass matrix at 2 K under size-selective excitation. Structures on the high energy edge are Rayleigh scattering of the excitation light. The bottom spectrum is for Z_{12} excitation. The inset shows the absorption spectrum and the excitation efficiency of R, OP(LO+TO) and P lines. (Pub.8)

Fig. 6   Stokes shifts of R and P lines versus confinement energy under size-selective excitation. The confinement energy is the difference between the excitation energy and the bulk exciton energy. Solid lines are the theoretical fits mentioned in the text. (Pub.8)

of the size. This structure is explained by the resonant exciton polaron luminescence, whose Stokes shift is caused by the exciton-LO phonon interaction as indicated by a solid line for R line in Fig. 6.[8]

The TO and LO lines are considered to be the exciton luminescence lines caused by the optical absorption followed by TO and LO phonon emission because of the constant Stokes shifts whose energies nearly coincide with those of TO and LO phonons in the bulk crystal. These structures in a glass matrix are observed more clearly than those in a NaCl matrix, indicating the difference in the exciton-LO phonon interaction between these two matrices. In the glass matrix it is suggested that the optical phonon is likely to be confined in a nanoparticle.

The structure P appears as a broad line which is decomposed from the other luminescence as indicated by a thin solid line. The intensity of the R, OP (TO+LO), and P lines as a function of excitation energy is shown in the inset of Fig. 5, where the maximum excitation efficiency of the R (OP) line is located almost at the Z_3 absorption peak (+TO phonon energy), and its width, as for the absorption, is determined by the size distribution. On the other hand, the excitation width for P line is narrower, and the maximum efficiency of the P line is realized when the Stokes shift is approximately equal to the LO or 2LO phonon energy. This indicates a second phonon resonance within the width of size distribution. As shown by a solid line for the P line in Fig. 6, the Stokes shift is almost a half the energy difference between the excitation laser and the bulk exciton peak. Therefore, the P line is caused by the result of the recombination of the 1 S

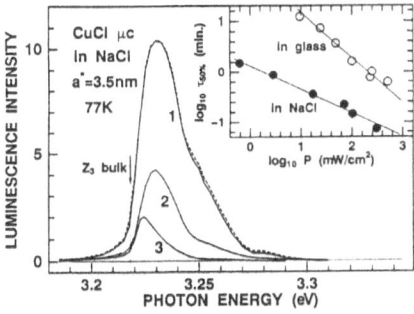

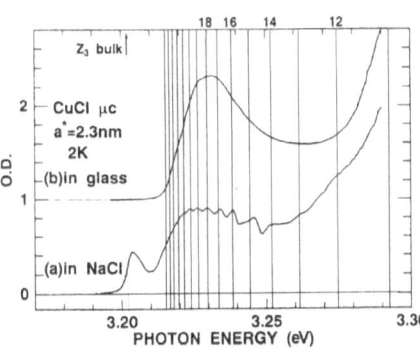

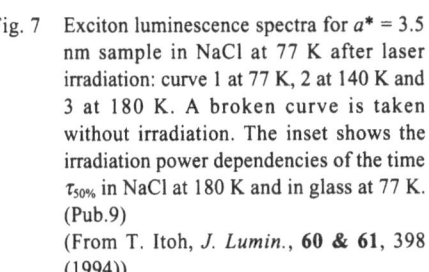

Fig. 7 Exciton luminescence spectra for $a*$ = 3.5 nm sample in NaCl at 77 K after laser irradiation: curve 1 at 77 K, 2 at 140 K and 3 at 180 K. A broken curve is taken without irradiation. The inset shows the irradiation power dependencies of the time $\tau_{50\%}$ in NaCl at 180 K and in glass at 77 K. (Pub.9)
(From T. Itoh, *J. Lumin.*, **60 & 61**, 398 (1994)).

Fig. 8 Exciton absorption spectra for $a*$ = 2.3 nm sample at 2 K: (a) in NaCl matrix and (b) in glass matrix. Vertical solid lines indicate the calculated exciton energies under the assumption of cubic nanoparticles. (Pub.9)
(From T. Itoh, *J. Lumin.*, **60 & 61**, 397 (1994)).

exciton which is selectively excited through the forbidden optical transition to the 1 P confined exciton state. To understand these results, the interaction of excitons with polar optical phonons is considered and the efficiency profile experimentally obtained can be theoretically reproduced.[8] This fact strongly supports the idea that the hybrid 1 S exciton-phonon complex is created through the mixing of 1 S and 1 P exciton states caused by P-type optical phonons.

3.4 Photofatigue effect

The luminescence intensity of the Z_3 exciton was found to decrease after exposure to UV laser light.[9,10] Solid curves in Fig. 7 show the change in the luminescence spectra for the samples of $a*$ = 3.5 nm in a NaCl matrix at 77 K after 10 minute exposure to the UV laser light of 3.35 eV and 200 mW/cm² at different temperatures; curves 1, 2 and 3 are at 77 K, 140 K and 180 K, respectively. The broken curve is the spectrum before the laser irradiation, showing almost no photofatigue effect on the luminescence at 77 K. Since little change occurs in the absorption spectra,[j] the decrease in the luminescence yield is suggested without any change in nanoparticle size. The decrease in the luminescence intensity is more pronounced on the higher energy side, that is, for smaller particle sizes. Therefore, the photofatigue effect is probably caused by some photo-induced change in crystal bonding at the interface region between nanoparticles and matrix through the trapping of carriers, resulting in the formation of nonradiative trapping centers for the excitons. As shown by closed circles in the inset of Fig. 7, the exposure time, $\tau_{50\%}$, necessary for the luminescence intensity to reach 50% of its initial value at 180 K is proportional to $P^{-0.5}$, where P is the irradiation power. Therefore, the photo-induced change for the formation of the exciton trapping center is considered to be caused by the capture of electron and/or hole thermally dissociated from the photo-generated exciton.

3.5 Dependence of the optical spectra on matrices

3.5.1 Shape of the nanoparticles

Figure 8 shows the absorption spectra of CuCl nanoparticles of $a* = 2.3$ nm embedded (a) in a NaCl and (b) in a glass matrices, respectively. Only in a NaCl matrix are oscillatory fine structures [9] observed between 3.22 and 3.28 eV around the Z_3 exciton absorption band, which is inhomogeneously broadened due to the wide distribution of the nanoparticle size. The energy positions of these structures are the same regardless of samples with different average nanoparticle sizes, suggesting the step-wise growth of CuCl nanoparticles in a NaCl matrix. Detailed analysis shows that the nanoparticle shape in a NaCl matrix is not spherical but cubic; the growth proceeds by steps of half the lattice constant of bulk CuCl single crystal of zinc-blende type, where the lattice constant is 0.54 nm. Each vertical thin solid line indicates the expected energy position of the Z_3 confined exciton of the lowest energy at each cube size, which is shown at the top in units of half the lattice constant. Such fine structures are not found for the nanoparticles in a glass matrix where no regular atomic arrangement is expected at the interface and the shape of the nanoparticles is more likely to be spherical due to surface tension in liquid phase growth. This result also suggests a difference in the chemical bonding at the interface region between these samples.

3.5.2 Exciton energy shift due to quantum size effect

In Fig. 9, exciton absorption spectra at 2 K are compared between samples in three different kinds of matrices and with the same average size of $a* = 1.5$ nm, that is, the same ΔE for the Z_3 exciton peak. In CuCl bulk crystal, the Z_3 and $Z_{1,2}$ exciton states are split from each other and the Z_3 exciton state is located on the lower energy side on account of the negative spin-orbit interaction and the exchange interaction in the valence band. The peak energies for the $Z_{1,2}$ band are found to be different between these samples: the highest in a NaCl matrix and the lowest in a glass matrix. [9] The onset of the absorption edge for NaCl is about 8 eV and is much larger than that for glass. Since the lowest Z_3 exciton state for bulk CuCl is 3.2 eV, the barrier height for the exciton confinement is roughly approximated to be infinite in a NaCl matrix, but it is still finite and small in a glass matrix. The influence of finite barrier height, which causes decrease not only in the exciton confinement energy but also in the redistribution of the exciton oscillator strength among the multi-components of the $Z_{1,2}$ exciton, may be the most pronounced for the $Z_{1,2}$ light exciton state located at the highest energy. The exciton barrier height for CuCl smoke is roughly estimated to be in the middle between those in glass and in NaCl, since the energy position of the $Z_{1,2}$ exciton peak is located in this order.

3.5.3 Photofatigue effect

Compared with the effect in a NaCl matrix, the photofatigue effect in a glass matrix is clearly observed even at 77 K. As shown in the inset of Fig. 7, the exposure time, $\tau_{50\%}$ in a glass matrix is almost proportional to P. This fact suggests that the exciton itself or a nonthermally dissociated electron and/or hole causes the

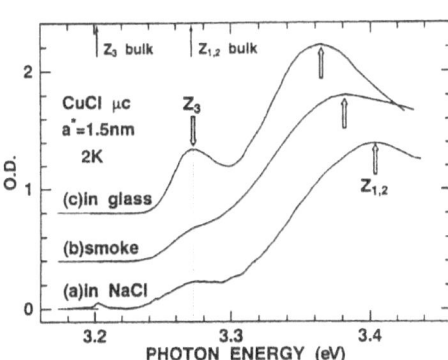

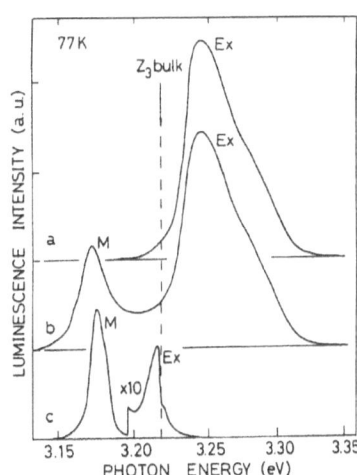

Fig. 9 Exciton absorption spectra for $a^* = 1.5$ nm: (a) in NaCl and (c) in glass, both at 2 K, (b) for smoke at 90 K, but redshifted by 20 meV. Peak positions for Z_3 and Z_{12} excitons are indicated by thick arrows. (Pub.9) (From T. Itoh, *J. Lumin.*, **60 & 61**, 398 (1994)).

Fig. 10 Exciton luminescence spectra for a sample in NaCl under band-to-band excitation, by (a) ordinary UV light, (b) intense laser of 100 kW/cm². (c) is the spectrum for a CuCl single crystal at intense laser excitation. (Pub.7) (From T. Itoh, *Optical Properties of Solids*, (K. C. Lee, P. M. Hui and T. Kushida ed.), p.103, World Scientific (1991)).

formation of the exciton trapping center. Although the difference in the photofatigue effect between these samples is mainly caused by the differences in the interface structure, chemical bonding, defects, etc., the effect of finite barrier height also accelerate the physical or chemical change in the interface region because of the increased probability of finding the confined exciton in this interface region. It is noted that the exciton-phonon interaction is also dependent on the matrices as already shown in 3.1 and 3.3. Therefore, various optical properties of confined excitons in nanoparticles are much influenced by the surrounding matrices.

3.6 Optical nonlinearities

When excitons are generated at high density in CuCl nanoparticles by strong laser irradiation, two excitons combine with each other to stably form a biexciton at low temperatures as in the case of bulk CuCl crystal. Figure 10 shows a comparison of the luminescence spectra, a) under ordinary UV light excitation, b) under nitrogen laser excitation for CuCl nanoparticles and c) for CuCl bulk crystal under the same laser excitation.[11] Since the optical nonlinearity of an exciton is caused by the deviation of the exciton character from the ideal Boson, that is, the existence of the mutual interaction among excitons, it is worth studying the size and pump power dependencies, and also the ultrafast dynamics of exciton and biexciton in order to clarify the origin and the mechanism of enhancement of optical nonlinearities in these nanoparticles.[f,g]

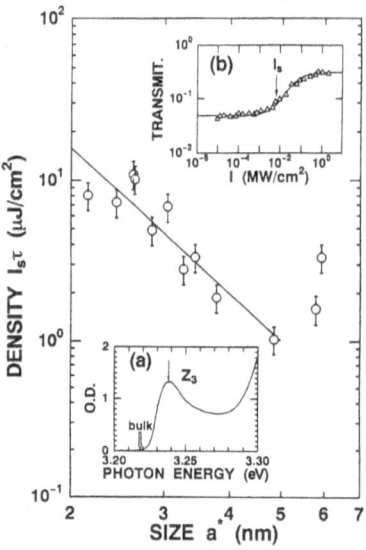

Fig. 11 Size-dependence of the saturation density of exciton for the Z_3 absorption. Open circles represent experimental results and a solid line represents $(a^*)^3$ dependence. Insets (a) and (b) are the Z_3 absorption spectrum and the intensity dependence of the transmittance at the absorption peak, respectively, for $a^* = 2.8$ nm sample in a NaCl matrix. (Pub. 13) (From T. Itoh, *Jpn. J. Appl. Phys.*, **34**, suppl.34–1, 2 (1995)).

3.6.1 Size dependence

The inset a) of Fig. 11 shows the Z_3 exciton absorption spectrum at 77 K for the nanoparticles with an average size of $a^* = 2.8$ nm embedded in a NaCl matrix. For the study of excitonic optical nonlinearity, the optical transmittance at the peak energy of the Z_3 exciton absorption was measured with the use of a ns dye laser beam [f)] and plotted as a function of the laser intensity as shown in the inset b).[12,13] Absorption saturation is clearly observed, mainly caused by the blueshift of the exciton peak as will be shown later. The dependence of transmittance on laser intensity was analyzed taking into account both the decrease of the light intensity inside the sample and the intensity independent term for the absorption coefficient. The result is shown by a solid line where the position of a vertical arrow indicates the derived saturation intensity I_s, which is found to be size-dependent; I_s is smaller for larger sizes except for those larger than 6 nm. Since the size-dependent exciton lifetime τ is shorter than the pulse width of the ns laser, [3)] one must transform the saturation intensity into the saturation density of exciton in order to discuss the size-dependence of the excitonic optical nonlinearities. The saturation density is proportional to the product $I_s\tau$, which is shown in Fig. 11 by open circles for different sizes. For $a^* < 5$ nm, the saturation density is almost inversely proportional to $(a^*)^3$ as indicated by a solid line. This fact means that the number of excitons per nanoparticle necessary to cause a fixed amount of optical nonlinear effect is independent of the size, that is, the existence of one exciton in a nanoparticle is enough to cause the saturation of exciton absorption, since the power density $I_s\tau$ of a few μJ/cm^2 corresponds to the formation of, at most, one exciton in a nanoparticle. This fact is probably explained by the fully coherent nature of the confined exciton and is distinctly different from the optical nonlinearity observed in the bulk crystal where the exciton density is an important

43

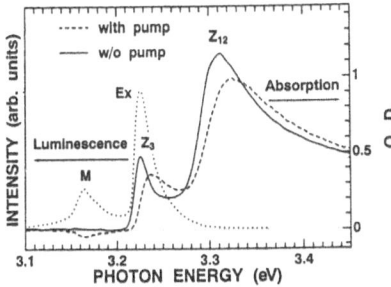 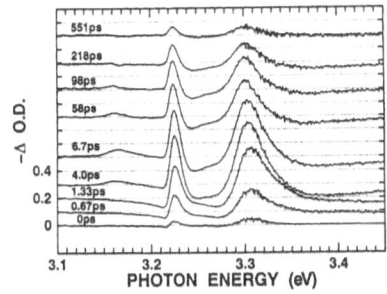

Fig. 12 Absorption spectra for a sample in NaCl without (solid curve) and with (broken curve) the sub-ps band-to-band excitation. The latter is taken at 10 ps after the excitation. Luminescence spectrum with the same excitation condition is also shown by a dotted curve. (Pub. 13)
(From T. Itoh, *Jpn. J. Appl. Phys.*, **34**, suppl.34–1, 2 (1995)).

Fig. 13 Difference spectra around the exciton absorption bands taken at different delay times. The time origin was taken at the time where the absorption started to change. (Pub. 13)
(From T. Itoh, *Jpn. J. Appl. Phys.*, **34**, suppl. 34–1, 2 (1995)).

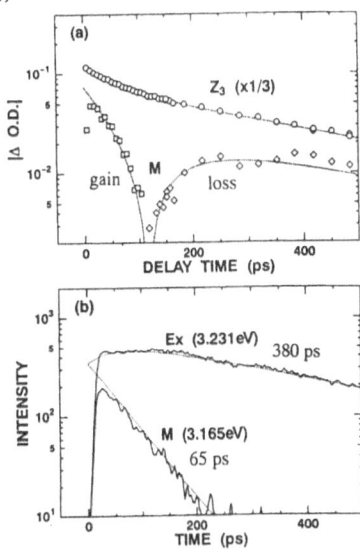

Fig. 14 (a) Delay-time dependencies of the absolute values of the bleaching at the Z_3 absorption peak (open circles), and the gain (squares) and loss (rhombi) around the M band region. Dotted lines are theoretical fits with two decay times. (b) Time responses of the exciton (Ex) and biexciton (M) luminescences. Dotted curves represent the theoretical fits mentioned in the text. (Pub. 13)
(From T. Itoh, *Jpn. J. Appl. Phys.*, **34**, suppl.34–1, 3 (1995)).

parameter. The experimental points in Fig. 11 deviate from the solid line for a^* > 5 nm; this may be closely related to the breakdown of the exciton superradiance through the partial disturbance of the exciton coherence observed in the same range of particle size.[3]

3.6.2 Exciton dynamics and its relation with optical nonlinearity

Figure 12 shows a comparison of the absorption spectra for the nanoparticles

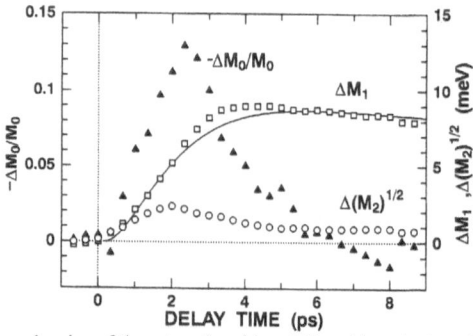

Fig. 15 Delay-time dependencies of the saturation, blue shift and broadening of the Z_3 absorption band obtained by moment analysis for the very early stage just after the sub-ps excitation. The solid line is a theoretical fit based on the cascade relaxation model described in the text. (Pub. 13) (From T. Itoh, *Jpn. J. Appl. Phys.*, **34**, suppl.34–1, 3 (1995)).

of $a^* = 4$ nm in a NaCl matrix at 77 K with and without a fs pump beam excitation, as indicated by a broken line and a solid line, respectively.[13–15,19,21,22] The spectrum with the pump beam excitation was taken at 10 ps after the excitation. Peak shift and broadening are found both for the Z_3 and Z_{12} exciton bands and the gain (negative absorption) appears at 3.165 eV where the biexciton luminescence M exists.[11] Figure 13 shows the difference spectra of the exciton absorption with and without pumping at different delay times after the excitation. The rise of the change in the Z_3 band is a little delayed compared with that in the Z_{12} band, and within 4 ps the absorption change becomes the maximum and lasts for more than 100 ps. At 3.165 eV, the loss observed at the very early stage changes to a gain after 2–3 ps and again becomes a loss after 100 ps. Under the same pump beam excitation as for the pump-and-probe measurement, luminescence was observed as shown by a dotted curve in Fig. 12, where the Z_3 exciton luminescence Ex and the biexciton luminescence M co-existed.[22]

Figure 14(a) shows the time responses of the absolute values of the absorption changes in optical density at the absorption peak (Z_3) by open circles and also at 3.165 eV (M) by open squares or rhombi for delay times longer than 10 ps. The former has two decay components of 60 ps and 380 ps, which coincide well with those for the M and Ex luminescence, 65 ps and 380 ps, respectively, as shown in Fig. 14(b). The decay time of the M band [22] has a value similar to that of the bulk crystal,[16] while the decay time of the confined exciton is governed by the exciton superradiance [3] and hard to compare with the bulk value.[17] For the Ex band, there exists a rise time of 65 ps at the early stage, which coincides with the biexciton luminescence decay time. This fact demonstrates the effective cascade relaxation from the biexciton state to the exciton state in a multi-exciton system. The theoretical fitting according to the above cascade model is shown by dotted curves in Figure 14 (a), which reproduce well the experimental data. The gain and loss observed at 3.165 eV are caused by the induced transition between the exciton and the biexciton states and the temporal behavior of the change from gain to loss around 100 ps is explained by the effective cascade relaxation from the biexciton (two excitons) to the ground state through one exciton state with longer lifetime. By taking into account Poisson's distribution, this result gives a value of two for the average number of excitons per nanoparticle, which is consistent with the average number estimated from the pump power.

In order to further analyze the time response of the exciton absorption change in the very early stage less than 10 ps after excitation, a moment analysis method was found to be useful to separate different kinds of contribution to the exciton absorption change.[14] The 0th, 1st and 2nd moments, M_0, M_1, and M_2 are calculated for the Z_3 absorption band. The changes in M_0, M_1 and $M_2^{1/2}$ represent bleaching, shift and broadening, respectively. Figure 15 shows the early-stage time responses of the absorption changes, $-\Delta M_0/M_0$, ΔM_1 and $\Delta M_2^{1/2}$ as indicated by triangles, squares and circles, respectively. The bleaching and broadening reach maxima within 2 ps after the excitation and the bleaching disappears within 6 ps, while the broadening has two decay components and remains even after 10 ps. The shift reaches maximum at 4 ps and lasts much beyond 10 ps as already mentioned, where the shift dominates over the broadening. The rise time of about 2 ps for the absorption change in the Z_3 band is considered to be the formation time of the Z_3 exciton under the band-to-band excitation. The rise time difference between the Z_3 and Z_{12} excitons is considered to be due to the cascade relaxation among the two exciton states with a successive decay time of ~1 ps, on which basis a theoretical fitting for the blue shift is shown by a solid line. Therefore, during the time of 2 ps after excitation, the main contribution is caused by free carriers (electrons and/or holes) or Z_{12} excitons. The saturation and fast part of the broadening are considered to be caused by the screening effect on the Z_3 exciton, while the blue shift may be caused by the exciton-exciton interaction and/or the phase space filling effect on the Z_3 excitons where they might be more likely to be Fermi particles in the confined system. Further investigation is necessary to clarify the mechanism of the blue shift.

4. Conclusions

In CuCl nanoparticles, the exciton confinement is realized and various kinds of optical properties of excitons characteristic of confined excitons have been found, including exciton-phonon interaction, matrix effect, optical nonlinearity and exciton dynamics. However, the following should be further investigated: (1) theoretical consideration of exciton-lattice interaction of nanoparticles whose exciton has mesoscopic character between local and nonlocal nature, (2) matrix effects which affect the nanoparticle shape, barrier height of quantum well, surface structure, dielectric effects, etc., (3) importance of the phase space filling effect and exciton coherence for optical nonlinearity.

Furthermore, some directions of future prospects in this field are as follows: (1) comparison with the results of the II-IV semiconductor nanoparticles, (2) use of two-photon or two-step absorption spectroscopies [18,20] to investigate excited states of confined excitons which are forbidden for one-photon absorption process, (3) application of methods such as single-molecule spectroscopy [k] or photon STM [l] to investigate the optical properties of a single nanoparticle [m] which are dependent on size and shape, (4) possibility of the laser action of Stokes shifted exciton luminescence with the aid of polaron effect, and (5) energy transfer between nanoparticles at high concentration,[n] which indicates the possibility of realizing a mesoscopic-particle crystal by means of nanoparticle array, etc. From these studies, further progress is expected in both basic and applied optical research work on semiconductor nanoparticles.

Acknowledgments

One of the authors (T. I.) would like to express his sincere thanks to Profs. T. Arai, K. Mihama, Y. Nishina and T. Goto for their useful discussions and encouragement. He is also grateful to Drs. Al. L. Efros and M. Rosen for their theoretical calculation. This work was partially supported by Grants in Aid from the Ministry of Education, Science, Sports and Culture. It was also supported by the exchange program of the Japan Society for the Promotion of Science.

PUBLICATIONS

1. T. Itoh, Y. Iwabuchi and M. Kataoka, *Phys. Stat. Sol.* (*b*), **145**, 567–577 (1988).
2. T. Itoh, Y. Iwabuchi and T. Kirihara, *Phys. Stat. Sol.* (*b*), **146**, 531–543 (1988).
3. T. Itoh, M. Furumiya, T. Ikehara and C. Gourdon, *Solid State Commun.*, **73**, 271–274 (1990).
4. T. Itoh and M. Furumiya, *J. Lumin.*, **48 & 49**, 704–708 (1991).
5. T. Itoh, T. Ikehara and Y. Iwabuchi, *J. Lumin.*, **45**, 29–33 (1989).
6. T. Itoh, *Ceramics*, 27, 508–514 (1992). (in Japanese)
7. T. Itoh, Proc. 1st. Taiwan-Japan Workshop on Solid State Optical Spectroscopy, 80–108, World Scientific (1991).
8. T. Itoh, A. I. Ekimov, C. Gourdon, Al. L. Efros and M. Rosen, *Phys Rev. Lett.*, **74**, 1645–1648 (1995).
9. T. Itoh, S. Yano, N. Katagiri, Y. Iwabuchi, C. Gourdon and A. Ekimov, *J. Lumin.*, **60 & 61**, 396–399 (1993).
10. T. Itoh, M. Furumiya and C. Gourdon, Proc. 2nd Taiwan-Japan Workshop on Solid-State Optical Spectroscopy, 208–212, Osaka, (1992).
11. T. Itoh, *Nonlinear Optics*, **1**, 61–69 (1991).
12. S. Yano, T. Goto and T. Itoh, *J. Appl. Phys.*, **79**, 8216–8222 (1996).
13. T. Itoh, S. Iwai, K. Edamatsu, S. Yano and T. Goto, *Jpn. J. Appl. Phys.*, **34**, Suppl. 34-1, 1–4 (1994).
14. K. Edamatsu, S. Iwai, T. Itoh, S. Yano and T. Goto, *Phys. Rev. B*, **51**, 11205–11208 (1995).
15. S. Yano, T. Goto, S. Iwai, K. Edamatsu and T. Itoh, *Jpn. J. Appl. Phys.*, **34**, Suppl. 34-1, 140–142 (1994).
16. T. Ikehara and T. Itoh, *Solid State Commun.*, **79**, 755–758 (1991).
17. T. Ikehara and T. Itoh, *Phys. Rev. B*, **44**, 9283–9294 (1991).
18. D. Fröhlich, M. Haselhoff, K. Reimann and T. Itoh, *Solid State Commun.*, **94**, 189–191 (1995).
19. K. Edamatsu, S. Yano, S. Iwai, T. Itoh, T. Goto and A. Ekimov, *J. Lumin.*, **66 & 67**, 406–409 (1996).
20. Y. Mimura, K. Edamatsu and T. Itoh, *J. Lumin.*, **66 & 67**, 401–405 (1966).
21. T. Itoh, S. Yano, S. Iwai, K. Edamatsu, T. Goto and A. Ekimov, *Mater. Sci. and Eng. A., MSA* **217 & 218**, 167–170 (1996).
22. K. Edamatsu, *J. Lumin.*, **70**, 377–385 (1996).

REFERENCES

a. Al. L. Efros and A. L. Efros, *Soviet Phys.-Semicond.*, **16**, 772 (1982).
b. A. I. Ekimov, Al. L. Efros and A. A. Onushchenko, *Solid State Commun.*, **56**, 921 (1985).
c. L. E. Brus, *J. Chem. Phys.*, **80**, 4403 (1984).
d. E. Hanamura, *Phys. Rev. B*, **37**, 1273 (1988); ibid. *B* **38**, 1228 (1988).
e. A. Nakamura, H. Yamada and T. Tokizaki, *Phys. Rev. B*, **40**, 8585 (1989).
f. Y. Masumoto, M. Yamazaki and H. Sugawara, *Appl. Phys. Lett.*, **53**, 1527 (1988).
g. T. Kataoka, T. Tokizaki and A. Nakamura, *Phys. Rev. B*, **48**, 2815 (1993).
h. S. Hayashi and K. Yamamoto, *J. Phys Soc. Jpn.*, **56**, 2229 (1987).
i. M. Kalm and Ch. Uihlein, *Phys. Stat. Sol.* (*b*), **87**, 575 (1978).
j. Absorption changes due to persistent hole burning have been recently reported in various semiconductor nanoparticles and summarized in the following reference and those cited therein, Y. Masumoto, *J. Lumin.*, **70**, 386 (1996).
k. S. A. Blanton, A. Dehestani, P. C. Lin and P. Guyot-Sionnest, *Chem. Phys. Lett.*, **229**, 317–322 (1994).
l. J. K. Trautman, J. J. Macklin, L. E. Brus and E. Betzig, *Nature*, **369**, 40 (1994).
m. S. A. Empedocles, D. J. Norris and M. G. Bawendi, *Phys. Rev. Lett.*, **77**, 3873 (1996).
n. C. R. Kagan, C. B. Murray, M. Nirmal and M. G. Bawendi, *Phys. Rev. Lett.*, **76**, 1517–1520 (1996).

5

Nonlinear Coherent Phenomena in Semiconductor Nanocrystals Embedded in Glass

Arao Nakamura and Takashi Tokizaki

Department of Applied Physics, Graduate School of Engineering, Nagoya University
Chikusa-ku, Nagoya 464-0814, Japan

Purpose of the present study

Clusters and nanocrystals of sizes smaller than several tens of nanometers exhibit interesting properties different from bulk crystals. We investigate linear and nonlinear optical response in semiconductor and metal nanocrystsals and amorphous small particles. By means of laser spectroscopy in the femto- and pico-second time region, we study relaxation dynamics of population and coherence of excited states in I-VII and II-VI semiconductor nanocrystals and Cu small particles. We elucidate quantum size effects on electron-hole pairs and large enhancement of third-order nonlinear susceptibility in semiconductor noncrystallines.

Contents of the present study

1. Introduction

In recent years interest is increasing in the development of materials and devices that are capable of processing light signals without converting them to electronic forms. The low dimensional semiconductor in which carriers are confined by the potential wall is one of the most promising materials because of the large optical nonlinearity and fast response time. In particular, the quantum well and superlattice structures of GaAs/AlGaAs have been intensively studied not only for optoelectronic devices but also for all-optical ones. In nanocrystals with radii of less than ~100 Å, electrons, holes and excitons are confined three-dimensionally by the matrix potential. Third-order susceptibilities $\chi^{(3)}$ ranging from 10^{-10} to ~3×10^{-8} esu were observed for CdSSe-doped glasses. [a–b] Larger enhancement of $\chi^{(3)}$ due to the quantum size effect has been theoretically predicted for larger nanocrystals or materials with a relatively small exciton Bohr radius. [f,g] Nonlinear absorption measurements and degenerate four-wave mixing measurements for CuCl nanocrystals have proved that the $\chi^{(3)}$ is enhanced to 10^{-6} esu depending on the crystallite size. [h,i,1–5]

The optical nonlinearity of small metal particles embedded in glass are also attractive for potential applications in nonlinear optical signal processing and optical devices because of their ultrafast response. Large picosecond optical nonlinearities have been shown for Au, Ag and Cu particles and the nonlinearity is attributed to the enhancement of the local field inside the particle and the

47

nonequilibrium electron heating at the surface plasmon resonance including the interband transition. [l,m,7)]

In this article, we review quantum size effects on photoexcited electron-hole pairs and excitons and large enhancement of optical nonlinearity in semiconductor nanocrystals. We give an experimental evidence for the mesoscopic enhancement of $\chi^{(3)}$ due to the size-dependent oscillator strength of the confined exciton. Furthermore, we demonstrate the subpicosecond time response of Cu nanocrystal-doped glass and quantum beats for confined excitons in CuBr nanocrystals by femtosecond nonlinear spectroscopy.

2. Experimental

CuCl nanocrystals used in the present study were obtained by heat treatment of borosilicate glasses doped with CuCl. The diffusive decomposition processes of supersaturated solid solution produce CuCl nanocrystals with radii of 15 to 80 Å in the volume fraction of 0.1 to 0.5 %. Cu nanocrystals with a radius of 40 to 320 Å in glass were prepared by a conventional melt and heat-treatment procedure using a $50BaO-50P_2O_5$ mole percent glass system. To measure absolute values of $\chi^{(3)}$ we used degenerated four-wave mixing (DFWM) with two incident beams of an excimer pumped dye laser. Femtosecond pump and probe experiments were performed to measure nonlinear response times in Cu-doped glass. The wavelength and duration of pumping pulses were 605 nm and 100 fs, respectively.

3. Results and discussion

3.1 Semiconductor nanocrystals

In the three-dimensional confinement of an electron-hole pair, there appear two limiting situations according to the ratio of the crystallite radius R to the Bohr radius a_B of exciton. For $R/a_B < 2$, an electron and a hole are individually confined. If the confinement potential is spherical and infinite, the lowest energy of an electron-hole pair is written by the following equation,

$$E = E_g + \hbar^2\pi^2/2\mu R^2 \tag{1}$$

where E_g and μ are the band gap and the reduced mass, respectively. For $R/a_B > 4$, the exciton is weakly confined. Since the confinement energy us less than the Coulomb energy, the character of the exciton as a quasiparticle is well conserved and the translational motion is confined. In this case, the lowest energy of exciton is given by the following equation,

$$E = E_g - E_{ex} + \hbar^2\pi^2/2MR^2 \tag{2}$$

where E_{ex} and M are the exciton binding energy and the translational mass, respectively.

Quantum size effects have been observed in various semiconductor nanocrystals in the glass matrix. Figure 1 shows absorption spectra of CdSe nanocrystals ($a_B = 45$ Å) measured at 2 K.[5)] In CdSe nanocrystals for $R < 31$ Å, absorption edge is shifted to the shorter wavelength side, and for radii of 21 and

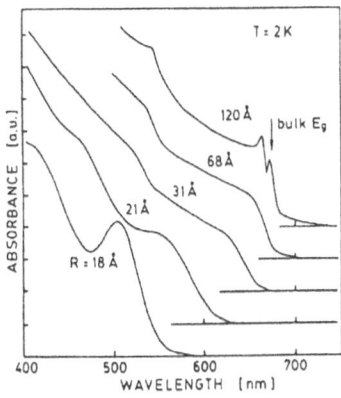

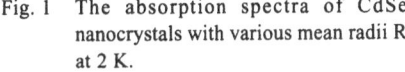

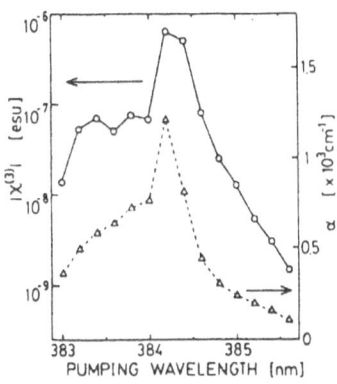

Fig. 1 The absorption spectra of CdSe nanocrystals with various mean radii R at 2 K.

Fig. 2 The absorption spectrum($\triangle$) and the absolute value of $\chi^{(3)}$ ($\bigcirc$) near the exciton energy of CuCl nanocrystals a the radius of 33 Å at 80 K.

18 Å we observe distinct absorption peaks. These peaks are due to the optical transition of the lowest quantized level of the A (or B) valence band to the lowest level of the conduction band. This behavior indicates the individual confinement of an electron and a hole. If the crystallite radius is increased to more than 100 Å, we can see the exciton peak due to the A exciton and B exciton and the blue shift of these peaks indicated the exciton confinement. The confinement effect on the translational motion of excitons was more clearly demonstrated for CuCl nanocrystals in which the exciton Bohr radius is relatively small to be 6.8 Å.[2,n]

Figure 2 represents the $|\chi^{(3)}|$ and absorption spectra measured for CuCl nanocrystals at 80 K.[2] The mean radius of the nanocrystals is 33 Å and the peak power density of the pumping beam is ~kW/cm². When the wavelength of the pumping laser is tuned at the absorption peak due to the exciton confined in nanocrystals, the resonant enhancement of $|\chi^{(3)}|$ is observed. The absorption spectrum exhibits a peak at the Z_3 exciton energy, which is shifted towards the shorter wavelength side because of the quantum confinement. The values of $|\chi^{(3)}|$ illustrated by open circles shows a peak at the absorption peak energy. The resonance enhancement is of two orders of magnitude.

Figure 3 shows a figure of merit $|\chi^{(3)}|/\alpha T_1$ as a function of R.[4] Here, α is the absorption coefficient and T_1 is the response time (lifetime of excitons). The figure of merit is increased with increase of R and subsequently decreased after reaching a maximum value $a \sim 45$ Å. The radius-dependence for $R < 45$ Å is approximately R^3. Consequently, we can obtain the higher polarizabilities with the faster response for $R < 45$ Å. The size-dependent behavior of the figure of merit suggests that the oscillator strength and/or homogeneous width depend upon the radius.

Next, we discuss the origin of the size-dependent figure of merit. Taking into account a two-level atomic model, $|\chi^{(3)}|/\alpha T_1$ is related to the homogeneous width Γ_h, the lifetime T_1 and the oscillator strength f_x as

$$\chi^{(3)}/\alpha T_1 = 1.3 \times 10^{17} \cdot \frac{n}{\left(n^2 + 2\right)^2} \cdot \left(\frac{e^2}{2m_0\omega}\right)^2 \cdot \frac{\hbar f_x}{\Gamma_h} \tag{3}$$

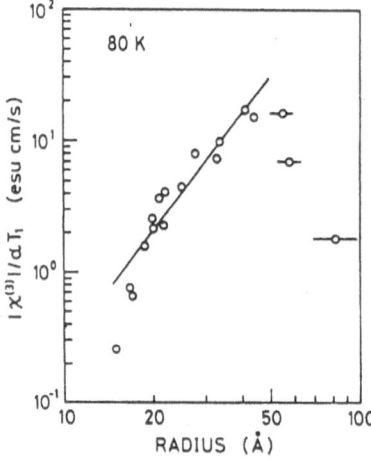

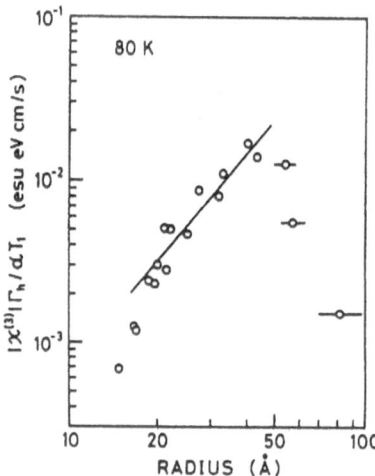

Fig. 3   The figure of merit $|\chi^{(3)}|/\alpha T_1$ as a function   Fig. 4   The size dependence of $|\chi^{(3)}|\Gamma_h/\alpha T_1$ ($\propto$
of crystallite radius. The straight line f_x) at 80 K. The straight line indicates
indicates the R^3 dependence. $R^{2.2}$ dependence.

where n is the refractive index. To investigate the contribution of homogeneous
width to the size-dependence of the figure of merit, we measured values of Γ_h for
the samples studied here by resonant luminescence experiments. Shown in Fig. 4
is the size-dependence of $|\chi^{(3)}|\Gamma_h/\alpha T_1$ obtained at 80 K.[4] According to Eq. (3) this
quantity is proportional to f_x. Therefore, this result exhibits clearly that the
oscillator strength is size-dependent and enhanced upon an increase of R for $R <$
40 Å. We obtain the dependence of $R^{2.2}$ from the least square fit of the radius
dependence for 19 Å $< R <$ 40 Å. This dependence is in good agreement with the
dependence of $R^{2.1}$, which was derived from the radiative lifetimes.[11] This
dependence, however, is smaller than the theoretically calculated one on the basis
of the giant oscillator strength effect on the confined exciton.[f] The main reason
for the smaller dependence is the finite confinement potential of the matrix glass
in which the gap is 4.3 eV, while the band gap of CuCl is 3.42 eV.

In what follows, we can estimate values of fx using the measured values of the
figure of merit and Γ_h. Taking the value of $|\chi^{(3)}|\Gamma_h/\alpha T_1$ to be $1.5 \times 10^{-3} \cdot$ esu $\cdot$ eV
$\cdot$ cm/s for $R = 15$ Å and 3.4×10^{-2} esu $\cdot$ eV $\cdot$ cm/s for 40 Å, we estimate $f_x \sim 0.24$
and 5.5 for 15 Å and 40 Å, respectively. If we compare the values with that of the
bulk exciton in CuCl crystal, which is $f_{z3} = 5.85 \times 10^{-3}$, the enhancement factor
f_x/f_{z3} is 41 to 940.

3.2 Cu nanocrystals in glass

A recent study has shown that copper nanocrystals in glass provides the
highest susceptibility ($\chi_m^{(3)} \sim 2 \times 10^{-6}$ esu) among other noble metals.[7] If the
relaxation process of the nonequilibrium electrons is determined by the electron-
phonon interaction as was proved for thin films, the response time is expected to
be the cooling time of hot electrons. It is of interest to measure it for such a
nonlinear material with high susceptibility.

Figure 5(a) shows the linear absorption spectrum of copper particles with a

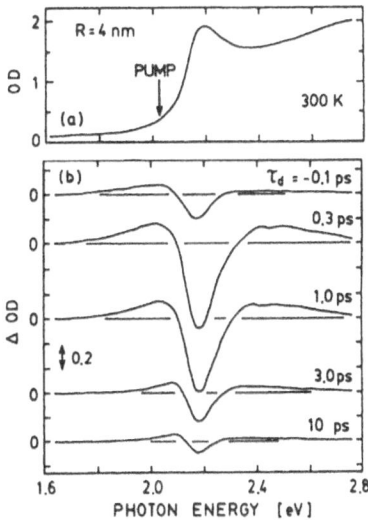

Fig. 5 (a) The linear absorption spectrum of copper nanocrystals with a radius of 40 Å at 300 K, (b) the differential absorption spectra for various delay times.

radius of 40 Å at 300 K. A surface plasmon peak is observed at 2.2 eV and a large shoulder is due to the transition from the d band. [8] Shown in Fig. 5(b) are differential absorption spectra for various delay times. The spectra exhibit a bleaching at the plasmon peak and the absorption increase emerges on both sides of the peak, which suggests broadening of the plasmon absorption band.

Figure 6 shows the time response measured at the early stage of time delay for different laser fluences. [8] The bleaching signal decreases within ~4 ps and the decay behavior is non-exponential depending on the pumping laser fluence. The bleaching at 4.0 mJ/cm² recovers in 4 ps, while the recovery time at 0.2 mJ/cm² is shortened to be ~1 ps. This behavior for the small particles is quite similar to the results of the thermomodulation observed for thin copper films by Elsayed-Ali et al. [o] As the laser fluence was changed from 11.4 mJ/cm² to 1.4 mJ/cm², the decay time decreased from 4 ps to 1 ps. If we define the response time by the time at which the signal decays to $1/e$, we obtain $\tau_{1/e}$ ~0.7 ps for 0.2 mJ/cm². This demonstrates that we can achieve a subpicosecond time response for the copper particles at the lower laser level.

We now examine the relaxation dynamics of the electron temperature using the usual electron-phonon coupling model.[p] The time evolution of the electron and the lattice effective temperatures, T_e and T_1, respectively, is described by the coupled differential equations:

$$C_e T_e \frac{\partial T_e}{\partial t} = \kappa \Delta^2 T_e - G(T_e - T_1) + P(t) \tag{4}$$

$$C_1 \frac{\partial T_1}{\partial t} = G(T_e - T_1) \tag{5}$$

where C_e (T_e) is the temperature-dependent electronic heat capacity, C_1 is the lattice heat capacity, κ is the thermal conductivity, G is the electron-phonon coupling constant, $P(t)$ is the excitation energy density per unit time and unit

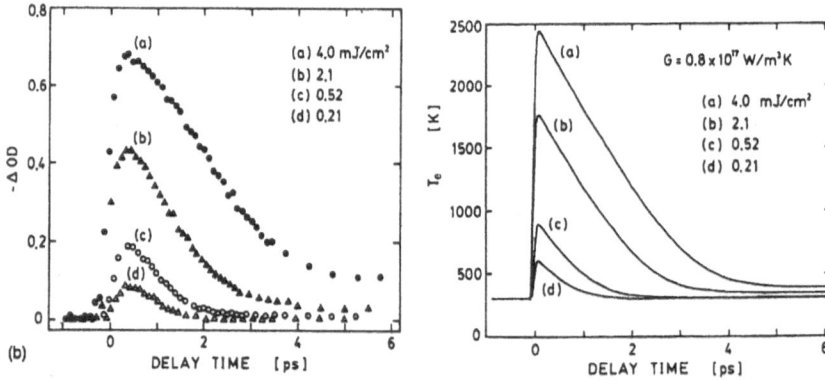

Fig. 6   The time evolution of the bleaching signal measured at 2.2 eV for various laser fluences.

Fig. 7   The calculated results of the electron temperature as a function of the delay time for measured laser fluences.

volume in a particle.

In Fig. 7 we show the calculated results of the electron temperature as a function of the delay time for the pumping fluences of 0.21, 0.52, 2.1, 4.0 mJ/cm² used in the experiments.[8] We used the values of C_e and C_l for the bulk Cu crystal and the value of $G = 0.8 \times 10^{17}$ W/m³ K, which is slightly smaller than that for the thin film ($\sim 1 \times 10^{17}$ W/m³ K).[o] The decay behavior depending on the pumping fluence can be well reproduced by this model.

3.3 Quantum beats in CuBr nanocrystals

In CuBr with zinc-blende structure the Z_{12} exciton consists of the multicomponent exciton states due to the light and heavy holes and a k-linear term in the Γ_8 valence band.[q] The confinement of the multicomponent excitons forms a multilevel system. In such a system we can expect quantum interference by simultaneous excitation of the states by a light pulse with a duration shorter than the reciprocal of the spliting frequency.[r] Here, we represent quantum beats which were observed by a newly developed method of DFWM experiments with femtosecond time resolution.

Experiments of DFWM with two-beam configuration were carried out using the second harmonics of a femtosecond mode-locked Ti:sapphire laser. The pulse duration and the spectral width are ~200 fs and 5 nm, respectively. The spectral width is large enough to cover two distinct peaks of the Z_{12} exciton band. The self-diffracted signal due to third-order nonlinear polarization is observed in the direction of $2\vec{k}_2 - \vec{k}_1 (2\vec{k}_1 - \vec{k}_2)$, where $\vec{k}_1$ and $\vec{k}_2$ are the wavevectors of two incoming beams. The decay of the diffracted signal with increasing delay time between the incoming pulses is a measure of the phase relaxation of excitons. In the case of the *spectrally resolved* DFWM (SR-DFWM), the self-diffracted signal light is analyzed using a monochromator with a band width of 0.2 nm.

Figure 8 shows the absorption spectrum of CuBr nanocrystals with radius R = 200 Å at 77 K.[9] The Z_{12} and Z_3 exciton bands are shifted to higher energy sides compared to the exciton energies of a bulk crystal, and the Z_{12} exciton band is split into two peaks. The dashed line shows a spectrum of pumping pulses,

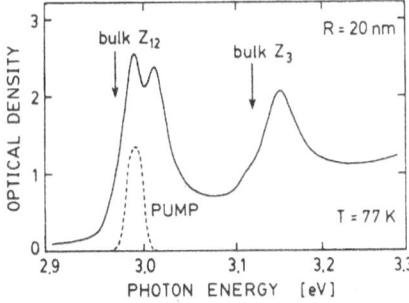

Fig. 8 The absorption spectrum of the CuBr
nanocrystal (R = 200 Å) in glass at 77K (solid
line) and the spectrum of the excitation pulses
(dashed line).

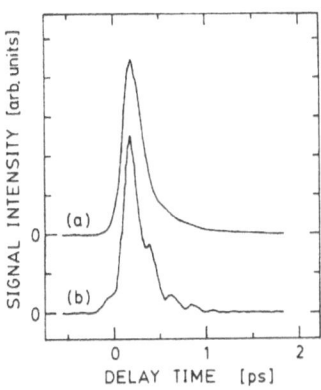

Fig. 9 The correlation curve of conventional
DFWM (a), and that of SR-DFWM
(b).

which covers both peaks. The correlation trace of the conventional DFWM is shown in Fig. 9(a). In this configuration, the self-diffracted signals were not spectrally resolved and all signals in the direction of $2\vec{k}_2 - \vec{k}_1$ were detected. The signal exhibits a decay behavior with a time constant of 0.35 ps, which indicates a finite optical dephasing time T_2 longer than the pulse duration. Assuming that the absorption profile is inhomogeneously broadened, the value of T_2 is obtained to be 1.4 ps, which corresponds to a homogeneous width of 0.5 meV, defined by the half width at half maximum of Lorentzian profile. Although both peaks of the Z_{12} exciton band are simultaneously excited by the femtosecond pulses, no oscillatory structure corresponding to quantum beats could be observed.

Next, we measured spectrally-resolved signals of DFWM using a monochromator. The correlation trace of the $2\vec{k}_2 - \vec{k}_1$ signal detected at a photon energy of 3.002 eV is shown in Fig. 9(b). An oscillatory structure superimposed onto the exponential decay is observed. The oscillation period is 220 fs, which corresponds to an energy of 19 meV. This energy coincides approximately with the splitting energy of the Z_{12} exciton band (23 meV). Therefore, we found that the quantum beat can be observed by SR-DFWM, while it is not observed if the whole spectrum of the self-diffracted signals is detected.

We discuss theoretically quantum interference effects in the three-level system with the inhomogeneous broadening. We carried out numerical calculations of signals which are measured by the conventional DFWM and SR-DFWM. As shown in the inset of Fig. 3, the three-level system consists of the states $|a>$, $|b>$ and $|g>$, and the transition energies between the ground state $|g>$ and the excited states, $|a>$ and $|b>$ are denoted to ω_a and ω_b, respectively. In order to describe the inhomogeneous broadening of the three-level system, we assume that ω_a and ω_b are proportional to a parameter x. We further assume that each three-level system has the same T_2 and no energy relaxation among three levels is included. Using the theory in Ref. s), the third-order polarization with the wavevector of $2\vec{k}_2 - \vec{k}_1$ at the time t for the three-level system with the parameter x is given by

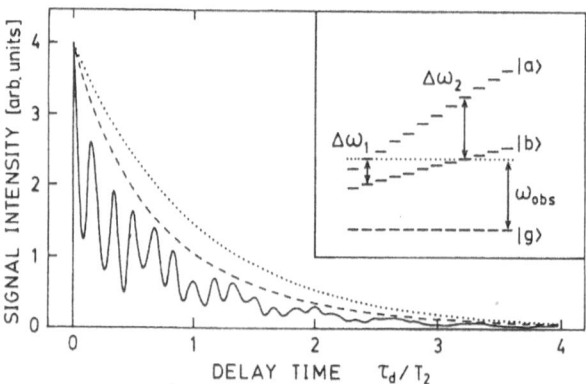

Fig. 10 The calculated correlation curve of the conventional DFWM for a three-level system (dashed line), a two-level system (dotted line) and the calculated correlation curve of SR-DFWM for three-level system. The inset shows the model of the three-level system with inhomogeneous broadening.

$$P_x^{(3)}(t) = C \, |\mu|^4 \sum_{n,m=a,b} \int dt_3 \int dt_2 \int dt_1 \Big\{ E_2(t_3) E_1^*(t_2) E_2(t_1) + E_2(t_3) E_2(t_2) E_1^*(t_1) \Big\}$$
$$\times \Big\{ 1 + \exp\big[-\{ \Gamma_{ab} + i\omega_{ab}(x) \}(t_3 - t_2) \big] \Big\} \tag{6}$$
$$\times \exp\Big[-\{ \Gamma + i(\omega_m(x) - \Omega) \}(t_2 - t_1) - \{ \Gamma + i(\omega_n(x) - \Omega) \}(t - t_3) \Big]$$

where E_1 and E_2 describe the electric fields of the pumping pulses, and μ and Γ are the dipole matrix element and the reciprocal of T_2, respectively. Ω is the carrier frequency of the pumping pulse. ω_{ab} and Γ_{ab} are the frequency difference and the reciprocal of the dephasing time between $|a>$ and $|b>$, respectively. The macroscopic polarization is obtained by the integration over the whole inhomogeneous broadening. Assuming the infinite broadening of the inhomogeneous width and the d function of the excitation pulse, the macroscopic polarization is written by

$$P^{(3)}(t, \tau_d) = C \int_{-\infty}^{+\infty} dx \, e^{-\Gamma t} \sum_{n,m=a,b} \exp\Big\{ -i(\omega_n(x) - \Omega)(t - \tau_d) + i(\omega_m(x) - \Omega)\tau_d \Big\}. \tag{7}$$

The correlation signal of the conventional DFWM is obtained by the time integration of the square of the polarization. For SR-DFWM, we pick up a frequency component ω_{obs} of the polarization by the time integration of $P^{(3)}(\tau, \tau_d)$ multiplied by $\exp(i\omega_{obs}t)$.

The dashed line in Fig. 10 shows the calculated correlation trace of the conventional DFWM for the inhomogeneously broadened a three-level system. The trace shows no beat structure or single exponential decay. The decay time is shorter than that for the two-level system with the same T_2 and infinite inhomogeneous width, as shown in Fig. 10 by the dotted line. The disappearance of the beating is explained in terms of the destructive interference effects of the electromagnetic waves with various beat frequencies which are due to quantum beats of three-level systems with different transition frequencies. Shown in Fig. 10 by the solid line is the correlation trace of SR-DFWM at the photon energy

$\hbar\omega_{obs}$ and clearly demonstrates quantum beats. The fluctuation of beat amplitudes noticed in the figure indicates the interference effect of the optical field with two beat frequency components, $\Delta\omega_1$ and $\Delta\omega_2$. Therefore, the theoretical calculations allow us to find that in the three-level system with inhomogeneous broadening we can observe quantum beats only by SR-DFWM. Comparing the calculated results with the experiments, an oscillatory structure with a single period was observed instead of the complicated beating behavior. This is due to the finite inhomogeneous broadening of the exciton system as well as the finite spectral width of the excitation pulses in the experiments.

Finally, we briefly discuss the origin of the splitting of the Z_{12} exciton band. Taking a simple picture of the quantum confinement of excitons, the confinement of the translational motion takes place at an exciton wavevector of $K = \pi/R$ on a dispersion curve. From the dispersion curves along the [110] direction, [9] we find that the energy splitting between the transverse heavy-hole exciton and "longitudinal" light-hole exciton is about 15 meV. In bulk crystal, the latter excitons are weakly dipole-active because of the mixing with the transverse excitons due to the k-linear term. However, the lack of translational symmetry in spherical nanocrystals may cause strong mixing, and this seems to be responsible for the observed splitting.

4. Summary

We have investigated the quantum size effects and the third-order optical nonlinearity of semiconductor and metal nanocrystals embedded in glasses. We have found that the giant oscillator strength effect on the confined exciton in CuCl nanocrystals gives rise to a size-dependent enhancement of the $\chi^{(3)}$. The quantum confinement of the exciton in the nanocrystal is externally imposed by the barrier potential. Therefore, an important and interesting feature of this system is to engineer the nonlinear optical properties associated with the oscillator strength changing the extent and the height of the barrier potential. The nonlinear time response of Cu nanocrystals has been investigated. The response time is as short as 0.7 ps, which depends on the laser fluences. Simulation has shown that we can attain a faster response for the lower pumping fluence. Furthermore, we have investigated quantum beats due to the multicomponent excitons in CuBr nanocrystals by means of the femtosecond DFWM. Quantum beats were not observed by conventional DFWM, but a clear beat structure could be measured by the SR-DFWM method. The three-dimensional confinement effect on the multicomponent excitons in CuBr nanocrystals gives rise to a three-level system with inhomogeneous broadening due to size distribution.

PUBLICATIONS

1. A. Nakamura, T. Tokizaki, T. Kataoka, N. Sugimoto and T. Manabe, Tech. Digest Int. Quantum Electronics Conf. Anaheim, p178 (1990).
2. A. Nakamura, T. Tokizaki, H. Akiyama and T. Kataoka, *J. Lum.*, **53**, 105 (1992).
3. T. Tokizaki, T. Kataoka, A. Nakamura, N. Sugimoto and T. Manabe, *Jpn. J. Appl. Phys.*, **32**, L782 (1993).
4. T. Kataoka, T. Tokizaki and A. Nakamura, *Phys. Rev.*, **B48**, 2815 (1993).
5. T. Tokizaki, H. Akiyama, M. Takaya and A. Nakamura, *J. Crys. Growth.*, **117A**, 603 (1992).
6. A. Nakamura, H. Yamada and T. Tokizaki, *Phys. Rev.*, **B40**, 8585 (1989).
7. K. Uchida, S. Kaneko, S. Omi, C. Hata, H. Tanji, Y. Asahara, A. J. Ikushima, T. Tokizaki and

A. Nakamura, *J. Opt. Soc. Am.*, **B11**, 1236 (1994).

8. T. Tokizaki, A. Nakamura, S. Kaneko, K. Uchida, S. Omi, H. Tanji, Y. Asahara, *Appl. Phys. Lett.*, **65**, 941 (1994).
9. T. Tokizaki, A. Nakamura, Proceedings of International Conference on Physics of Semiconductors p.2019 (1995).
10. A. Nakamura, Y. L. Lee, T. Kataoka and T. Tokizaki, *J. Lumin.*, **60&61**, 376 (1994).
11. H. Hosono, Y. Abe, Y. L. Lee, T. Tokizaki and A. Nakamura, *Appl. Phys. Lett.*, **61**, 2747 (1992) .

REFERENCES

a. R. K. Jain and R. C. Lind: *J. Opt. Soc. Am.*, **73**, 647 (1983) .
b. P. Roussignol, D. Ricard, J. Lukasik and C. Flytzanis: *J. Opt. soc. Am.*, **B4**,5 (1987).
c. L. H. Acioli, A. S. L. Gomes and J. R. Rios Leite: *Appl. Phys. Lett.*, **53**, 1788 (1988).
d. J. Yumoto, H. Shinojima, N. Uesugi, K. Tsunetomo, H. Nasu and Y. Osaka, *Appl. Phys. Lett.*, **57**, 2393 (1990).
e. N. Finlayson, W. C. Banyai, C. T. Seaton, G. I. Stegeman, M. O'Neil, T. J. Cullen and C. Ironside, *J. Opt. Soc. Am.*, **B6**, 675 (1989).
f. E. Hanamura, *Phys. Rev.*, **B37**, 1273 (1988).
g. T. Takagahara: *Phys. Rev.*, **B39**, 10206 (1989).
h. Y. Masumoto, M. Yamazaki and H. Sugawara, *Appl. Phys. Lett.*, **53**, 1527 (1988).
i. B. L. Justus, M. E. Seaver, J. A. Ruller and A. J. Campillo, *Appl. Phys. Lett.*, **57**, 1381 (1990).
j. D. Ricard, P. Roussignol and C. Flytzanis, *Opt. Lett.*, **10**, 511 (1985).
k. M. J. Bloemer, J. W. Haus and P. R. Ashley, *J. Opt. Soc. Am.*, **B7**, 790 (1990).
l. F. Hache, D. Ricard, C. Flytanis, *J. Opt. Soc. Am.*, **B3**, 1647 (1986).
m. F. Hache, D. Ricard, C. Flytanis and U. Kreibig, *Appl. Phys.*, **A47**, 347 (1988).
n. A. I. Ekimov, Al. L. Efros, M. G. Ivanov, A. A. Onushchenko and S. K. Shumilov, *Solid State Commun.*, **69**, 565 (1989).
o. H. E. Elsayed-Ali, T. B. Norris, M. A. Pessot and G. L. Eesley, *Phys. Rev. Lett.*, **58**, 1212 (1987).
p. S. I. Anisimov, B. L. Kapeliovich and T. L. Pere'man, *Sov. Phys. JETP*, **39**, 375 (1975).
q. Y. Nozue, *J. Phys. Soc. Jpn.*, **51**, 1840 (1982).
r. T. Tokizaki, A. Nakamura, Y. Ishida, T. Yajima, I. Akai and T. Karasawa, *Ultrafast Phenomena VII*, (C. B. Harris, E. P. Ippen, G. A. Mourou and A. H. Zewail ed.) p.253, Springer Verlag (1990).
s. T. Yajima and Y. Taira, *J. Phys. Soc. Jpn.*, **47**, 1620 (1979).

6

Atomic Bonding and Excitons in Ultrathin PbI$_2$ Crystallites

Takenari Goto[a], Shingo Saito[a], Hitoshi Tanaka[a], Michiyoshi Tanaka[b], Satoshi Sasaki[c] and Noboru Miura[c]

[a] Department of Physics, Graduate School of Science, Tohoku University
   Aoba-ku, Sendai 980-8578, Japan
[b] Research Institute for Scientific Measurements, Tohoku University
   Aoba-ku, Sendai 980-8577, Japan
[c] Institute for Solid State Physics, University of Tokyo
   Minato-ku, Tokyo 106-0032, Japan

Purpose of the present study

We aim to study quantum confinement of an exciton, atomic bonding changes and surface state in ultrathin crystallites of nanometers thickness. In particular, it is clarified how an effective mass approximation can be applied in quantum confinement of exciton translational motion, and where deviation originates. We study also the effects of size confinement and surrounding medium on the internal motion of the exciton. We further report on the phonons characteristic of the ultrathin crystallite and the exciton-phonon interaction.

Contents of the present study

1. Introduction

One-dimensional confinement of electrons and holes has been extensively studied in multiquantum well of GaAs, and interpreted well according to the features of particles confined in quantum well potential. On the other hand, study on confinement of exciton translational motion within a two-dimensional space has not been conducted until the beginning of the present study.[a] To study quantum confinement of excitons in two-dimensional space, a semiconductor having large exciton binding energy is useful, because such an exciton is stable even at elevated temperatures. As excitons with large binding energy have a small Bohr radius, an ultrathin crystallite with thickness comparable to the exciton radius is necessary for the study of size quantization effect. The present experiments were performed using an ultrathin crystallite of PbI$_2$ which has a layer crystal structure and a direct allowed exciton in the visible region.

The PbI$_2$ single crystal has various polytypes. The simplest structure is of the 2H type and belongs to spatial symmetry D_{3d}^3. The crystal structure of 2H-PbI$_2$ is shown in Fig. 1. Open and closed circles show the positions of Pb and I, respectively. The PbI$_2$ crystal is constructed from a unit layer with a thickness of 7 Å and is composed of three layers I-Pb-I. The unit layer is bonded by a very weak van der Waals force.

The electronic band structure has been calculated by Schlüter and Schlüter.[b]

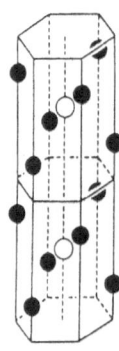

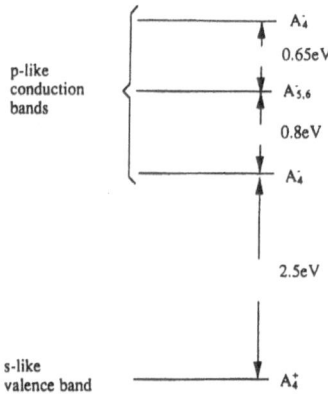

Fig. 1   Crystal structure of 2H-PbI$_2$. Open and closed circles represent positions of Pb and I.

Fig. 2   Energy levels near the bottom of the conduction and the top of the valence band at point A of the Brillouin zone in the 2H-PbI$_2$ crystal.

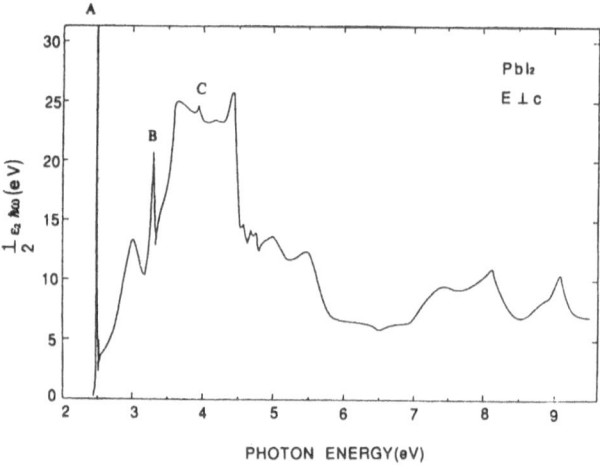

Fig. 3   Optical conductivity spectrum for polarization $E_\perp c$ at 4.2 K obtained by Kramers Kronig analysis from the reflectivity spectrum.
(From C.Gahwiller and G.Harbeke, *Phys.Rew.*, **185**, 1143 (1969)).

According to their calculation, a direct gap is located at point A of the Brillouin zone. An energy diagram near the band gap is shown in Fig. 2, taking a spin-orbit interaction into account. Electrons near the top of the valence band and the bottom of the conduction band consist of electrons of Pb. The valence band orbital is s-like and has symmetry A_4^+. On the other hand, the conduction band orbital is p-like and the energy level splits into three levels by a spin-orbit interaction and a crystal field. These orbitals have symmetries A_4^-, $A_{5,6}^-$, A_4^-, in that order from the bottom level. Figure 3 shows an optical conductivity spectrum calculated from a reflectivity spectrum using Kramers-Kronig relation by Gahwiller and Harbeke.[c] Sharp lines indicated by A, B, C are assigned as due to excitons corresponding to A_4^-, $A_{5,6}^-$, A_4^- conduction levels, respectively. The broad band near 3 eV is not due to an exciton but a band-to-band transition and the spectrum reflects joint density of states.

2. Sample preparation

The ultrathin PbI_2 crystallite used in the present study comprises microcrystallites embedded in polymer. The polymer is ethylene-methacrylic acid copolymer (E-MAA) purchased from Du Pont-Mitsui Polychemicals co., Ltd. Pendant molecules of the carbon backbone chain are carboxyl groups. This polymer transmits light of wavelength between 240 nm and 3000 nm.

The preparation method is similar to Mahler's method [d] for making PbS fine particles in polymer. Powder of $Pb(CH_3COO)_2 \cdot 3H_2O$ and the polymer was vacuum sealed in a glass ampoule, and then kept at 170°C for several days. By this treatment, two hydrogens of the carboxyl group were changed to Pb. Rotating the glass ampoule at 170°C, a transparent polymer film was made on the inner wall of the ampoule. Polymer film of thickness of about 0.1 mm was exposed to HI gas for one hour at 50°C, turning orange in color. The size of the PbI_2 crystallite was controlled by heat treatment in which the film was heated at various temperatures down to 90°C and maintained for less than 3 hours.

3. Shape of the PbI_2 microcrystallites

The polymer film was sliced perpendicular to the surface to a thickness of 100 nm. the photograph of this film was taken by TEM. For this measurement, the film was cooled down to 100 K to avoid temperature rise by electron beam irradiation. An example of TEM photographs is shown in Fig. 4. The direction of the polymer chain was arranged after making a film by flowing it along the inner wall of the glass tube. The platelet-like PbI_2 microcrystallite embedded in the polymer was grown along the polymer chain, *i.e.* the surface of the microcrystallite was parallel to the stretched direction of the polymer. Hence, the electron beam is parallel to the surface of the microcrystallite.

Taking the crystal orientation into account, the edge of the platelet-like crystallite is seen in this photograph. These microcrystallites were known to be a single crystal from the electron diffraction pattern. By measuring the thicknesses of these microcrystallites, layer numbers were estimated and shown in this figure, because the unit layer is 0.7 nm thick. The surface size of the microcrystallites ranges from 10 nm to 20 nm in this case. The surface size of all samples ranges from 5 nm to 50 nm, and becomes larger as the thickness increases.

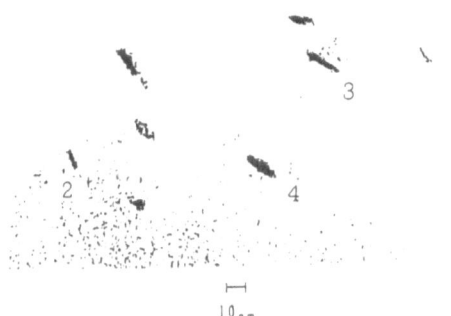

Fig. 4   TEM photograph of the PbI_2 microcrystallites embedded into polymer at 100 K. A layer number is indicated near a crystallite. The c-axis is parallel to the photograph.

4. Experimental results and discussion

4.1 Size quantization effect on translational motion of exciton and on a rigid layer mode phonon in ultrathin PbI₂ crystallites

Figure 5 shows absorption spectra of polymer films including PbI_2 microcrystallites with various thermal treatments at 4.2 K. The average thickness of the PbI_2 microcrystallites incrases in order from the lowest curve. In the uppermost spectrum of the film containing the thickest microcrystallites, there appear narrow absorption bands at 2.53 and 2.37 eV, which are close to the energies of A and B exciton bands, respectively, of Fig. 3. Hence, the 2.53 eV and 2.37 eV bands are assigned to be due to excitons associated with A_4^- and $A_{5,6}^-$ conduction electrons, respectively, by analogy with the bulk crystal. The broad band near 3.1 eV E is observed also in Fig. 3, and hence, is assigned as due to the band-to-band transition. With decreasing average thickness, some peaks appear on the high energy side of the 2.53 eV band, and the lower energy peak decreases. These peaks are considered to be associated with the microcrystallites of different layer numbers. The peak energy becomes higher as the layer number decreases, because the exciton shows blue shift for size quantization effect. The layer number is decided from the resonant Raman experiment as stated later. Figure 6 shows an expanded absorption spectra near the A exciton band. In this figure, there clearly appear some peaks due to excitons generated in ultrathin microcrystallites with different layer numbers.

Next, we introduce resonant Raman scattering of these samples in order to study phonons characteristic of ultrathin crystallites and to determine the layer number. The sample was excited by light from Stilben 420 and Coumarin 480 dye lasers pumped by UV lights from a cw Ar⁺ laser, the scattered light was detected by a photomultiplier after passing through a Jovin Yvon U-1000 double monochromator. The excitation energies are shown by arrows in Fig. 6.

The resonant Raman spectra of samples #1, 2 and #3, 4 at 77 K are shown in Figs. 7 and 8, respectively as a function of the Stokes shift. The upper four curves of Fig. 7 are Raman spectra of sample #1 in excitation light with energies corresponding to band α. The lower four curves are the spectra of sample #2 in

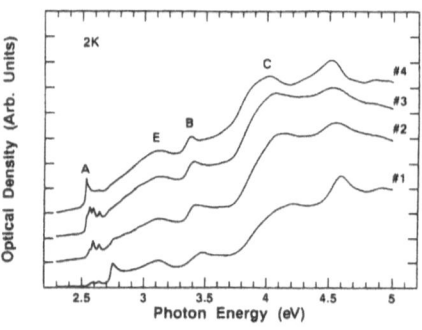

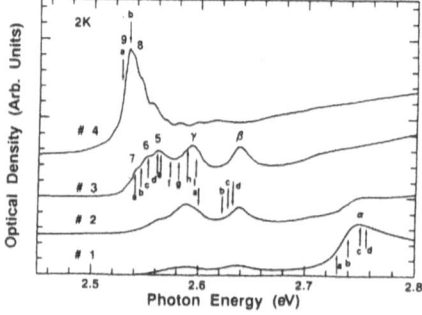

Fig. 5   Absorption spectra of the polymer films after different thermal treatments. Average thickness increases from the bottom to the top curve.

Fig. 6   Detailed absorption spectra near the A band of different samples at 2 K. Arrows show the excitation energies for the Raman scattering spectra.

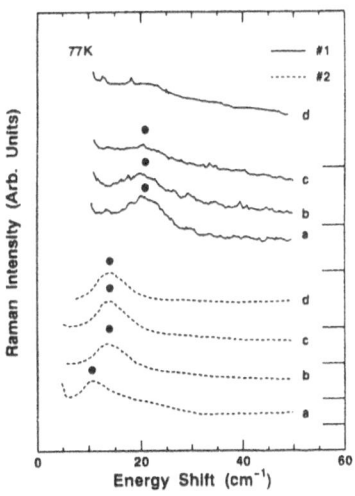 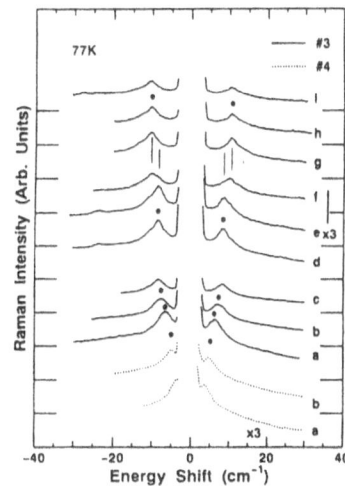

Fig. 7   Raman spectra of samples 1 and 2 at 77 K.   Fig. 8   Raman spectra of samples 3 and
The letters on the righthand side represent   4.
excitation energy shown in Fig. 6.

the excitation energies of bands β and γ. The upper nine curves and lowest two curves of Fig. 9 are Raman spectra of samples #3 and #4, respectively.

A Raman line appears with small energy shift, and the energy shift becomes smaller as the excitation light energy decreases. Even if the excitation energy changes within the absorption band, the Raman shift remains unchanged, as seen in the second, third and fourth curves from the top of Fig. 7, for example. This means that this Raman line is characteristic of the layer number and does not depend on the lateral size of the ultrathin crystallites. (The absorption band width can sometimes be caused by inhomogeneity in the lateral size of ultrathin crystallites.)

In the layer semiconductor, a phonon associated with the Raman line with the small Stokes shift is of a rigid layer mode. There are two kinds of phonons, compressional and shear modes.

Here, we assume that the rigid unit layers vibrate with each other, but unlike bulk crystal, the layer number is finite. As shown in Fig. 9, a linear chain model is adopted to interpret the Raman line. The mass of the unit layer is m and the force constant between the layers is K. The positions of the layer are indicated by x_i ($i = 1, 2, 3, ... n$). The equations of the layer motions are written as

$$m\frac{d^2x_i}{dt} = -K(x_i - x_{i-1}) - K(x_i - x_{i+1}), \quad i = 1, 2, ... n. \tag{1}$$

Using parameters K_c = 7.0 N/m and K_s = 1.67 N/m [e] which are the force constants of the compressional and shear modes, respectively, and the mass parameter $m = 1.66 \times 10^{-27}$ kg in the bulk, the phonon energies of compressional and shear modes are calculated from eq. 1 and shown with open circles and triangles, respectively, in Fig. 10. Only the energy of the longest wavelength phonon is shown as a function of the inverse thickness $1/L$ of the ultrathin crystallites. The layer number is indicated in the upper abscissae. Experimental phonon energies obtained from the Raman shift are plotted with closed circles.

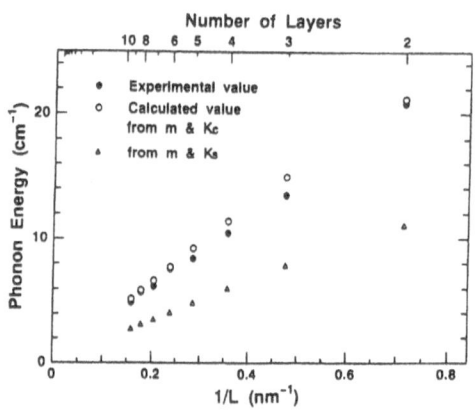

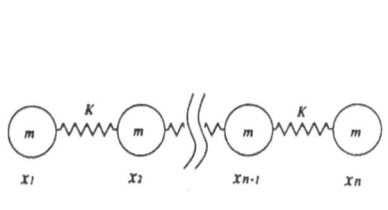

Fig. 9   Linear chain model of a rigid layer mode in the crystal with the finite layers n. $x_1, x_2,...x_n$ show positions of the rigid layers, m and K are the mass and force constant of a simple harmonic oscillator, respectively.

Fig. 10   Phonon energies calculated from eq. 1 using parameters $m = 1.66 \times 10^{-27}$ kg, $K_c = 7.0$ N/m and $K_s = 1.67$ N/m are shown with open circles and triangles, respectively, as a function of $1/L$. Experimental phonon energies obtained from the Raman shifts are shown by closed circles. At the upper edge, the number of the layers is indicated.

Here we assume that the highest energy band α is associated with 2-layer crystallites, because the phonon mode can be produced only in crystallites of more than 2-layers and the exciton energy is largest in 2-layer crystallites. The exciton peaks β, γ with the lower energies are assigned as absorption bands of the crystallites with 3, 4 layers. The peaks with further lower energies are also assigned as absorption bands of the crystallites of 5, 6, 7, 8, 9 layers in order from the highest peak.

As seen in Fig. 10, the experimental phonon energies coincide with those of the compressional mode phonon. This means that the compressional mode phonon with the longest wavelength couples well with the specially confined exciton. Raman scattering of the bulk crystal is forbidden for the compressional mode but allowed for the shear mode from group theoretical analysis.[f] This is inconsistent with the above result that it is not the shear mode but the compressional mode that is observed in the Raman spectrum.

In this experiment, however, the excitation energy is resonant to the exciton energy. Furthermore, the microcrystallite has symmetry different from that of the bulk crystal because of the finite layers. Hence, the selection rule for the bulk crystal cannot be applied.

Let us consider the reason why the compressional mode with the longest wavelength couples with the specially confined exciton. The vibrational direction of the compressional mode is the same as that of the exciton confinement. Both the confined exciton and the deformation potential induced by the lowest energy phonon are expressed by a standing wave with the same length $2L$ and the same phase. Therefore, a deformation potential type interaction may be strong. On the other hand, the higher energy phonons have wavelengths smaller than that of the exciton, and hence, the interaction with the exciton may become weaker.

Next, we will discuss the exciton energy of the ultrathin microcrystallites. As seen in Figs. 5 and 6, the exciton band energy becomes larger with decreasing layer number. This effect is considered to be due to spatial confinement of the exciton in the c-direction. In the simple model where the exciton is confined in the one-dimensional quantum well with infinite potential, the exciton energy E in an effective mass approximation is expressed by

$$E = E_B + \frac{\hbar^2}{2M} \cdot \frac{\pi^2 p^2}{L^2} \quad p = 1, 2, 3, ..., \tag{2}$$

where E_B is an exciton formation energy in the bulk and M an exciton mass, L a width of the potential well, and p a quantum number of the confinement.

Equation 2 with the exciton translational mass of the bulk $M = 1.0m_0$,[g] an adjustable parameter, $E_B = 2.524$ eV and $p = 1$ is shown by a solid straight line in Fig. 11, where E vs. $1/L$ is shown. As the exciton energy in the bulk depends on the polytype, an adjustable parameter is used as E_B. On the other hand, the broken line represents eq. 2 when the exciton reduced mass $M = 0.18m$ [h] is used and the band gap energy is used instead of E_B.

The solid line shows good fit to the experimental peak energies shown by the open circles, in the range $1/L^2 < 0.1$. This coincidence strongly suggests that the translational motion of the exciton is quantized in crystallites of more than five layers. In contrast, the exciton energies in the 2-, 3- and 4-layer crystallites are higher than those given by the solid line. This deviation may be caused mainly by transfer from "exciton confinement" to "electron-hole confinement." This may be intimately related to the ratio or the effective Bohr radius, 1.9 nm,[j] to the potential well width, as pointed by Kayanuma.[i]

The absorption band B near 3.4 eV also shifts to the higher energy side as the crystal thickness decreases, as shown in Fig. 5. The amount of the shift, however, is smaller than that of the first exciton band. The exciton translational mass is estimated to be $2.2m_0$ from eq. 2. This means that the mass of the exciton consisting of the A_4^- conduction electron is about twice as large as that of the $A_{5,6}^-$ electron, i.e. the dispersion of the A_4^- electron is smaller than that of the $A_{5,6}^-$ electron.

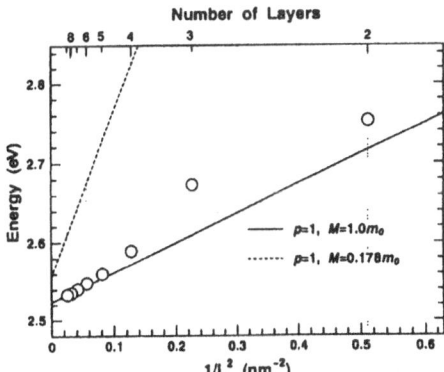

Fig. 11 The peak energies of the absorption bands in Fig. 6 are shown as a function of $1/L^2$. Open circles show the experimental energies. Solid and broken lines represent eq. 2 with parameters $p = 1$, $M = 1.0m_0$ and $p = 1$, $M = 0.178m_0$, respectively.

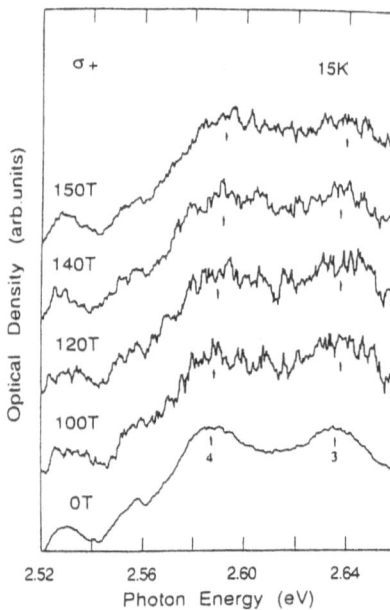

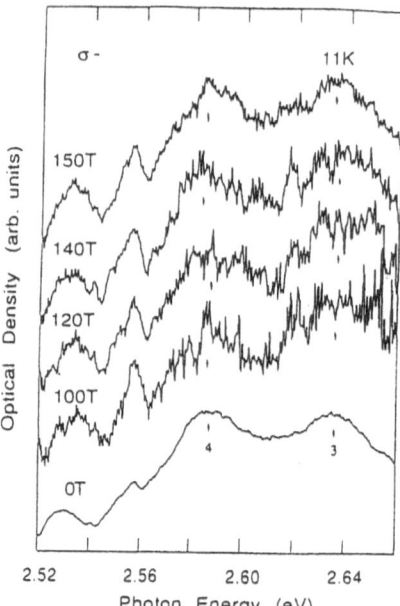

Fig. 12 Absorption spectra of a sample containing 3- and 4-layer crystallites for σ^+ polarization in Faraday configuration in ultra high magnetic fields at 15 K.

Fig. 13 Absorption spectra for σ^- polarization at 11 K.

The broad band near 3.1 eV in Fig. 5 is independent of the layer number. This fact supports the Schlüter's assignment that this band is associated with band-to-band transition.

4.2 Size effect on internal motion of the exciton

To study exciton internal motion in ultrathin crystallites we measured diamagnetic shift of the exciton absorption band in ultra high magnetic fields. The sample was set at the center of one turn coil and cooled by helium gas. Pulsed magnetic fields of up to 150 T were applied to the sample. White light from Xe flash lamp was transmitted through the sample, analyzed by a monochromator and detected by a streak camera. Figures 12 and 13 show the exciton absorption spectra of the sample containing 3- and 4-layer crystallites for σ^+ and σ^- circular polarizations, respectively, in Faraday configuration at 15 K and 11 K, respectively.

The magnetic field for each curve is shown on the lefthand side and the number on the lowest curve shows the layer number of the crystallites corresponding to the exciton bands. In this polymer film, crystallites with 3- and 4-layers are embedded. Structures below 2.56 eV appear owing to insufficient cancellation because of too complicated spectrum of the Xenon flash lamp used as a light source.

Here we discuss the diamagnetic shift of the PbI₂ microcrystallite. The diamagnetic shift $\Delta E_{d//}$ for the magnetic field B_z parallel to the c-axis is expressed by

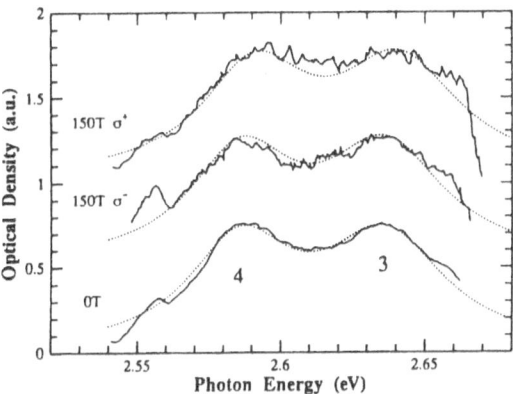

Fig. 14 Absorption spectra of a sample containing 3- and 4-layer crystallites for σ^+ and σ^- polarization at the magnetic field of 150 T and a spectrum without magnetic field. Solid and dotted curves show the experimental and simulated spectra, respectively.

$$\Delta E_{d//} = \sigma B_z^2. \tag{3}$$

In the configuration $B_{\perp c}$, however, the exciton state is not an eigen state for circularly polarized light. In fact, the apparent diamagnetic shift of the bulk is less than half.[j] Taking account of the apparent shift of the bulk, the diamagnetic shift for B perpendicular to the c-axis $\Delta E_{d\perp}$ may be written as

$$\Delta E_{d\perp} = p\sigma (B_x^2 + B_y^2), \tag{4}$$

where $p = 0$ and $p = 0.25$ for σ^+ and σ^- polarization, respectively.

The Zeeman splitting ΔE_z is given by

$$\Delta E_z = \pm (1/2)g \, \mu_B \, B, \tag{5}$$

where signs + and − correspond to σ^+ and σ^-, respectively. Here, it is assumed by analogy with the bulk that ΔE_z is isotropic. On the other hand, the oscillator strength of the lowest energy exciton for $E_{//c}$ polarization is known to be one fourth that for $E_{\perp c}$.[b] It is observed from the electron micrograph that the average lateral size of the microcrystallite is about ten times as large as the thickness.

Assuming that the absorption band shape is Lorentzian and the c-axis of the microcrystallite are randomly oriented, the absorption spectra in the magnetic fields of zero and 150 T for σ^+ and σ^- orientations have been calculated and are shown by the dotted lines in Fig. 14.

Adjusting the calculated curves to the experimental solid curves, we obtain the diamagnetic shift σ as 3×10^{-7} eV/T^2 for excitons created in both 3- and 4-layer crystallites. The value of σ is smaller compared to diamagnetic coefficients 6.1 and 7.0×10^{-7} eV/T^2 for 3-and 4-layer crystallites, respectively, calculated by Basterd et al.[k] on the basis of an effective mass approximation. The smaller value may originate from dielectric screening effect,[l] because the PbI$_2$ microcrystallite is embedded in polymer with a smaller dielectric constant, that is, the exciton wave function shrinks as the Coulomb interaction between electron

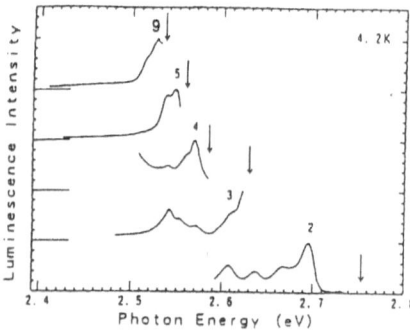
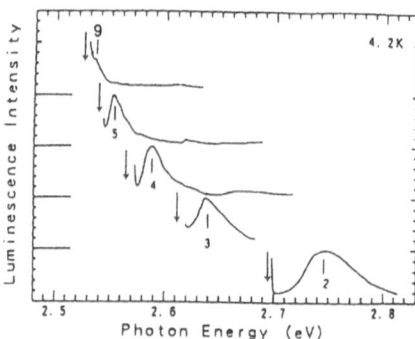

Fig. 15 Luminescence spectra near the absorption edge at 4.2 K when the sample is excited by light with the peak energy of the exciton absorption band. Numbers indicate layer number, and arrows the excitation energy.

Fig. 16 Excitation spectra of the luminescence indicated by the number in Fig. 15. Numbers indicate layer number and arrows the luminescence energy.

and hole becomes stronger outside the crystal. From the diamagnetic constant, the exciton binding energy is estimated to be 200 meV, which is 3.3 times as large as the bulk value.

Next, we discuss the exciton binding energy estimated from analysis of the luminescence spectrum. Fig. 15 shows the luminescence spectra under selective excitations at 4.2 K. The layer number of the microcrystallite in which the luminescence occurs is shown near the luminescence line. An arrow indicates the excitation light energy which equals the peak energy of the exciton absorption band. Each luminescence line has a Stokes shift from the corresponding absorption band and the shift becomes smaller as the layer number increases. Figure 16 shows the excitation spectra of these luminescences at 4.2 K.

An arrow indicates the energy of the detected luminescence, which is indicated by a number in Fig. 15. Also in this figure, it is seen that the Stokes shift decreases with increasing layer number and approaches that of the I, bound exciton, 7.4 meV. This luminescence becomes weaker when the temperature rises, and disappears at 77 K. These facts suggest that the luminescence line is caused by radiative annihilation of the I_1 bound exciton. The I_1 bound exciton is assigned as an exciton bound to a neutral donor by Skolnick and Bimberg.[h] The binding energy of the exciton bound to a neutral donor is known to be proportional to that of a free exciton.[m] The proportional factor is 8.5 for the I_1 bound exciton. If we assume that this relation holds also in an ultrathin crystal, we can estimate the binding energy of the free exciton in an ultrathin crystal. As the exciton Bohr radius is comparable to the crystal thickness, this assumption is ambiguous, and hence, theoretical calculation is needed for the exact relation. The relation between the estimated binding energy and a_B/L is shown by open circles in Fig. 17. Here, a_B is the exciton Bohr radius of the bulk, which has been obtained as 1.9 nm by Nagamune et al.[j] Triangles in Fig. 17 show the theoretical binding energy calculated by Hanamura et al.,[l] taking the dielectric screening effect into account. Dielectric constants of the PbI₂ crystal and the polymer, 8.5 and 2.67 respectively, are used in the theoretical calculation.

The experimental values are nearly equal to the theoretical values. The

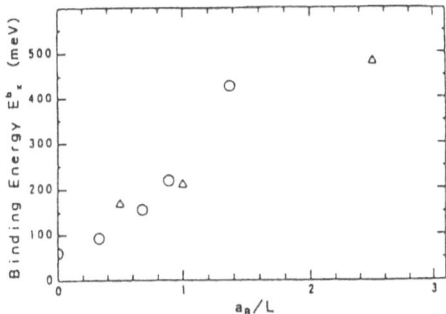

Fig. 17 Binding energies of the free exciton vs. a_B/L. Circles and triangles show the experimental and theoretical values, respectively.

experimental binding energies of excitons in 3- and 4-layer crystallites, 220 meV and 160 meV, are close to that estimated from the diamagnetic shift 200 meV. These coincidences suggest that the exciton wave function shrinks not only by confinement of excitons in the ultrathin crystallite but also by confinement in dielectric medium surrounding the film resulting in enhancement of the exciton binding energy.

4.3 Scattering of the exciton in ultrathin microcrystallites

In order to study the physical origin of the exciton absorption band shape, the homogeneous bandwidth was measured by a pump-probe method. The upper part of Fig. 18 shows the exciton absorption spectrum of 3- and 4-layer crystallites at

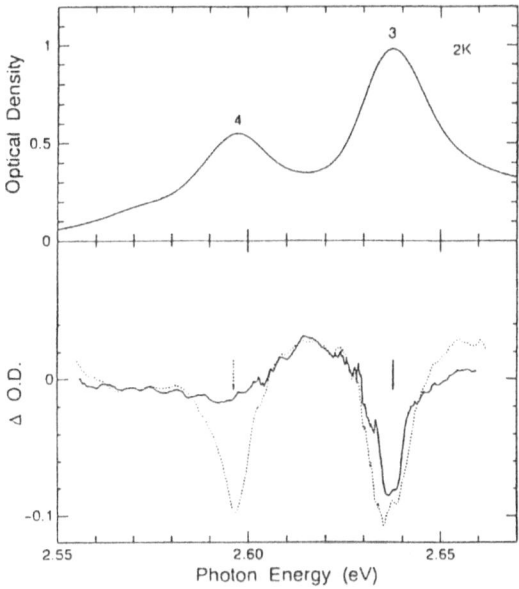

Fig. 18 The upper part shows the absorption spectrum of 3- and 4-layer crystallites at 2 K. The lower part shows the absorption change ΔOD when the samples are irradiated by light with the peak energy of the exciton band in 3-layer crystallites. A dotted curve shows ΔOD when the pump energy is located at the exciton band peak in 4-layer crystallites.

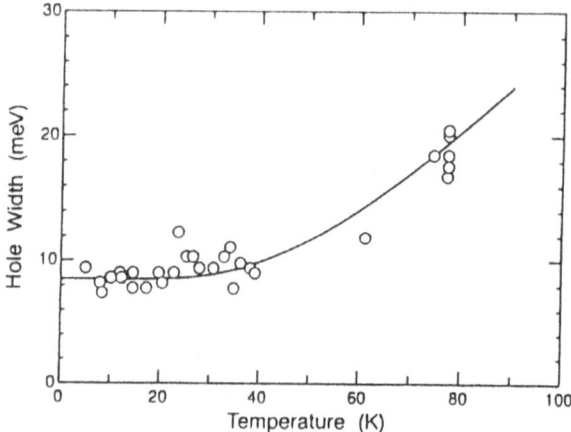

Fig. 19 Temperature dependence of the hole width. Circles show the experimental hole width and the
solid curve the theoretical hole width calculated using eq. 3.

2 K. In the lower part is shown the optical density change ΔOD when the
crystallites are excited by pump light with 10 ns pulse width. The pump light had
a spectral width of less than 0.1 meV and the intensity was about 400 kW·cm⁻².
Solid and dotted curves show the ΔOD when the 3- and 4-layer crystallites are
excited, respectively. The pump light energy is indicated by an arrow.

   When the 3-layer crystallite is excited, the absorption band of the 4-layer
crystallite also changes as well as that of the 3-layer crystallite. When the 4-
layer crystallite is excited, the absorption band of the 3-layer crystallite also
changes. This suggests that the microcrystallite temperature rises by pump light
irradiation. This effect is ensured from the pump power dependence of the hole
burning spectrum. The hole width as a function of temperature is shown by open
circles in Fig. 19. The solid curve represents the formula

$$\Gamma = 8.5 + 76\{\exp\left(\hbar\omega_{LO} / k_B T\right) - 1\}^{-1} \tag{6}$$

   where $\hbar\omega_{LO}$ is the LO phonon energy, 13.9 meV, and k_B the Boltzman
constant.

   From coincidence between the experimental points and the solid curve, it is
concluded that the exciton is scattered by impurities, defects and/or the surface
below 40 K and by LO phonons above 40 K.

4.4 Size effect on the atomic bonding

   We measured exciton absoption spectra at different hydrostatic pressures at
77 K using diamond and sapphire anvil cells.

   Figure 20 shows absorption spectra at 77 K at different pressures the values
of which are indicated on the lefthand side. The layer number is shown near the
band peak. Fig. 21 also shows the spectra of another sample containing 3-, 4-, 5-
layer crystallites. Each absorption band shifts to the lower energy side with
increasing the pressure. The peak energies as a function of the pressure are shown
in Fig. 22. Closed circles show the relation in the bulk. The layer numbers are

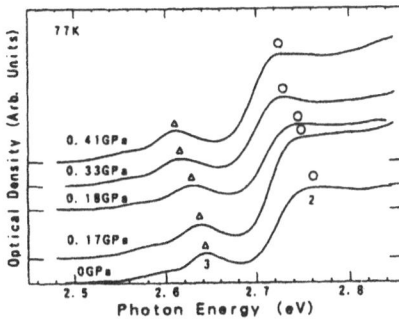

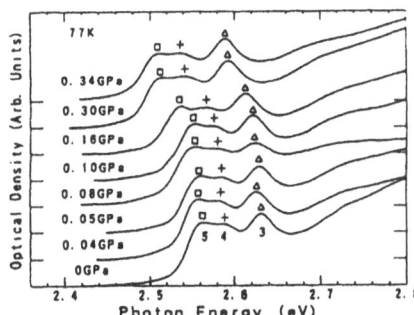

Fig. 20 Pressure dependence of the absorption spectra of the sample containing 2- and 3-layer crystallites at 77 K. The pressure is indicated on the lefthand side of each curve.

Fig. 21 Pressure dependence of the absorption spectra of the sample containing 3-, 4-, 5-layer crystallites.

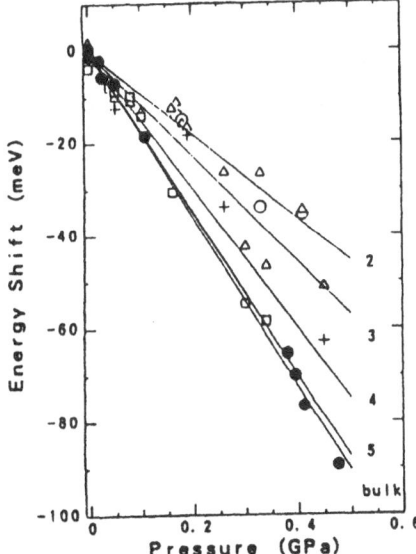

Fig. 22 Peak energy shift of the exciton band vs. the pressure in the crystallites with the different layer numbers. Closed circles show the energy shift of the bulk measured by Powell.[n]

indicated on the righthand side. With increasing layer number, the negative pressure coefficient increases and approaches the bulk value.

This phenomenon is interpreted as follows. The top of the valence band consists of the antibonding state between a 6s orbital of Pb and a $5p_z$ orbital of I,[b] hence the energy level is expected to rise with decreasing the lattice constant. In the ultrathin crystallite, the distance between iodine atoms is larger than that of the bulk, so that the covalent bond between Pb and I becomes weak and ionicity becomes strong. This results in decrease of the bonding-antibonding splitting. Accordingly, the decreasing rate of the band gap with the pressure becomes smaller as the layer number decreases.

Such a tendency is deduced also from esonant Raman scattering. Fig. 23

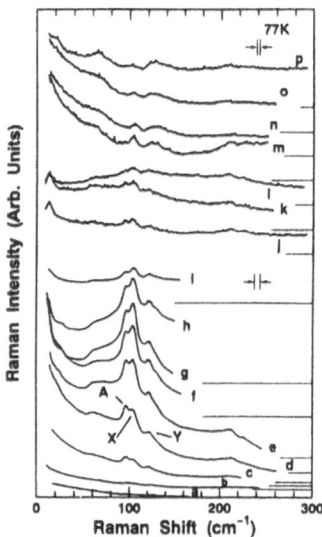

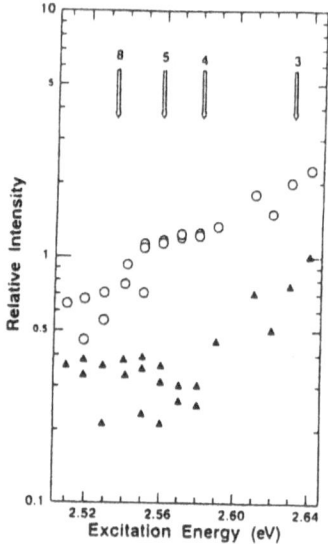

Fig. 23 The Raman spectra of the sample containing 2-,3-, 4-layer crystallites excited by light with different energies. The excitation energies are: a, 2.52 eV; b,2.53 eV; c, 2.54 eV; d, 2.55 eV; f, 2.57 eV; g, 2.58 eV; h,2.59 eV; i, 2.61 eV; j, 2.62 eV; k, 2.63 eV; l, 2.64 eV; m,2.67 eV; n, 2.69 eV; o, 2.70 eV; p, 2.73 eV.

Fig. 24 Dependence of the X, Y Raman intensities normalized to the A Raman intensity on the excitation energy. Open circles and closed triangles represent the Raman intensities of X and Y lines, respectively. Numbers in the upper part of this figure show the layer number.

shows the Raman spectra under the excitation lights with the energies of the exciton bands in the 2-, 3-, 4-, 5-layer crystallites at 77 K. The excitation energies of curves k, g, e, d equal to the exciton peak energies of the 2-, 3-, 4-, 5-layer crystallites, respectively. At 98 cm⁻¹, there appears a Raman line associated with an A mode phonon in the bulk. On the higher energy side of the A line, two new lines characteristic of ultrathin crystallites are observed. These lines are named X and Y and have energies of 105 and 122 cm⁻¹, respectively.

The intensities of the X and Y lines normalized to that of the A line are shown with circles and triangles, respectively, in Fig. 24. With decreasing layer number, both Raman intensities increase. Hence, these lines are considered to be associated with the surface phonon. So-called surface phonon, however, generally has smaller energy compared to the bulk phonon. In PbI₂, atomic bonding between I and Pb is expected to be stronger in the ultrathin crystallite from the stress effect of the exciton energy, as stated earlier.

This expectation agrees well with the observation of the new Raman lines on the higher energy side of the bulk phonon.

Summary

1) The exciton translational motion perpendicular to the c-axis is size quantized in the ultrathin PbI₂ crystallite.
2) The compressional mode phonon of the rigid layer in the finite layer crystallite appears in the resonant Raman spectrum because of strong exciton-phonon interaction.

3) The exciton wave function shrinks in the ultrathin crystallite not only by the size confinement but also by the dielectric confinement.
4) The exciton is scattered by imperfections or the surface below 40 K and by LO phonons above 40 K.
5) Atomic bonding between I and Pb becomes stronger in the ultrathin crystallite compared to that of the bulk.

Acknowledgments

We thank M. Saito and F. Sato of Tohoku University for technical assistance. This work was partially supported by a Grant in Aid for Scientific Research from the Ministry of Education of Japan.

PUBLICATIONS

1. T. Goto, S. Saito and M. Tanaka, *Solid State Commun.*, **80**, 331 (1991).
2. T. Goto and S. Saito, SPIE **1675** *Quantum Well and Superlattice Physics IV*, 128 (1992).
3. T. Goto and H. Tanaka, *Solid State Commun.*, **89**, 17 (1994).
4. S. Saito and T. Goto, *Solid State Commun.*, **84**, 1043 (1992).
5. S. Saito and T. Goto, *J. Luminescence*, **58**, 127 (1994).
6. S. Saito and T. Goto, *Phys. Rev.*, **B52**, 5929 (1995).

REFERENCES

a. Z. K. Tang, A. Yanase, T. Yasui and Y. Segawa, *Phys. Rev. Letts.*, **30**, 1431 (1993).
b. I. Schlüter and M. Schlüter, *Phys. Rev.*, **89**, 1652 (1974).
c. ch. Gähwiller and G. Harbeke, *Phys. Rev.*, **185**, 1141 (1969).
d. Y. Wang, A. Suna, W. Mahler and R. Kasowski, *J. Chem. Phys.*, **87**, 7315 (1987).
e. W. M. Sears, M. L. Klein and J. A. Morrison, *Phys. Rev.*, **B19**, 2305 (1979).
f. R. Loudon, *Adv. Phys.*, **13**, 423 (1964).
g. T. Hayashi, *J. Phys. Soc. Jpn.*, **55**, 2043 (1986).
h. M. S. Skolnick and D. Bimberg, *Phys. Rev.*, **B18**, 7080 (1978).
i. Y. Kayanuma, *Phys. Rev.*, **B38**, 9797 (1988).
j. Y. Nagamune, S. Takeyama and N. Miura, *Phys. Rev.*, **B40**, 8099 (1989).
k. G. Basterd, E. E. Mendez, L. L. Chang and L. Esaki, *Phys. Rev.*, **B26**, 1974 (1982).
l. E. Hanamura, N. Nagaosa, M. Kumagai and T. Takagahara, *Mat. Sci. and Eng.*, **B1**, 215 (1988).
m. R. E. Halsted, *Physics and Chemistry of II-VI Compounds*, (M. Aven and J. S. Prener, eds.), pp. 385, North-Holland, New York (1967).
n. M. J. Powell, *Phyl. Mag.*, **B38**, 71 (1978); *ibid.*, **B38**, 5565 (1980).

7

Optical Properties of Mesoscopic Particles Prepared by Sputtering Method

Shinji Hayashi and Keiichi Yamamoto

Department of Electrical and Electronics Engineering, Faculty of Engineering, Kobe University
Nada-ku, Kobe 657-0013, Japan

Purpose of the present study

The first step toward a systematic study of mesoscopic particles is to develop an appropriate technique which allows us to prepare a sufficient amount of stable mesoscopic particles and to control the size of the particles. One of the purposes of the present study is to establish a technique for preparing the mesoscopic particles embedded in a solid matrix with size ranging from a few nanometers (microcrystals) to less than one nanometer (clusters). Another purpose of this study is to acquire various types of optical data to investigate the size effects on the electronic and vibrational states in these mesoscopic particles. This article describes and discusses results of our recent studies on Raman spectra of Si clusters embedded in SiO_x matrices and impurity-doped Si microcrystals.

Contents of the present study

1. Introduction

We have been using the rf cosputtering technique to prepare mesoscopic particles embedded in solid matrices. This technique enables us to prepare mesoscopic particles of varying average size ranging from several nanometers (microcrystals) to less than one nanometer (clusters).[1,2] In our previous publications, we described the rf cosputtering method and presented the optical data for various samples prepared. We discussed the confinement of optical phonons [1] and acoustic phonons [3,4] into microcrystals, effects of microcrystals on the phonon modes of SiO_2 matrices,[5] photoluminescence and Raman spectra of clusters of group IV elements embedded in SiO_2,[6-11] and light emission from SiO_2 films containing metallic particles induced by a tunneling current.[12] We discussed also some optical properties of microcrystals prepared by gas-evaporation method, *i.e.*, the infrared absorption form the surface oxide layers of Si particles,[13] Raman scattering from WO_3 [14] and MoO_3 [15] microcrystals in relation to phase transitions and crystal structures. Very recently, we initiated the study of electrical transport in SiO_2 films containing clusters of group IV elements.[16] In this article, in order to complement the previous publications, we report exclusively the results of Raman studies on Si clusters formed in Si-rich SiO_2 films and Si microcrystals doped with B atoms.

2. Raman spectra of Si clusters embedded in SiO$_2$

2.1 Introduction

In our previous work,[7] we demonstrated that SiO$_2$ films containing excess Si atoms (Si-rich SiO$_2$ films), prepared by rf cosputtering of Si and SiO$_2$, exhibit a broad PL peak at almost the same photon energy (~1.7 eV) as the red-PL peak of porous Si. From various experimental results, including those of infrared (IR) absorption and Raman scattering measurements and transmission electron microscopy (TEM), we suggested that Si structures responsible for the observed PL peak are not Si nanoparticles several nm in size, but Si clusters smaller than ~2 nm.

In an attempt to confirm the existence of Si clusters in Si-rich SiO$_2$ films and explore the Raman spectra of relatively large Si clusters, we performed systematic Raman studies of Si-rich SiO$_2$ films. Raman spectra were measured for as-deposited Si-rich SiO$_2$ films having various concentrations of excess Si atoms as well as for the films annealed in high vacuum. The observed overall spectral shapes are in good qualitative agreement with the spectra of vibrational density of states for Si$_{33}$ and Si$_{45}$ clusters calculated by Feldman, Kaxiras and Li.[a] The present results confirm our previous suggestion that the visible PL is due to Si clusters smaller than ~2 nm and not due to well-grown nanoparticles larger than ~2 nm.[7]

2.2 Experimental

Si-rich SiO$_2$ films were prepared by an rf cosputtering method similar to that used in our previous work.[1,2] Small pieces of Si wafers $5 \times 15 \times 0.5$ mm^3 in size (Si targets) were placed on a pure SiO$_2$ target (99.99 %, 4 inch in diameter) and cosputtered in Ar gas of 2×10^{-2} Torr with an rf power of 200 W, using an ANELVA SPF210HS magnetron sputtering apparatus. The films were deposited onto either Si or sapphire substrates. The substrate was not intentionally heated during the deposition. Therefore, the substrate temperature was kept lower than about 100°C. In this method, the number of excess Si atoms in SiO$_2$ is roughly proportional to the number of Si targets used. In the present work, the number of Si targets was varied from 0 to 24. Hereafter, the number of Si targets is used to specify the samples and the samples are denoted as Si-N, where N is the number of Si targets. In order to keep the total number of excess Si atoms in the films roughly constant, the film thickness was varied from ~0.2 to ~2 μm. That is, for higher excess Si concentrations, thinner films were deposited. Thermal annealing of the films was performed in a vacuum of 3×10^{-6} Torr for 30 min at various temperatures (200–800°C).

Raman measurements were carried out in a 90° delete scattering configuration at room temperature. The spectra were recorded by a Spex-Ramalog 5M spectrophotometer, equipped with a double monochromator, a R943-02 photomultiplier (Hamamatsu Photonics) and a photon counting system. The excitation source was the 488.0 nm line of an Ar$^+$ laser and its power was about 100 mW on the sample. When the Raman spectra are measured in air, Raman signals of air in the low frequency range (< 200 cm^{-1}) are usually superimposed

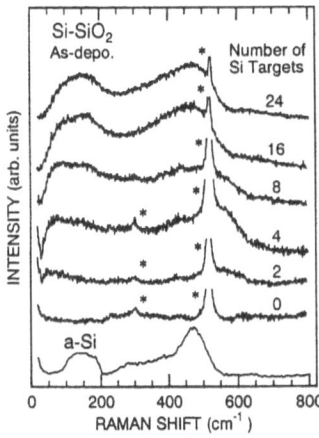
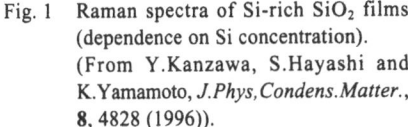
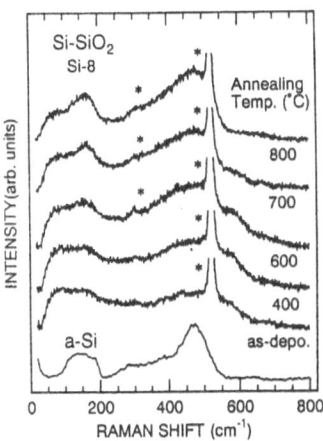

Fig. 1 Raman spectra of Si-rich SiO₂ films (dependence on Si concentration). (From Y.Kanzawa, S.Hayashi and K.Yamamoto, *J.Phys,Condens.Matter.*, **8**, 4828 (1996)).

Fig. 2 Raman spectra of Si-rich SiO₂ films (dependence on annealing temperature). (From Y.Kanzawa, S.Hayashi and K.Yamamoto, *J.Phys,Condens.Matter.*, **4**, 4828 (1996)).

on the signals from the sample. In order to eliminate the Raman lines of air, all measurements were carried out by spraying Ar gas on the sample.

2.3 Results and discussion

Figure 1 shows Raman spectra of the as-deposited samples with various Si concentrations. On the right-hand side of each spectrum, the number of Si targets is given. A spectrum of a pure SiO₂ film (~2 μm thick) deposited onto a Si substrate is also shown. These spectra were obtained by subtracting the tails of the photoluminescence signals appearing in the high frequency region of the raw spectra. The figure also includes a Raman spectrum of an a-Si film prepared by sputtering only a Si wafer. The sharp peak at 520 cm⁻¹ and weak peak at 300 cm⁻¹ (marked with * in the figure) are due to the phonon modes of the Si substrate. Since the sample is not strongly absorbent at the excitation wavelength of 488.0 nm, the sharp peak at 520 cm⁻¹ is excited by the incident laser beam reaching the substrate. For the samples deposited on the sapphire substrates, the sharp peak at 520 cm⁻¹ did not appear and only the broad peaks were observed. Therefore, the Raman signals from the sample layer exhibit no sharp structures.

For the pure SiO₂ film (number of Si targets = 0), we cannot see the Raman peaks except for the substrate peaks. However, for samples containing excess Si atoms, broad Raman peaks are observed and the spectrum changes depending on the Si concentration. For the sample Si-2, although they are not strong, two broad Raman components centered around 70 cm⁻¹ and 550 cm⁻¹ are clearly seen. Note that the Raman signal around 550 cm⁻¹ extends up to 600 cm⁻¹. As the Si concentration increases (Si-4, Si-8), the intensity of the low-frequency component around 70 cm⁻¹ increases and a component located around 150 cm⁻¹ appears and grows. At the same time, the high-frequency component around 550 cm⁻¹ grows and Raman signals in the intermediate frequency range (from 200 to 400 cm⁻¹) appear. As the Si concentration increases further (Si-16, Si-24), Raman signals extending up to 600 cm^{-1} disappear, a component around 480 cm^{-1} becomes dominant and the low-frequency component around 150 cm^{-1} further grows. It should be noted that for low Si concentrations (Si-2, Si-4, Si-8), the intensities of the low-frequency components are comparable to that of the high-frequency component. However, as the Si concentration increases (Si-16, Si-24), the high-frequency component around 480 cm^{-1} becomes stronger than the low-frequency component around 150 cm^{-1}.

The annealing temperature dependence of the Raman spectrum for sample Si-8 is shown in Fig. 2 together with the spectrum of an a-Si film. On the right-hand side of each spectrum, the annealing temperature, T_a, is given. The sharp peak at 520 cm^{-1} and the weak peak at 300 cm^{-1} (marked with * in the figure) are again due to the Si substrate. In order to prove that these peaks really come from the Si substrate even after annealing the sample, the sample Si-8 deposited on a sapphire substrate was annealed at 800°C. The sample deposited on the sapphire substrate no longer exhibited a sharp peak at 520 cm^{-1} or a weak peak around 300 cm^{-1}, but exhibited several sharp peaks due to the sapphire substrate. The same results were obtained for samples annealed at other temperatures. It is now very clear that the present sample exhibits no sharp Raman lines even after annealing.

In Fig. 2 we see that the spectrum remains almost unchanged upon annealing at 400°C. However, as the annealing temperature is increased further, the low-frequency component around 150 cm^{-1} grows and the splitting of the low-frequency component becomes apparent. The high-frequency shoulder seen around 550 cm^{-1} first increases in intensity and then decreases, and finally disappears at $T_a = 800$°C. For low annealing temperatures, $T_a = 400$°C, 600°C, the intensities of the low-frequency components are comparable to that of the high-frequency component, but for higher annealing temperatures, the high-frequency component around 480 cm^{-1} grows and becomes stronger than the low-frequency components. Comparing Fig. 2 with Fig. 1 we note that the spectral changes caused by annealing are very similar to those caused by the increase in the excess Si concentration, although the splitting of the low-frequency component is less pronounced in the concentration dependence shown in Fig. 1.

It should be noted here that the Raman signals observed for the present samples are not due to the vibrational modes of Si-O-Si bonds in SiO_2. It is well known that the Raman spectrum of amorphous SiO_2 (a-SiO_2) exhibits a broad feature extending from about 50 to 600 cm^{-1} with a pronounced peak around 420 cm^{-1}. However, the Raman efficiency of a-SiO_2 is not so large and consequently, in the case of thin films, the films should be sufficiently thick (thicker than several micrometers) to give rise to detectable Raman signals. In fact, for our sputter-deposited pure SiO_2 film of about 2 μm in thickness (sample Si-0 in Fig. 1), the Raman signals attributable to the SiO_2 film were hardly detected. Since the Raman spectrum presently observed changes depending on the concentration of excess Si atoms and the annealing temperature, the present Raman signals are thought to come from Si-Si vibrations introduced by excess Si atoms.

Although the present spectra, in particular those of the as-deposited sample Si-24 and sample Si-8 annealed at 800°C, resemble somewhat that of a-Si, a close comparison reveals that the present spectra differ considerably from that of a-Si. As described above, almost all spectra of as-deposited and annealed samples exhibit a high-frequency shoulder around 550 cm^{-1} extending up to 600 cm^{-1}, while the spectrum of a-Si shows a relatively sharp cutoff around 500 cm^{-1}.

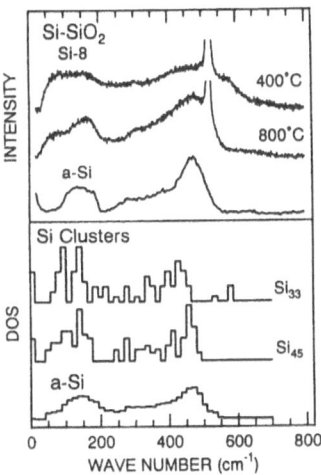

Fig. 3   Comparison of experimental spectra with calculated DOS spectra of Si_{33} and Si_{45} clusters.
(From Y.Kanzawa, S.Hayashi and K.Yamamoto, *J.Phys,Condens.Matter.*, **4**, 4828 (1996)).

Furthermore, the low-frequency component located around 70 cm^{-1} is not present in the spectrum of a-Si. The intensity of the low-frequency component relative to that of the high-frequency component in the present spectra is larger than that of a-Si.

It is known that Si microcrystals several nm in size can be grown in Si-rich SiO$_2$ films by annealing the films at high temperatures.[7] If the present samples contain Si microcrystals several nm in size, they would exhibit a sharp Raman peak around 520 cm^{-1} similar to those reported in our previous paper.[7] Since such a sharp peak was not observed for the present samples, we can rule out the existence of Si microcrystals. The cross-sectional TEM observation of the present samples did not show lattice fringes corresponding to Si microcrystals, thus confirming the Raman results. The absence of well-grown Si microcrystals in our present samples is due to lower initial concentrations of excess Si atoms and lower annealing temperatures compared to the previous studies. To our knowledge, spectra similar to those shown in Figs. 1 and 2 have not been reported so far for c-Si, a-Si, nanostructures of Si and SiO$_x$ films.

We point out here that the spectral features presently observed agree qualitatively with those of the vibrational DOS of Si clusters calculated by Feldman *et al.*[a] They calculated the DOS spectra for the magic-number clusters Si_{33} and Si_{45}, whose geometric structures were constructed based on the reconstructed structure of Si surface. The structure of Si_{33} was constructed based on the Si(111) 7 × 7 surface, while that of Si_{45} on the Si(111) 2 × 1 surface. Fig. 3 compares the calculated DOS spectra with our typical Raman spectra of the samples Si-8 annealed at 400°C and 800°C. In the figure the calculated DOS spectrum and the Raman spectrum of a-Si are also included. Due to the potential used, calculated frequencies of the vibrational modes are somewhat higher than the values which will be measured experimentally. To facilitate the comparison between theory and experiment, the frequency scale of the calculated DOS spectra was so adjusted that the calculated DOS spectrum of a-Si agrees well with the experimental Raman spectrum, *i. e.*, the original frequency scale was reduced by a factor of 0.88 to bring the high-frequency DOS peak of a-Si at the experimental value of 480 cm^{-1}.

In Fig. 3 we can see that characteristic features of the experimental spectrum of $T_a = 400°C$ agree very well with those of calculated DOS spectrum of Si$_{33}$. A striking feature in the DOS spectra of Si$_{33}$ is the appearance of high-frequency split-off modes (at 540 and 585 cm^{-1} after rescaling), which are located at frequencies higher than the highest bulk mode in c-Si and a-Si. These split-off modes correspond to adatom vibrations similar to those in Si (111) 7 × 7 surface. A high-frequency shoulder around 550 cm^{-1} appearing in the experimental spectrum is thought to be the manifestation of these split-off modes. Furthermore, the calculated DOS spectrum shows a low-frequency component around 70 cm^{-1} in addition to a component around 150 cm^{-1}. The component at 70 cm^{-1} does not appear in the DOS spectrum of a-Si. In the experimental spectrum of $T_a = 400°C$, Raman signals attributable to these components are seen, although the splitting of the components is not so clear as in the DOS spectrum. In the DOS spectrum, the components around 70 and 150 cm^{-1} are stronger than the high-frequency components around 45 m^{-1}, in good qualitative agreement with the experimental spectrum.

In the DOS spectrum of Si$_{45}$ cluster, the high-frequency split-off modes do not appear, because the structure assumed is based on the (2 × 1) reconstruction of Si (111) surface and there is no adatom analogous to those in Si$_{33}$. The low-frequency component around 70 cm^{-1} becomes weaker than the component around 150 cm^{-1}. Furthermore, the high-frequency component around 480 cm^{-1} becomes stronger than the low-frequency components. These changes in the DOS spectrum agree fairly well with those in the Raman spectrum observed by raising T_a from 400°C to 800°C. Since the Raman spectrum is determined by not only the DOS spectrum but also the matrix elements, the argument based only on the DOS spectrum is not sufficient. Moreover, the calculations of Feldman et al. [a] were limited to only Si$_{33}$ and Si$_{45}$ clusters. However, the good qualitative agreement between theory and experiment seen in Fig. 3 allows us to conclude that the present Raman spectra originate from the Si clusters formedin the films.

The presently observed spectra are rather broad and the split-off modes are not well separated from the rest of the spectra. This may be due to the size distribution of the clusters. The appearance of the split-off modes in the spectra may not directly indicate the dominance of the Si$_{33}$ cluster in our samples. Although theoretical analysis is still lacking for Si clusters other than Si$_{33}$ and Si$_{45}$, cluster structures analogous to the Si(111) 7 × 7 surface and corresponding adatom vibrations may be possible in Si clusters of various sizes. What was presently observed as a high-frequency shoulder may be the Raman signals from various adatom vibrations, with frequencies slightly different from each other, in clusters of various sizes within a size distribution. However, the average size of the clusters is thought to increase with increasing excess Si concentration and annealing temperature.

3. Raman scattering from B-doped Si microcrystals

3.1 Introduction

Although the quantum size effects arising from the confinement of electrons

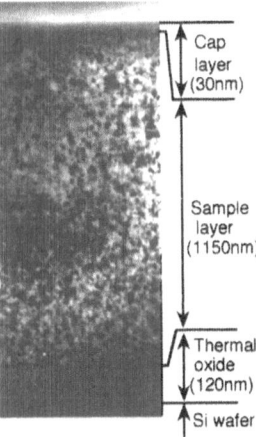

Fig. 4 Cross-sectional TEM image of the sample.

and holes in microcrystals have been widely studied so far, few experimental studies have been conducted to achieve the impurity doping of microcrystals in a controlled way to search for the quantum size effects on impurity states. In a previous study,[17] we demonstrated the feasibility of nitrogen doping into GaP microcrystals by means of gas evaporation in a mixture gas of N_2 and Ar. However, the N-doped GaP microcrystals obtained were too large (~100 nm) for the quantum size effects to occur. The gas-evaporation technique does not allow us to prepare much smaller microcrystals. In the present study, we attempted to prepare much smaller Si microcrystals and dope them with B atoms by applying the rf cosputtering method. Results of Raman measurements clearly demonstrate the success of the doping.

3.2 Experimental

The present samples were prepared by cosputtering the targets of SiO_2, B_2O_3 and Si. The targets used were a Si wafer (4 in in diameter), and small pieces of fused quartz plates and pellets of B_2O_3 placed on the Si wafer. The sputtering was performed in Ar gas of 2×10^{-2} Torr with an rf power of 100 W, using an ANELVA SPF210HS magnetron sputtering apparatus. Si wafers covered with thermal oxide layers 100 nm thick were used as the substrates. After deposition of the cosputtered film, the sample was covered with a thin SiO_2 cap layer, which was also deposited by sputtering. In order to grow B-doped μc-Si embedded in SiO_2 matrices, the samples were annealed at 1100°C for 30 min under N_2 gas flow.

Raman measurements were performed in a back-scattering configuration at room temperature. The spectra were recorded by a Jobin Yvon U1000 double monochromator, equipped with a R943-02 photomultiplier (Hamamatsu Photonics) and a photon counting system. The excitation source was two lines of an Ar-ion laser (457.9 and 514.5 nm) and two lines of a Kr-ion laser (647.1 and 676.5 nm).

Samples for cross-sectional TEM observation were prepared by standard procedures including the mechanical and Ar-ion thinning techniques. The TEM

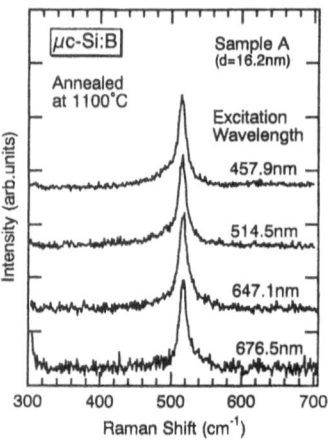

Fig. 5 Raman spectra of B-doped Si microcrystals.

observation was made using a JEM-200CX electron microscope operated at 200 kV.

3.3 Results and discussion

Figure 4 shows the cross-sectional TEM picture of a sample prepared with 18 pieces of the SiO_2 target and 4 pellets of B_2O_3 placed on the Si wafer. We can see clearly the multilayer structure of the sample, *i.e.*, the Si substrate, 120 nm-thick thermal oxide layer, 1150 nm-thick sample layer and the 30 nm-thick cap layer. In the sample layer,the grown μc-Si (dark patches) are dispersed rather uniformly. The average size of μc-Si obtained from pictures taken with a higher magnification is 16.2 nm. The electron diffraction patterns obtained indicated the diamond structure of μc-Si.

The Raman spectra obtained for the sample shown in Fig. 4 are presented in Fig. 5. In this figure, the dependence of the spectrum on the wavelength of the excitation light, λ_{ex}, is presented. For $\lambda_{ex} = 457.9$ nm, an asymmetric peak with a tail at the low-frequency side is seen. The spectral shape is very similar to those previously reported for sufficiently small Si microcrystals in which confinement of phonon occurs. As the wavelength of the excitation light becomes longer, the low-frequency tail gradually disappears and a tail at the high-frequency side appears. For $\lambda_{ex} = 676.5$ nm the peak is highly asymmetric with a tail at the high-frequency side. It should be stressed that changes in the spectrum similar to those seen in Fig. 4 were observed only for the samples prepared with B_2O_3 targets, and never observed for the samples prepared without B_2O_3 targets.

Raman spectra of bulk Si crystals doped with a large amount of B atoms were studied in detail more than 20 years ago. [b] It was reported that heavy doping results in the asymmetric shape of the Raman peak with a dip at the low-frequency side and a tail at the high-frequency side. The high-frequency tail becomes very much pronounced for longer excitation wavelengths. The asymmetry in the spectral shape was interpreted in terms of Fano-type interference between the continuum electronic transitions in the valence bands and the discrete optical phonon scattering. The qualitative features in the spectral shape presently observed for our μc-Si are in good qualitative agreement with those of heavily B-doped bulk Si crystals. From the good qualitative agreement with the bulk data and a fact that asymmetry was never observed for samples prepared without B_2O_3 targets, we can conclude that B-doping into μc-Si was successfully achieved. As far as we know, this is the first report on impurity doping of μc-Si.

PUBLICATIONS

1. M. Fujii, S. Hayashi and K. Yamamoto, *Jpn. J. Appl. Phys.*, **30**, 687 (1991).
2. S. Hayashi, E. Nishimae and K. Yamamoto, *Z. Phys.*, **D26**, S228 (1993).
3. M. Fujii, T. Nagareda, S. Hayashi and K. Yamamoto, *Phys. Rev.*, **B44**, 6243 (1991).
4. M. Fujii, T. Nagareda, S. Hayashi and K. Yamamoto, *J. Phys. Soc. Jpn.*, **61**, 754 (1992).
5. M. Fujii, M. Wada, S. Hayashi and K. Yamamoto, *Phys. Rev.*, **B46**, 15930 (1992).
6. S. Hayashi, Y. Kanzawa, M. Kataoka, T. Nagareda and K. Yamamoto, *Z. Phys.*, **D26**, 144 (1993).
7. S. Hayashi, T. Nagareda, Y. Kanzawa and K. Yamamoto, *Jpn. J. Appl. Phys.*, **32**, 3840 (1993).
8. S. Hayashi, M. Kataoka and K. Yamamoto, *Jpn. J. Appl. Phys.*, **32**, L274 (1993).
9. Y. Kanzawa, S. Hayashi and K. Yamamoto, Proc. 22nd Int. Conf. on Physics of Semiconductors, Vancouver, 2031 (1994).
10. S. Hayashi, M. Kataoka, Y. Kanzawa and K. Yamamoto, Proc. 22nd Int. Conf. on Physics of Semiconductors, Vancouver, 2023 (1994).
11. S. Hayashi, M. Kataoka, H. Koshida and K. Yamamoto, *Surf. Rev. Lett.*, **3**, 1095 (1996).
12. S. Hayashi, A. Kato and K. Yamamoto, *Solid State Commun.*, **89**, 563 (1994).
13. S. Hayashi, S. Kawata, H. M. Kim and K. Yamamoto, *Jpn. J. Appl. Phys.*, **32**, 4870 (1993).
14. S. Hayashi, H. Sugano, H. Arai and K. Yamamoto, *J. Phys. Soc. Jpn.*, **61**, 916 (1992).
15. H. M. Kim, T. Fukumoto, S. Hayashi and K. Yamamoto, *J. Phys. Soc. Jpn.*, **63**, 2194 (1994).
16. Y. Inoue, S. Hayashi and K. Yamamoto, *Surf. Rev. Lett.*, **3**, 1059 (1996).
17. H. M. Kim, S. Hayashi and K. Yamamoto, *Jpn. J. Appl. Phys. Suppl.*, **34-1**, 40 (1994).

REFERENCES

a. J. L. Feldman, E. Kaxiras and X.-P. Li, *Phys. Rev.*, **B44**, 8334 (1991).
b. F. Cerdeira, T. A. Fjeldly and M. Cardona, *Phys. Rev.*, **B8**, 4734 (1973).

8

Mesoparticle Buried in Semiconductor
—Silicon in Gallium Arsenide—

Masashi Suezawa

Institute for Materials Research, Tohoku University
Aoba-ku, Sendai 980-8577, Japan

Purpose of the present study

We show various properties associated with change of occupation sites of Si in GaAs due to annealing. Si in GaAs is amphoteric impurity. It works as a donor and an acceptor when it occupies the Ga atom site and As atom site, respectively. One of the interesting feature of Si in GaAs is that it mainly occupies the Ga atom site (Si_{Ga}) at high temperature and some move to As atom site (Si_{As}) at low temperature. Hence we assume that Si clusters, *i.e.* semiconductor buried in another semiconductor, are probably generated in highly Si-doped GaAs after annealing at low temperature since the probability for the existence of the nearest neighbor Si atoms pair becomes high after annealing at low temperature. We annealed GaAs doped with various concentrations of Si and measured photoluminescence spectrum, its excitation spectrum, polarization and optical absorption spectrum (detection of localized vibration of Si atom) to investigate the effect of site change of Si atoms due to annealing.

Contents of the present study

1. Experiment [a,1–7)]

Specimens used in this experiment were single crystalline GaAs doped with Si between $1 \times 10^{17} – 6 \times 10^{19}$ (0.013 %) atoms·cm^{-3} grown by horizontal Bridgeman method. We call the specimen which contains 6×10^{19} Si atoms·cm^{-3} P crystal hereafter since it contains Si precipitates at as-grown state.

All crystals were slowly cooled after single crystal growth and hence imposed complex heat treatment. In order to investigate annealing effect, we first annealed all of the crystals at 1200°C to eliminate annealing effect during crystal growth. We call this annealing solution treatment since the distribution of defects and impurities is expected to become homogeneous due to the above annealing. Specimens were annealed at appropriate temperatures after solution treatment. All of the annealing was done under optimum [b,c)] arsenic pressure to reduce the evaporation of As atoms from specimens during annealing. We sealed specimens together with As in evacuated quartz capsules and annealed them using two furnaces, one for specimens and the other for As. We rapidly cooled the specimens after annealing.

In the measurement of normal photoluminescence (PL) spectrum, we excited specimens with Ar ion laser (514.5 nm line) and detected PL with a Ge-photodetector. In the measurement of excitation spectrum of PL, we used Ti-sapphire tunable laser for excitation. Specimens were immersed in liquid helium.

Free electron concentration, which is determined mainly from the difference of the number of Si atoms at a Ga atom site and that at an As atom site, was determined from the measurement of optical reflectance at room temperature. The concentration of Si at each atomic site can be determined from the measurement of optical absorption associated with atomic vibration of Si. Mass of Si atom is much smaller compared to that of Ga and As atom. Hence it has localized modes of vibration and we can identify Si at Ga and As atom sites and also pairs of Si_{Ga}, Si_{As} and vacancy from the measurement of frequency of localized vibration. [d] Those vibrational frequencies are around 380 cm^{-1} and not observable when free carrier density is high. We irradiated specimens with 15 MeV electrons to reduce free electrons by being captured by acceptors generated by high energy electrons and measured localized vibration related to Si at 6 K with the use of an FT-IR spectrometer.

2. Results and discussion

2.1 Changes in occupation sites of Si atom and of free electron density due to annealing

Figure 1 shows the dependence of free electron density on the annealing temperature of specimens from P crystal and non-P crystal, which does not include Si precipitates at as-grown state and has a carrier density of around 3×10^{18} cm^{-3}. In the case of P crystal, the free electron density becomes as high as about 8×10^{18} cm^{-3} after solution treatment in contrast with that at as-grown state, 3×10^{18} cm^{-3}. This means that many Si atoms change to isolated Si_{Ga} due to solution treatment.

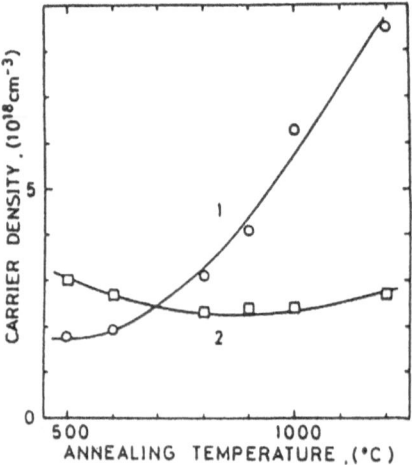

Fig. 1   Temperature dependence of free carrier density at room temperature of highly Si-doped GaAs. Curves 1 and 2 correspond to specimens of precipitation region and non-precipitation region, respectively. Free carrier density was determined from the measurement of optical reflectance.

(From M.Suezawa, A.Kasuya, Y.Nishino and K.Sumino, *J. Appl. Phys.*, **69**, 1622 (1991)).

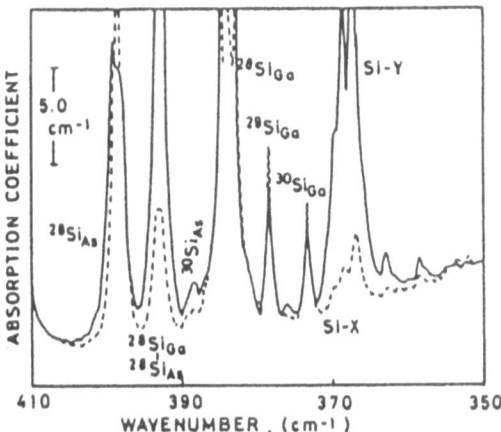

Fig. 2   Optical absorption spectra after annealing of specimen at precipitation region for 20 h at
1200°C solution treatment: dashed line) and 600°C (solid line). Inserted letters indicate the
responsible defects for vibrational absorption peaks. Measuring temperature was 6 K.
(From M.Suezawa, A.Kasuya, Y.Nishino and K.Sumino, *J. Appl. Phys.*, **69**, 1622 (1991)).

But this is much smaller compared with the total number of Si atoms included in
P-crystal. Hence many Si atoms do not occupy Ga atom sites even after solution
treatment. The free electron density decreases with decrease in annealing
temperature. As shown in Fig. 2, this is due to site change of Si atom from the Ga
atom site to the As atom site and also due to pairing with vacancy. On the
contrary, the free electron density of non-P crystal does not depend on the
annealing temperature. One interesting result is that curves 1 and 2 intersect as
shown in Fig. 1. If site change of Si atom occurs with a simple reaction, for
example one in which the reaction rate is proportional to the concentration, such
intersection of two curves cannot be expected to occur. There may be a complex
process for the site change of Si atoms.

Figure 2 shows the optical absorption spectra of localized vibration of Si-
related defects measured at 6 K. The responsible states [d] of Si atom are shown in
the figures. Figure 2 shows that of P crystal. After solution treatment, intensity due
to Si_{Ga} is very strong. In contrast with this, the intensity of Si_{As}, Si_{Ga}-Si_{As} pair and
Si-Y becomes very strong and that of Si_{Ga} becomes weak after annealing at 600°C.
This shows that many Si atoms change occupation sites and form pairs with other
defects (Defect Y is believed to be a Ga vacancy). The increase of the density of
Si_{Ga}-Si_{As} pair suggests an increase of Si clusters if we consider only the interaction
between 1st neighbor atoms. These behaviors probably explain the behavior of
Fig. 1. In the case of non-P crystal, optical absorption spectra are very similar after
solution treatment and 600°C annealing. This is consistent with Fig. 1 in which
free electron density is almost independent of annealing temperature.

2.2 Study of annealing effect by photoluminescence measurement

2.2.1 Photoluminescence spectrum

Above results clearly show that Si atom changes occupation site and that
pairs of Si_{Ga}-Si_{As} and Si-vacancy are generated after low temperature annealing.

Optical absorption method is very powerful to detect the state of Si atoms as shown above but its sensitivity is not high. We measured PL spectrum in order to know the behaviors of Si atoms and defects depending on the annealing treatment, which are not detected by optical absorption measurement in the infrared region. We measured PL spectrum of specimens doped with Si of wide concentration range, not only in the precipitation region but also in the non-precipitation region, in order to know the effects of annealing on the specimens of various concentration of Si. We found that PL spectra had different dependence on annealing temperature in the three regions of Si concentration, $i.e.$, region 1 is below 7×10^{17} Si atoms·cm^{-3}, region 2 is between 7×10^{17} and 3×10^{18} and region 3 is above 3×10^{18}. Si atoms precipitate after annealing at low temperatures in the case of region 3. The reason for division to regions 2 and 3 in high concentration is intuitively easy to understand since precipitation of Si occurs in region 3 and not in region 2. However, we cannot discuss this in detail since we do not know the solubility of Si in GaAs as a function of temperature. We have no idea how to explain the division into regions 1 and 2 at low concentration of Si. We show PL spectra in these three regions in the following.

PL spectra in the specimen of region 3 are shown in Fig. 3. The first characteristic feature is that a broad PL peak appears at a point higher than the band-gap energy after annealing at high temperatures, and it disappears after annealing at temperatures lower than 1000°C. The former is explained by Burstein-Moss effect, which is observed in degenerate system of high density of carriers. The latter phenomenon is difficult to understand since the carrier density is high enough to degenerate even after annealing at 800°C. We have not yet succeeded in explaining the latter. Probably we should introduce another factor, for example scattering of carriers by impurities, other than Burstein-Moss effect. The second is that the peak position of P-12 shifts to lower energy as the annealing temperature becomes lower. Figure 4 shows quantitative relation between peak position of P-12 and annealing temperature. The peak position of P-12 continuously shifts to lower energy as the annealing temperature becomes lower.

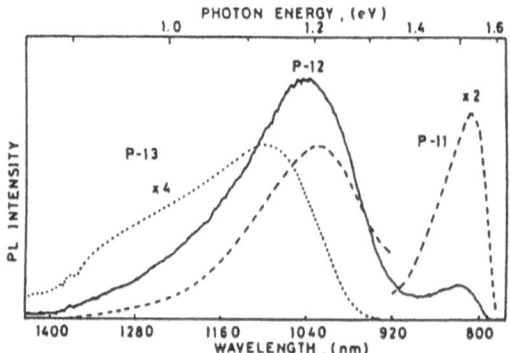

Fig. 3   Photoluminescence spectra after annealing of specimens in precipitation region at various temperatures. Dashed line, solid line and dotted line are spectra after annealing at 1200°C, 1000°C and 600°C, respectively. Annealing duration at each temperature was 20 h. Labels (*e.g.* P-11) are the name of luminescence peaks. Numerals (*e.g.* ×2) mean that real intensity should be multiplied by this factor.
(From M.Suezawa, A.Kasuya, Y.Nishino and K.Sumino, *J. Appl. Phys.*, **69**, 1622 (1991)).

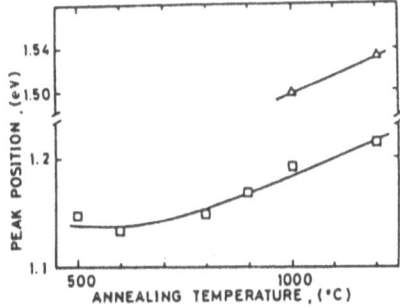

Fig. 4 Dependence of PL peak position of specimens in region 3 on annealing temperature. Symbols of triangle and square are for P-11 and P-12, respectively.
(From M.Suezawa, A.Kasuya, Y.Nishino and K.Sumino, *J. Appl. Phys.*, **69**, 1622 (1991)).

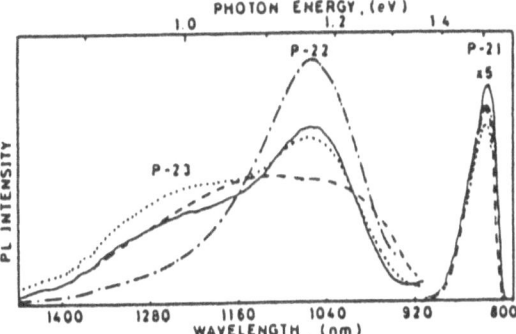

Fig. 5 Dependence of PL spectrum of specimens in region 2 on annealing temperature. Solid, dashed, dotted and chained lines are specimens after solution treatment, delete 1000°C, 900°C and 500°C, respectively.
(From M.Suezawa, A.Kasuya, Y.Nishino and K.Sumino, *J. Appl. Phys.*, **69**, 1622 (1991)).

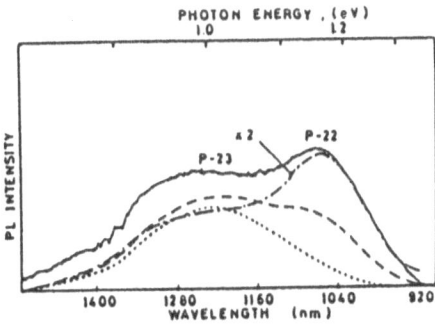

Fig. 6 Dependence of PL spectrum of the specimen in region 2 after annealing at 700°C on the excitation energy. Chained, solid, dashed and dotted lines are spectra excited by 774, 839, 860 and 888 nm light, respectively.
(From M.Suezawa, A.Kasuya, Y.Nishino and K.Sumino, *J. Appl. Phys.*, **69**, 1622 (1991)).

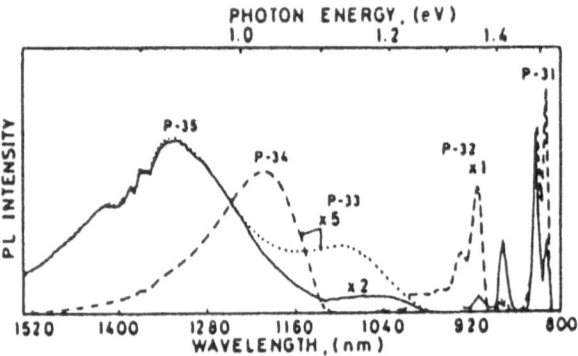

Fig. 7   Dependence of PL spectra of specimens in region 1 on annealing temperature. Solid, dashed and dotted lines are spectra in the specimens after annealing at 1200°C, 900°C and 500°C, respectively.
(From M.Suezawa, A.Kasuya, Y.Nishino and K.Sumino, *J. Appl. Phys.*, **69**, 1622 (1991)).

This means that the defect responsible for P-12 does not have a definite structure but a variable structure, for example cluster, depending on the annealing temperature. The second-neighbor pair of Si_{Ga} and V_{Ga} (Gallium vacancy) was proposed[e] to be the defect responsible for P-12, but the above result suggests that such a model does not hold.

PL spectra in the specimen of region 2 are shown in Fig. 5. The first characteristic feature is that there is a broad PL peak P-21 at near band-gap energy. Contrary to that of region 3, this peak is always observed and does not depend on the annealing temperature. The second is that the intensities of peaks P-22 and P-23 strongly depend on the annealing temperature as shown in Fig. 5. Figure 6 shows the excitation energy dependence of peaks P-22 and P-23. Both peaks are observed when excited by high energy light, around and above band-gap energy, but only P-23 is observed when excited by small energy light. These results clearly show that responsible energy levels for the two PL peaks are very different even though the peak positions themselves are close to each other.

PL spectra in the specimens of region 1 are shown in Fig. 7. The first characteristic feature is that the peak position of deep PL depends on the annealing temperature in a discrete, not continuous, manner. Peaks P-34 and P-35 are observed after annealing at high temperature (1200°C) or low temperature (*e.g.* 500°C) and P-34 is observed after annealing at intermediate temperature (*e.g.* 900°C).

2.2.2 Excitation spectra of PL

Above results of the dependence of PL spectra on annealing temperature are very different from those doped with Zn or Te, both of which have definite occupation sites, *i.e.*, Ga and As atom sites, respectively, not depending on the annealing temperature. Hence we believe that the above results come mainly from the change in occupation sites depending on the annealing temperature. In fact we observe optical absorption due to vacancy-Si pair after annealing at low temperature as shown in section 2.1. Vacancy is indispensable for the site change of Si atoms and it probably pairs with Si after site change. There may be many

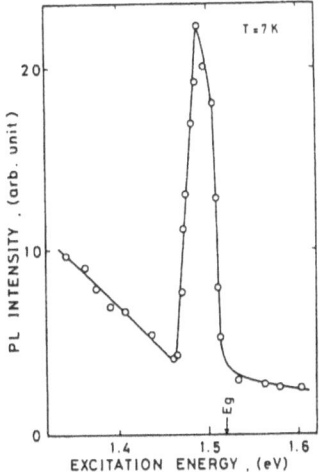

Fig. 8   Excitation spectrum of P-34 which is observed in specimen 1 after annealing at 900°C. E_g is the band-gap energy.
(From M.Suezawa, A.Kasuya, Y.Nishino and K.Sumino, *J. Appl. Phys.*, **76**, 1622 (1991)).

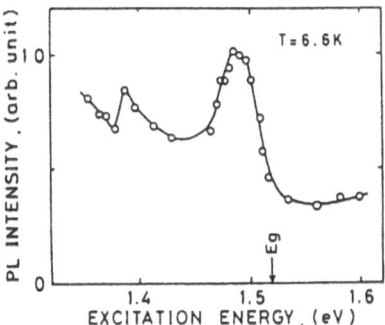

Fig. 9 Excitation spectrum of P-35 observed in specimen 1 after annealing at 500°C.
(From M.Suezawa, A.Kasuya, Y.Nishino and K.Sumino, *J. Appl. Phys.*, **76**, 1622 (1991)).

kinds of defects in Si doped GaAs. In order to find the energy levels responsible for PL, we measured excitation spectra of PL lines.

Figure 8 shows the excitation spectrum of peak P-34 observed in the specimen in region 1 after annealing at 900°C. There is a sharp peak (E_{DA}) near the band-gap energy and a broad background component below about 1.46 eV. We calculate E_{DA} assuming Eq. (1) for E_{DA}

$$E_{DA} = E_g - (E_D + E_A) + e^2/\varepsilon r_{DA} \tag{1}$$

where E_g is the band-gap energy, E_D and E_A are levels of donor and acceptor, respectively, and r_{DA} their average distance. This agrees well with the observed value. We therefore conclude that donor Si (Si_{Ga}) and acceptor Si (Si_{As}) are responsible for peak P-34. Both or one of these probably pairs with a defect, *e.g.* a vacancy, and capture an electron and/or a hole which will relax to deep levels. Radiative recombination occurs after relaxation. That is the reason for the large

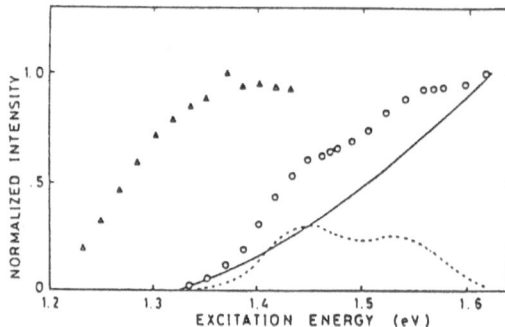

Fig. 10 Excitation spectra of P-22 (circle) and P-23 (triangle) which are observed in the specimen of
region 3 after annealing at 500°C. Solid and dotted lines are fitted line due to Lucovsky
model and the difference between smoothed lines fitted to circles and solid lines, respectively.

difference between E_{DA} and peak energy of P-34. There is a background
component in the excitation spectrum as shown in Fig. 8. Probably this is due to
other deep levels but we do not know its nature.

Figure 9 shows the excitation spectrum of peak P-35 in a specimen of region
1 annealed at 500°C. There are two peaks. The higher energy peak corresponds to
that in Fig. 9. Hence Si donor and Si acceptor are responsible for peak P-35.
There is another peak at lower energy (about 1.4 eV). We think that this is
probably related to V_{Ga} since it is expected to be generated associated with the site
change of Si from the Ga atom to the As atom site due to annealing at low
temperature.

The excitation spectra of specimens in regions 2 and 3 are very similar.
Figure 10 shows the excitation spectra of peaks P-12 (circles) and P-13 (triangles)
observed in the specimen in region 3 after annealing at 500°C. The excitation
spectrum of P-12 can be decomposed to that of solid and dotted lines. The solid
line is determined by Lucovsky's model, which is known to be well applicable to
the optical transition between a localized level and continuous state, such as the
conduction and valence band. The threshold energy of continuous state is about
1.32 eV. It means that there is a discrete energy level at this energy. There are two
broad peaks, at 1.45 and 1.53 eV, in the dotted spectrum. It means that these
peaks correspond to transition between localized levels. Peak widths are broad
probably because of high density of conduction electrons. The peak at 1.45 eV
corresponds to transition between two localized states in the band gap. On the
other hand, the peak position of another peak, 1.53 eV, is larger than the band gap
energy. We interpret that this peak corresponds to transition of electrons between
a localized state in the band gap and resonance state in the conduction band.
Such resonance state is found in the case of DX center in GaAs.

2.2.3 Polarization

As described in section 2.2.1, the position of peak P-12 in the specimen in
region 3 shifts depending on the annealing temperature. This behavior may be
interpreted in terms of cluster as responsible defect for this PL peak. Recently,
Muto et al. [1] reported their results of specimens in region 3 observed by high
resolution electron microscope. They found thin spheroidal precipitates, i.e.

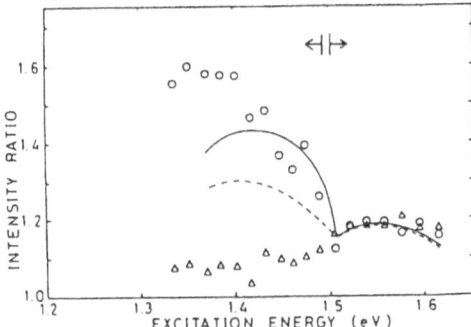

Fig. 11 Excitation energy dependence of polarization of P-12 observed after annealing of specimen in region 3 at 500°C (circles and triangles) and P-22 observed after annealing of specimen in region 2 at 900°C (solid and dashed line). The vertical axis is the ratio of the strongest and the weakest intensities. Circles and solid line are intensity ratios excited by light polarized along [$1\bar{1}1$]. Triangles and dashed line are intensity ratios excited along the [001] direction.

clusters, of Si on the {111} plane. If such precipitates are responsible for the PL, PL is probably polarized since the responsible defect for PL has a spatially anisotropic shape. We observed polarization characteristics of PLs at 1080 nm (P-12, P-22) and 1240 nm (P-13, P-23) in specimens in regions 2 and 3 after annealing at 900°C and 500°C, respectively. Excited and observed surface was (110) plane since it included crystallographic directions of low indices.

Figure 11 shows polarization characteristics. The polarization directions of excitation were [$1\bar{1}1$] (circle and solid line) and [001] (triangle and dashed line). The vertical axis is the PL intensity ratio of the strongest and weakest PL. The abscissa is the excitation energy. The arrow in the figure is the band gap energy. The direction of the strongest PL depends on the excitation energy, namely [$\bar{1}12$] when excited above the band gap energy and [$1\bar{1}1$] when excited below the band gap energy. This is interpreted to be due to the existence of many paths for the relaxation of electrons and holes to emit PL when excited by high energy photons but limited paths when excited by low energy photons. Circles and triangles are data from the specimen in region 3 and solid and dashed lines are from the specimen in region 2. Both specimens show similar polarization when excited along the [$1\bar{1}1$] direction but they showed very different behavior when excited along the [001] direction. Table 1 shows a summary. The 1240 nm (P-13, P-23) lines from specimens in regions 3 and 2 show similar behavior but the 1080 nm (P-12, P-22) lines show very different behavior. We therefore conclude that silicon cluster is responsible for the 1080 nm line and randomly distributed defects are responsible for the 1240 nm line. Some polarization characteristics can be explained if we assume Si cluster to be Si precipitates observed by high resolution electron microscopy.

The 1080 nm line in specimens in regions 2 and 3 have similar excitation spectra but have very different polarization characteristics. This probably suggests that responsible defects themselves are the same but their spatial distributions are different.

Table 1 The relation between direction of excitation and that of the strongest PL and the intensity ratio at 1.4 eV

Excitation Direction	Photoluminescence			
	1080 nm line		1240 nm line	
	P-500	NP-900	P-500	NP-900
$[1\bar{1}1]$	$[1\bar{1}1]$ 1.55	$[1\bar{1}1]$ 1.4	$[\bar{2}21]$ 1.1	$[\bar{2}21]$ 1.1
$[001]$	nonpolar	$[\bar{1}12]$ 1.3	$[\bar{2}21]$ 1.1	$[\bar{2}21]$ 1.1
$[\bar{1}12]$	$[\bar{2}21]$ 1.8	$[\bar{1}12]$ 1.5	$[\bar{2}21]$ 1.5	$[\bar{2}21]$ 1.25
$[\bar{1}10]$	$[\bar{2}21]$ 1.55	$[1\bar{1}1]$ 1.3	$[\bar{2}21]$ 1.4	$[\bar{2}21]$ 1.25

3. Summary

We investigated the properties of Si-doped GaAs after annealing at various temperatures noting the change in occupation sites of Si depending on the annealing temperature. Photoluminescence spectra and their excitation spectra strongly depend on the concentration of Si and annealing temperature. This suggests that many kinds of defects are generated associated with the change of occupation sites of Si due to annealing. We have proposed some kinds of defects responsible for some PL lines.

Acknowledgment

The author thanks Profs, A. Kasuya, Y. Nishina and K. Sumino for their collaboration in this research.

PUBLICATIONS

1. M. Suezawa, A. Kasuya, Y. Nishina and K. Sumino, Proc. 20th Internat. Conf. Physics of Semicon. (eds. E. M. Anastassakis & J. D. Joannopoulos, World Scientific, 1990) 678.
2. M. Suezawa, A. Kasuya, Y. Nishina and K. Sumino, *J. Appl. Phys.*, **69**, 1618 (1991).
3. M. Suezawa, A. Kasuya, Y. Nishina and K. Sumino, *Materials Sci. Forum*, **83-87**, 953 (1992).
4. M. Suezawa, A. Kasuya, Y. Nishina and K. Sumino, *J. Appl. Phys.*, **73**, 3035 (1993).
5. M. Suezawa, A. Kasuya, Y. Nishina and K. Sumino, *J. Appl. Phys.*, **76**, 1164 (1994).
6. M. Suezawa, A. Kasuya, Y. Nishina and K. Sumino, MRS Symp. Proc. **262**, 543 (1992).
7. M. Suezawa, A. Kasuya, Y. Nishina and K. Sumino, *Materials Sci. Forum*, **143-147**, 1269 (1994).

REFERENCES

a. As a review, E. W. Williams and H. B. Bebb, *Semiconductors and Semimetals*, eds. R. K. Willardson and A. C. Beer, **8**, 321, Academic Press (1972).
b. J. Nishizawa, H. Otsuka, S. Yamakoshi and K. Ishida, *Jpn. J. Appl. Phys.*, **13**, 46 (1974).
c. J. Lagowski, H. C. Gatos, C. H. Kang, M. Skowronski, K. Y. Ko and D. G. Lin, *Appl. Phys. Lett.*, **53**, 892 (1986).
d. R. T. Chen and W. G. Spitzer, *J. Electrochem. Soc.*, **127**, 1607 (1980).
e. E. W. Williams, *Phys. Rev.*, **168**, 922 (1968).
f. S. Muto, S. Takeda, M. Hirata, K. Fujii and K. Ibe, *Phil. Mag.*, **A66**, 257 (1992).

9

FIR Spectroscopy of Fine Particles under High Pressure

Takao Nanba[a] and Mitsuhiro Motokawa[b]

[a] Department of Physics, Faculty of Science, Kobe University
Nada-ku, Kobe 657-8501, Japan
[b] Institute for Materials Research, Tohoku University
Aoba-ku, Sendai 980-8577, Japan

Purpose of the present study

The change in properties of materials under high pressure is one of the most interesting phenomena and it is important to know the difference of the pressure effect between bulk and fine particles. We have been working on phonon spectra of alkali halides microcrystals and exciton spectra of CdS fine particles in FIR region applying high pressure by diamond anvil cell (DAC). The purpose of the present study is to observe the phase transitions of fine particles from the pressure dependence of spectra. In this report, we describe the experimental results of pressure dependence of surface phonon spectra of alkali halide fine particles.

Experimental methods

To study the structural phase transition of microcrystals caused by pressure, measurements of pressure dependence of Raman spectroscopy and FIR spectroscopy of phonon modes are important. FIR spectroscopy in particular provides more information on crystals which have high symmetries and are inactive for Raman spectroscopy. High pressure experiment of FIR spectroscopy, however, is usually difficult because the usual blackbody radiation used in a normal FIR spectrometer does not provide light strong enough for the small specimen in DAC. On the other hand, synchrotron radiation has a high brightness compared to the thermal light sources. We developed a method of spectroscopy in the FIR region by use of synchrotron radiation and have been doing high pressure experiments up to 3.2 GPa at the electron storage ring facility, UVSOR, of the Institute for Molecular Science at Okazaki. Details of the apparatus for transmission measurement under high pressure in the FIR region are described in references a and b.

Contents of the present study

Pressure effect on surface phonon mode of NaCl microcrystal

1. Introduction

As the size of an ionic crystal decreases below the micrometer level, a so-called surface phonon mode appears because of the important role of atoms on the surface.[c] The fundamental mechanism of an infrared absorption due to such a surface phonon mode of the spherical microcrystals embedded in a nonabsorbing medium at atmospheric pressure ($P = 0$) has been developed as a continuum theory by Genzel and Martin,[d,e] and up to now the optical properties of microcrystals with not only a spherical but also an irregular shape have been widely investigated experimentally and theoretically. The pressure dependence of the TO phonon mode in the bulk state of NaCl has been measured by Lowndes and Rastogi [f] in a low pressure range below 0.7 GPa and analyzed from the anharmonicity in lattice dynamics. The behavior of the surface phonon of the microcrystal under high pressure, however, has not yet been studied. We made experiment under high pressure to clarify the pressure dependence of the surface phonon energies of a spherical and a cubic microcrystal of NaCl. In order to interpret the obtained results, the continuum model at $P = 0$ was extended to the case of a sample under high pressure. In the extended continuum model, the following two contributions are taken into consideration for the continuum model at $P = 0$, the blue energy shifts of the bulk TO phonon under pressure and the increase in the refractive index of the grease which works as a pressure transmitting medium.

It was confirmed by electron microscopic observation that the shape of NaCl microcrystals obtained by a conventional gas evaporation technique showed was spherical and that their size distribution was within 0.5–2.0 μm. On the other hand, the cubic microcrystals of NaCl obtained by grinding normal NaCl crystals were 1–3 μm. The size of these microcrystals is much smaller than the wavelength of the incident far-infrared light (40–400 μm). Therefore, a long wavelength approximation for the analysis of a surface phonon holds well in the present study. This means that the observed phonon absorption is due to the lowest mode among the polarizations of the surface phonons of the microcrystal. The pressure dependence of thin evaporated films of NaCl was also measured as a reference and compared with those of the surface phonons of microcrystals. The thickness of the evaporated film was about 1.21 μm.

NaCl crystal undergoes a structural phase transition from a rocksalt structure (B1) to a cesium chloride (B2) at about 30 GPa at room temperature. This experiment covers only the B1 phase of NaCl. The thickness and the diameter of the pinhole of the stainless steel gasket of DAC were 0.2 and 0.7mm, respectively. The double-sided interferogram pattern was measured by an inteferometer of the Martin-Pupplet type (Specac Co., U. K.) combined with a germanium bolometer (Infrared Lab., USA) which was operated at liquid helium temperature. The spectral intensities were obtained by a Fourier transformation of the interferogram. The spectral resolution was 4 cm^{-1}. The transmission intensities of the bare grease (I_0) and the mixture comprising the microcrystals and the grease (I) were recorded alternatively. It was difficult to measure simultaneously both spectral intensities, I_0 and I, at the same pressure. We then obtained the transmission spectrum (I/I_0) of the sample by using the spectrum of the grease under almost the same pressure as I_0. The slight difference in pressure gave rise to an error which appeared as a small noisy structure in the spectrum. As a sensor of the pressure induced in DAC, ruby chips were immersed together with the mixture in the cell. The calibration of the applied pressure to the sample was done by measuring the energy shift of the fluorescence R1 line of ruby which was excited by an argon ion laser. The fluorescence was recorded by an HR-320 monochromator (Jobin-Ybon Co., France) after each transmission measurement.

2. Experimental results and discussion

The transmission spectra of a spherical and a cubic microcrystal of NaCl are shown in Fig. 1 together with that of a thin film for comparison at (a) atmospheric pressure and at (b) 0.8 GPa. The large dip in each spectrum in the figures corresponds to the absorptions due to the surface phonons of the microcrystals and to the bulk TO phonon mode of the thin film, respectively. The energy of the bulk TO phonon agreed with the result obtained by Lowdnes and Rastogi. [f] In the microcrystals, the bandwidths of the absorptions were about 50 cm^{-1} and they were broader than that of the bulk. At $P = 0$ (a), the peak energy of the absorption were 198 cm^{-1} and 184.8 cm^{-1} for the microcrystals with a spherical and a cubic shape, respectively, and 165.4 cm^{-1} for the bulk. At $P = 0.8$ GPa (b), they shifted to 208 cm^{-1}, 193 cm^{-1} and 174.8 cm^{-1} for the spherical, cubic microcrystal and the bulk, respectively.

The peak positions of the observed absorption arc plotted as a function of the applied pressure in Fig. 2. The broken lines indicate the least square fitting curves. The magnitude of the shifts from the values at $P = 0$ for the two kinds of microcrystals was small, but different from that of the bulk.

In this paper, we extended the continuum model to a scheme under pressure in order to interpret the differences in peak energy shifts with pressure caused by

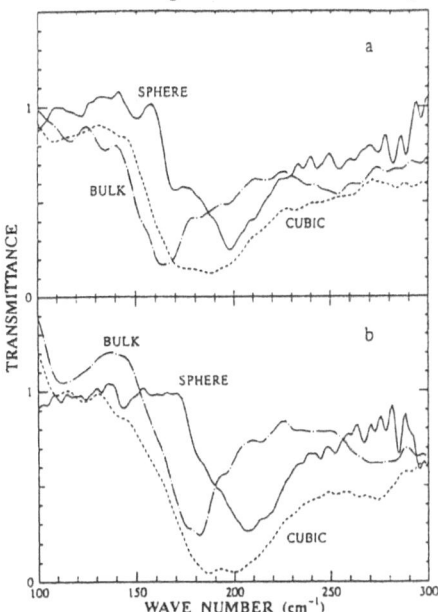

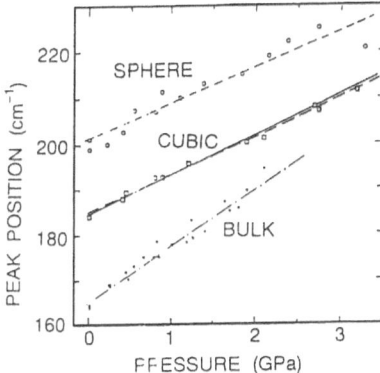

Fig. 1 Transmission spectra at room temperature of NaCl. The transmittance over unity us caused by the slight difference in the pressure applied to samples in obtaining the spectral intensities. (From T.Nanba, *J.Phys.Soc.Jpn.*, **63 (10)**, 3886 (1994)).

Fig. 2 Peak positions of the absorptions of a spherical microcrystal (open circles), a cubic microcrystal (open squares) a bulk (closed squares), respectively, as a function of the pressure applied to the samples. (From T.Nanba, *J.Phys.Soc.Jpn.*, **63 (10)**, 3886 (1994)).

the shape of the microcrystals.

2.1 Surface phonon at atmospheric pressure

According to the continuum model [d,e)] for the composite medium consisting of fine spherical microcrystals with a volume fraction f and a nonabsorbing medium surrounding the spheres, the characteristic energy of the surface phonon (ω_s) of the microcrystal is approximately given by

$$\varepsilon_1(\omega_s) = -\varepsilon_m^0 (2 + f)/(1 + 2f). \tag{1}$$

where $\varepsilon_1(\omega_s)$ is a real part of a dielectric function of the material forming a sphere and ε_m^0 is the dielectric constant of the material surrounding the sphere. In the case of small f, this equation leads to the simple equation,

$$\varepsilon_1(\omega_s) = -2\,\varepsilon_m^0 \tag{2}$$

which gives a Frolich mode with an uniform polarization within a sphere.

On the other hand, a surface phonon mode of a cubic microcrystal has been calculated by Gelder [g)] and Fuchs. [h)] Six surface phonon modes are obtained for a cubic microcrystal as the lowest order mode in the polarization which correspond to a Frolich mode of a small sphere. Their weight in the contribution to the absorption power of an incident light was calculated. The boundary condition which gives rise to the resonance energies for the six modes of a cubic microcrystal is given by,

$$\varepsilon_1(\omega_s^i) = -a_i\,\varepsilon_m^0 \tag{3}$$

where a_i is a constant coefficient corresponding to each surface phonon mode ($i = 1,2,...,6$) which possesses a weight fraction of ω_i in the contribution to the absorption power of light. Fuchs showed that the largest contribution to the absorption comes from the $i = 1$ mode ($a_1 = 3.68$ and $\omega_1 = 0.44$), the second large contribution comes from the $i = 2$ mode ($a_2 = 2.37$ and $\omega_2 = 0.24$), and so on. [h)] In the following analysis to interpret qualitatively the energy shifts of the surface phonon of a cubic microcrystal with pressure, for simplicity we refer to only the strongest $i = 1$ mode among the six surface phonon modes in a cubic microcrystal.

$\omega_{TO} = 164$ cm^{-1} and $\gamma = 19$ cm^{-1} are obtained from the peak position and the full width of the absorption spectrum for bulk in Fig. 1(a). $\varepsilon_m^0 = 2.25$ was estimated from the refractive index (n) of the grease measured by the interference fringes which appeared in the transmission spectrum of the bare grease in the cell in the far-infrared region. Then we obtained the value of 196.5 cm^{-1} as the resonance energy ω_s^0 for a spherical microcrystal by eq. (2). This value coincides well with the measured value of 198 cm^{-1} shown in Fig. 1. On the other hand, the energy ω^p for the lowest mode of a cubic microcrystal was calculated to be 183.3 cm^{-1}, while the measured one was 184.8 cm^{-1}. This good agreement between the two values means that the continuum model is applicable to the present analysis.

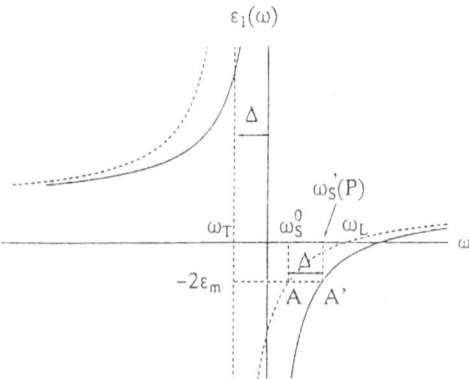

Fig. 3   Interpretation of the resonance condition for a surface phonon of a spherical microcrystal
under pressure according to eq. (2). Dashed and solid curves correspond to real parts of a
dielectric function without a damping at atmospheric and under pressure respectively.
(From T.Nanba, *J.Phys.Soc.Jpn.*, **63(10)**, 3889 (1994)).

2.2 Surface phonon under pressure

The position and the shifts of the surface phonon under pressure for a
spherical microcrystal are explained in Fig. 3. The surface phonon denoted by ω_s^0
at atmospheric pressure is located at the cross point A of the two broken lines,
$\varepsilon_1(\omega_s)$ and $-2\,\varepsilon_m$. Under pressure, on the other hand, the TO phonon energy
increases due to the compression of the volume, as seen in Fig. 2. This
corresponds to the blue shift of the whole dielectric function curve denoted by a
solid curve, $\varepsilon_1(\omega)$, by Δ as shown in Fig. 3. Then, the cross point A will shift to
the point A′, where the resonance energy is defind by ω's (P), by the same amount
Δ with the energy shift of the TO phonon due to the compression. This means that
the energy shift of the surface phonon is always Δ and consequently, independent
of the shape of the microcrystal. This scheme clearly goes against the obtained
experimental results. Experiment showed that the energy shifts of the surface
phonons were different from that of the bulk TO phonon.

In order to interpret consistently the obtained results, we first consider a
possible scheme for the case under pressure. When pressure is applied to the
specimen, it transmits the medium surrounding the microcrystals and the dielectric
constant, ε_m^0, increases to ε_m (the small increment is given by $\delta(P)$). Due to such
an increase in the dielectric constant, the cross point at which the resonance
occurs for a spherical microcrystal changes from A to A″ under pressure, as
shown in Fig. 4. Now the resonance energy is given by ω_s (P) instead of ω_s′ (P).
Consequently the energy shift for a spherical microcrystal at P amounts to Δ′
(P), which is given by ω_s (P) $= \omega_s^0 + \Delta$′ (P). The magnitude of Δ′ is always
smaller than the energy shifts Δ of the TO phonon (Δ′ $< \Delta$).

In the case of a cubic microcrystal, the resonance at $P = 0$ occurs at B where
the energy of the surface phonon is give by ω_c^0. The value of the coefficient, a_1
$= 3.68$, in eq. (3) for a cube, which is larger than the factor 2 in eq. (2) for a
sphere, gives the relation $\omega_c^0 < \omega_s^0$. By similar reasoning in the case of a sphere,
the resonance point for a cube under pressure changes from B to B″ at $P = 0$ due
to the same amount of the increase of Δ in the dielectric constant, In this case, the

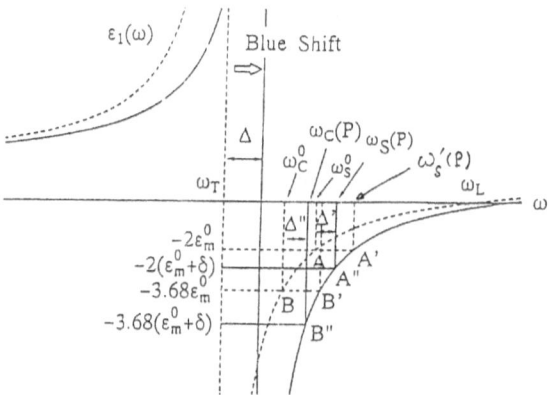

Fig. 4   Possible resonance conditions with an increase of δ in the dielectric constant of the surrounding medium of microcrystals by the compression in the frame work of the extended continuum model. The resonance energy is given by $\varepsilon_1(\omega_s^0) = 3.68\,\varepsilon_m$ for a cubic microcrystal. (From T.Nanba, *J.Phys.Soc.Jpn.*, **63(10)**, 3886 (1994)).

resonance occurs at ω_c (P). The resultant energy shift for a cubic microcrystal at P amounts to Δ'' (P) which is given by ω_c (P) $= \omega_c^0 + \Delta''$ (P). The magnitude of Δ'' is always smaller than Δ and between Δ and Δ' $(\Delta' < \Delta'' < \Delta)$ because the curvature of the dielectric function, $\varepsilon_1(\omega)$, of the microcrystal is steeper at B$''$ than at A$''$ even with the same increment of Δ in the dielectric constant, ε_m.

From the above interpretation, the expected resonance energies and the energy shifts of the surface phonons of microcrystals are schematically drawn in comparison with the bulk TO phonon in Fig. 5. At $P = 0$, the surface phonon of the spherical microcrystal appears at ω_c^0, which has the highest energy. The surface phonon of the cubic microcrystal also appears at the higher energy side of ω_{TO}^0 by $\Delta\omega''$. The relation of $\omega_{TO}^0 < \omega_c^0 < \omega_s^0$ always holds, as seen in Fig. 4. The difference in the curvature of the dielectric function at the cross points of A$''$ and B$''$ causes the shape-dependent energy shift of the surface phonon of the microcrystal, as well as the increase of δ in the dielectric constant of the medium. In this way, a continuum model which was extended in the present paper to the case under pressure explains qualitatively the shape dependence of the energy shifts of the surface phonons of the microcrystals under pressure.

Now we analyze quantitatively the results obtained for the peak energy shifts of the absorption of microcrystals under the scheme of the extended continuum model, which is given previously and shown in Figs. 4 and 5. The energy of the surface phonon in the lowest mode under pressure can be calculated according to eqs. (2) and (3), respectively, for a spherical and a cubic microcrystal, if we know the $\varepsilon_1(\omega)$ of microcrystal and the ε_m of the surrounding material under pressure. We use the following function as the dielectric function of NaCl,

$$\varepsilon_1(\omega) = \varepsilon_\infty + (\varepsilon_0 - \varepsilon_\infty)\, \omega_{TO}^2/(\omega_{TO}^2 - \tilde{\omega}^2) \qquad (4)$$

where ε_0 and ε_∞ are a static and a high frequency dielectric constant of NaCl, and $\tilde{\omega}^2 = \omega^2 + i\omega\gamma$, where γ is a damping constant in a Lorentz oscillator for the phonon and corresponds to the full width at the half maximum of the absorption band.

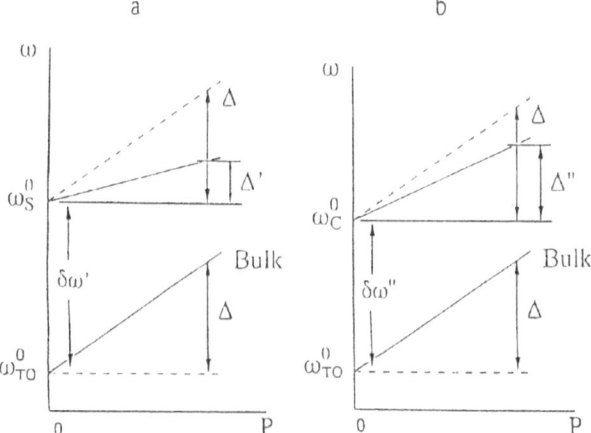

Fig. 5   Interpretation of the difference between the energy shifts of the surface phonons for a
spherical (a) and a cubic (b) microcrystal under pressure.
(From T.Nanba, *J.Phys.Soc.Jpn.*, **63(10)**, 3886 (1994)).

For the sample under pressure we assume that similar equations at $P = 0$
hold for the resonance conditions (2) and (3) and the dielectric function (4). In this
case, however, the quantities appearing in these equations change with pressure.
They are expressed as $\omega_{TO}(P)$, $\gamma(P)$ and $\varepsilon_m (P)$. Here, $\omega_{TO} (P) = \omega_{TO}^0 + \Delta (P)$ and
$\varepsilon_m (P) = \varepsilon_m^0 + \delta (P)$. Therefore, eqs. (2) and (3) become

$$\varepsilon(\omega_s^0 + \Delta'(P)) = -2(\varepsilon_m^0 + \delta'(P)) \text{ for a sphere,} \tag{5}$$

and

$$\varepsilon(\omega_c^0 + \Delta''(P)) = -3.68(\varepsilon_m^0 + \delta(P)) \text{ for a cube,} \tag{6}$$

respectively, under pressure. From these results, we can conclude that the
proposed extended continuum model works for the sample under pressure, if the
observed energy shifts for both types of microcrystals coincides with $\Delta' (P)$ and
$\Delta'' (P)$, respectively, appearing in eqs. (5) and (6). For this purpose, we need
$\delta(P)$, which appears in common in the right hand side of each equation.
Unfortunately, however, we failed to measure the refractive index of grease under
pressure because of the lack of the interference fringe in the transmission spectrum
of the grease in the diamond anvil cell. This seems to be because the refractive
index of grease under pressure is close to that of a diamond. Therefore we
analyzed our data in the following way. First, we determined $\delta(P)$ at one pressure
so that $\omega_s^0 + \Delta'(P)$ in eq. (5) coincided with the observed energy shifts of the
surface phonon for a sphere, and next examined if $\Delta''(P)$ which was obtained by
$\delta(P)$ used in eq. (6), could reproduce the measured energy shifts of the surface
phonon for a cubic microcrystal under pressure. If the present scheme works in
common for the samples under pressure, we can expect the measured energy
shifts for the cubic microcrystals to coincide with $\Delta'' (P)$ in eq. (6). First, we
determined $\delta (P)$ so that $\Delta' (P)$ in eq. (5) coincided with the observed energy shifts
for a sphere under pressure. These are shown in the broken curve marked
"SPHERE" in Fig. 2. At 1 GPa, for example, $\delta (1 \text{ GPa})$ was found to be 0.39 by
this procedure. This change is an increase of 17% of the value ε_m^0 at atmospheric

pressure. This value is a reasonable change under the pressure of 1 GPa. Next, we estimated Δ'' (1 GPa) as 9.0 cm^{-1} by eq. (6) using both values 0.39 as δ (1 GPa) and 12.5 cm^{-1} as Δ (1 GPa), which is the energy shifts of the bulk TO phonon of NaCl at 1 GPa. The obtained value of 9.0 cm^{-1} as Δ'' (1 GPa) is in good agreement with the measured energy of 9.3 cm^{-1} for the cubic microcrystal under the pressure of 1 GPa. Subsequently, we obtained Δ'' (P) at different pressures by this procedure under the assumption that the increment $\delta(P)$ in the dielectric constant ε_m^0 is proportional to the applied pressure. A least square fitting curve for values of Δ'' (P) obtained at different pressures is shown by a solid line marked "CUBIC" in Fig. 2. The expected curve for Δ'' (P) of the cubic microcrystal reproduced the measured data well. The small difference between the two curves seems to be caused mainly by neglecting the small contribution from the higher order modes in the polarization of the surface phonons in a cube to the absorption. This good agreement for a cubic microcrystal between the measured energy shifts (a broken line) and the expected ones under pressure (a solid line) indicates the validity of the present analysis by the extended continuum model.

PUBLICATION

1. T. Nanba, T. Matsuya and M. Motokawa, *J. Phys. Soc. Jpn.*, **63**, 3886 (1994).

REFERENCES

a. T. Nanba, *Rev. Sci. Instrum.*, **60**, 1680 (1989).
b. T. Nanba and M. Watanabe, *J. Phys. Soc. Jpn.*, **58**, 1535 (1989).
c. H. Frohlich, *Theory of Dielectrics*, 2nd ed., p.163, Oxford University Press, Oxford (1958).
d. L. Genzel and T. P. Martin, *Phys. Stat. Sol.(b)*, **51**, 91 (1972).
e. L. Genzel and T. P. Martin, *Surf. Sci.*, **34**, 33 (1973).
f. R. P. Lowndes and A. Rasrogi, *Phys. Rev.*, **B14**, 3598 (1976).
g. A. P. von Gelder, J. Holvat, J. H. M. Stoelinga and P. Wyder, *J. Phys.*, **C5**, 2757 (1972).
h. R. Fuchs, *Phys. Rev.*, **B11**, 1733 (1975).

10

Nanostructure Semimagnetic Semiconductors

Yasuo Oka, Hiroshi Okamoto, Kohei Yanata and Masaaki Takahashi

Research Institute for Scientific Measurements, Tohoku University
Aoba-ku, Sendai 980-8577, Japan

The properties of semimagnetic semiconductors with nanometer-scale structures (nanostructures) are described. Superlattices and microcrystals of the semimagnetic semiconductors have been grown successfully by epitaxy and sputtering methods and have shown remarkably enhanced magneto-optical responses. The enhanced optical properties of nanostructures are interpreted by the confined excitonic states interacting with the magnetic ion spins involved in the lattice. Possible applications of these nanostructure materials are discussed.

1. Introduction

Semimagnetic semiconductors are mixed crystals of compound semiconductors in which cation ions are partially substituted by magnetic ions. A variety of materials can be synthesized by a combination of a host material (typically a II-VI compound semiconductor such as CdTe, CdSe, ZnTe or ZnSe) and magnetic ions (Mn, Fe or Co). Besides the optical and electrical properties of host lattices, these materials have magnetic properties due to the magnetic ions involved and show novel effects such as a giant Zeeman splitting, a magnetic polaron effect and a spin glass phase.[a–b]

We have fabricated microcrystals and superlattices of semimagnetic semiconductors and studied the properties of quantum confined electronic states in the quantum well where the electrons are interacting with the magnetic ions. These semimagnetic nanostructures show new optical properties and high possibilities for novel magneto-optical devices. In this paper we discuss several possibilities of applications in the nanostructure semimagnetic semiconductors.[1–3]

2. Properties of semimagnetic semiconductors

Single crystals of semimagnetic semiconductors can be synthesized by the Bridgman method. $Cd_{1-x}Mn_xTe$, $Cd_{1-x}Mn_xSe$ and $Cd_{1-x}Fe_xSe$ are typical examples which can be grown by the Bridgman method using quartz ampules. The $Cd_{1-x}Mn_xTe$ crystal is a zinc blende structure. For high x the material approaches MnTe, although MnTe with zinc blende structure is not reported as a bulk crystal. The zinc blende MnTe has been grown recently as epitaxial films on GaAs substrates.[c–e] The band gap energy of the zinc-blende MnTe is 2.98 eV, which coincides with the extrapolating band-gap energy of $Cd_{1-x}Mn_xTe$ to $x = 1$. A III–V compound based semimagnetic semiconductor, $In_{1-x}Mn_xAs$, has also been

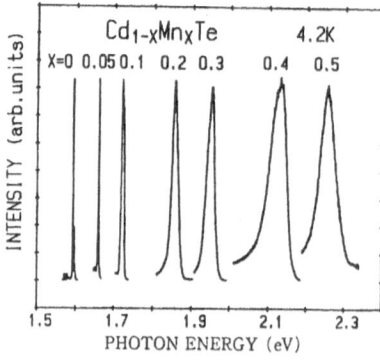

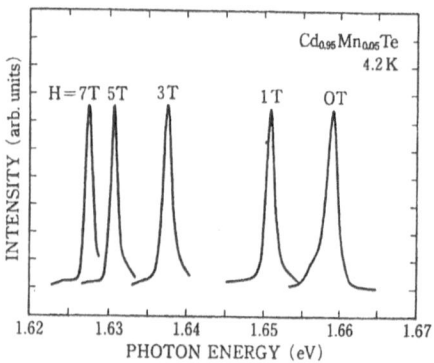

Fig. 1 Luminescence spectra of $Cd_{1-x}Mn_xTe$.

Fig. 2 Zeeman shift of the exciton luminescence in $Cd_{1-x}Mn_xTe$ ($x = 0.05$).

reported.[f] Rare earth ions such as Eu and Er can be introduced as magnetic ions in semimagnetic semiconductors. Therefore the semimagnetic semiconductors show a large variety of material properties.

Luminescence spectra of excitons in $Cd_{1-x}Mn_xTe$ single crystals with $x = 0 - 0.5$ are shown in Fig. 1. The exciton energy increases with increase of Mn mole fraction x, which is due to increase in the band gap energy.[1] Figure 2 shows the magnetic field effect of the exciton luminescence in $Cd_{1-x}Mn_xTe$ with $x = 0.05$, where the exciton band shows the Zeeman shift of 32 meV at a magnetic field of 7 T. The effective g value is given to be 360. Therefore the Zeeman splitting of the exciton ground state in the $Cd_{0.95}Mn_{0.05}Te$ crystal is 180 times as large as that of pure CdTe. The huge magnetic splitting called "the giant Zeeman effect" is caused by the exchange interaction of the band electron with the magnetic ions.

3. Nanostructure semimagnetic semiconductors

The semimagnetic semiconductor nanostructure can be classified into two types: Fig. 3(a) is the case of the electron system confined in a well in which magnetic ions also exist, while Fig. 3(b) corresponds to the system with magnetic ions outside the well. In system (a) the confined electron in the well interacts directly by the exchange interaction with the magnetic ions. On the other hand, in system (b), the electron in the well can only interact with the magnetic ions outside the well by the tunneled portion of the wavefunction. The difference in these two types of structure produces distinct properties in these semimagnetic nano-systems.

Microcrystals of semimagnetic semiconductors have successfully been grown by the simultaneous sputtering of $Cd_{1-x}Mn_xSe$ and SiO_2 and heat treatments. In transmission microscopy, the $Cd_{1-x}Mn_xSe$ microcrystals of 50–200 Å diameter are observed in the SiO_2 amorphous matrix and show a clear image of the lattice. This is the first report of synthesis of semimagnetic semiconductor quantum dots.[2–5]

Semimagnetic semiconductor superlattices can be synthesized by using molecular beam epitaxy,[e–h] cluster beam epitaxy,[d] and hot wall epitaxy. We have fabricated $CdTe/Cd_{1-x}Mn_xTe$ and $Cd_{1-x}Mn_xTe/ZnTe$ superlattices by the hot wall epitaxy method.[6–9] Bulk crystals of CdTe, MnTe and ZnTe have been used as

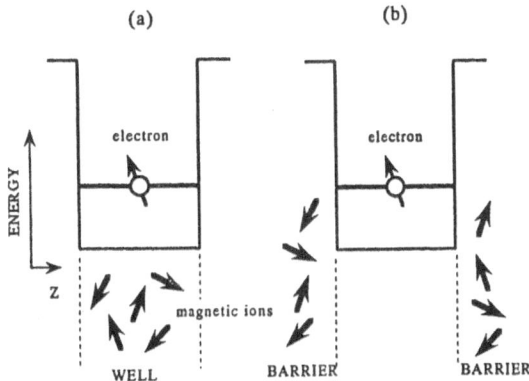

Fig. 3 Semimagnetic semiconductor nanostructures.

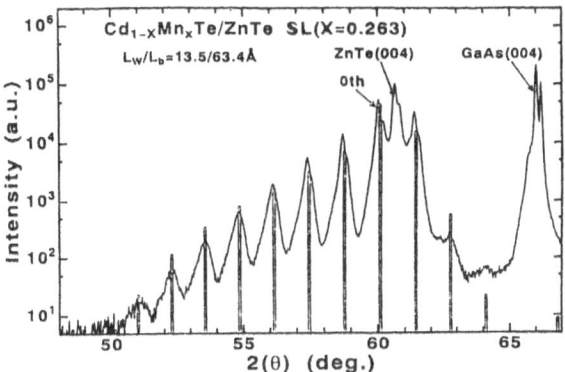

Fig. 4 X-ray diffraction pattern of the $Cd_{1-x}Mn_xTe/ZnTe$ superlattice.

source materials in the epitaxy and the $Cd_{1-x}Mn_xTe$ and CdTe (ZnTe) layers are alternately grown to make superlattices. ZnTe buffer layers are initially grown on GaAs substrates with thickness of 1 μm before the starting growth of the superlattices. Figure 4 shows the X-ray diffraction pattern of the $Cd_{1-x}Mn_xTe/ZnTe$ superlattice.[8] Satellite lines to the 7th order in X-ray diffraction are observed, displaying evidence of an excellent periodicity of the superlattice. These superlattices are strained superlattices due to the difference in the lattice constant of the composite materials. Therefore the electronic states in the superlattices are caused by the strain effect. The superlattices of $Cd_{1-x}Mn_xTe/ZnTe$ have been synthesized for the first time in our study.

4. Exchange interaction and Zeeman effect

Semimagnetic semiconductors show many interesting magnetic and optical properties. These are induced by the exchange interaction of the band electron with the magnetic ions involved. The exchange interaction is expressed as follows:

$$H_{ex} = \sum_i \alpha_i s \cdot S_i, \tag{1}$$

where s, S_i are the spins of the band electron and the magnetic ion, α_i is the exchange constant of the interaction.

The exchange interaction between magnetic ions is as follows:

$$H_{ex}' = \sum_{ij} J_{ij} S_i \cdot S_j. \tag{2}$$

The interaction among the magnetic ions changes significantly with increasing magnetic ion concentration x, which affects the magnetic properties of the materials. For the increase of x, the magnetic ion spin state changes from an isolated spin state to pair and triplet states and to a larger cluster spin state. The variation in the spin state affects the magnetization process of the materials. In semimagnetic semiconductors with higher concentration of magnetic ions, a spin glass phase appears at low temperatures. Galazka determined the phase diagram of the spin glass in $Cd_{1-x}Mn_xTe$ by measuring the transition temperature from the paramagnetic phase to the spin glass.[i]

Optical spectrum related to the exciton state in the semimagnetic semiconductor shows marked change in magnetic fields, as can be seen in Fig. 2. The change is due to the Zeeman effect of the exciton state, which is caused by the exchange interaction of the exciton with the magnetic ions contained. The interaction is given as follows:

$$\Delta E(H) = N_0(\alpha - \beta)x \langle S_z \rangle, \tag{3}$$

where N_0 is the density of cation, $\alpha(\beta)$ the exchange constant of an electron (hole) with the magnetic ions, x the mole fraction of the magnetic ions, and $\langle S_z \rangle$ the averaged spin component of the magnetic ion along the external magnetic field. $\langle S_z \rangle$ can be expressed in the low concentration regime of magnetic ions by using the Brillouin function $B_J(\eta)$,

$$\langle S_z \rangle = B_{5/2}(5g\mu_B H/2k_B T), \tag{4}$$

where g is the g value of the d electron in the magnetic ions, μ_B, k_B and T are the Bohr magneton, Boltzmann constant and temperature of the lattice. The exchange constant $N_0(\alpha - \beta)$ can be determined from the Zeeman shift ($= \Delta E/2$) of the peak energy of the exciton luminescence by using eqs. (3) and (4). In the high concentration region of magnetic ions, the effects of pairs and higher clusters interacting with the exciton must be taken into account besides the isolated ions since the antiferromagnetic interaction between the magnetic ions becomes significant. It is interesting to study how these magnetic and optical properties vary in the quantum wells, superlattices and quantum dots from those of the bulk crystal.

5. Luminescence of nanostructure semimagnetic semiconductors

Figure 5 displays the luminescence spectrum of the $CdTe/Cd_{1-x}Mn_xTe$

superlattice synthesized by hot wall epitaxy. The superlattice is composed of well layers of CdTe and barrier layers of $Cd_{1-x}Mn_xTe$, where the thickness of the well (L_w) and the barrier (L_b) are 150 Å and the Mn concentration x is 0.24. A band denoted as QX around 1.60 eV is the exciton luminescence, while at 1.97 eV the luminescence appears due to the barrier exciton in the $Cd_{1-x}Mn_xTe$ layers, BX. The magnetic field shift of these luminescence band is shown in Fig. 6. The BX band shows the Zeeman shift of 21 meV at 7 T, from which the effective g value of the BX band can be deduced to be g_{eff} = 104. The Zeeman shift of the QX band at 7 T is 3 meV, which corresponds to g_{eff} of 15.

In this superlattice, the exciton in the quantum well of the CdTe layer, QX, interacts with Mn ions in the $Cd_{1-x}Mn_xTe$ barrier. However the probability of penetration of the QX wavefunction to the barrier layer due to the tunneling effect is only 10^{-3}. Therefore the strength of the exchange interaction of the QX exciton with Mn ions in the barrier due to the tunneling effect cannot give the observed g value. In this case the Mn ions penetrate into the well layer by diffusion from the barrier and the QX exciton directly interacts with the Mn ions in the well. From the g_{eff} value of QX the diffusion in the Mn ions through the interface is estimated to be 10 Å.

Figure 7 shows the magnetic field effect of the exciton luminescence in the $Cd_{1-x}Mn_xTe/ZnTe$ superlattice, where x = 0.16 and L_w and L_b are 28 Å and 53 Å respectively.[9] Since, in this superlattice, the QX luminescence arises in the $Cd_{1-x}Mn_xTe$ layer, the Zeeman shift of the QX band becomes a large value, 16 meV at 7 T, which corresponds to g_{eff} of 80 for QX. The large effective g value of the exciton is the evidence of the direct exchange interaction of the exciton with the Mn ions in the well. The exciton luminescence peak energy is shown in Fig. 8 as a function of the well thickness. In the superlattice system with x ~0.16, the energy of QX is 1.90 eV for the wide well of L_w = 45 Å, while for the narrower well of L_w = 6 Å the exciton energy increases to 2.35 eV. This is evidence of the

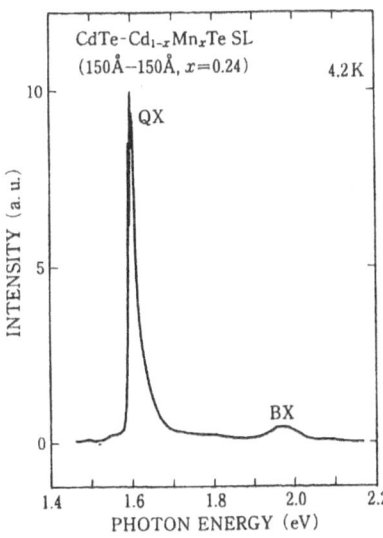

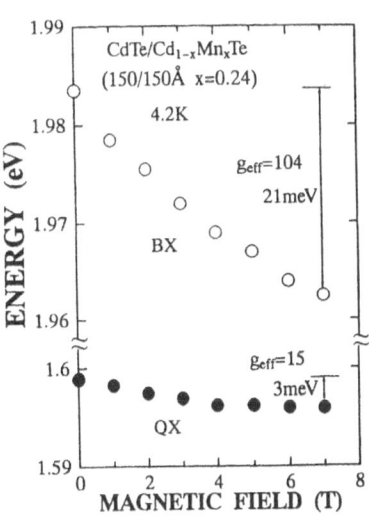

Fig. 5 Luminescence due to the quantum well exciton (QX) and the barrier exciton (BX) in the CdTe/$Cd_{1-x}Mn_xTe$ superlattice.

Fig. 6 Zeeman shifts of the quantum well exciton (QX) and the barrier exciton (BX).

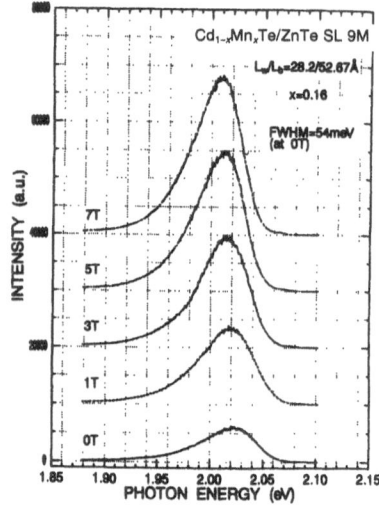

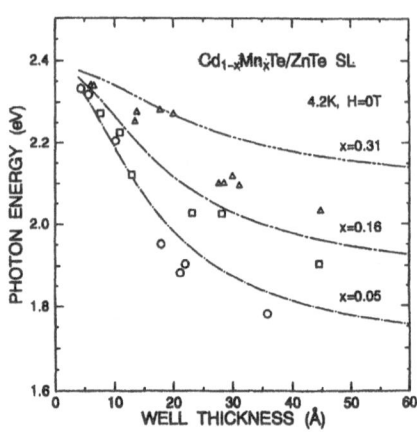

Fig. 7   Luminescence spectra of the Cd$_{1-x}$Mn$_x$ Te/ZnTe superlattice.

Fig. 8   Exciton energy of the Cd$_{1-x}$Mn$_x$Te/ZnTe superlattice as a function of the well width.

quantum confinement effect. The dash-dotted lines show the energy of the QX calculated by the Kronig-Penney model.

6. Excitonic magnetic polaron in nanostructures

As we have shown in the previous section, the exciton in semimagnetic semiconductors is affected by the exchange interaction of magnetic ions. On the other hand the spins of the magnetic ions are polarized within the area of the exciton Bohr radius due to this exchange interaction. Figure 9 schematically shows the exciton interacting with magnetic ions, which is called "the excitonic magnetic polaron." [10,11]

To form the bound state of the magnetic polaron, the exciton polarizes the neighboring magnetic-ion spins. Formation of the excitonic magnetic polaron takes place during the gain for the self energy of the magnetic polaron, while the

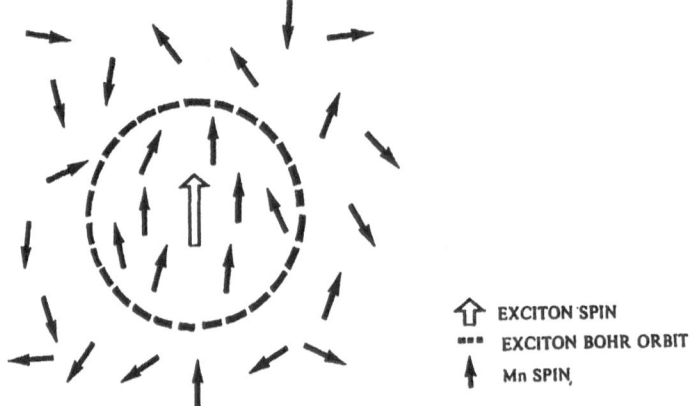

Fig. 9   Excitonic magnetic polaron.

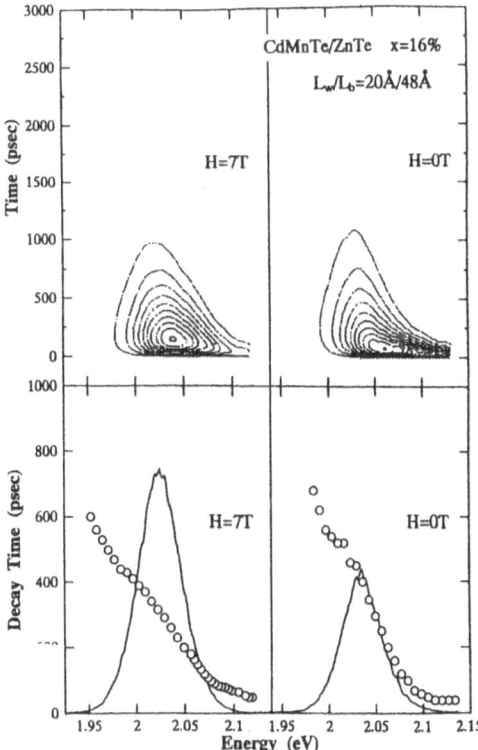

Fig. 10 Time characteristics and the decay time of the exciton luminescence in the $Cd_{1-x}Mn_xTe/ZnTe$ superlattice.

created excitonic magnetic polaron annihilates by the radiative or non-radiative recombination processes. These changes are the dynamical process of the excitonic magnetic polarons. In semimagnetic semiconductor nanostructures, we must consider the magnetic polaron state for the reduced dimensional exciton confined in the quantum wells. In the low dimensional exciton system the formation of the excitonic magnetic polarons will be different from that of the bulk crystal.

To elucidate the formation and annihilation of the excitonic magnetic polaron in the nanostructure semimagnetic semiconductors, we study the time variation of the exciton luminescence.[11] Figure 10 shows the time variation of the exciton luminescence in the $Cd_{1-x}Mn_xTe/ZnTe$ superlattice. Top figures in Fig. 10 display the time variation of luminescence by a contour map of the emission intensity,while the bottom figures show the lifetime of the exciton corresponding to the exciton luminescence energies.[12,13] At $H = 0$ T (right hand side figures), a fast decay component with the decay time of 50–100 ps arises in the higher part of the luminescence band and at around 2.04 eV the decay time increases to 400 ps. In a much lower energy region of the exciton luminescence band, the decay time of the exciton becomes 600 ps. The fast-decay components appeared in the higher energy region display the formation process of the excitonic magnetic polaron by aligning the Mn-ion spins in the $Cd_{1-x}Mn_xTe$ quantum well. In the lower energy region, the excitonic magnetic polaron shows the localization process by lowering the energy in the inhomogeneous quantum well due to the fluctuation of the interfaces.

In the left hand side of Fig. 10 at 7 T, the first decay component in the higher energy region almost disappears, as can be seen from the contour map. Since most of the Mn-ion spins are aligned by the external field of 7 T, the exciton cannot further align the Mn spins. This effect can be seen as the disappearance of the transient formation process of the excitonic magnetic polaron in the finite time region. In this superlattice the lifetime of the exciton in the quantum well is 400 ps at the peak energy of the luminescence band, which is considerably shorter than that of the bulk crystal of $Cd_{1-x}Mn_xTe$ (1-2 ns). The reason for the shortening of the lifetime in the superlattice is due to the increase of the oscillator strength and also partly due to the increase of the non-radiative recombination process caused by the lattice defects in the strained superlattices.

The $CdTe/Cd_{1-x}Mn_xTe$ superlattices are the system where the exchange interaction of the exciton with Mn ions takes place indirectly through the superlattice interfaces. On the other hand the $Cd_{1-x}Mn_xTe/ZnTe$ superlattice contains Mn ions in the well layers and shows a large magneto-optical effect due to the direct exchange interaction of the exciton with Mn ions. We can also expect novel systems of semimagnetic semiconductor superlattices other than the type mentioned above.[j] $Zn_{1-x}Mn_xSe$ shows decrease in band-gap energy by increasing Mn concentration x from 0 to 0.02. The band gap increases for the further increase of x. Therefore in the $Zn_{1-x}Mn_xSe/ZnSe$ superlattice with $x = 0.04$, the gap energies of the $Zn_{1-x}Mn_xSe$ and $ZnSe$ layers coincide with each other. The Zeeman splitting of the exciton level in the $Zn_{1-x}Mn_xSe$ layer in the magnetic field is large, while the splitting in the $ZnSe$ layer is small. The higher energy level with the spin-up configuration in the $Zn_{1-x}Mn_xSe$ layer becomes an unbound state due to the lack of barrier potential, while the spin-down state in the $ZnSe$ layer is unstable due to the existence of the lower spin-down state in the $Zn_{1-x}Mn_xSe$ layer. As a result each layer of the superlattice contains one type of spin in the magnetic field, i.e., the spin-down state is stable in the $Zn_{1-x}Mn_xSe$ layer and the spin-up state is only stable in the $ZnSe$ layer. A superlattice in which the up and down spin states of the carriers are separated into the two layers, is called "a spin superlattice." The spin superlattices have interesting electrical and optical properties.[j]

7. Semimagnetic semiconductor quantum dots

Figure 11 shows the transmission microscope photograph of the $Cd_{1-x}Mn_xSe$ microcrystal ($x = 0.15$) grown in SiO_2 matrix. [5,7,14,15] The microcrystals are prepared by a sputtering method and subsequent heat treatment. The microcrystal of 150 Å diameter in the SiO_2 amorphous matrix shows the (110) lattice fringes without significant lattice defect. The lattice constant of the microcrystal is 4.3 Å, from which the Mn concentration x in the microcrystal could be determined. The crystal structure of the microcrystal is wurtzite, which is the same as that of the bulk crystal.

Luminescence spectrum of the $Cd_{1-x}Mn_xSe$ microcrystal is shown in Fig. 12. The microcrystals with $x = 0.15$ and with an average diameter of 200 Å show the exciton luminescence band peaked at 1.97 eV and a Zeeman shift of 16 meV at 3.5 T. The g value of the exciton in the microcrystal can be estimated to be 100-140. Since the exciton Bohr radius in the $Cd_{1-x}Mn_xSe$ bulk crystal is 32 Å, the

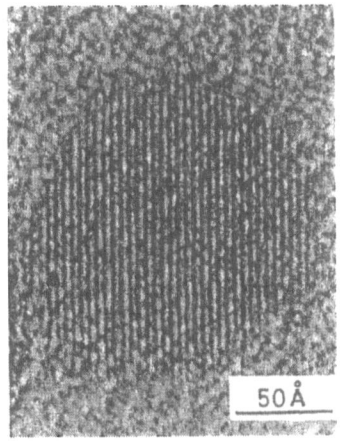

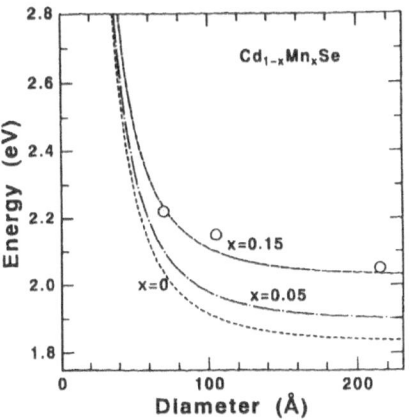

Fig. 11 Lattice image of the $Cd_{1-x}Mn_xSe$ quantum dot grown in the SiO_2 amorphous film.
(From K. Yanata and Y. Oka, *J. Appl. Phys.*, **73(9)**, 4596 (1993)).

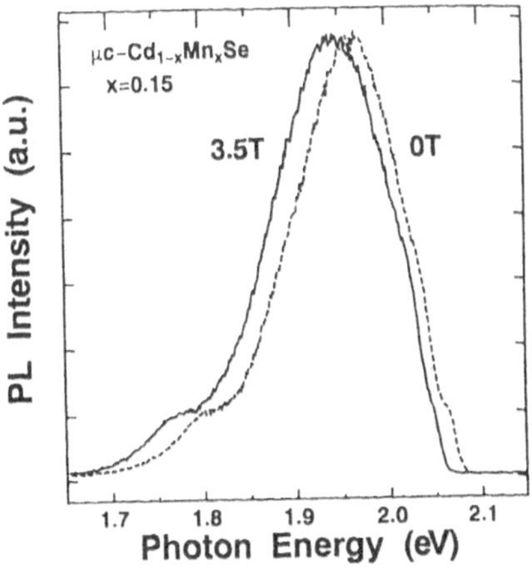

Fig. 12 Exciton luminescence spectra of the $Cd_{1-x}Mn_xSe$ quantum dot.

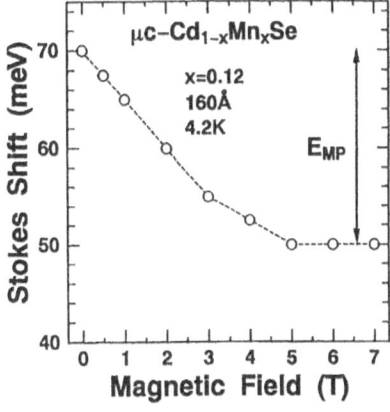

Fig. 13 Size dependence of the exciton energy in the $Cd_{1-x}Mn_xSe$ quantum dot.
(From K. Yanata and Y. Oka, *J. Appl. Phys.*, **73(9)**, 4596 (1993)).

Fig. 14 Excitonic magnetic polaron energy E_{MP} in the $Cd_{1-x}Mn_xSe$ quantum dot.
(From K. Yanata and Y. Oka, *Superlattices Microstruct.*, **15(3)**, 233 (1994)).

microcrystal with diameter of 60–70 Å is similar in size as the exciton. Figure 13 shows the size dependence of the exciton energy measured from the luminescence peak. For the decrease in diameter, the exciton energy increases due to the increase of the quantum confinement. Figure 14 shows the magnetic field dependence of the Stokes shift of the exciton luminescence obtained by the selective excitation of the exciton absorption band. The reduction of the Stokes shift due to the magnetic field, E_{MP} corresponds to the binding energy of the excitonic magnetic polaron. The existence of the excitonic magnetic polaron in the semimagnetic semiconductor quantum dot is clarified for the first time in this work. The lifetime of the exciton in the quantum dots is plotted in Fig. 15, which indicates the

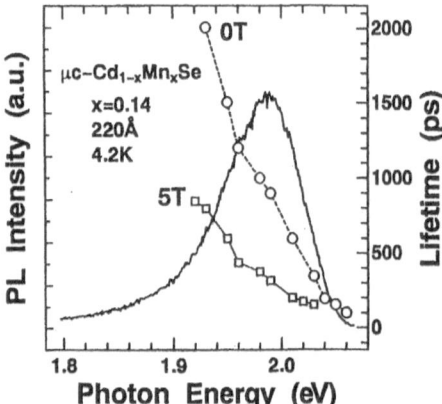

Fig. 15 Exciton lifetime in the $Cd_{1-x}Mn_xSe$ quantum dot.
(From K. Yanata and Y. Oka, *Superlattices Microstruct.*, **15**(3), 233 (1994)).

dynamical process of the excitonic magnetic polaron formation.[14,15]

8. Mn-ion luminescence and Faraday effect

Luminescence due to the d-electrons in Mn^{2+} ions in II–VI compound semiconductors has been studied for long years. This is caused by the $^4T_1 \rightarrow {}^6T_1$ transition of the $3d^5$ electrons. Some of the results are applied to practical devices such as electroluminescence displays by the $Zn_{1-x}Mn_xS$ system. In $Cd_{1-x}Mn_xTe$ the orange luminescence of the Mn ions by the d-electron transition appears for high x under the band to band excitation. We can clarify the energy transfer process from the exciton to the Mn ions.[1] For $x \sim 0.4$ the Mn luminescence intensity is 50 times that of the exciton luminescence.

Recent photoemission spectroscopy study using synchrotron radiations shows that the d-electron level of Mn ion is located at the energy position of 3.5 eV below the top of the valence band and the d-level is significantly mixed with the valence band.[16–18] The efficient energy transfer from the electron-hole state of the host semiconductor to the Mn ions is caused by this strong mixing of the valence band and the d-electron state. In the nanostructure semimagnetic semiconductors, the d-electron state will be influenced by the low dimensional confinement effect of the host band electrons.

The giant Zeeman effect causes large Faraday rotation in semimagnetic semiconductors.[19] Crystals of $Cd_{1-x}Mn_xTe$ and $Zn_{1-x}Mn_xTe$ with 1 mm thickness show the large Faraday effects with a rotation of the polarization plane by more than 180×5 degrees for the incident photon energy near the band gap energy. In the superlattice of semimagnetic semiconductors, significant increase in oscillator strength arises due to the quantum confinement of the electronic state. Therefore we can expect efficient Faraday effect of semimagnetic nanostructures in the wavelength region from infrared to ultraviolet.

9. Application of nanostructure semimagnetic semiconductors

The novel properties of semimagnetic semiconductors can be applicable to practical use as follows. First, the efficient Faraday effect can be used to fabricate Faraday rotators or optical isolators. We can obtain these optical devices in the infrared to ultraviolet light region by using various crystals and compositions of the semimagnetic semiconductors. The optical isolators in the red (630 nm) and infrared (980 nm) region are actually fabricated by using $Cd_{1-x}Mn_xTe$ and $Cd_{1-x-y}Hg_xMn_yTe$.[k] In the near future optical communication and optical storage technologies will be developed in the blue and ultraviolet wavelength regions, and the optical isolators in these energy regions will become important. A waveguide-type device will be efficient for the isolator using superlattices.

Light-emitting devices are a good application for the semimagnetic semiconductor materials. Light-emitting diodes and laser diodes in various wavelength regions can be fabricated by using the various band gap materials of semimagnetic semiconductors. To realize these devices, full control of the materials for the n- and p-types is required. Recent development of impurity doping for ZnSe-based blue laser diodes will be helpful.[1] Since the emission wavelength can be varied by the external magnetic field in these light-emitting devices with semimagnetic nanostructures, we can fabricate magnetically tunable light-emitting devices. Therefore the semimagnetic semiconductor superlattices and microcrystals will be applicable to unique optical devices which cannot be obtained from non-magnetic semiconductor materials. [20]

A system of ZnS with Mn^{2+} ions has been used for the electroluminescence display. Recently it has been reported that the microcrystals of $ZnS:Mn^{2+}$ show markedly efficient luminescence of Mn ions. The result is interpreted by the confinement and boundary effects in the microcrystal.[m] Therefore nanostructure semiconductors can be expected to be used as efficient electroluminescence materials since the activators of Mn ions are involved in the host lattice.

In conclusion the nanostructure semimagnetic semiconductors have remarkable functional properties, where the electron system is affected by the quantum confinement and the exchange effect. Further application as new magneto-opto-electronic materials will become possible if we extend the freedom of the material control by developing epitaxy and microfabrication techniques.

This research is partly supported by a Priority Area Research Program for "Physical and Chemical Properties of Mesoscopic Materials" by the Ministry of Education, Science and Culture, Japan. The authors are grateful to I. Souma and T. Sato for their helpful collaboration. The study of semimagnetic semiconductor superlattices was also made in collaboration with Professor H. Fujiyasu, Shizuoka University.

PUBLICATIONS

1. Y. Oka, *Oyobutsuri*, **57**, 894 (1988) [in Japanese].
2. Y. Oka, *Oyojiki-Gakkaishi*, **17**, 869 (1993) [in Japanese].
3. Y. Oka, *Materials Science*, **31**, 16 (1994) [in Japanese].
4. Y. Oka, Proc. 4th Japan-Korea Joint Symposium, p26, Cheju Island, Korea (1994).
5. K. Yanata, K. Suzuki and Y. Oka, *J. Appl. Phys.*, **73**, 4596 (1993).
6. I. Souma, K. Izumisawa, Y. Oka, H. Fujiyasu, H. Noma, Bulletin of the Research Institute for Scientific Measurements, Tohoku University **38**, 15 (1989) [in Japanese].
7. K. Suzuki, M. Nakamura, I. Souma, K. Yanata, Y. Oka, H. Fujiyasu and H. Noma, *J. Crystal Growth,* **117**, 881 (1992).
8. M. Takahashi, S. Muto, K. Suzuki and Y. Oka, *Physics of Semiconductors*, p879, World Scientific (1992).
9. M. Nogaku, M. Takahashi, Y. Oka, Abstracts of the 48th Japan Physical Society Meeting, **Vol.**

2, 184 (1993) [in Japanese].

10. Y. Oka, *Bull. Phys. Soc. Jpn.*, **43**, 705 (1988) [in Japanese].
11. Y. Oka, K. Ishikawa, I. Souma and M. Nakamura, *Physics of Semiconductors*, p1939, World Scientific (1990).
12. H. Okamoto, K. Suzuki, K. Tsuzuki, M. Takahashi and Y. Oka, *Jpn. J. Appl. Phys.*, 32 Suppl. 32-3,746 (1993).
13. H. Okamoto, K. Thuzuki, M. Nogaku, T. Hisatsugu, M. Takahashi and Y. Oka, *Physics of Semiconductors*, World Scientific (1994) **Vol. 3**, 2525 (1994).
14. K. Yanata and Y. Oka, *Superlattice and Microstructures*, **15**, 233 (1994).
15. K. Yanata and Y. Oka, *Jpn. J. Appl. Phys.*, **34**, 164 (1994).
16. M. Taniguchi, A. Fujimori, M. Fujisawa, T. Mori, I. Souma and Y. Oka, *Solid State Commun.*, **62**, 431 (1987).
17. M. Taniguchi, K. Soda, I. Souma and Y. Oka, *Phys. Rev.*, **B46**, 15789 (1992).
18. Y. Ueda, M. Taniguchi, T. Mizokawa, A. Fujimori, I. Souma and Y. Oka, *Phys. Rev.*, **B49**, 2167 (1994).
19. I. Souma, T. Sato, Y. Oka, Abstracts of the Japan Physical Society Meeting (October, 1993), **Vol. 2**, 194 (1993) [in Japanese].
20. Y. Oka, *Quantum Theory of Solids and Devices*, Kogaku Kenkyusha (1993) [in Japanese].

REFERENCES

a. J. K. Furdyna and J. Kossut, *Semiconductors and Semimetals*, **Vol. 25**, "Diluted Magnetic Semiconductors," Academic Press, New York (1988).
b. M. Jain, *Diluted Magnetic Semiconductors*, World Scientific, Singapore (1991).
c. S. M. Durbin, J. Han, M. Kobayashi, D. R. Menke, R. L. Gunshor, Q. Fu, N. Pelekanos, A. V. Nurmikko, D. Li, J. Gonsalves and N. Otsuka, *Appl. Phys. Lett.*, **55**, 2087 (1989).
d. H. Anno, T. Koyanagi and K. Matsubara, *J. Cryst. Growth*, **117**, 816 (1992).
e. K. Ando, K. Takahashi, T. Okuda and M. Uemura, *Phys. Rev.*, **B46**, 12289 (1992).
f. H. Ohno, H. Munekata, T. Penney, S. von Molnar and L. L. Chang, *Phys. Rev. Lett.*, **68**, 2664 (1992).
g. N. Otsuka, *Oyobutsuri*, **56**, 1148 (1987) (in Japanese).
h. H. Akinaga, T. Abe, K. Ando, K. Ando, Y. Yoshida, K. Uchida, S. Sasaki and N. Miura, *Phys. Rev.*, **B47**, 15954 (1993).
i. R. R. Galazka, S. Nagata and P. H. Keesom, *Phys. Rev.*, **B22**, 3344 (1980).
j. N. Dai, H. Luo, F. C. Zhang, N. Samarth, M. Dobrowolska and J. K. Furdyna, *Phys. Rev. Lett.*, **67**, 3824 (1991).
k. K. Onodera, M. Kimura, T. Masumoto, *Tokin Technical Review*, **21**, 33 (1994) [in Japanese].
l. M. A. Haase, J. Qiu, J. M. DePuydt and H. Cheng, *Appl. Phys. Lett.*, **59**, 1272 (1991).
m. R. N. Bhargava, D. Gallagher, X. Hong and A. Nurmikko, *Phys. Rev. Lett.*, **72**, 416 (1994).

11

Kubo Effects in Small Particles of Metals

Shun-ichi Kobayashi, Hidenori Goto and Shingo Katsumoto

Department of Physics, Faculty of Science, University of Tokyo
Bunkyo-ku, Tokyo 113-0033, Japan

Abstract

In this paper, we discuss the NMR properties in particles. It also includes some historical review. Emphasis is placed on level quantization, level statistics, finiteness of systems, single electron charging energy, etc. These properties are current topics in the field of mesoscopic systems.

1. Introduction

1.1 Historical background

Kubo's pioneering paper[a] was the first to point out the anomalies that occur when the volume of particles is so small that the energy levels are discrete. Hence these anomalies are called Kubo effects.[b] The concept that, in a piece of metal, the energy spectrum of conduction electrons is continuous and the Fermi function is applicable, breaks down when the size of the piece is reduced to 100 Å at temperatures below 1 K, where the level spacing usually exceeds the broadening. The spacing between adjacent orbital levels, on average, is $\delta = 1/N(0)V$, where $N(0)$ and V are the density of states at the Fermi level and the volume of the particle, respectively. Because the discrete levels are a consequence of the electron standing wave in the particle, broadening takes place when electrons are inelastically scattered by thermal phonons, other electrons or magnetic impurities, which have their own degrees of freedom. The discreteness is smeared out when the inelastic lifetime τ_{inel} is shorter than h/δ that is, the time required for the electron to sweep the entire volume of the particle.

The electronic properties of small particles are governed not only by the average spacing δ but also by the distribution of the spacing. Let $P(\varepsilon)\,d\varepsilon$ be the probability of finding the next level between interval of ε and $\varepsilon+d\varepsilon$. It can be shown that, in the limit of small ε, $P_n(\varepsilon) \approx \varepsilon^n/\delta^{n+1}$ with $n = 0,1,2,4$, depending on the symmetry of the system. In the limit of large ε, all $P_n(\varepsilon)$'s fall off as $\exp(-(\varepsilon/\delta)^2)$. The following four parameters are important.

τ_δ ; the time necessary for establishing a level $(= h/\delta)$

τ_A ; the time necessary for twisting the phase of the orbital to turn by 2π in the presence of the vector potential $(\tau_A \propto H^2V)$

τ_{so} ; the time necessary for turning the spin direction by 2π with the spin-orbit interaction

τ_z ; the time necessary for turning the spin direction by 2π in a magnetic field (τ_z = $g\mu_B H$ for large τ_{so}, and $\tau_z \propto \tau_{so}H^2$ for small τ_{so})

When τ_δ is much smaller than the other three, $n = 1$ (in terms of the universality classes, orthogonal). Similarly, $n = 2$ for τ_A (unitary), $n = 4$ for τ_{so} (symplectic), and $n = 0$ for τ_A (poissonian). $P_n(\varepsilon_z)$ determines the thermal properties of the system at low temperatures.

Kubo was also the first to point out the importance of the charging energy in small particles.[a] To take out one electron from a particle, the charging energy e^2/d, which amounts to 10^3 K for $d = 100$ Å, is required. Consequently, the charge neutrality should be strictly observed, say, at 1 K, and the statistics to be applied are not grand-canonical but canonical, *i.e.*, the Fermi function is not applicable. This was the first time attention was paid to the charging energy associated with a single electron, which has become a current topic in mesoscopic tunneling junctions.[c]

Nuclear magnetic resonance is a powerful method for detecting Kubo effects, because it can gather information, both static and dynamic, of conduction electrons, without attaching electrodes to particles. What NMR can detect are the Knight shift, which is directly proportional to the real part of the spin susceptibility of conduction electrons at zero frequency, and the spin-lattice relaxation time the imaginary part at the resonance frequency.

The static susceptibility of metal particles deviates from the temperature-independent Pauli susceptibility below $T = \delta/k_B$. As $P(\varepsilon)$'s peak at around $\varepsilon \approx \delta$ except for $n = 0$, the spin susceptibility of an "even" particle, in which the number of electrons is even, tends toward zero as $T \to 0$, while that in "odd" diverges.

This feature is modified by a spin-orbit interaction. It mixes the up- and down-spin eigenstates, and the mixing is dependent on the magnetic field. As a consequence, in the limit that τ_{so} is much shorter than the other time parameters, the susceptibility of "even" particles revives (the mechanism is entirely analogous to the van Vleck paramagnetism and to the residual Knight shifts in heavy metal superconductors). The residual susceptibility at $T = 0$ of "even" particle is given as $\chi_{even} = 2\mu_B(2\eta - \eta^2)N(0)$, where $\eta = \tau_d/\tau_s$ measures the degree of mixing between spin eigenstates. η is proportional to d^2.[d]

1.2 Early experiments

1.2.1 Knight shift in normal metals

Copper is a metal with a moderate spin-orbit interaction. The first evidence of the Kubo effect was reported by Knight for copper particles [5] (an experiment preceding this on Li particles [f] was exciting but not convincing enough). The samples were island-shaped metal particles which were naturally formed in vacuum deposition onto room-temperature substrates.

The Knight shifts, which are directly proportional to the static spin susceptibilities, measured at the peak positions of NMR absorption, were smaller than that of pure Cu. The line shapes were broad and asymmetric having longer tails at the low magnetic field side. The low-field tail was assigned to the signals from "odd" particles and the peak from "even" particles. The broadening is partly due to the second-order quadrupole effect and partly to the Kubo effect.

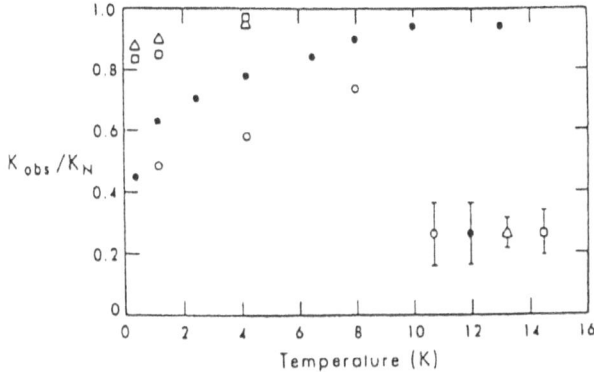

Fig. 1 The temperature dependence of the Knight shift of Cu particles. [e]
○,25 Å; ●,40 Å; △,100 Å; □,110 Å.

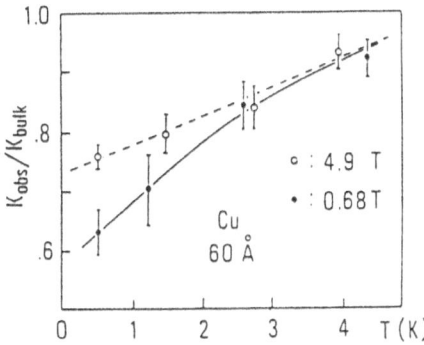

Fig. 2 The Knight shift of Cu particles versus temperature, in high and low magnetic fields. [1]

Figure 1 shows the temperature dependence of the Knight shifts for samples with four different sizes. The residual shift is smaller for smaller particles, supporting the validity of the above expression of χ_{even}. The value of η estimated by using the g-shift of bulk Cu is about 0.6 for $d = 100$ Å, which is close to η of sample No. 3 in Fig. 1. The system is conjectured to be almost orthogonal but slightly symplectic. The fact that the slopes of data points in Fig. 1 at $T \approx 0$ are finite suggests that the levels are not equally spaced, although it is hard to identify values of n.

The universality classes of the electron system in Cu particles can be switched by a high magnetic field from one to the other. Figure 2 shows the Knight shift of 60 Å Cu particles in low and high magnetic fields. [1] The parameters are: $\delta = 9.3$ K, $h/\tau_{\text{so}} = 1.9$ K and $h/\tau_z = 4.1$ K in 4.9 T. Therefore, the levels of up and down spin bands are nearly uncorrelated, the system being close to the case where $n = 0$. The recovery of the shift in a high field here is an evidence of this change.

1.2.2 Spin lattice relaxations in normal metals

While the Knight shift (or the static susceptibility) is a thermally equilibrium property, the nuclear spin-lattice relaxation (more precisely, the relaxation from nuclear spins to conduction electron system) reflects a dynamical feature of the level discreteness. A rapid relaxation of nuclear spins in metals takes place

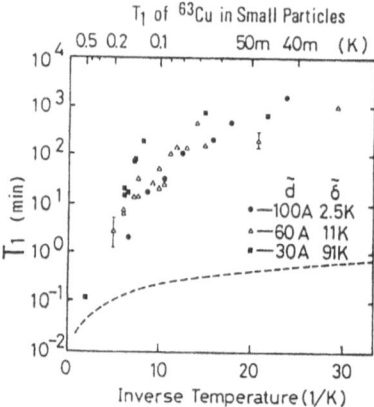

Fig. 3   The temperature dependence of spin lattice relaxation time in Cu particles.[3] The dashed line indicates Korringa's relation, $T_1 T = 1.27$ Ks.

through the contact hyperfine interaction with the conduction electrons, *i.e.*, a nuclear spin and a conduction electron spin mutually flip and the difference in the Zeeman energies is carried away by the change in the kinetic energy of the conduction electron. This process, known as the Korringa mechanism, is expected to be largely suppressed when the energy spectrum of conduction electrons is discrete, because of difficulty in satisfying the energy conservation throughout the process.[2]

In Fig. 3, the spin lattice relaxation time T_1 of ^{63}Cu in particles measured in 6.83 kOe is shown.[3] T_1 in particles is several order of magnitude longer than the bulk value shown by a dashed curve. The rather weak dependence on size may be attributed to some non-electronic relaxation processes.

Because the volume of particles is so small, the entire space is filled with Friedel oscillations generated at the surface. This inhomogeneous density of electrons couples to the electric quadrupole moment of nuclei and broadens the NMR. Thus it is difficult to saturate a whole spectrum of NMR in the magnetic field as is usually done in the relaxation measurements. Instead, it was found that the nuclear polarization can be completely lost when the field is once switched off to zero. This happens because the spin system experiences microscopically non-equilibrium states due to the quadrupole interaction. The data in Fig. 3 were obtained by making use of this depolarization.

1.2.3 The Knight shifts in superconducting particles

The most fundamental problem of superconducting particles may be the critical size for the occurrence of superconductivity. An atom of aluminum is certainly net superconductive. Then how many atoms are necessary for superconductivity? The problem is analogous to the superconductivity in heavy nuclei, *i.e.*, a few body problem with attractive interaction. The answer is not given yet, although the NMR properties of Al particles as small as 40 Å ($\delta \gg \Delta$, Δ; the BCS energy gap) still exhibit superconductive features.

In Fig. 4, the Knight shift of Sn particles extrapolated to T = 0 is plotted against the mean diameter d.[4] Here we can see an interesting competition among the three energy parameters. The shift in bulk Sn is zero because of the singlet

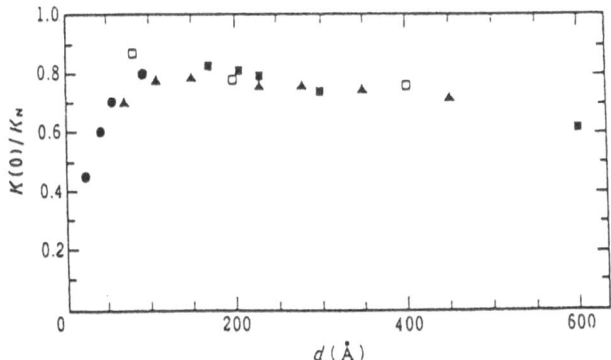

Fig. 4 The residual Knight shift of Sn particles versus the particle diameter.[4]

pairing of the Cooper pairs, and the gap is insensitive to d i.e., $\Delta \propto d^0$. When particle size is reduced, the surface scattering mixes the spin states via spin-orbit interaction, the pairing mates are no longer pure up or down states, and the spin susceptibility revives. The parameter relevant to the mixing is $h/\tau_{so} \propto d^{-1}$. With further reduction of the size, the discreteness of the levels governs the susceptibility. Because the mean spacing is $\delta \propto d^{-3}$, the susceptibility again decreases (to be precise, in "even" particles). The value of d at which the peak appears agrees with the condition $\tau_{so}/\tau_d = 1$.

2. Recent experiments

In this section, we describe the results which were obtained in the program supported by a Grant-in-Aid for Scientific Research on Priority Areas, Mesoscopic Particles from the Ministry of Education, Science and Culture.

2.1 Spin-orbit interaction; effect on the Knight shift

In heavy metal particles or particles with heavy impurities, the spin-orbit interaction mixes the spin up and down states. It affects the spin susceptibility through two mechanisms. First it causes the reduction of the magnetic moment, which appears as the g-shift of the conduction electrons. Second, the wave functions perturbed by the spin-orbit interaction change in the magnetic field and the expectation value of the z-component of spins becomes non zero. Therefore, the spin susceptibility of even particles is finite at T = 0. This is analogous to the non-vanishing spin susceptibility in heavy superconductors. It was shown that the spin susceptibility of even particles is given as

$$\chi = 2\mu_B^2(2\eta - \eta^2)N(\varepsilon_F) \tag{1}$$

where μ_B is the Bohr magneton, $N(\varepsilon_F)$ the density of states at the Fermi level, and η the degree of mixing of spin states which is given as

$$\eta = h/\delta\tau_{so.} \tag{2}$$

Here τ_{so} is the spin orbit scattering time, which is roughly estimated as

$$\tau_{so} \approx (\alpha Z)^{-4}\tau_0, \tag{3}$$

where α, Z and τ_0 are the material constant, the atomic number and the electron mean free time, respectively. Because $\tau_0 = d/v_F$, the residual susceptibility is larger for larger particles.

This dependence of the residual susceptibility on the particle size was observed in Cu particles through the Knight shift, which is proportional to the spin susceptibility.

We carried out experiments to verify the theory directly by varying the strength of the spin-orbit interaction, instead of the particle size. The spin-orbit interaction, which is represented by τ_0/τ_{so} in the above discussion, increases rapidly with increasing atomic number. Therefore, it can been enhanced when a heavier element is alloyed to the particles .

The samples were ensembles of fine Cu particles containing Au impurities. The mean diameter of the particles was 60 Å. The mean spacing δ for this diameter is about 9.3 K. The sources for vacuum evaporation were prepared by melting two metals together in a plasma jet furnace. The rest of the process was the same as that described in ref. 1.

To measure the Knight shift, a bridge spectrometer with an rf hybrid junction was used for the steady-state detection of NMR signals. A one-shot ^{3}He cryostat was used to cool the samples to 0.5 K. The shifts are plotted against the temperature in Fig. 5. The shifts refer to the values of bulk alloys of corresponding concentration. The value of pure Cu particles taken from ref. 1 are also plotted. We attribute the observed shifts to those of the even particles. (For the reason, see ref. 1) The gradual restoration of the shift toward the Pauli value with increasing concentration of Au is clear. Because the shift recovers the bulk value in the 3.1 % sample, we estimate its τ_{so} from eq. (2) to be h/δ, i.e., $\tau_{so} = 5.2 \times 10^{-12}$ s.

The value of τ_{so} can be estimated from eq. (3) . We assume that α is the same for pure Cu particles and Cu-Au particles, and that the effective Z is the concentration-weighted sum of Z^4 of Cu and Au as

$$Z_{eff}^4 = (1 - c)Z_{Cu}^4 + cZ_{Au}^4 \tag{4}$$

where c is the concentration of Au. Then τ_{so} is given as

$$\tau_{so} = (\alpha Z_{eff})^{-4} \tau_{so}. \tag{5}$$

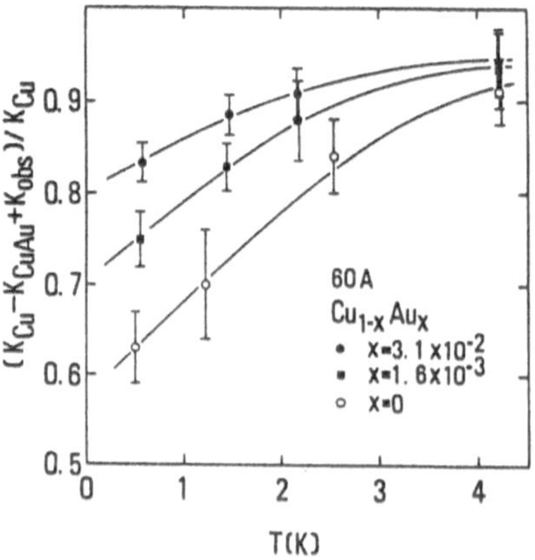

Fig. 5   The temperature dependence of the Knight shifts of Cu-Au alloy particles (solid symbols), and pure Cu particles (open symbols).[5]

Using α evaluated from the NMR data for pure Cu particles and $c = 3.1 \times 10^{-1}$, we obtain

$$\tau_{so} = 4.8 \times 10^{-12}\,\mathrm{s}. \tag{6}$$

This agrees fairly well with the value obtained above. Hence we conclude that we observed the recovery of spin susceptibility due to the spin-orbit interaction.

The spin susceptibility anomaly in small particles can also be suppressed by making the spin scattering time τ_s shorter than h/δ, although the nature of spin-orbit scattering and that of spin scattering differ substantially, i.e., the former is elastic while the latter, inelastic.

Furthermore, one must note the following. Since inelastic scatterings by thermal phonons and the electron-electron interact ion are direct perturbation to the orbital states, they naturally broaden the kinetic energy levels. Therefore, the anomalies disappear when inelastic scattering time is much shorter than h/δ. On the other hand, the spin scattering by localized impurity spins, which is certainly inelastic, is perturbation to the spin state, and it is not self-evident that spin scattering time $\tau_s \ll h/\delta$ is the condition to suppress the anomalies. We carried out experiments to verify whether this is valid.

2.2 Spin scattering; effect on the Knight shift

The samples are ensembles of Cu fine particles containing Mn impurities. The mean diameter of particles is 60 Å, and $\delta = 9.3$ K. The samples were prepared in the following manner. First, a mother alloy containing 3.4 at% of Mn was prepared by melting Cu and Mn metals in a plasma jet furnace. The mother was diluted successively by adding Cu to obtain desired concentrations. The rest of the process is the same as that described in ref. 1. We prepared three samples and the nominal concentrations are listed in Table 1. Because of a large difference in vapor pressures of Cu and Mn, the Mn concentration in the final product of particles can be substantially different from that of the starting alloy. We analyzed the concentration of Mn in particles using an X-ray micro-analyzer, and found that the Mn concentration of the highest concentration sample is about three times larger than the nominal concentration. For less-concentrated samples, the Mn signals were below the noise level, so that the concentrations were estimated by using the ratio observed for the most concentrated sample. The values are given in Table 1. The NMR detection system was the same as the one used for Cu-Au alloys.

The observed Knight shifts are plotted in Fig. 6, with the data for pure Cu particles taken from ref. 1. A gradual restoration of the shift toward the Pauli value is seen. Cu-Mn alloy is one of the well-known spin glasses. The glass transition temperature T_g for 100 ppm Mn is about 0.2 K. Therefore, samples #1 and #2 are

Table 1 List of Cu-Mn alloy samples [6]

	Nominal Mn concentration	Analyzed Mn concentration
#1	600 ppm	2000 ppm
#2	50	200
#3	5	20

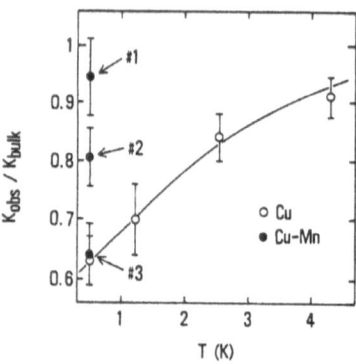

Fig. 6   The Knight shifts of Cu-Mn alloy particles (solid symbols), and pure Cu particles (open symbols).[6]

in glass states at 0.5 K, if the smallness of the size does not change T_g. Generally, in magnetically ordered metals, the mutual flip of the conduction electron spin and the localized spin is not possible, so that the presence of the magnetic impurities does not necessarily mean a short τ_s. However, spin glass is exceptional. The glass state allows many spins to be nearly free because of its complexity, and τ_s can still be very short.

Actually, τ_s in Cu-Mn alloys was measured in weakly localized Anderson films, which consisted of lightly oxidized 30 Å particles. In Fig. 7, we plotted the relationship between the Mn concentration and the values of τ_s taken from ref. 7. The horizontal line represents h/δ and the arrows show the Mn concentrations of the present samples. The result that the eye-guide line for τ_s's intersects h/δ near the concentration of sample #1 supports the idea that the relevant spin scattering time to destroy the level discreteness is h/δ.

We can explain this result as follows. To establish sharp energy levels in a particle, an electron must travel the entire space within it, and collect complete information of the potential. In a particle, the motion of the electron is ballistic, so the electron can sweep volume of $\pi\lambda_F^2 v_F dt$ in a time dt. The time required to see the entire space V in the particle is then $\tau = V/\pi\lambda_F^2 v_F$. It can be easily shown that this τ is equal to h/δ. If the phase memory of the electron, regardless of the orbital

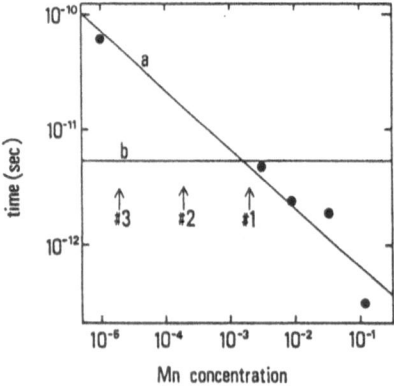

Fig. 7   Mn concentration dependence of two characteristic time scales.[6] a: τ_s of Cu-Mn alloy films obtained by the Anderson localization experiments. b: τ_δ of pure Cu particles with diameter of 60 Å.

state or the spin state, is lost before this τ is finished, the orbital levels cannot be established, resulting in broadening of the levels. As a result, we obtain the condition $\tau_s \ll h/\delta$.

Thus, as long as the Knight shift is involved, spin-orbit and spin scatterings qualitatively give the same results. However, a difference will appear in the spin-lattice relaxation rate of Cu NMR.

2.3 Spin-orbit and spin scattering; effects on the nuclear relaxation

The energies of the initial and final states in the Korringa process are different because of the difference in the magnetic moments of a nucleus and an electron. In bulk metals this difference is carried away as the change in electron kinetic energy. However, in fine particles, the discrete nature of the levels obstructs the energy conservation. Actually we showed that the rate in pure Cu particles is anomalously slow at low temperatures because of the level discreteness which suppresses the Korringa relaxation mechanism. We extended the relaxation time measurement of ^{63}Cu to Cu-Mn and Cu-Au particles to highlight the difference between elastic and inelastic scatterings.

The samples are identical to the highest concentrated ones used in the Knight shift measurements. The samples were cooled to 50 mK in a mixing chamber of a dilution refrigerator. The method of relaxation measurement is the same as that we used for pure Cu particles. The measured T_1's are plotted against temperature in Fig. 8. In pure Cu particles, T_1 increases exponentially with decreasing temperature and saturates when $T < 100$ mK. Data of T_1 for the pure case in Fig. 8 are taken from ref. 3. The result of Cu-Au particles is similar to that of the pure case, though the concentration of Au is high enough to restore full susceptibility. This means that the levels are not broadened by spin-orbit interaction that is strong enough to completely mix up and down spin states. The difference between pure Cu and Cu-Au at the lowest temperatures is not significant because relaxation paths other than those of the Korringa type, *e.g.*, a direct coupling to magnetic

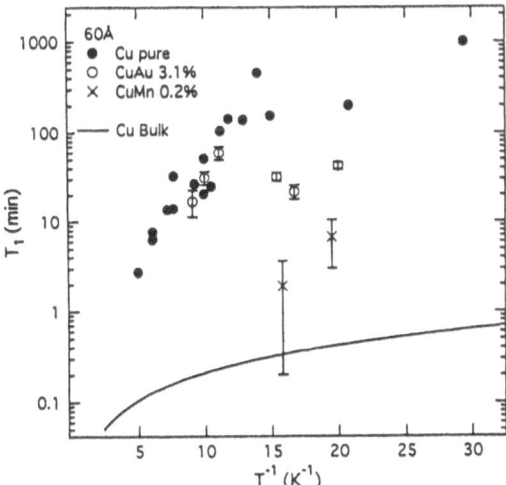

Fig. 8   The temperature dependence of the nuclear spin lattice relaxation time of ^{63}Cu in pure Cu, Cu-Au and Cu-Mn small particles. [8] The solid line indicates Korringa's relation, $T_1 T = 1.27$ K s.

impurities or a quadrupolar coupling to phonons, can depend on preparation conditions and on lattice properties of the alloys.

On the other hand, T_1 in Cu-Mn particles is much shorter than that in pure Cu particles. There are two possibilities in interpreting this result. One is that the levels, originally discrete, are smeared out by spin scattering. In this case, the energy spectrum of electrons is continuous and the relaxation rate is the same as that in a bulk metal. The other is that ^{63}Cu spins relax to Mn local spins through spin diffusion. This type of relaxation has been studied in bulk metals with magnetic impurities. In the present case, however, the spin diffusion is largely suppressed by the large quadrupolar inhomogeneity at the atomic scale. Therefore, we attribute the recovery of T_1 towards the bulk value to the level smearing due to the spin scattering.

3. Summary

NMR experiments have clearly detected Kubo effects in small particles of metals. The symmetries of the system (the universality classes) and the nature of scatterings have been controlled and the their physical significance demonstrated. For more detailed studies, *e.g.*, direct determination of the level distribution functions, further elaborate experiments are necessary. Other properties, such as specific heat, optical response and static susceptibility, are important to understand the whole physics of particles. As for superconductivity, much work remains to be done.

Acknowledgment

The authors would like to thank R. Kubo, W. D. Knight, W. P. Halperin, J. Sone, E. Simanek, A. Kawabata and his collaborators for valuable discussions.

PUBLICATIONS

1. S. Kobayashi and S. Katsumoto, *J. Phys. Soc. Jpn.*, **56**, 2256 (1987).
2. S. Kobayashi, *J. de Phys.*, **38**, C2-121 (1977).
3. T. Goto, F. Komori and S. Kobayashi, *J. Phys. Soc. Jpn.*, **58**, 3788 (1989).
4. Y. Fukagawa, S. Kobayashi and W. Sasaki, *J. Phys. Soc. Jpn.*, **51**, 1095 (1982).
5. S. Kobayashi, H. Goto and S. Katsumoto, *J. Phys. Soc. Jpn.*, **61**, 1856 (1992).
6. S. Kobayashi, H. Goto and S. Katsumoto, *J. Phys. Soc. Jpn.*, **61**, 762 (1992).
7. F. Komori, S. Kobayashi and W. Sasaki, *J. Phys. Soc. Jpn.*, **52**, 4306 (1983).
8. H. Goto, S. Katsumoto and S. Kobayashi, *J. Phys. Soc. Jpn.*, **62**, 1439 (1993).

REFERENCES

a. R. Kubo, *J. Phys. Soc. Jpn.*, **17**, 975 (1962).
b. See for example the following reviews.
    W. P. Halperin, *Rev. Mod. Phys.*, **58**, 533 (1986).
    S. Kobayashi, *Phase Transitions*, **24–26**, 463 (1990).
c. See for example, *Z. für Phys.*, **B85** No.3 (1991) Special Issue on Single Charge Tunneling.
d. J. Sone, *J. Phys. Soc. Jpn.*, **42**, 1457 (1977).
e. P. Yee and W. D. Knight, *Phys. Rev.*, **B11**, 3261 (1975).
f. C. Taupin, *J. Phys. Chem. Solids*, **28**, 41 (1967).

12

Growth of Crystalline Composite Films of Au-MoS$_2$ and CaF$_2$-MgO

Kazuhiro Mihama [a], Saburo Iwama [a] and Nobuo Tanaka [b]

[a] Department of Applied Electronics, Daido Institute of Technology
  Minami-ku, Nagoya 457-0811, Japan
[b] Department of Applied Physics, School of Engineering, Nagoya University
  Chikusa-ku, Nagoya 464-8603, Japan

Purpose of the present study

Metal-ceramics composite films, such as "cermet", have been studied as functional materials. However, structural details have not yet been clarified. They are polycrystalline or amorphous structures in most cases. In the present study, the structure of composite films is examined using single crystalline ceramics and metals.

1. Introduction

For several years, we have been studying composite films of metal (Au [1], Fe [2,3], Ti [4]) embedded in MgO prepared by the simultaneous deposition technique on cleavage surfaces of sodium chloride in vacuum. The characteristics of the composite films are as follows. [1] Fine metal particles grow epitaxially inside MgO single crystalline films. [2] For Au-MgO composite films, extremely thin gold particles whose diameter is about 2 nm, show negative temperature coefficient of resistivity in most cases. [3] For Fe-MgO composite films α-iron particles initially formed in MgO film were transformed into γ-iron, the high temperature phase, by heat treatment at 500°C for 2 hours, in spite of the fact that the transition temperature is 906°C for bulk iron. Both α- and γ-iron particles showed ferromagnetic characteristics at low temperatures, and the M–T curve was similar to that of a spin glass.

The present paper concerns the growth and characterization of Au-MoS$_2$ and CaF$_2$-MgO composite films. The former is an example of metal crystallites embedded in MoS$_2$, a layered material and the latter, ionic particles in MgO.

2. Experimental

Au-MoS$_2$ composite films were prepared by RF magnetron sputtering (200 W) in 20 Pa, Ar atmosphere onto cleavage surfaces of mica or molybdenite at 300°C substrate temperature. MoS$_2$, 100 mm$\phi \times$ 6 mm, was used as the target, and gold wires of 0.3 mmϕ were mounted on the MoS$_2$ target. The quantity of Au embedded in MoS$_2$ films was controlled by the number of gold wires on the target. The growth rate of the composite films was 5 nm/min.

CaF$_2$-MgO composite films were prepared by the simulteneous deposition of CaF$_2$ and MgO onto cleavage surfaces of sodium chloride at 150°C to 400°C in vacuum of around 5×10^{-4} Pa, the evaporation being carried out by resistive heating and electron beam heating, respectively. The mean thickness of CaF$_2$ was 1 to 10 nm and 40 nm, respectively. The deposition rate of MgO was around 10 nm/min. Cleavage surfaces of sodium chloride were used as substrates at 150°C to 380°C. In the simultaneous deposition, CaF$_2$ deposition was carried out in the middle of MgO deposition to ensure the embedding of the CaF$_2$ crystallites inside the MgO film. The deposition of CaF$_2$ on a MgO film was also carried out for reference.

3. Results and discussion

3.1 Au-MoS$_2$ composite films

Figure 1 shows an electron micrograph and the corresponding diffraction pattern of a Au-MoS$_2$ composite film formed on mica at 300°C with five gold wires. The MoS$_2$ film grows with (00.1) fiber orientation and lattice constant a_0 was 0.262 nm about 5% of shrinkage in comparison with that of normal MoS$_2$. The shrinkage observed also for MoS$_2$ grown on molybdenite may be due to a lack of sulphur atoms in a MoS$_2$ lattice formed by sputtering. The gold particles grown inside the MoS$_2$ film grew epitaxially. The size of gold particles can be controlled by the number of gold wires mounted on the target; the size is around 1 to 2 nm for a single gold wire. The size of the gold particles remains unchanged by heat treatment at 300°C for 6 hours. In the diffraction pattern, a weak ring pattern assigned to the cubic system ($a_0 = 0.48$nm) of an unknown substance is always recognized when cosputtering of MoS$_2$ and gold is carried out. The unknown substance can be observed as extremely thin flakes in the electron micrograph, as indicated by arrows in Fig. 1. The unknown substance is possibly related to Ag$_3$AuS$_2$, a cubic system reported by Smit et al.[a]

Figure 2 shows an electron micrograph and the corresponding diffraction pattern of Au-MoS$_2$ composite film grown on molybdenite under the same sputtering conditions as Fig. 1. In this case, the sputtered MoS$_2$ film grows epitaxially on the molybdenite substrate. In the micrograph, Moiré fringes of 1.7 nm due to (220) of Au and molybdenite can be observed in addition to those of 1.4 nm due to the unknown substance and (10.0) of molybdenite.

Figure 3 shows a high resolution electron micrograph of Au-MoS$_2$ composite

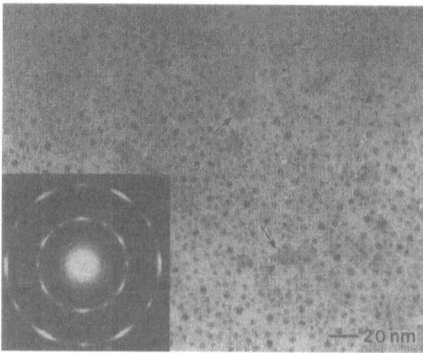

Fig. 1 Au-MoS$_2$ composite film grown on mica.

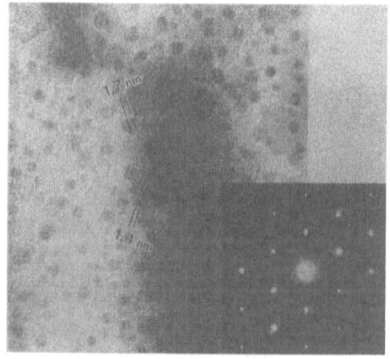

Fig. 2   Au-MoS$_2$ composite film grown on molybdenite.

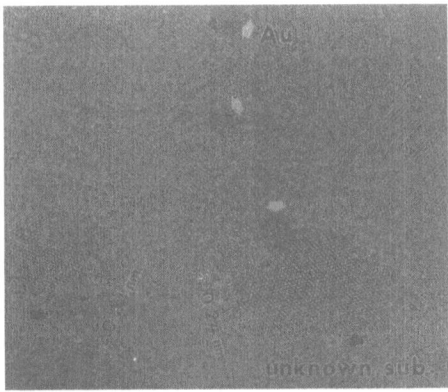

Fig. 3 High resolution image of Au-MoS₂ composite film.

film prepared with a single gold wire, in which gold particles of around 2 nm in size with a round shape are recognized as a dark contrast. Lattice fringes of MoS_2 as well as the unknown substance are observed. In the present Au-MoS_2 composite films, gold particles are situated in the sputtered MoS_2, but not in the interlayer of MoS_2 film, because no strain contrast caused by intercalation can be observed.

3.2 CaF₂-MgO composite films

For the CaF_2 composite film, CaF_2 crystallites grow with (111) fiber orientation in MgO film with (001) orientation at 180°C substrate temperature. Fig. 4 shows an electron diffraction pattern of CaF_2-MgO composite film grown at 300°C, the thickness of CaF_2 and MgO being 10 nm and 40 nm, respectively. Three kinds of epitaxial orientation can be observed: A predominant orientation of CaF_2 crysallites is (111) orientation, and (002) and (112) orientations are also observed. For CaF_2 crysallites grown on MgO films, nearly the same orientation can be observed. Sometimes weak CaO (200) spots are also detected, probably due to electron irradiation damage. In the diffraction pattern, a considerable intensity of (200) spots, which have nearly zero intensity for a stoichiometric CaF_2 crystal, is recognized. The high intensity may be due to a lack of fluorine atoms in CaF_2 lattice formed inside the MgO film. The (200) spots can also be recognized in the

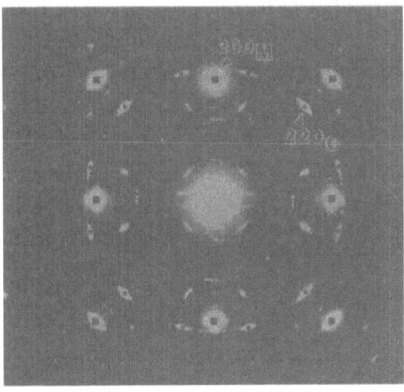

Fig. 4 Electron diffraction pattern of CaF₂-MgO composite film.

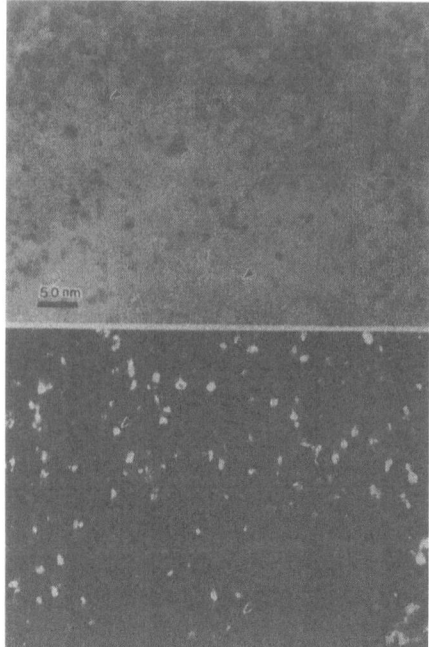

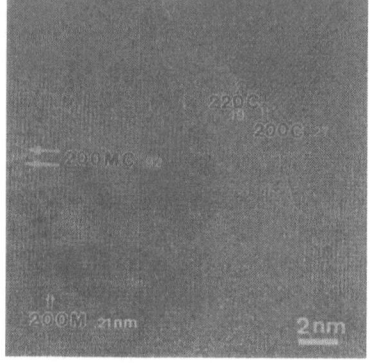

Fig. 5   Bright and dark field images of CaF₂-MgO   Fig. 6   High resolution image of CaF₂-MgO
composite film grown at 300°C.   composite film grown at 300°C.

diffraction pattern of CaF_2 grown on MgO films, although the intensity is a little weaker than that of the composite films.

Figure 5 shows bright and dark field images taken with the (220) spots of CaF_2 of CaF_2-MgO composite films grown at 300°C, the thickness of CaF_2 and MgO being 6 nm and 40 nm, respectively. In the micrographs, CaF_2 crystallites are dispersed uniformly inside the MgO film, the size being 10 to 20 nm with an indefinite shape. After heat treatment of the film at 300°C for three hours in vacuum, the orientation and the shape remain unchanged. With increasing mean thickness of CaF_2, the size of CaF_2 crystallites increases. For a mean thickness of 10 nm of CaF_2, growth of crystallites 20 to 30 nm in size is observed.

At 380°C substrate temperature, the orientation of CaF_2 crystallites becomes uniquely (001) orientation and (200) spots are also observed. In the lattice image, it is very difficult to distinguish the periphery of CaF_2 crystallites by image contrast and this becomes possible only by means of lattice spacing. Fig. 6 shows a lattice image of CaF_2-MgO composite film grown at 300°C, in which 0.21 nm of MgO (200) spacing and 0.92 nm of Moiré fringes due to (200) of MgO and CaF_2 are observed in the central region and a CaF_2 crystallite with 0.27 nm of (200) spacing and 0.19 nm of (220) spacing is observed in the upper right hand corner.

PUBLICATIONS

1. N. Tanaka, N. Nagao and K. Mihama, *Ultramicroscopy*, **25**, 241 (1988).
2. N. Tanaka, K. Kimoto, F. Yoshizaki and K. Mihama, *Trans. J. I. M.*, **31**, 588 (1990).
3. F. Yoshizaki, N. Tanaka and K. Mihama, *J. Electron Microsc.*, **39**, 459 (1990).

4. M. Nagao, N. Tanaka and K. Mihama, *Jpn. J. Appl. Phys.*, **26**, L216 (1987).

REFERENCES

a. T. J. M. Smit *et al.*, *J. Solid State Chem.*, **2**, 309 (1970).

13

Growth of Au-Cu Alloy Particles on and in MgO Films

Kazuhiro Mihama [a] and Nobuo Tanaka [b]

[a] Department of Applied Electronics, Daido Institute of Technology
  Minami-ku, Nagoya 457-0811, Japan
[b] Department of Applied Physics, School of Engineering, Nagoya University
  Chikusa-ku, Nagoya 464-8603, Japan

Purpose of the present study

Formation of long period antiphase boundaries has been understood by the electronic structure of the alloy. The behavior of the formation of the structure is studied for the size effect as well as the growth inside MgO crystalline films.

Content of the present study

1. Introduction

Alloy particles of Au-Cu 50/50 have been studied by means of electron diffraction and electron microscopy to clarify the size effect on the formation of AuCu II phase. [1] The present study concerning the growth and structure of AuCu alloy particles grown on and in MgO single crystalline films is carried out with respect for size effect on the formation of AuCu II meso-particles. The composite films were prepared by successive deposition and simultaneous deposition [2] of MgO and the alloy (atomic composition 1:1) on cleavage surfaces of NaCl in a vacuum of around 10^{-4} Pa at around 400°C substrate temperature. The mean thickness of MgO and the alloy was 40 nm and a few nm, respectively. The specimens thus prepared including structure images were observed by transmission electron microscopy.

2. Experimental results

2.1 Growth on MgO single crystalline films

As-grown alloy particles grown on MgO single crystalline films at 400°C shows generally AuCu I phase. For the formation of AuCu II phase, further heat treatment at 400°C for about 15 hours is required. Figure 1 shows AuCu I particles in which lattice fringes of 0.28 nm due to (110) ordered spots in addition to Moire' fringes of 2.34 nm due to (200) of MgO and the alloy are recognized. AuCu II particles were found after 17 hours of heat treatment. In the micrograph, 110 fringes of 0.28 nm spacing are shifted by half period at the antiphase boundary apart from about 2 nm, as was already observed. [1]

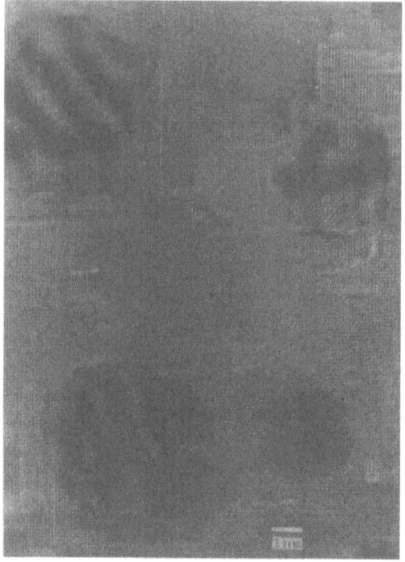 

Fig. 1 AuCu particles on MgO (001) film.   Fig. 2 AuCu particles embedded in MgO film.

2.2 Growth in MgO single crystalline films

As-grown alloy particles in MgO single crystalline films show AuCu I phase with parallel orientation, the size being around 2 nm. Above 400°C substrate temperature, alloy particles show the disordered phase, but they transform into the AuCu I phase with heat treatment of several hours at 400°C. Figure 2 shows AuCu I particles embedded in MgO single crystalline film. In the micrograph, growth traces with a square of MgO crystallite around the alloy particles can be observed. However, the details of contrast due to alloy particles, for example, the existence of antiphase domain boundaries, are barely detectable.

REFERENCES

1. K. Mihama, *J. Phys. Soc. Jpn.*, **31**, 1677 (1971).
2. N. Tanaka, M. Nagao and K. Mihama, *Ultramicroscopy*, **25**, 241 (1988).

14

In situ Observation of Spontaneous Alloying in Nanometer-Sized Atom Clusters

Hidehiro Yasuda and Hirotaro Mori

Research Center for Ultra-High Voltage Electron Microscopy, Osaka University
Suita-shi, Osaka 565-0871, Japan

Abstract

Alloying behavior of solute atoms into nanometer (nm)-sized clusters has been studied through *in situ* deposition experiments using a double-source evaporator installed in the specimen chamber of an electron microscope. Isolated nm-sized clusters were first prepared on a supporting film, and solute atoms were then evaporated onto the same film kept at ambient temperature. Upon depositing, solute atoms quickly dissolved into clusters, and solid solution or compound clusters were successfully formed. Solute diffusivity in clusters was estimated to be many orders of magnitude faster than that in the corresponding bulk materials. In this work, causes and mechanisms of such a spontaneous alloying have been systematically investigated by means of transmission electron microscopy (TEM).

1. Introduction

Interest in studies on ultrafine particles (hereafter designated as atom clusters or simply clusters) has been rapidly growing in recent years because the clusters often exhibit properties and structures which are quite different from those in the corresponding bulk materials.[a,b] Quite recently, it was found that spontaneous alloying takes place in nm-sized atom clusters even at ambient temperature.[1] For example, instantaneous dissolution of copper atoms takes place in nm-sized gold clusters at room and reduced temperatures and solid solution clusters were successfully formed. The copper diffusivity in gold clusters was estimated to be at least nine orders of magnitude faster than that in bulk gold.[2,3]

This paper is a review of the research on spontaneous alloying in nm-sized clusters by means of transmission electron microscopy (TEM), and is organized as follows. In section 2, equipment for *in situ* observation in a TEM is described. In section 3, cluster-size dependence on spontaneous alloying and the alloying process are characterized. In section 4, effect of surface-facet structure of clusters on spontaneous alloying is investigated. In section 5, temperature rise of clusters during *in situ* observation of spontaneous alloying is estimated. In section 6, a general relation between initial cluster sizes and spontaneous alloying is discussed. In section 7, results are summarized.

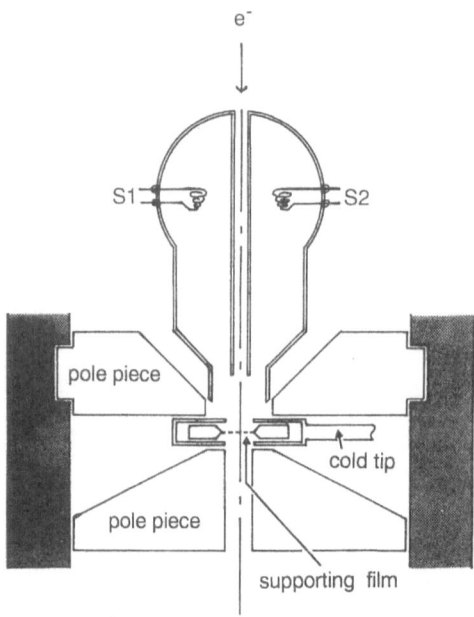

Fig. 1   Schematic illustration of a double-source evaporator installed in the specimen chamber of a TEM. The source heaters S1 and S2 consist of two spiral-shaped tungsten filaments. (From H. Mori, H. Yasuda and K. Fujii, *Surf. Rev. Lett.*, **3**, 1177–1179 (1996)).

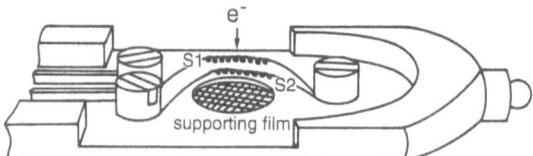

Fig. 2   Schematic illustration of a miniature double-source evaporator set at the tip of a side-entry specimen holder for a transmission electron microscope. (From H. Mori *et al.*, *Philos. Mag. Lett.*, **69**, 279–283 (1994)).

2. Experimental procedures

Preparation of clusters and subsequent deposition of solute atoms onto the clusters were carried out using a double-source evaporator in an electron microscope. Figure 1 shows a schematic illustration of a double-source evaporator installed in the specimen chamber of an 200 kV electron microscope. The evaporator consists of two spiral-shaped tungsten filaments. The distance between the filaments and the supporting film (substrate) for atom clusters was approximately 100 mm. Fig. 2 is the schematic illustration of a miniature double-source evaporator set at the tip of a side-entry specimen holder of a 300 kV electron microscope. The evaporator also consists of two spiral-shaped tungsten filaments. The length and diameter of the filaments are as small as 3 mm and 25 μm, respectively. The distance between the filaments and the supporting film is approximately 2 mm. An amorphous carbon film was used as the supporting film and was mounted on a molybdenum grid.

Using these evaporators, source 1 was first evaporated from the filament (S1) to produce from a few to several nm-sized clusters on a supporting film.

Next, solute atoms were evaporated from the other filament (S2) onto the same film kept at ambient temperature. The flux of these depositing atoms was on the order of $10^{17} - 10^{18}$ $m^{-2} \cdot s^{-1}$. Changes in the morphology and structure of the clusters associated with deposition of solute atoms were examined by means of bright-field images (BFIs), dark-field images (DFIs), high-resolution images (HRIs) and selected area electron diffraction patterns (SAEDs).

The microscope used for conventional observation was a Hitachi H-800TEM, operating at an accelerating voltage of 200 kV. This microscope was equipped with a turbo-molecular pumping system and a liquid N_2-cooled anti-contamination device to achieve a base pressure below 5×10^{-5} Pa in the specimen chamber. The electron flux used was approximately 1.5×10^{20} $e \cdot m^{-2} \cdot s^{-1}$, which is quite low. The microscope used for the observation of the atomistic structure of the clusters was a Hitachi H-9000NAR high-resolution electron microscope (HREM), operating at an accelerating voltage of 300 kV. Point-to-point resolution of the microscope was better than 0.18 nm. The base pressure was below 5×10^{-6} Pa in the specimen chamber. The electron flux used was approximately 3.1×10^{24} $e \cdot m^{-2} \cdot s^{-1}$.

3. *In situ* observation of spontaneous alloying of copper atoms into gold clusters

3.1 Cluster-size dependence of alloying behavior [4)]

A typical example of spontaneous alloying in gold clusters of approximately 4 nm in mean diameter is depicted in Fig. 3. Figs. 3(a) and (a') show a BFI of as-produced gold clusters on a supporting film and the corresponding SAED, respectively. The Debye rings in the SAED are indexed as those of face-centered cubic (fcc) gold. Figs. 3(b) and (b') show a BFI of clusters after copper deposition and the corresponding SAED, respectively. With copper deposition, the clusters

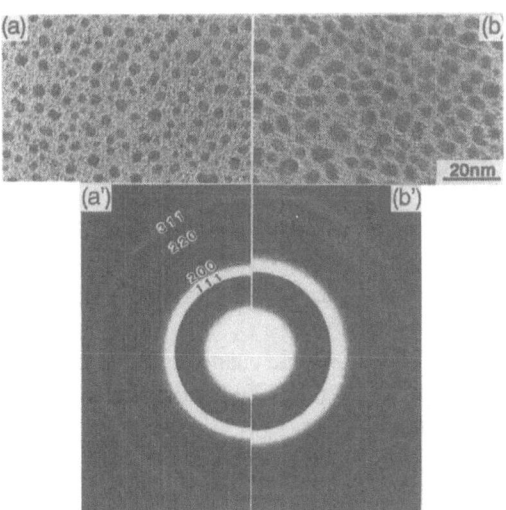

Fig. 3   Typical example of spontaneous alloying in 4 nm-sized gold clusters.(a) BFI of as-produced gold clusters on a supporting film, (b) BFI of clusters after copper deposition. (a') and (b') are SAEDs corresponding to (a) and (b), respectively.
(From H. Yasuda and H. Mori, *Z. Für. Phys. D.*, **31**, 131–134 (1994)).

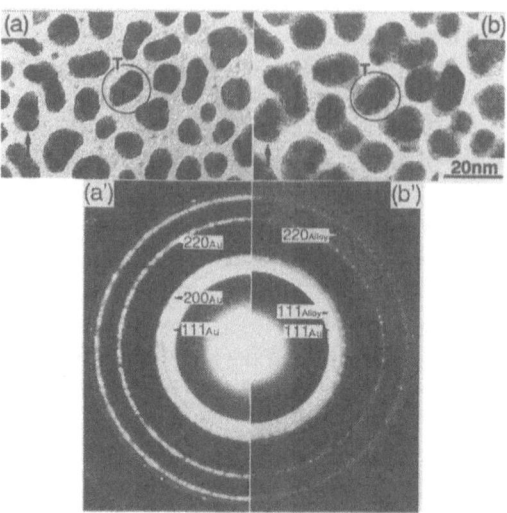

Fig. 4   Typical example of alloying in 10 nm-sized gold clusters. The arrows indicate a fixed position.
(a) A BFI of as-produced gold clusters on a supporting film. (b) A BFI of the same area as
in (a) after copper deposition. (a′) and (b′) are SAEDs corresponding to (a) and (b),
respectively. A twinned structure is seen in cluster T in (a). It seems that the twinned structure
is retained even after copper deposition (b).
(From H. Yasuda and H. Mori, *Z. Für. Phys. D.,* **31**, 131–134 (1994)).

increased from approximately 4 to 6 nm in mean diameter, as seen from Figs. 3(a)
and (b), and at the same time the diameter of the Debye rings also increased
somewhat, as seen from Figs. 3(a′) and (b′). This indicates that when vapor-
deposited copper atoms came in contact with gold clusters, they quickly dissolved
into gold clusters to form Au-Cu solid solution clusters with somewhat reduced
lattice constants. Analysis of the magnitude of the increase in the diameter of the
Debye rings revealed that the copper concentration of the Au-Cu solid solution
clusters is approximately 40–50 at% Cu[c].
   A simple relationship

$$x = (Dt)^{1/2}$$

can be used as a rough estimate for the diffusion coefficient D, where t is the time
needed to achieve appreciable diffusion of solute atoms over a distance x [d]. With
$x = 2.0$ nm (half the average diameter of the gold clusters shown in Fig. 3(a)) and
$t = 20$ s (the time needed to take photographs after the atom beam was turned off),
a value of 8×10^{-19} m²·s⁻¹ is obtained for D. This value gives a guide for the lower
limit of the diffusivity of copper in gold clusters, since the alloying takes place in
less than 20 s. On the other hand, the diffusion coefficient of copper in bulk gold
$D_{Cu \to Au}$ can be written (in m²·s⁻¹) as

$$D_{Cu \to Au} = (1.2 \times 10^{-7}) \exp(-14300/T)$$

where T is the temperature.[e] Extrapolation of this relationship to room temperature
gives a value of 2.4×10 m²·s⁻¹ for $D_{Cu \to Au}$ at 300 K. Thus it can be said that the
observed copper diffusivity in gold clusters is at least nine orders of magnitude
faster than that in bulk gold.

135

Figure 4 shows a typical example of alloying in gold clusters of approximately 10 nm in mean diameter. Figs. 4(a) and (a′) show a BFI of as-produced gold clusters on a supporting film and the corresponding SAED,respectively. Figs. 4(b) and (b′) are a BFI of the same area after copper deposition and the corresponding SAED, respectively. The arrows in Fig. 4 indicate a fixed position. With copper deposition the mean cluster size increased from approximately 10 nm to 13 nm, as seen from a comparison of Fig. 4(a) with 4(b). The Debye rings in the SAED in Fig. 4(b′) can be divided into two groups: one is rings of fcc gold, and the other is broad fcc rings with somewhat large diameters. This indicates that with copper deposition two phases were produced: one is pure gold and the other is Au-Cu solid solution. Analysis of the magnitude of the increase in the diameter of the Debye rings revealed that the copper concentration of the Au-Cu solid solution is approximately 55–65 at%Cu. An interesting observation pertinent to Fig. 4 is as follows: Cluster T in Fig. 4(a) contains twins and the twin structure in this cluster is retained even after copper deposition, as seen from a comparison of cluster T in Fig. 4(a) with that in Fig. 4(b). Namely, no drastic changes in the microstructure were induced with copper deposition. This fact suggests that the rapid alloying which resulted in the formation of Au-(55–65)at% Cu alloy took place not via a melting process,[f-i] but via a purely solid-state process.

In an attempt to elucidate the spatial arrangement of the two phases (i.e. Au-(55–65)at%Cu and pure gold) produced on the supporting film, microstructures in individual clusters were examined by dark-field electron microscopy. Examples of the results are depicted in Fig. 5. Figs 5(a) and (b) are a BFI and a DFI of the same clusters, respectively. The DFI was taken by setting the objective aperture on a part of the Au 111 Debye ring. Clusters A to C in Figs. 5(a) and (b) are enclosed by dashed lines. From a comparison of clusters A to C in Fig. 5(a) with those in 5(b), it is evident that pure gold is present preferentially at the central region of individual clusters. This suggests that in gold clusters of approximately 10 nm in mean size, spontaneous alloying takes place only at a shell-shaped region beneath the free surface of individual clusters and no alloying occurs at the central region of clusters.

Figure 6(a) shows an example of as-produced gold clusters on a supporting

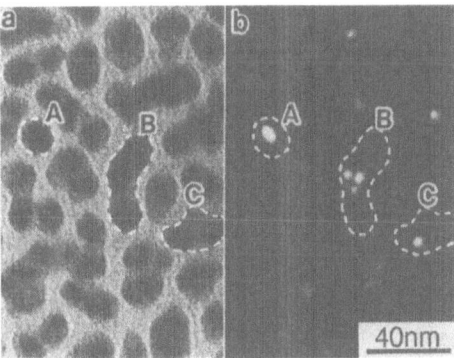

Fig. 5   (a) BFI of clusters that were formed by depositing copper onto 10 nm-sized gold clusters. (b) DFI of the same area as in (a) taken by setting the objective aperture on a part of the Au 111 Debye ring. In (b) regions of pure gold appear bright.
(From H. Yasuda and H. Mori, *Z. Für. Phys. D.*, **31**, 131–134 (1994)).

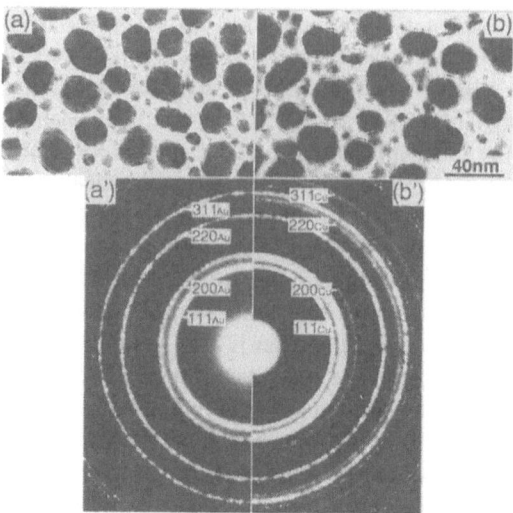

Fig. 6   (a) BFI of as-produced, 30 nm-sized gold clusters on a supporting film, (b) BFI of clusters after copper deposition. (a') and (b') are SAEDs corresponding to (a) and (b), respectively. In (b) there appear Moiré patterns in some clusters, suggesting that gold clusters are wrapped with a film of copper.
(From H. Yasuda and H. Mori, *Z. Für. Phys. D.*, **31**, 131–134 (1994)).

film. The mean size of the clusters is approximately 30 nm. A BFI of the clusters after copper deposition and the corresponding SAED are depicted in Figs. 6(b) and (b'), respectively. In the SAED, all the Debye rings can be indexed as those of fcc gold and copper. In Fig. 6(b), there appear Moiré patterns in some clusters, suggesting that copper film is present on gold clusters. All these facts indicate that in gold clusters of approximately 30 nm in mean size, such spontaneous alloying as observed in 4 nm-sized gold clusters did not take place and only a two-phase mixture of pure gold and copper was produced on a supporting film by depositing copper. The formation of the unalloyed, two-phase mixture is consistent with what is predicted by the copper atom diffusivity in bulk gold.[e)]

Through the present experiments it becomes evident that spontaneous alloying becomes more difficult with increasing cluster size. A schematic illustration of cluster-size dependence of spontaneous alloying of copper in gold clusters is

Initial gold cluster (nmφ)	④	10	30
After copper deposition	Alloy	Au / Alloy	Cu Au

Fig. 7   Schematic illustration of cluster-size dependence of spontaneous alloying of copper in gold clusters.

shown in Fig. 7. In gold clusters of approximately 4 nm in mean size, a rapid dissolution of copper atoms took place and homogeneously mixed Au-Cu alloy clusters were formed. In gold clusters of approximately 10 nm in mean size, rapid alloying of copper took place only at a shell-shaped region beneath the free surface of individual clusters and pure gold was retained at the central region of clusters. In gold clusters of approximately 30 nm in mean size, no rapid alloying of copper was induced, judging from analyses of BFIs and SAEDs.

3.2 Process of spontaneous alloying [5]

The alloying behavior of copper atoms into gold clusters has been studied by HREM in an attempt to understand the atomistic process involved in the spontaneous alloying.

Figures 8(a) and (b) depict HRIs of the same gold cluster. These two images of cluster before copper deposition were obtained in slightly different reflecting conditions; the image in Fig. 8(b) was taken with the sample tilted at an angle of the order of 10^{-2} radian after the image in Fig. 8(a) was taken. From these images it is evident that the cluster is a multiply-twinned crystal in nearly five-fold orientation; the cluster is composed of five twins, and the [110] direction, which is common to all five twins, is nearly parallel to the electron beam. The twinned structure in the cluster is schematically illustrated in Fig. 8(c). The almost vertical lattice fringes which appear in twins X and Y in Fig. 8(b) are (111) lattice fringes. HRIs of the cluster after copper deposition are depicted in Figs. 8(d) and (e).

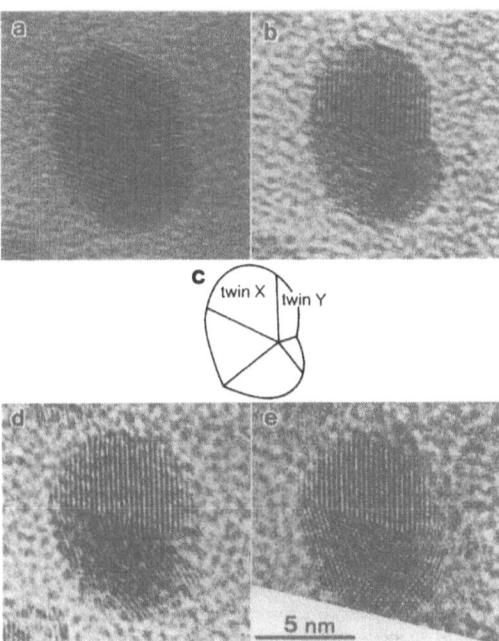

Fig. 8   HRIs of the same cluster. Images in (a) and (b) were obtained before copper deposition, and those in (d) and (e) after copper deposition. A schematic illustration of twin boundaries in the cluster is depicted in (c).
(From H. Mori *et al.*, *Philos.Mag.Lett.*, **69**, 279–283 (1994)).

Again these two images were taken under slightly different reflecting conditions. From an electron diffraction experiment, the copper concentration of the Au-Cu solid solution clusters containing this cluster was estimated to be approximately 30 at% Cu. It is evident from Figs. 8(d) and (e) that with copper deposition no changes were induced in the twinned structure of the cluster; the cluster remained composed of five twins and assumed the same nearly five-fold orientation.

If a melting process is responsible for the spontaneous alloying, then it is postulated that the arrangement and orientation of grains in a cluster would be completely different between conditions before and after solute atom (copper atom) deposition. However, the present experimental results are in contradiction with this postulate, since with copper deposition no changes were induced in the arrangement and orientation of grains in the cluster. This observation provides evidence for the idea that the spontaneous alloying takes place not via a melting process,[f-i] but via a purely solid-state process.

4. Effect of the surface-facet structure on spontaneous alloying [6]

The alloying behavior of gold in amorphous antimony (a-Sb) clusters has been investigated in order to see whether spontaneous alloying takes place also in amorphous clusters in which no surface-facet structures are involved.

A typical sequence of spontaneous alloying of gold atoms into antimony clusters at ambient temperature is shown in Fig. 9. Figs. 9(a) and (a′) show a BFI of as-produced antimony clusters on a supporting film and the corresponding SAED, respectively. The size of the antimony clusters is, on average, 10 nm in diameter. In the SAED (Fig. 9(a′)), only halos are recognized. This fact indicates that approximately 10 nm-sized antimony clusters prepared by the vapor-

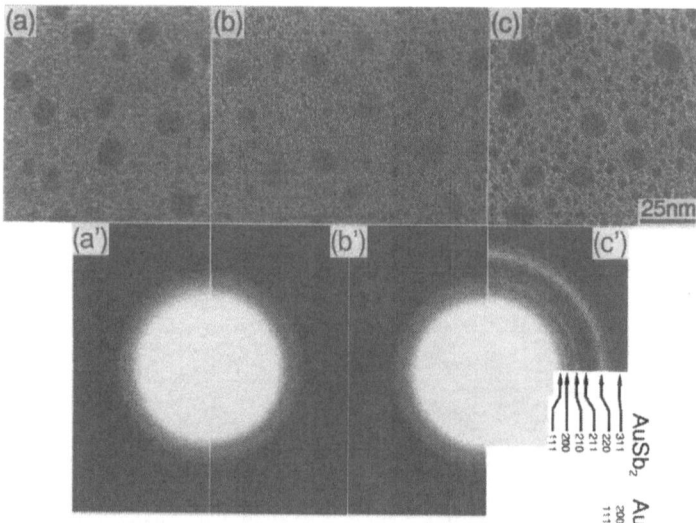

Fig. 9   Typical sequence of spontaneous alloying of gold atoms into antimony clusters at ambient temperature. (a) BFI of as-produced a-Sb clusters on a supporting film and (a′) the corresponding SAED. (b) BFI of clusters after gold deposition and (b′) the corresponding SAED. (c) BFI after additional gold deposition and (c′) the corresponding SAED, in which Debye rings can be consistently indexed as those of AuSb₂, superimposed on those of fcc gold. (From H. Yasuda and H. Mori, *Surf. Rev. Lett.*, **3**, 1167–1170 (1996)).

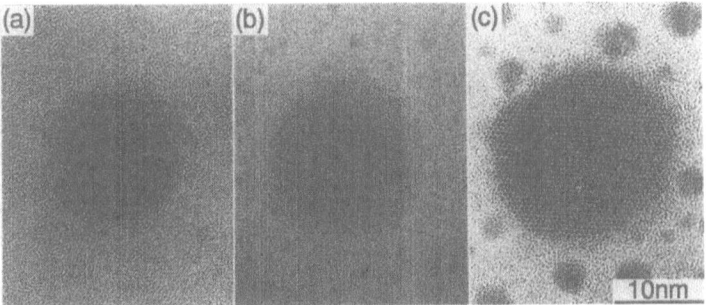

Fig. 10 (a) Example of HRIs of an as-produced a-Sb cluster. (b) Example of HRIs of an a-(Sb-Au) alloy cluster formed after the first gold deposition. Gold clusters of approximately 1.5 nm in the mean diameter can be seen on the supporting film. (c) Example of HRIs of an $AuSb_2$ cluster formed after the second gold deposition. The beam direction for the $AuSb_2$ cluster is [111]. Gold clusters of approximately 4 nm in the mean diameter can be seen on the supporting film.
(From H. Yasuda and H. Mori, *Surf. Rev. Lett.*, **3**, 1167–1170 (1996)).

deposition have an amorphous structure. The value of the scattering vector K (= $4\pi \sin \theta / \lambda$) for the first halo is approximately 20.5 nm^{-1}. Figs. 9(b) and (b′) show a BFI of clusters after gold deposition and the corresponding SAED, respectively. As seen in Fig. 9(b), the mean diameter of the clusters has increased from approximately 10 to 11 nm after gold deposition. It is also noted that small gold clusters with a mean diameter of approximately 3 nm have been formed on the supporting film after gold deposition. In the SAED (Fig. 9(b′)), halos are recognized, superimposed on the diffuse 111 Debye ring of fcc gold. The value of K for the first halo is again approximately 20.5 nm^{-1}. It is then considered that when gold is vapor-deposited onto about 10 nm-sized a-Sb clusters, gold atoms quickly dissolve into the a-Sb clusters and amorphous Sb-Au (a-(Sb-Au)) alloy clusters are formed. Figs. 9(c) and (c′) show a BFI of clusters after an additional gold deposition and the corresponding SAED, respectively. With this second deposition of gold, the mean size of alloyed clusters has increased from approximately 11 to 12 nm in diameter. As illustrated in the SAED (Fig. 9(c′)), the Debye rings can consistently be indexed as those of $AuSb_2$, which has the C2 structure with a lattice constant of $a_0 = 0.66_6$ nm.$^{\text{j)}}$ In the SAED, Debye rings of fcc gold should also be present. It is, however, difficult to discriminate Debye rings of fcc gold from those of $AuSb_2$ because the intensity of the former is much weaker than that of the latter and in addition the 111 and 200, rings of fcc gold almost overlap with the 220 and 311 rings of $AuSb_2$ respectively. It is evident from Figs. 9(c) and (c′) that a-(Sb-Au) alloy clusters have changed into $AuSb_2$ clusters with the second deposition of gold. It is assumed that after the second deposition of gold the gold concentration in alloyed clusters has reached the stoichiometric composition of the compound $AuSb_2$. All these observations shown in Fig. 9 clearly indicate that a fraction of vapor-deposited gold atoms have come in contact with a-Sb clusters on supporting film and they have quickly dissolved into a-Sb clusters to form clusters of either an a-(Sb-Au) alloy or the compound $AuSb_2$ depending on the gold concentration in clusters.

An atomistic scale observation of spontaneous alloying has been carried out by HREM. Fig. 10(a) shows an example of HRIs of an as-produced a-Sb cluster. The cluster shows contrast similar to the "salt and pepper" contrast characteristic

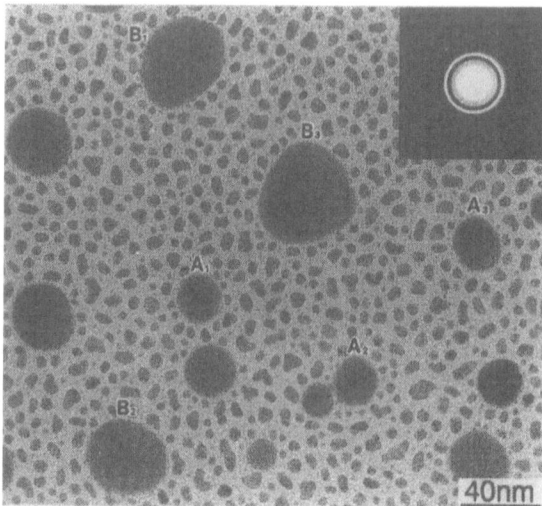

Fig. 11 BFI showing the cluster-size dependence of spontaneous dissolution of gold into a-Sb clusters. (From H. Yasuda and H. Mori, *Surf. Rev. Lett.*, **3**, 1167–1170 (1996)).

of an amorphous structure. Fig. 10(b) shows an example of HRIs of an a-(Sb-Au) alloy cluster formed after the first gold deposition. In the cluster image contrast similar to that in an a-Sb cluster (*i.e.* Fig. 10(a)) appears, showing that a similar amorphous structure is retained in a-Sb containing a small amount of gold. Tiny clusters with a mean diameter of approximately 1.5 nm, dispersed rather uniformly on the supporting film are gold clusters. Fig. 10(c) shows an example of HRIs of an $AuSb_2$ cluster formed after the second gold deposition. It was confirmed by an optical diffraction experiment of the cluster image that the incident beam direction is [111] of the $AuSb_2$ cluster. Small clusters of approximately 4 nm in the mean diameter, dispersed on the supporting film are again gold. HREM observations also revealed that every $AuSb_2$ cluster is a single crystalline particle, as illustrated in Fig. 10(c). Through the experiments, it became evident that an a-Sb cluster can absorb an small amount of gold and remain amorphous at nonstoichiometric compositions but crystallizes into a single crystalline $AuSb_2$ when the gold content reaches the stoichiometric composition (*i.e.* 1/3) in the course of continued gold deposition.

In order to see the cluster-size dependence of spontaneous dissolution, a-Sb clusters with rather wide size distribution, ranging from approximately 15 to 40 nm in diameter, were prepared on a supporting carbon film and gold was then vapor-deposited onto them. Fig. 11 shows a BFI of clusters after gold deposition and the corresponding SAED (inset). In the image, somewhat large clusters (such as clusters A_1, A_2, A_3, B_1, B_2 and B_3) are gold-deposited Sb clusters whereas the approximately 5 nm-sized clusters that are rather uniformly distributed on the film are gold clusters. It should be noted here that in clusters B_1, B_2 and B_3, which are more than 30 nm in diameter, approximately 2 nm-sized gold particles (arrowed) are present on the surface of each cluster, whereas in clusters A_1, A_2 and A_3, which are approximately 15 nm in diameter, no such gold particles remain on the surface. The size of gold particles that remain on the surface of clusters B_1 to B_3 (*i.e.* ~2 nm) is smaller than that of gold clusters on the supporting film (*i.e.* ~5 nm). In the SAED, Debye rings of both $AuSb_2$ and fcc gold appear, superimposed on amorphous halos. These facts indicate that in approximately 15 nm-sized a-Sb clusters almost all gold atoms which have come into contact with clusters can be dissolved into clusters while in clusters larger than approximately 30 nm in diameter only a fraction of them can be dissolved into clusters and undissolved gold atoms congregate on the surface to form gold particles.

In this experiment, it was confirmed that spontaneous alloying takes place in amorphous clusters which have no surface-facet structures. This shows that the presence of surface-facet structures is not prerequisite for spontaneous alloying. Moreover, it was confirmed that the ease with which spontaneous alloying takes place decreases with increasing cluster size. This tendency is consistent with what is observed in crystalline clusters.[4] All these facts suggest that the origin and mechanism of spontaneous alloying are common to both amorphous and crystalline clusters.

5. Temperature rise of clusters during *in situ* observation of spontaneous alloying

The magnitude of temperature rise in clusters is discussed in order to see whether or not such rapid dissolution can be ascribed to any temperature rise in clusters. The temperature of clusters would be increased by (1) electron beam heating, (2) heat of condensation, (3) heat of mixing, and (4) impingement of flying solute atoms with a kinetic energy of the order of kT.

A model calculation was made to analyze the temperature rise in Au-Cu alloy clusters. Using the value of 1.5×10^{20} e·m^{-2}·s^{-1} for the electron flux, a value of 8 K is obtained by calculation for the temperature rise in nm-sized gold clusters. It is reasonable to estimate that the magnitude of the temperature rise in gold clusters due to the beam heating is of the order of 10 K in conventional observation. This estimation is not so different from that by Takayanagi *et al.*[k] The cohesive energy and atom-accumulation rate of copper for a gold cluster are 336 kJ·mol^{-1} [l] and 20 s^{-1}, respectively. With the use of these values, a value of 2.8 $\times 10^{-3}$ K is obtained by calculation for the temperature rise in clusters due to the heat of condensation. A value of the heat of mixing for an $Au_{0.5}Cu_{0.5}$ alloy at 300 K is 8.75 kJ·mol^{-1} [m] and the atom-accumulation rate is 20 s^{-1}. Using these values, a value of 2×10^{-4} K is obtained by calculation for the temperature rise in clusters due to the heat of mixing. When a copper atom directly impinges on a gold cluster and is incorporated into the cluster, the kinetic energy of the copper atom will be converted into heat in the cluster. The average kinetic energy of copper atoms is given by 2 kT, where k is the Boltzmann constant. The value for T in the above equation is 1358 K (the melting temperature of copper). The rate of direct impingement of copper atoms on a gold cluster was approximately 7 s^{-1}. Using of these values, the value of 4×10^{-4} K is obtained by calculation for the temperature rise in clusters due to the impingement of flying copper atoms.

Of the four items mentioned above, the beam heating brings about the highest temperature rise (*i.e.* ~10 K). It was concluded that the spontaneous alloying in clusters is not an artifact originating from the temperature rise in clusters but is an intrinsic property of clusters.

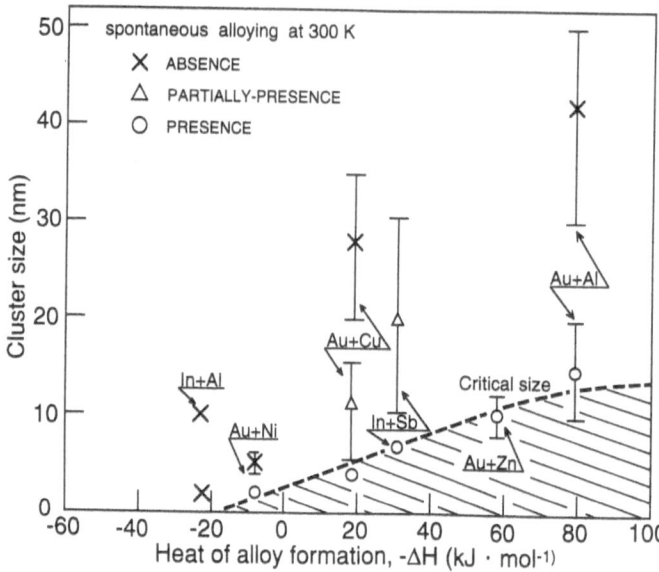

Fig. 12 Critical size of initial clusters, below which alloy clusters are successfully formed by spontaneous alloying as a function of the heat of alloy formation.

6. Generality in spontaneous alloying

It has been confirmed in previous experiments by the authors' group that solid solution and compound clusters are formed by spontaneous alloying in numerous binary systems. [1-17] Fig. 12 shows the critical size of initial clusters below which alloy clusters are successfully formed by spontaneous alloying at 300 K as a function of the heat of alloy formation. The circles, triangles and crosses indicate the presence, partially-presence and absence of spontaneous alloying, respectively. In the Au-Cu,[4] In-Sb,[12] Au-Zn[13] and Au-Al[16] systems in which the heat of formation is negative, spontaneous alloying takes place with ease. Also in the Au-Ni system in which the heat of formation is small positive, spontaneous alloying takes place but the critical size is decreased. On the other hand, in the In-Al system in which the heat of formation is large positive, no spontaneous alloying takes place. In the hatched region, spontaneous alloying takes place to form homogeneously mixed alloy clusters. The critical size of initial clusters increases with increasing heat of alloy formation. This result suggests that in nm-sized atom clusters the atom displacements driven by the free energy difference take place with ease. A definitive study on the mechanism of the atom displacements is in progress.

7. Summary

The results are summarized as follows.
(1) Spontaneous alloying becomes more difficult with increasing cluster size.
(2) Spontaneous alloying takes place not via a melting process but via a solid-state process.
(3) The presence of surface-facet structures is not prerequisite for spontaneous alloying.

(4) Spontaneous alloying is not an artifact originating from the temperature rise but an intrinsic property of clusters.
(5) The critical size of initial clusters below which alloy clusters are successfully formed by spontaneous alloying increases with increasing negative heat of alloy formation.

PUBLICATIONS

1. H. Mori, M. Komatsu, K. Takeda and H. Fujita, *Phil. Mag. Lett.*, **63**, 173 (1991).
2. H. Mori, M. Komatsu, K. Takeda, H. Yasuda and H. Fujita, Proc. of Special Symposium on Advanced Materials- III Nagoya, 144 (1991).
3. H. Yasuda, H. Mori, M. Komatsu, K. Takeda and H. Fujita, *J. Electron Microsc.*, **41**, 267 (1992).
4. H. Yasuda and H. Mori, *Z. Phys. D*, **31**, 131 (1994).
5. H. Mori, H. Yasuda and T. Kamino, *Phil. Mag. Lett.*, **69**, 279 (1994).
6. H. Yasuda and H. Mori, *Surf. Rev. Lett.*, 3, 1167 (1996).
7. H. Yasuda and H. Mori, Proc. 13th International Congress on Electron Microscopy, 355, Paris, July 17-22 (1994).
8. H. Yasuda, H. Mori, M. Komatsu and K. Takeda, *J. Appl. Phys.*, **73**, 1100 (1993).
9. H. Yasuda, H. Mori, K. Takeda and H. Fujita, *Defect and Diffusion Forum*, **95-98**, 697 (1993).
10. H. Mori and H. Yasuda, Proc. 3rd Japan International SAMPE Symposium Dec. 7-10, 1270 (1993).
11. H. Yasuda, H. Mori, M. Komatsu and K. Takeda, *Appl. Phys. Lett.*, **61**, 2173 (1992).
12. H. Yasuda, H. Mori, T. Muraki and T. Sakata, *Z. Phys. D*, **31**, 209 (1994).
13. H. Yasuda and H. Mori, *Phys. Rev. Lett.*, **69**, 3747 (1992).
14. H. Yasuda and H. Mori, Proc. 13th International Congress on Electron Microscopy, 367, Paris, July 17-22 (1994).
15. H. Mori and H. Yasuda, *Intermetallics*, **1**, 35 (1993).
16. H. Mori and H. Yasuda, *J. Microsc.*, 180, 33 (1995).
17. H. Mori, H. Yasuda and K. Fujii, *Surf. Rev. Lett.*, 3, 1177 (1996).

REFERENCES

a. R. P. Andres, R. S. Averback, W. L. Brown, L. E. Brus, W. A. Goddard III, A. Kaldor, S. G. Louie, M. Moscovits, P. S. Peercy, S. J. Riley, R. W. Siegel, F. Spaepen, Y. Wang, *J. Mater. Res.*, **4**, 704 (1989).
b. W. P. Halperin, *Rev. Modern Phys.*, **58**, 533 (1986).
c. W. B. Pearson, *A Handbook of Lattice Spacings and Structures of Metals and Alloys*, Pergamon, London (1958).
d. C. R. Barrett, W. D. Nix and A. A. Tetelman AA, *The Principles of Engineering Materials*, 150, New Jersey, Prentice-Hall (1973).
e. O. Kubaschewski, *Trans. Faraday Soc.*, **46**, 713 (1950).
f. L. D. Marks and P. M. Ajayan, *Ultramicrosc.*, **20**, 77 (1986).
g. P. M. Ajayan and L. D. Marks, *Phys. Rev. Lett.*, **60**, 585 (1988).
h. P. M. Ajayan and L. D. Marks, *Phys. Rev. Lett.*, **63**, 279 (1989).
i. S. Iijima and T. Ichihashi, *Phys. Rev. Lett.*, **56**, 616 (1986).
j. P. Villars and L. D. Calvert, *Pearson's Handbook of Crystallographic Data for Intermetallic Phases*, 1231, ASM, Metals Park, Ohio (1985).
k. M. Mitome, Y. Tanishiro and K. Takayanagi, *Z. Phys. D*, **12**, 45 (1989).
l. C. Kittel, *Introduction to Solid State Physics*, 2nd ed., 99, John Wiley and sons, New York (1956).
m. R. Hultgren, P. D. Desai, D. T. Hawkins, M. Gleiser and K. K. Kelley, *Selected Values of the Thermodynamic Properties of Binary Alloys*, 332, American Society for Metals, OH (1973).

15

Cluster-Cluster Aggregation of Calcium Carbonate at the Air/Water Interface

Tomoo Nakayama, Akio Nakahara and Mitsugu Matsushita

Department of Physics, Chuo University
Bunkyo-ku, Tokyo 112-0003, Japan

Purpose of the present study

To elucidate the structure of clusters that are formed by aggregation of fine particles, we experimentally studied the cluster-cluster aggregation of calcium carbonate ($CaCO_3$) particles and examined fractal dimensions of these clusters, aggregation process, cluster-size distribution function and dynamic scaling.

Contents of the present study

1. Introduction

The aggregation process is one of the most common phenomena in nature and can be seen in various fields of science and technology. For example, flocculations of smoke particles, colloids, aerosols, etc. are regarded as the cluster-cluster aggregation (CCA) process. These aggregation processes are classified into two types: fast and slow. Resultant clusters obtained through each process are now known to have different fractal dimensions, different time dependence of mean cluster-radius [a] and different cluster size distribution functions.[b] Under the fast aggregation condition, which corresponds to diffusion-limited CCA, the fractal dimension D of clusters is approximately equal to 1.43 (spatial dimension $d = 2$). In this case the time dependence of the mean cluster radius R can be described by a power-law relationship. The cluster-size distribution function n_s is almost flat with a cutoff at large cluster sizes and a depletion of very small clusters. On the other hand, under the slow aggregation condition, which corresponds to reaction-limited CCA, D is approximately given by 1.55 ($d = 2$), the mean cluster radius shows exponential growth and the distribution function exhibits a power-law form with a cutoff at large cluster size. It is known that the cluster-size distribution function with size s at time t, $n_s(t)$, can be scaled by the form $n_s \sim s^{-\theta} f(s/t_z)$ under fast aggregation. [c] And if the number of the particles is conserved, the scaling exponent θ is known to be 2. Fine particles of calcium carbonate ($CaCO_3$) have been found to aggregate and form clusters in two dimensions at the air/water interface. We study the cluster structure, aggregation dynamics, cluster statistics and so on to determine into which type of process this simple but real system should be classified.

2. Experimental Procedures

When a fresh aqueous solution of calcium hydroxide $(Ca(OH)_2)$ is poured into a flat glass dish, $Ca(OH)_2$ reacts with CO_2 in the air and forms calcium carbonate $(CaCO_3)$ fine particles at the air/water interface. This reaction is described as follows :

$$CO_2 + H_2O \longrightarrow H_2CO_3$$

$$H_2CO_3 + Ca(OH)_2 \longrightarrow CaCO_3 + 2H_2O.$$

Resultant particles of $CaCO_3$, whose diameter is about 2–3 µm, float on the surface of the solution (due to its surface tension), collide with each other and aggregate to form clusters. Clusters also move around and aggregate to produce still larger clusters. As a result, clusters grow in two-dimensional space (the air/water interface).

Experimental procedures are as follow: 5.0 ml of a fresh aqueous solution containing $Ca(OH)_2$ was poured into a petri dish with a diameter of 60 mm. We recorded the aggregation process of clusters of $CaCO_3$ particles on video tape using a microscope of 64 × magnification. We carried out digital image analysis using an image processor TVIP-4100 (NIPPON AVIONICS Co.,Ltd.) which has a frame grabber of 512 × 480 pixel resolution. One pixel corresponds to 3.5 µm × 3.5 µm in our data. The concentration of the $Ca(OH)_2$ solution was varied as 0.3 g/l, 0.2 g/l, 0.15 g/l and 0.1 g/l, respectively, because when the concentration is smaller than 0.1 g/l, particles do not aggregate enough and when the concentration is greater than 0.3 g/l, gelation occurs immediately after we pour the solution into the petri dish. We repeated the experiments six times under each condition.

3. Experimental Results and Discussion

3.1 Cluster structure

Figures 1(a)–(c) show snapshots of $CaCO_3$ clusters at time t = 6, 15 and 24 min, respectively, after initiation of aggregation. As time passes, the number of clusters decreases and their size becomes larger. In Fig. 2 is the photograph of an aggregated cluster at higher magnification. Particles tend to aggregate densely on shorter length scales. Using the box-counting method, we measured the fractal dimension D of $CaCO_3$ clusters. The results are summarized at Table 1.

D has little dependence on the $Ca(OH)_2$ concentration. The $CaCO_3$ clusters have many loops. These are little observed in the patterns obtained by CCA simulation. The values of D are larger than that of fast aggregation (diffusion-

Table 1

Concentration(g/l)	0.10	0.15	0.20	0.30
D	1.48 ± 0.07	1.49 ± 0.04	1.48 ± 0.06	1.46 ± 0.06

Fractal dimension of $CaCO_3$ clusters, D, for various values of the initial concentration of $Ca(OH)_2$ solution, C. Note that D has little dependence on C.

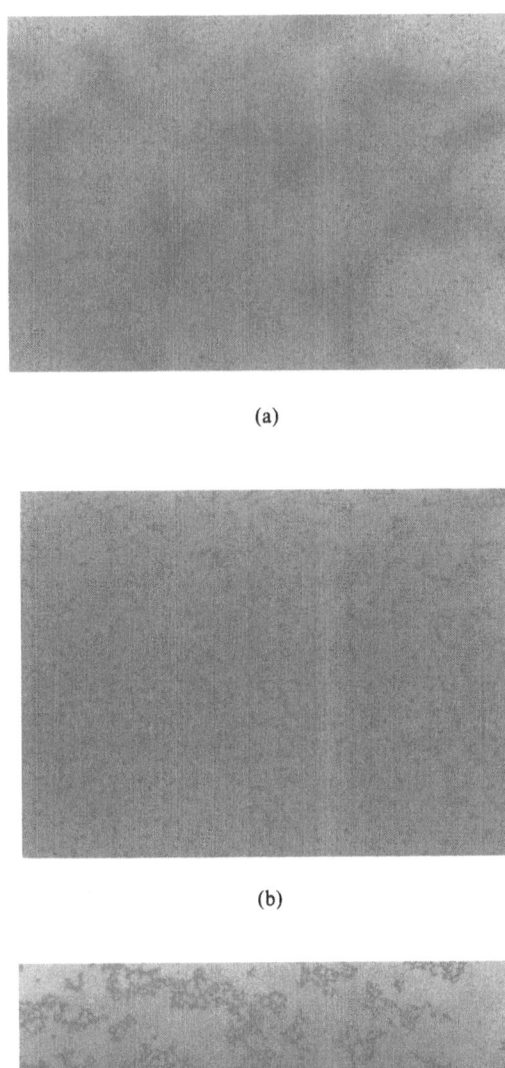

(a)

(b)

(c)

Fig. 1 Snapshots of aggregates of CaCO$_3$ clusters at time t = 6 (a), 15 (b), and 25 min (c), respectively. Initial concentration of Ca(OH)$_2$ is 0.3 g/l. The side length of the photographs is 2 mm.

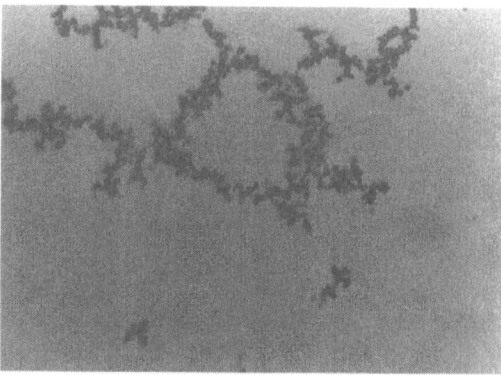

Fig. 2   Photograph of a cluster of CaCO₃ particles taken at higher magnification. Initial concentration of Ca(OH)₂ is 0.3 g/l. Note that rather compact structures are seen in very short length scales, in contrast to ramified structures in larger length scales. The side length of photographs is about 0.2 mm.

limited CCA) clusters ($D \cong 1.43$) but smaller than that of slow aggregation clusters (reaction-limited CCA; $D \cong 1.53$). These values are close to that obtained from numerical simulations by Meakin and Jullien.[d] In their simulations, after two clusters have made contact with each other, one cluster is allowed to rotate around the contact particle in the other cluster until the second contact is formed. They obtained $D = 1.475 \sim 1.50$ for these simulations. Patterns obtained from these simulation look similar to the CaCO₃ clusters of our experiments. The experiments by Skjeltrop [e] corresponds to these simulations. He used monodisperse polystyrene spheres dispersed in water and confined to essentially one layer between two parallel glass plates. He obtained $D = 1.49 \pm 0.05$ as the fractal dimension of polystyrene clusters grown through diffusion-limited CCA. He also noted the existence of structural readjustment in clusters. We therefore assert that the values of D which are larger than those obtained for the fast aggregation clusters are also due to the rotational structural readjustment. In our experiment we actually observed the phenomenon in which after two clusters collide to form one cluster they rotate around the contact point till they touch at another point.

3.2 Aggregation dynamics

Cluster aggregation can be classified into two types and the time dependence of the mean-cluster radius R can be described as follows.

$$R \sim t^{\zeta} \qquad\qquad\qquad \text{(fast aggregation)}$$

$$R \sim e^{\alpha t} \qquad\qquad\qquad \text{(slow aggregation)}$$

To investigate which aggregation process fits the aggregation of CaCO₃ clusters, we show the evolution of the mean-cluster radius R as a function of time t in Fig. 3. We find that the mean-cluster radius R obeys a power-law in time as $R \sim t^{\zeta}$ with $\zeta \cong 0.78$. We can now summarize that the aggregation process of CaCO₃ clusters observed in the present experiments belong essentially to the fast aggregation category. The reason for this conclusion is as follows. First we

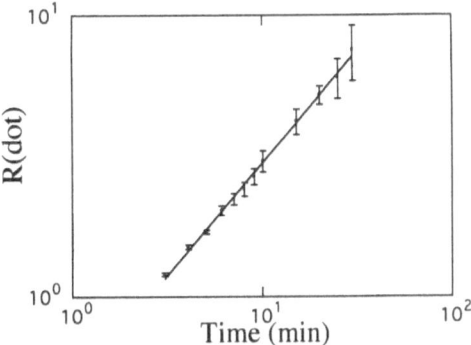

Fig. 3   Mean cluster size R is plotted double-logarithmically as a function of time t. Note the power-law dependence $R \sim t^{\zeta}$ with $\zeta \cong 0.78$.
(From T.Nakayama, A.Nakahara and M.Matsushita, *J.Phys.Soc.Jpn.*, **64**, 1117 (1995)).

observed that clusters aggregate and form larger ones as soon as they touch each other. This implies that the diffusion of clusters limits the aggregation process. Secondly, the mean cluster radius grows as $R \sim t^{\zeta}$. This is characteristic of fast aggregation (diffusion-limited CCA). Moreover, our values of fractal dimension D agree with the value obtained from numerical simulations of diffusion-limited CCA with structural readjustment.

3.3 Size distribution function and dynamic scaling

Figure 4 shows the size distribution function of $CaCO_3$ clusters. We found that the distribution functions look bell-shaped in the larger size region and monotonically decrease in the smaller size region, that is, cluster size distribution function consists of two parts. With lapse of time, the number of smaller clusters decreases and that of larger clusters increases.

In diffusion-limited CCA, it is known that the size distribution function can be scaled as follows:

$$n_s(t) \sim s^{-\theta} f(s/t^z),$$

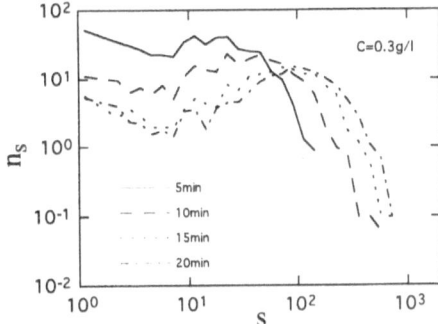

Fig. 4   Cluster-size distribution functions $n_s(t)$ are plotted double logarithmically as a function of cluster size s at $t = 5, 10, 15$ and 25 min, respectively.
(From T.Nakayama, A.Nakahara and M.Matsushita, *J.Phys.Soc.Jpn.*, **64**, 1117 (1995)).

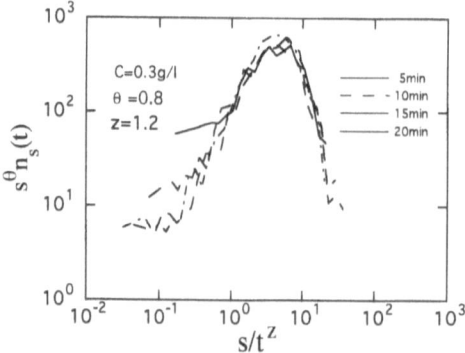

Fig. 5   Scaled cluster-size distribution function $s^\theta n_s$ versus s/t^z for the data shown in Fig. 4 with $\theta =$
0.8, $z = 1.2$. Note the prominent data collapse seen in the larger cluster-size region.
(From T.Nakayama, A.Nakahara and M.Matsushita, *J.Phys.Soc.Jpn.*, **64**, 1117 (1995)).

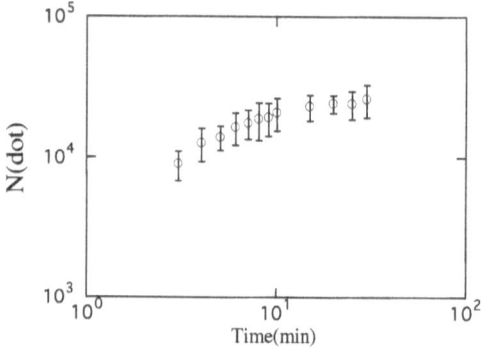

Fig. 6   Time dependence of the total number of CaCO$_3$ particles N per area 1 mm^2. Note that N
saturates at a finite value after about 10 min.
(From T.Nakayama, A.Nakahara and M.Matsushita, *J.Phys.Soc.Jpn.*, **64**, 1117 (1995)).

and when the total number of particles is independent of time, the scaling
exponent θ is 2. Fig. 5 shows the scaled cluster-size distribution $s^\theta n_s$ as a function
of scaled cluster-size $x = s/t^z$. We have found that the case with $\theta = 0.8$ and z
$= 1.2$ shows the best data collapse.

In order to explain the large difference between theoretical and experimental
θ values, we measured the time evolution of CaCO$_3$ particles N, as shown in Fig.
6.

We found that N in fact increases early in the experiment, then gradually
approaches a finite value later on. Therefore we conclude that there is another
reason for such a small value of θ. As recognized in Fig. 5, although distribution
functions are well scaled in the larger cluster-size region, they are not scaled in
the smaller size region. We tried to examine the scaling of distribution functions
in only the larger size region, namely in the bell-shaped part. Although the mass
conservation holds as a whole, the number of particles contained in the
monotonically decreasing part (smaller size region) of the distribution functions
decreases with time, while the number of particles in the bell-shaped part
increases. We believe that the mass transfer observed in the cluster-size space may
explain why the value of θ becomes smaller than 2.

4. Conclusion

From above results we conclude the following:
1) CaCO$_3$ clusters have self-similarity and the fractal dimension is given by D = 1.46 ~ 1.49.
2) Time dependence of the mean-cluster radius R can be described by a power-law form. Therefore, the CaCO$_3$ clusters are generated through a fast aggregation (diffusion-limited CCA) process.
3) Cluster-size distribution function obeys the dynamic scaling in the form $n_s(t)$ ~ $s^{-\theta} f(s/t_z)$. These results show following relations:
   1) $s \sim R^D$
   2) $R \sim t^\zeta$
   3) $s \sim t^z$

We can, therefore, derive a scaling law of the form $D\zeta = z$ among the three exponents. Taking into account the fact that D is about 1.48 and ζ is about 0.79, one finds that the value of $D\zeta$ is in good agreement with that of z.

PUBLICATIONS

1. M. Matsushita, S. Ouchi and K. Honda, *J. Phys. Soc. Jpn.*, **60**, 2109 (1991).
2. M. Ohgiwari, M. Matsushita and T. Matsuyama, *J. Phys. Soc. Jpn.*, **61**, 816 (1992).
3. S. Isogami and M. Matsushita, *J. Phys. Soc. Jpn.*, **61**, 1445 (1992).
4. M. Yasui and M. Matsushita, *J. Phys. Soc. Jpn.*, **61**, 2327 (1992).
5. S. Isogami and M. Matsushita, *J. Phys. Soc. Jpn.*, **62**, 2200 (1992).
6. T. Matsuyama, R. M. Harshey and M. Matsushita, *Fractals*, **1**, 302 (1993).
7. J. Yokoyama, Y. Kitagawa, H. Yamada and M. Matsushita, *Physica*, **A 204**, 789 (1994).
8. S. Isogami and M. Matsushita, *J. Phys. Soc. Jpn.*, **63**, 2919 (1994).

REFERENCES

a. D. A, Weitz, J. S. Huang, M. Y. Lin and J. Sung, *Phys. Rev. Lett.*, **54**, 1416 (1985).
b. D. A. Weitz and M. Y. Lin, *Phys. Rev. Lett.*, **57**, 2037 (1986).
c. T. Vicsek and F. Family, *Phys. Rev.*, **B31**, 564 (1985).
d. P. Meakin and R. Jullien, *J. Phys.* (Paris), **46**, 1543 (1985).
e. A. T. Skjeltorp, *Phys. Rev. Lett.*, **58**, 1444 (1987).
f. T. Nakayama, A. Nakahara and M. Matsushita, *J. Phys. Soc. Jpn.*, **64**, 1114 (1995).

16

Electronic State of Ultrafine Particles Suspended in Liquid Media

Keisaku Kimura

Department of Material Science, Faculty of Science, Himeji Institute of Technology
Ako-gun, Hyogo 678-1231, Japan

Aim of this research

Suspensions, which are systems composed of small particles suspended in liquid, have many advantages in the study of ultrafine particles (UFPs). The first one is the maintenance of particles in high density free from contact with each other in a liquid. The second is versatility in the handling process. We can imagine this merit when compared with samples in a solid. In this case, inactive liquid should be selected. The third advantage is that one can control the extent of particle contact depending on the kind of liquid. As described below, we have produced various particle assemblies, including completely isolated particles to fractal flocculates and amalgamated cluster aggregates.

One object of small particle physics is to see the change in properties that occurs in the process from particle to bulk. There are two ways to access this process. One is to observe the size change of a single particle. Much research effort is devoted to this approach. The other is to see the sequential process of connecting many particles. In the latter case, the particle assembly is not necessarily the same as the bulk in its solid state properties. We can expect to have noble materials different from both single particle and bulk. However, studies along these lines are rare in the literature. We will describe first from the view point of size effect of a single particle the magnetic properties of metallic ultrafine particles (Section 1) and the ^{1}H high resolution solid state NMR of organic nanoparticles (Section 2). Then from the viewpoint of network formation, we discuss photocoagulation of Au ultrafine particles and the photoexcited van der Waals attractive force in addition to the ultrafine particles of solid electrolytes.

1. Magnetic properties of metallic ultrafine particles

In order to evaluate the quantum size effect in light of the electronic properties of matter, a non-contact method is commonly used. In a memorial paper by Kubo,[a] he pointed out the important role of the magnetic susceptibility in a study of the quantum size effect. The Pauli paramagnetism, a typical characteristic of metallic properties, is reduced at low temperature corresponding to the energy of E_F divided by the number of electrons in a particle. However, this behavior is sometimes masked if the sample is contaminated by a small amount of impurities. If the metal element is monovalent or an odd spin system, particles inherently contain odd or even number electrons in which the number of odd particles to the even particles is the same. Consequently, the interpretation of the experiment is not straightforward. Therefore, using a divalent metal element removes this complexity and makes it easy to check whether a sample is contaminated by paramagnetic impurities or not.

Magnesium is a divalent metal having paramagnetic moment in the bulk in contrast to the diamagnetic moment in the atomic state. Hence we may expect the conversion of the sign of magnetism from paramagnetic to diamagnetic in a small-size region. This property of magnesium is very convenient for the study of the size effect. We have already reported the spin paramagnetism of Mg particles of less than 5 nm. [1] We found that the paramagnetism from conduction electrons is a function of size, temperature and shape of particle. The prediction of the random matrix theory must be modified incorporating the shape effect in order to quantitatively explain the magnetic behavior of nano-size materials. We report here the result of magnetic measurement of Mg particles larger than those of the former experiment.

Ferromagnetism is also an interesting topic in the study of size effect. We summarized the data reported hithertofore to obtain the phase diagram of iron particles as a function of size, temperature and magnetic moment. The characteristic features of this diagram from atoms to bulk is also reported.

1.1 Magnetic measurement of magnesium ultrafine particles

In this study, the field dependence of magnetization of Mg small particles up to 5 T is investigated in order to obtain detailed information about the size effect. In the former study, it was reported that (1) the magnetic susceptibility of particles less than 5 nm in size decreases with decreasing temperature, (2) the susceptibility at room temperature has a tendency to increase with reducing size. The Mg ultrafine particles were prepared by the matrix isolation method. [b] Mg tips were heated and sublimed in high-purity He gas (1–100 Torr) and the formed Mg fumes were trapped onto the cryogenic-temperature hexane matrix. After warming at room temperature, the colloidal solution was stored in a storage bottle.

The sample particles were classified by electron microscopic observation depending size as follows.
(A) typical size of around 20 nm.
(B) a mixture of 10–30 nm UFPs and 500–1000 nm microcrystals.
(C) very broad size distribution of 10–300 nm.
Therefore, the size tendency can be classified in the order of (A) < (C) < (B).

Figure 1 shows the magnetization curves at room temperature for the sample (A), (B), (C). No significant difference from room temperature was observed for the measurements at temperatures of 2 K and 4.2 K. As seen in the figure, every sample showed the same deviation from the bulk sample at high field over 1 T. This diamagnetic shift increases with decrease in the size of particle. However, the magnetization becomes larger than that of the bulk below 1 T. This behavior is consistent with that of the previous report. [1] Since the average energy gap (0.3 K) at the Fermi level for particles 20 nm in size is much smaller than 300 K (room temperature), this abnormalty cannot be ascribed to the quantum size effect in spin magnetism. We believe that this is due to orbital quantization as described below.

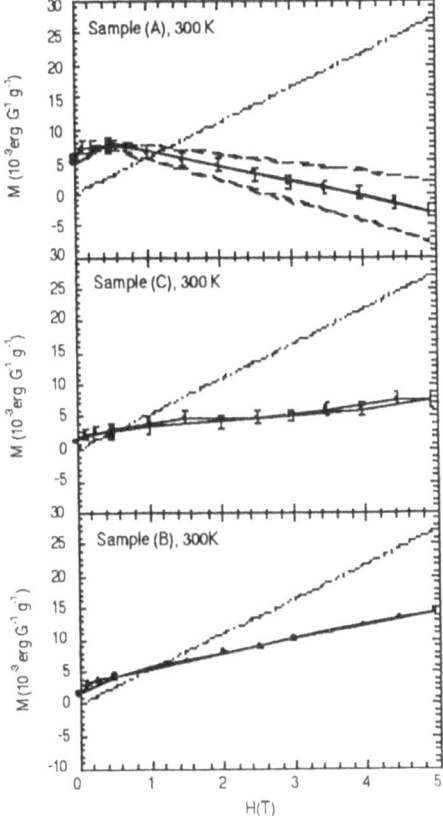

Fig. 1   Magnetization curves of sample (A) at 300, 4.2 and 2 K. Dotted lines for bulk. Broken lines in the upper panel are systematic errors.

1.2 Orbital magnetism of free electrons in a finite system

In this section, we explain the diamagnetic shift observed in the high field for the ultrafine Mg particles. The relevant parameters in question are the average energy gap $\sim E_F/N$, kT and Zeeman energy $\mu_B H$. Let us consider the Hamiltonian incorporating explicitly the effect of magnetic field H as

$$\hat{H} = \frac{1}{2m}\left(\frac{\hbar}{i}\nabla - \frac{e}{c}A\right) \pm \mu_B H.$$

Applying this Hamiltonian to the basis set (3D standing wave) being the eigenfunction at $H = 0$, we calculate the free energy of the free electron system after diagonalization of the matrix. Magnetization M is given as

$$M = -\partial F/\partial H.$$

Magnetic susceptibility is therefore

$$\chi = -\partial^2 F/\partial H^2.$$

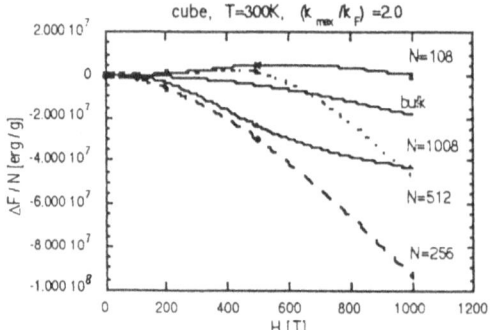

Fig. 2 Field dependence of free energy.

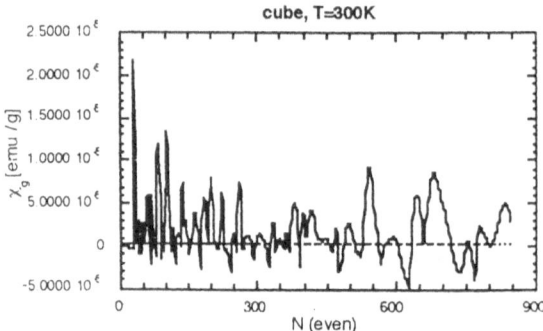

Fig. 3 Magnetic susceptibility as a function of N.

Figure 2 shows the results of the calculation. It is found that the magnetic susceptibility is not necessarily constant with field. Figure 3 shows the dependence of magnetic susceptibility on the number of electrons in a particle. The susceptibility for the even electron particles was found to deviate from bulk value even at room temperature.

1.3 Magnetic properties of small iron particles

It is very important not only from the view point of understanding the origin of ferromagnetism but also from that of technical application to the high performance magnet to comprehend the differences and common features of bulk and cluster ferromagnetic materials. There are many studies on magnetite [3] and other compound ferromagnets as well as on ferromagnetic metals such as iron, cobalt and nickel. Here we take the magnetic moment of iron cluster as an example of metallic ferro-magnetism.

Recently, many reports have appeared on the magnetic moment of iron cluster.[c-f] From Stern-Gerlach's experiment on iron cluster, it was concluded that the magnetic moment in dimer and trimer is larger than that in bulk.[c-e] On the contrary, de Heer et al. reported that the moment was very small and seemed to disappear for small field limit.[f] Taking into account the fact that a particle itself rotates in the magnetic field in the beam experiment and if not, that the magnetic moment itself fluctuates by the thermal energy, caution must be used in examining these results.

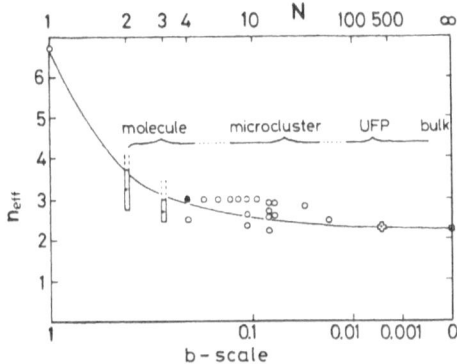

Fig. 4 Magnetic moment (Bohr unit) of Fe cluster. The ordinate indicates the number of atoms in a cluster.
(From K. Kimura, *Phys. Lett.*, **A158**, 86 (1991)).

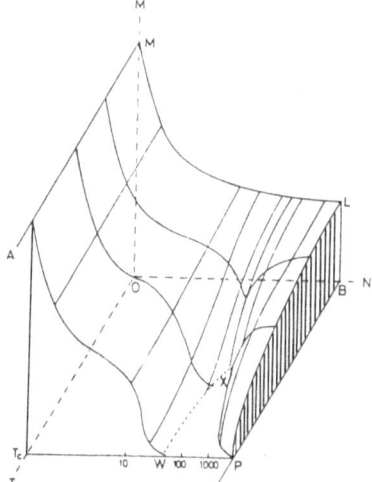

Fig. 5 Phase diagram of Fe cluster. LBP, bulk phase. ML line is the same as for presented in Fig. 4. the O-T line indicates temperature axis.
(From K. Kimura, *Phys. Lett.*, **A158**, 87 (1991)).

Starting from the view point that the properties of clusters and small particles must be continuously connected to bulk, we develop a phase diagram of magnetic moment as a function of size and temperature. [4] At 0 K, the magnetic moment per atom converted from the volume magnetic moment of cluster is plotted in Fig. 4. The solid fitting curve was confirmed by recent experiments. [8] For bulk materials, the temperature dependence of spontaneous magnetization is well known both from experiment and theory. The size dependence on Curie temperature (T_c) is analytically represented by its asymptotic shape by molecular field approximation. For atoms and molecules, it can be regarded as constant because of the large energy gap (larger than 1 eV) between ground electronic state and excited state. There is no phase transition in the rigorous meaning for finite clusters. When we trace a minimal given moment as a function of temperature, we obtain a curve which indicates threshold magnetization analogous to the finite size rounding of T_c. In summary, we obtain a phase diagram as shown in Fig. 5 in which $T_c = 770°C$ and the hatched region expresses bcc bulk iron that becomes fcc iron above

900°C. As seen in the figure, magnetic properties of cluster iron does not monotonically change to that of bulk but shows different temperature dependence on the region divided by a crevasse structure. Recent results obtained by Bucher *et al.*[h)] appear to support our prediction.

2. Ultrafine particle NMR (UFP-NMR)

2.1 High resolution solid state ¹H NMR of organic nanoparticles

NMR spectral lines of solid-state samples are, in general, broad due to the presence of anisotropic interactions such as a magnetic dipole-dipole interaction, chemical shielding interaction, quadrupole interaction, etc. It is inevitable to reduce the broad line width in order to get useful NMR parameters such as chemical shifts of individual nuclei, multiplet structure of absorption peak, etc. Rapid and random motion of nuclei in a solid suppress these broadening effects, This is well known as motional narrowing. The present author has reported that Brownian motion of nanometer-sized particles dispersed in a liquid phase provides well resolved high-resolution NMR spectra for solid state samples, which has been named "UFP-NMR".[5)] Recently, this UFP-NMR method has been applied to nanometer-sized organic substances.[6,7)] UFP of terephthalic acid (TPA) dispersed in carbon tetrachloride (CCl_4) was chosen as a test material for UFP-NMR because TPA is chemically and physically stable. Low solubility of TPA to conventional solvents is also important because a strong signal from molecularly dissolved TPA may be obtained in the case of good solubility.

Although various preparation techniques have been developed for metallic and inorganic substances, very few studies have been reported for the preparation of organic substances. TPA-UFP was prepared by the gas evaporation technique combined with the matrix isolation method specially modified for organic nanoparticles.[7)] TPA-UFP dispersion is provided based on an island formation of deposit on a cold matrix by the vacuum evaporation method. It was not able to disperse TPA-UFP by a single component among those tested. TPA was too soluble in some solvents or deposited to sediment in other solvents. A dispersing reagent such as deuterated acetone (acetone-d_6) or deuterated methanol (methanol-d_4) which covers TPA-UFP is required. Among various solvents tested, CCl_4 was the best as the dispersing medium for these dispersing agents, because solubilities of TPA to CCl_4, CCl_4/ acetone-d_6 mixture and CCl_4/ methanol-d_4 mixture are low enough to stabilize UFP-state in addition to suppressing molecular signals of TPA. ¹H NMR measurements were carried out by a conventional liquid state FT NMR spectrometer operated at a frequency of 400 MHz. Colloidal and liquid samples were sealed in a 5-mm conventional NMR sample tube with 17 Hz sample spinning rate. All chemical shifts were quoted from tetramethylsilane. The solid state ¹H NMR was measured by a CRAMPS method with a BR-24 pulse-sequence. The average size of particles was determined by the dynamic laser scattering (DLS) method (Ohtsuka Electronics; ELS-800) and was *ca.* 10 nm.

Figure 6 shows the spectra of ¹H NMR from aromatic proton region of TPA in various states after accumulation of 500 signals. The colloidal dispersion of TPA-UFP/acetone-d_6/CCl_4 provides six peaks. The full widths at half maxima of six peaks were on the order of 0.1 ppm, *i.e.* 40 Hz. Six similar signals with an additional peak positioned at 8.15 ppm appeared in the case of the TPA-

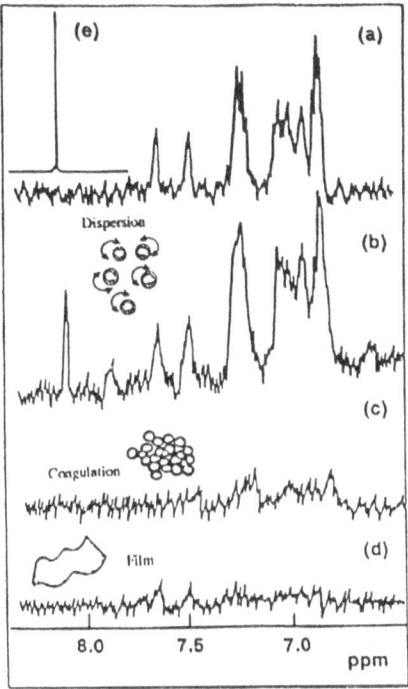

Fig. 6 ¹H NMR spectra of aromatic proton region of some TPA preparations taken by a conventional liquid state FT NMR. (a) TPA-UFP/acetone-d₆/CCl₄, (b)TPA-UFP/methanol-d₄/CCl₄, (c)TPA-UFP aggregates in CCl₄, (d)TPA-UFP film in CCl₄, (e)TPA/acetone-d₆ solution.

UFP/methanol-d₄/CCl₄ system (Fig. 6(b)). The original signal from aromatic proton of TPA dissolved in acetone-d₆ is referenced in Fig. 6(e). Because of higher solubility of TPA in methanol than in acetone, a small amount of TPA is supposed to dissolve into liquid phase in the TPA-UFP/methanol-d₄/CCl₄ system. The signals observed in the TPA-UFP/dispersing reagent/CCl₄ systems were significantly diminished or completely lost both when the dispersing reagents were not used on the preparation process and when TPA was deposited at more than 200 nm thickness. In both cases, DLS measurement displayed the existence of solely large particles. In addition, no signal was observed from TPA saturated acetone-d₆ diluted with CCl₄ in this region. These results strongly suggest that the signals originated from the dispersed TPA-UFPs in CCl₄.

In order to identify the signal origin, the solvent of TPA-UFP/acetone-d₆/CCl₄ was evaporated to obtain UFP powders of TPA, which were then dissolved again into aceone-d₆. During this process, UFPs are believed to disperse into more fine assemblies than UFPs in their present size, namely, cluster states. Figure 7 shows the spectral change of TPA due to various processes in order of size: bulk, UFP, cluster and molecule from the top down. In this figure, the bulk broad peaks converge into the molecular peak through. UFP and cluster states and the peaks from the UFPs shift to a lower field in spectrum c, the degree of shifts is depending on each peak. This is the first time the development of molecules to bulk through UFP state via an NMR window has been observed.

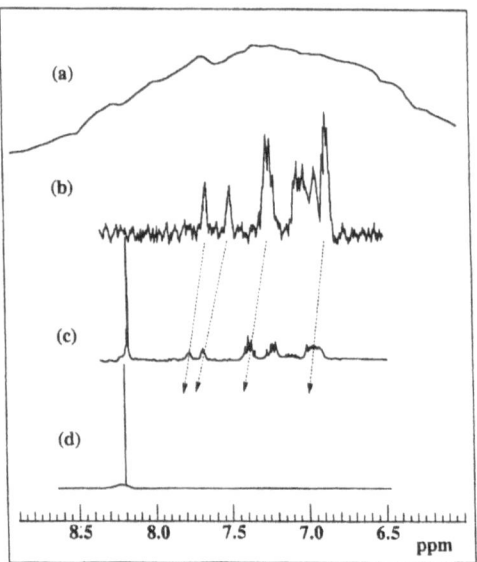

Fig. 7   Spectral change with decrease in size of TPA-UFP. (a) bulk spectrum from CRAMPS, (b)
TPA-UFP, (c)cluster(sharp line at 8.2 ppm) from dissolved molecules, (d) free molecule. Note
the well-resolved spectral lines of the cluster preparation (c).
(From K. Kimura, *Bull. Chem. Soc. Jpn.*, **68**, 2155 (1995)).

3. Solid electrolyte ultrafine particles

Solid electrolytes are characterized by high ionic conductivity in solid phase.
It has recently been reported that mixtures of a solid electrolyte with submicron
particles of an insulating material generally show a larger increase in ionic
conductivity than the pure, homogeneous system. It is therefore interesting to
examine the size effect of solid electrolyte itself on the conductivity. In this
study, attention is focused on the micron and/or submicron size level electrolyte
prepared by an evaporation method and the transport properties of the compaction
pellet of the well-defined small particles of solid electrolyte materials are
investigated.

The fine particles were dropped into a 4 mm-diameter steel die together with
the organic solvents. After the sublimation of solvents, particles were further
dried under vacuum and pelletized at a pressure of 0.2–0.4 GPa. In order to
examine the kind of charge carrier, the electric conductivity was measured both
in the conventional electric current mode and an ion-blocking electrode mode. The
ac-conductivity measurements were also performed using RC-oscillator and a
single-phase lock-in-amplifier. The observed impedance of the sample was
corrected from the parallel open-circuit impedance of the measurement system.

The dc-conductivity of the pellets of as-grown CuBr and CuCl samples are on
the order of 10^{-1} and 10^{-3} $\Omega^{-1} \cdot cm^{-1}$, respectively. It was found that hole conduction
is dominant in these samples. After washing with acetone, conductivities of these
pellets were significantly decreased. Applying the polarity exchange experiment,
it was concluded that the ionic conduction was not dominant in the CuBr pellets
but was dominant in the CuCl pellets in this study.[8]

Arrhenius plots of the ac-conductivity of AgI pellets are shown in Fig. 8.

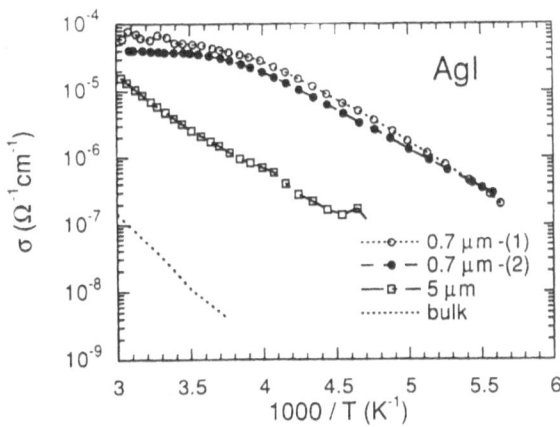

Fig. 8   Plot of conductivity of the polycrystals of AgI (0.7, 5 μm) microcrystals. The samples indicated by 0.7 μm -(1) and 0.7 μm-(2) denote respectively pellets of 0.33 and 0.17 cm in thickness. Bulk data are referred to the literature (J. B. Boyce, T. M. Hayes and J. C. Mikkelsen, Jr, *Phys. Rev.*, **B23**, 2876 (1981)).
(From K. Kimura, *Sung. Rev. Lett.*, **3**, 43 (1996)).

Typical particle sizes of two samples are 0.7 and 5 μm, respectively. Ionic conduction is significant in both samples. Large enhancement effect was observed in the small-size particles, possibly be due to an internal structural disorder of the sample.

4. Photocoagulation of ultrafine Au particles

It is well known that metallic ultrafine particles dispersed in glass sometimes reveal beautiful coloration. This phenomenon was well explained by the classical Mie theory over 80 years ago. In this theory, it is assumed that particles are dispersed homogeneously in the medium. In contrast to this assumption, ultrafine particles often coagulate, agglomerate or flocculate depending on the solution conditions and in extreme cases, particles coalesce with each other through a contact point. These effects affect the visible spectrum such as peak position, bandwidth and peak profile. When the coalescence process is promoted to give a metallic reflectance, this process can be regarded as the development of particle to bulk. The author and his team accidentally found the coagulation phenomenon of Au nanoparticles dispersed in organic liquids by the illumination of light.

4.1 Fractal coagulation of Au nanocolloids by illumination

Nanometer-sized gold particles were prepared by the gas flow-solution trap method.[1] The dispersing solvent is 2-propanol and average size of starting particles is 8 nm. On irradiation of a high-pressure mercury lamp, the color of dispersion changed from wine red to dark blue, otherwise it remained wine red for years.[9] TEM observation shows that dispersed particles agglomerated into fractal cluster concurrent with the spectral change of increase in 750 nm band and decrease in 523 nm plasmon band. Fractal dimensional analysis revealed that the dimension of the system change from 0 D (isolated particles), 1 D (linear assembly) to 1.7 dimension as time passed. The dimension of 1.7 roughly agrees

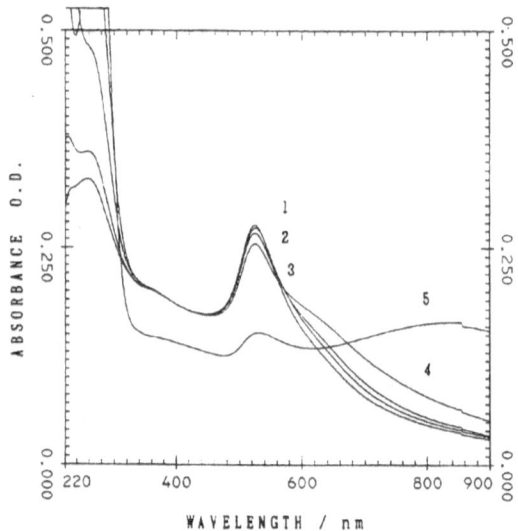

Fig. 9   Spectral change of Au colloids as a function of illumination time. 1: without illumination, 2; 3; 4; and 5; are for 4, 8, 12 and 16 h after irradiation.
(From K.Kimura, *J.Phys.Chem.*, **98**, 2144 (1994)).

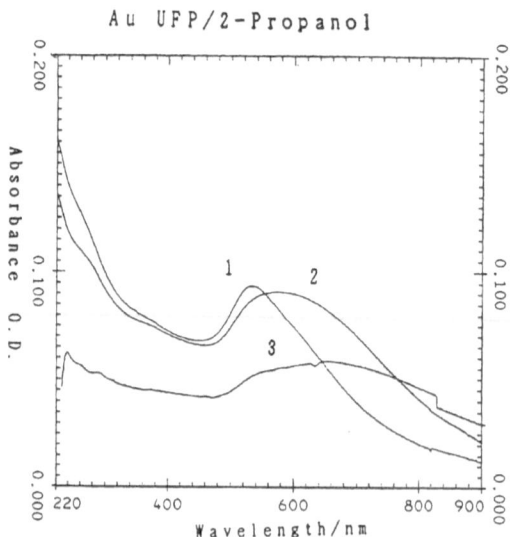

Fig. 10   Spectral change with addition of salt. 1; without NaCl, 2; 4.3×10^{-6} M, 2; 8.6×10^{-4} M just after addition of NaCl.
(From K.Kimura, *J.Phys.Chem.*, **98**, 2144 (1994)).

with the value of 5/3 by virtue of the theory of diffusion-limited aggregation. Figure 9 shows the spectral change of Au UFP dispersions as a function of time.

4.2 Electrolyte effect[10]

The change in dispersion by the illumination of light resembles the effect of the addition of salt. The stability of dispersion is most sensitive to the concentration of the electrolyte in dispersion because ions affect the Debye length around the particle sphere. Following the Schultze-Hardy rule, the more the electrolyte concentration increases, the less is the stability of the sols. Fig. 10 shows the effect of the addition of NaCl on the dispersion in the absorption spectrum as a function of salt concentration. Sudden coagulation started at the instant of addition of NaCl solution. With increase in the amount of NaCl, the original spectrum 1 (wine red) changes to 2 (dark blue) and then to 3 (blue black), and the sols are finally deposited. That is, the initial peak maximum at 520 nm shifted toward the red wavelength region with increased concentration of salt. This tendency is similar to the irradiation time dependence on photocoagulation.

4.3 Coagulation by plasmon excitation[10]

Following the standard theory of colloids, the stability of sols is governed by the balance between van der Waals attraction and Coulombic repulsion of charged particles. If the charge on the particles is lost by the irradiation of light (photoneutralization of charged particles), coagulation takes place. To examine this possibility, an experiment on the wavelength dependence of irradiation light was conducted since deionization is very effective in the UV-region. Having illuminated UV-cut light to the dispersion, a new shoulder in the long-wavelength region appears after 11 h of irradiation, and this tendency is more clear after 22 h. It is obvious that the coagulation proceeds even with visible light energy. Figure 11 shows TEM pictures of particles at various steps. Unirradiated samples (C) show well-isolated particles in contrast to the chain-like or fractal structures in (B). The result of the addition of salt is also shown in the same figure as a comparison. The size of the starting particles (8 nm) did not change by the

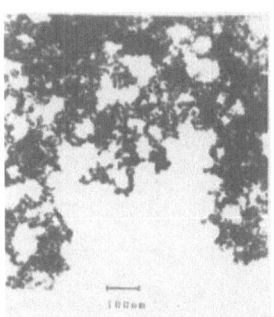 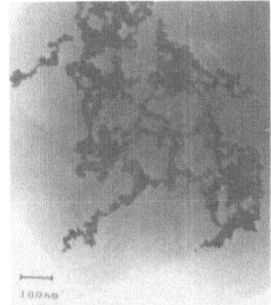 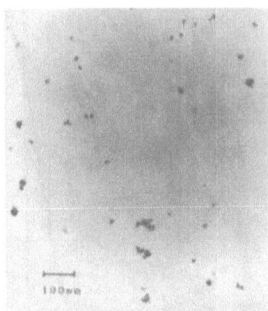

(A) Coagulation induced by the addition of salt   (B) Coagulation by UV-cut irradiation   (C) Reference system

Fig. 11
(From K. Kimura, *J. Phys. Chem.*, **98**, 2145 (1994)).

illumination with UV-cut light in contrast to the marked increase in the case of UV irradiation where the size increased to 27 nm after 16 h irradiation.

5. Photoenhanced van der Waals attraction force

The stability of a sol system is governed by the balance between van der Waals attractive force and electrostatic repulsive force according to the DLVO (Deryagin-Landau-Verway-Overbeek) theory, the received standard theory. Below it is demonstrated that the photocoagulation can be induced by the photoenhanced van der Waals attraction assuming that the interparticle interaction is caused by a gigantic plasma resonance oscillation of free electrons.

5.1 Mechanism of photoenhanced van der Waals force

At Mie resonance frequency, the free electron system in a metallic particle feels forced vibration at the resonance frequency causing large dipole oscillation. This large oscillation-induced interparticle attractive force in the same particle occurs likewise in the normal van der Waals force. The calculation proceeds following below steps.[11]
1) Derivation of a dielectric function of metals under electromagnetic field.
2) Expression of the interparticle interaction as a function of dielectric function.
3) Comparison of the interaction potential of the forced oscillation with that in dark condition.

First, normal van der Waals interaction energy, U_d, is expressed by the size of particles. Then the interparticle energy under irradiation, U_{irr}, is given by a function of the external field. Finally, we get the ratio of these energies as

$$U_d = -\frac{3R^6}{4d^6}\hbar\omega_M, \quad U_{irr} = -\frac{4R^6}{9d^6}\left(-\frac{R}{D_M}\right)^2 E^2\alpha_0, \quad U_{irr}/U_d = \frac{16E^2\alpha_0}{27\hbar\omega_M}\left(-\frac{R}{D_M}\right)^2,$$

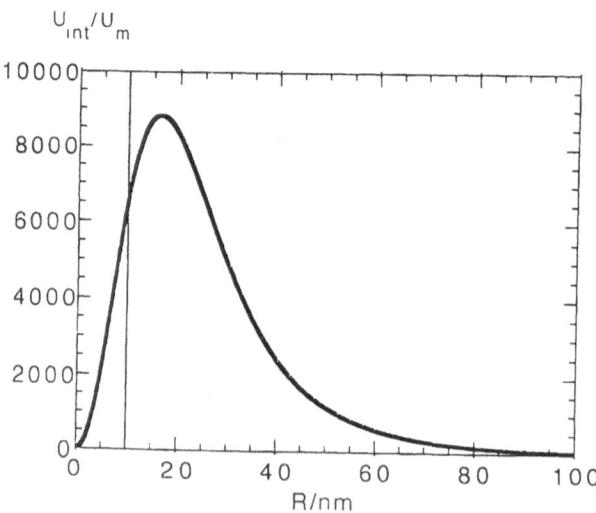

Fig. 12 Size dependence of the interaction energy. Parameters used: $R = 10$ nm, $\lambda_M = 500$ nm. (From K. Kimura, *Bull. Chem. Soc. Jpn.*, **69**, 323 (1996)).

where R is radius of particles, d interparticle distance, $\hbar\omega_M$ Mie resonance energy, E field strength of electromagnetic wave, D_M effective distance of spilling-out electrons into medium (~1 Å)[i], α_0 polarizability of particles. As a result, the interaction is a function of the square of the size of particles and the power of a field. When the dephasing effect that can be incorporated by expanding the damping term in R/λ (λ; wave length) is taken into account, the interparticle interaction is also expressed by the size or particles.[12] Figure 12 stands for the interparticle interaction energy normalized to the molecular value (U_m) as a function of the size of particles. The interaction energy has its maximum at about 20 nm and gradually decays for larger sizes. Hence the system must be of mesoscopic scale for a large enhancement to be observed.

6. Control of dispersion for metallic nanocolloids

In the study of the size effect of ultrafine metallic particles, it is desirable to use isolated particles in order to obtain information on the electronic state of a single particle. However, sometimes particles in contact with each other are used, leading to complicated problems concerning the particle-particle interface contacts. Therefore, it is very important to control the dispersion-coagulation of the nanoparticles in a liquid. The author has reported on several methods for obtaining stable dispersions of nanometer-sized particles in pure organic liquids.[13] These methods were based on the evaporation of metals in flow gas stream combined with a trapping medium for particles such as cryogenic matrix or cooled organic liquid. So far, these techniques have been applied to more than a hundred combinations of metals and liquids.

6.1 Preparation of contact-free Ag nanocolloids by aerosol process

Highly concentrated state (10^{21}–10^{22} particles/m^3) can be achieved in hexane as a solvent for particles with 2.7 nm-diameter without aggregation by the gas flow-cold trap method combined with cationic surfactants such as (2-dodecylhexadenyl)trimethylammonium chloride (g-C28TAC).[14] The preservation of the particle size distribution in the concentrating and drying process has been confirmed by electron microscopy. However, the particle size changes when other kinds of surfactants such as anionic ones are used. Contact-free silver clusters 1 nm in diameter were prepared in hexane with g-C28TAC by a modified gas cold trap method.

6.2 Isolation and agglomeration of metallic nanocolloids

The gas flow-cold trap method was also applied to obtain nanoparticles of palladium carbide, a meta-stable phase with carbon atoms on interstitial sites of metallic lattice.[15] This was synthesized by a single process *in situ* reaction in contact with acetone vapor. The resultant particles are in the state of colloidal sols. The average size was determined to be 5 nm by transmission electron microscopy. Zinc ultrafine particles were prepared by the gas evaporation-matrix isolation method in hexane and acetone.[16] The colloidal state changed as flocculation or precipitation depending on the solvent used. The contact of particles was checked by ESR spectroscopy, which reflects the degree of contact on the spectra.

6.3 Preparation of Si nanocolloids

Silicone nanoparticles are a current topic due to their potentiality photoluminescence. The nanometric Si was prepared by an Ar ion sputtering method combined with a solution trap method[17] using 2-propanol as the dispersing reagent. The author observed, for the first time, absorption spectra of Si nanoparticle dispersions overlapping a fine structure in the visible/UV region which can be correlated with the Si-O-Si stretching vibration. The spectra depended strongly on the size of particles.

PUBLICATIONS

1. a: K. Kimura and S. Bandow, *Phys. Rev.*, **B37**, 4473 (1988).
   b: K. Kimura, *Mat. Res. Soc. Symp. Proc.*, **272**, 193 (1992).
2. T. Ida and K. Kimura, *Z. Phys.*, **D26**, S140 (1993).
3. S. Bandow and K. Kimura, *Z. Phys.*, **D19**, 271 (1991).
4. K. Kimura, *Phys. Lett.*, **A158**, 85 (1991).
5. K. Kimura and N. Satoh, *Chem. Lett.*, 271 (1989).
6. N. Satoh and K. Kimura, *Chem. Lett.*, 2155 (1994).
7. N. Satoh, A. Naito and K. Kimura, *BulL. Chem. Soc. Jpn.*, **68**, 2151 (1995).
8. T. Ida, H. Saeki, H. Hamada and K. Kimura, *Surf. Rev. Lett.*, **3**, 41 (1996).
9. H. Hasegawa, N. Satoh, K. Tsujii and K. Kimura, *Z. Phys.*, **D20**, 325 (1991).
10. N. Satoh, H. Hasegawa, K. Tsujii and K. Kimura, *J. Phys. Chem.*, **98**, 2143 (1994).
11. K. Kimura, *J. Phys. Chem.*, **98**, 11997 (1994).
12. K. Kimura, *Bull. Chem. Soc. Jpn.*, **69**, 321 (1996).
13. a: N. Satoh, S. Bandow and K. Kimura, *J. Colloid and Interface Sci.*, **131**, 161 (1989).
    b: N. Satoh and K. Kimura, *Bull. Chem. Soc. Jpn.*, **62**, 1758 (1989).
14. S. Tohno, M. Itoh and K. Kimura, *J. Chem. Soc. Jpn.*, (in Japanese). (11) 1027 (1994).
15. T. Yamamoto, M. Adachi, K. Kawabata, K. Kimura and H. W. Hahn, *Appl. Phys. Lett.*, **63**, 1 (1993).
16. K. Kimura and S. Bandow, *J. Colloid & Interface Sci.*, **171**, 356 (1995).
17. Y. Zhu and K. Kimura, *Chem. Lett.*, 643 (1995).

REFERENCES

a. R. Kubo, *J. Phys. Soc. Jpn.*, **17**, 975 (1962).
b. N. Wada and M. Ichikawa, *J. Appl. Phys.*, **15**, 755 (1976).
c. A. J. Cox, J. G. Louderback, S. E. Apsel and L. A. Bloomfield, *Phys. Rev. Lett.*, **71**, 923 (1993).
d. D. C. Douglass, J. P. Bucher and L. A. Bloomfield, *Phys. Rev. Lett.*, **68**, 1774 (1991).
e. L. A. Bloomfield, *ISSPIC7 Abstracts*, p.32 (1994).
f. W. A. de Heer, P. Milani and A. Chatelain, *Phys. Rev. Lett.*, **65**, 488 (1990).
g. W. A. de Heer, *ISSPIC7 Abstracts*, p.31 (1994).
h. J. P. Bucher, D. C. Douglass and L. A. Bloomfield, *Phys. Rev. Lett.*, **66**, 3052 (1991).
i. A. Liebsch, *Phys. Rev.*, **B48**, 11317 (1993).

17

Electron Correlation within Fine Particles and Quantum Dots

Kiyoshi Kawamura, Mikio Eto, Katsuki Amemiya, Takayuki Mizuno, Fumiko Yamaguchi and Norikazu Urata

Department of Physics, Faculty of Science and Technology, Keio University
Kohoku-ku, Yokohama 223-8522, Japan

Purpose of the present study

The physics of mesoscopic systems has developed rapidly since the start of the present program. In particular, the boundary between the physics of fine particles and the physics of quantum dots has disappeared. In this situation, the purpose of the present study has shifted gradually and continuously, since theoretical study is always adapted to the worldwide trend of researches. This feature of theoretical study is very different from that of experimental studies which are strongly influenced by the cost of research facilities. Nevertheless, we have continuously kept the viewpoint that we should study the correlation among coherent electrons within conductors of mesoscopic size, *i.e.* the size between macroscopic and microscopic matter to provide a new type of image concerning conductor materials.

Contents of the present study

1. Introduction

The theory of metallic fine particles began with the paper by Kubo[a] in 1962, in which he pointed out that (i) discreteness of energy level distribution affects the thermodynamic properties of small particles at helium temperature and (ii) canonical ensemble should be employed instead of grand canonical ensemble because the Coulomb interaction is as large as the thermal energy. These two points are related by two key words which play an important role in the physics of mesoscopic conductors, that is, the coherence of electron waves and the charging energy of a small conductor. Coherence of the electron wave in conductors of mesoscopic size has drawn attention ever since the Aharonov-Bohm effect was discovered in the electron conduction of mesoscopic systems. When the coherence is important, we must have precise knowledge about wave functions in individual conductors for which physical quantities are measured. In fact the samples used in experimental studies often have a complicated shape instead of the cubic boxes in the theory of bulk conductors and the wave function is influenced by boundary conditions.

In sections 2 and 3 the dependence of the Hall resistance on the shape of Hall elements is studied. In such studies we have solved the Schrödinger equation for electrons in the presence of a magnetic field within the Hall elements of quite complicated shape. We give below two examples in which we have employed the boundary element method, which is an algorithm to solve the Schrödinger equation with arbitrary geometry of boundaries. We believe our study to be the first and best application of such an algorithm to the study of electron conduction in mesoscopic systems.

When the perturbation theory is valid for analyzing influences of the Coulomb interaction, we employ the Hartree-Fock approximation. Although a more sophisticated method like the local density functional method is often used in condensed matter physics, we believe that the Hartree-Fock approximation is sufficient to analyze certain manybody effects in mesoscopic systems.

We will study optical absorption in sections 4 and 5. We find the Hartree-Fock approximation powerful for explaining the experiments and that it provides a guiding principle for future experimental studies. The study of the density of states within a quantum dot in the same approximation presented in section 6 is also useful in the study of strong Coulomb interactions.

In sections 7 and 8 nonequilibrium currents through a quantum dot are discussed. When electrons enter a small conductor which can accommodate a small number of electrons, the Hartree-Fock approximation is not valid. This is because the number of electrons is an integer but the number is not necessarily an integer in the Hartree-Fock approximation. In such cases, the Hubbard model and the Anderson model have been employed to study the electron system in a nonperturbational way. We will present in sections 7 and 8 typical examples of studies of I–V characteristics in terms of the Anderson model in the Hubbard approximation.

We will apply the Friedel sum rule to studies of quantum states of electrons within a quantum dot. The Friedel sum rule was applied for the first time by Kawabata [n] to analyze the influence of the Coulomb interaction on linear conductance of a symmetric quantum dot. Since an actual quantum dot is more or less asymmetric, the study of the influence of the asymmetry is important. We will present in section 9 an extension of the Friedel sum rule to asymmetric scatterers.

We will consider the coherence of electronic waves and the Coulomb correlation in various scenes but we have not succeeded in the study of strong Coulomb correlation in arbitrary shape of mesoscopic conductors. This remains a problem for future studies.

2. Electrons in Hall elements [1]

The Hall effect of a bulk conductor is well known and its theory is almost established except for some problems concerning the quantum Hall effect. However, there remain various unsolved problems about the Hall resistance of conductors of mesoscopic size. In particular, the explanation of observed Hall resistance using the quantum electronic states within Hall elements is open to question, since the computation of wave functions in the presence of a magnetic fields is difficult.

Figure 1(a) represents a Hall element of the simplest geometry. The four branches of the cross will be called the wave guides, hereafter. A voltage V appears in the vertical direction when a current I flows in the horizontal direction. The ratio $R_H = V/I$ is called the Hall resistance. In experiments on the Hall

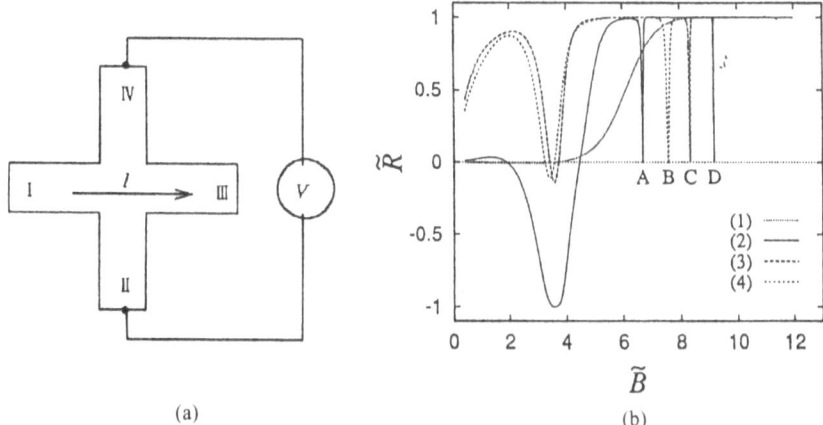

(a)   (b)

Fig. 1   (a) A Hall element of perfect cross geometry. Wave guides I to IV are defined. (b) Scaled Hall
resistance versus scaled magnetic induction for a Hall element of a perfect cross geometry (1),
with a round junction in which the corners of a junction are cut off (2) [See Fig. 3(a)], and
with an antidot at a center of the round junction (3 and 4) [See Fig. 3(b)].

resistance, a volt meter is connected with the wave guides II and IV. However,
when we qualitatively discuss the Hall resistance, it is useful to imagine that a
current can also flow in the vertical direction. Then, when a magnetic field is
applied from the back to the surface of the paper, the Hall resistance has a normal
sign if electrons coming from the wave guide I go out to the wave guide II after
a right turn. If, on the other hand, electrons go out into the wave guide IV for a
certain reason, the Hall resistance has a negative sign.

The Hall resistance R_H can be computed by invoking the Landauer-Büttiker
formula[b)]

$$R_H = \frac{h}{e} \times \frac{T_R - T_L}{e(T_R + T_P)^2 + (T_L + T_P)^2}, \tag{1}$$

when electrons come from the wave guide I, T_R, T_L, T_F in this formula represents
the transmission probability into the wave guides II, IV and III, respectively.

These transmission probabilities can be computed from the solution of the
Schrödinger equation under the boundary condition that electrons come in the
junction exclusively from the wave guide I. The computed Hall resistances are
plotted in Fig. 1(b) for four types of Hall elements.

In Fig. 1(b), as well as in the following discussion. the Hall resistance and the
magnetic induction are scaled by (h/e^2) and (h/ed^2), where d is the width of the
four electronic wave guides, and scaled quantities are indicated by tildes.

The scaled Hall resistance R_H of the perfect cross geometry shown in Fig. 1(a)
is almost quenched (see the curve in Fig. 1(b)) when the magnetic field is weak,
while its value jumps to unity when the magnetic field becomes stronger than a
certain value.

The quenching of the Hall resistance is experimentally known. The behavior
that the scaled Hall resistance takes the value (h/e^2) divided by an integer is
called the quantization of the Hall resistance.

Contours of the probability distribution of electrons are presented in Figs. 2(a)

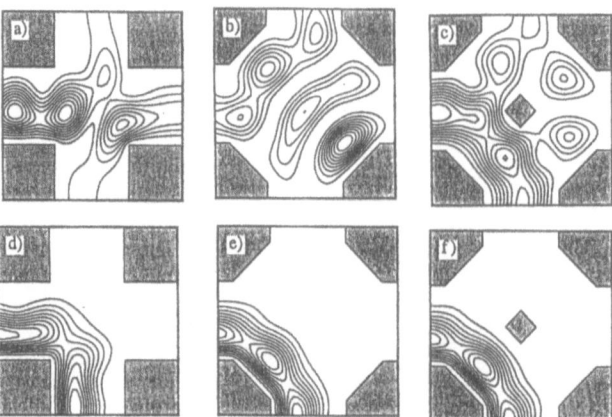

Fig. 2   The probability densities within Hall junctions when the Hall resistance is suppressed (a), negative valued (b), and enhanced (c). The Hall resistance is quantized, electronic streams are not disturbed by the junctions (d to f).

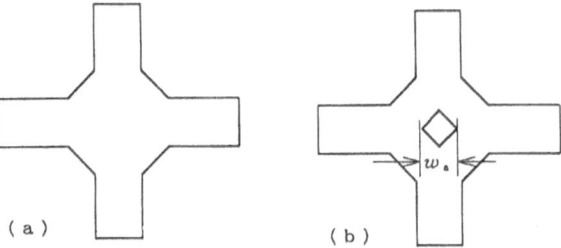

Fig. 3   A model of a device for which the Hall resistance is negative valued (a) and enhanced (b).

and (d), to interpret these behaviors of the Hall resistance in terms of the wave function. Figure 2(a) represents contours when the Hall resistance is quenched in the presence of a weak magnetic field ($\tilde{B} = 4.0$). In this case, electrons which flow from the wave guide I cannot turn the corner and enters the wave guide III in the forward direction. The wave function for the case of $\tilde{B} = 10.0$ is represented in Fig. 2(d), which implies that electrons can completely turn at the corner since the cyclotron radius is very small.

When the corners of the junction of the perfect cross are cut off as shown in Fig. 3(a), the Hall resistance behaves as the curve (2) in Fig. 1(b). The Hall resistance becomes negative when $\tilde{B} = 4.0$. Such a negative Hall resistance was experimentally observed by Ford et al.[f] The curve also exhibits a sharp dip, which we will discuss later.

The probability density of electrons in the Hall element shown in Fig. 3(a) in a magnetic field of $\tilde{B} = 3.6$ is represented in Fig. 2(b). Electrons which come in from the wave guide on the left-hand side are reflected by the wall on the right-below and go out to the wave guide IV on the top. The Hall resistance is negative in this case. When $\tilde{B} = 10.0$, the Hall resistance is quantized because of the complete turn of electrons, as shown in Fig. 2(d).

Furthermore, an antidot, on which electrons cannot exist, is placed at the center of the junction as Fig. 3(b). The Hall resistance of such elements is represented by curves (3) and (4) in Fig. 1(b), which are the Hall resistance of

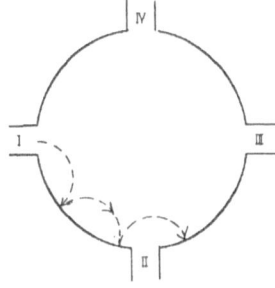

Fig. 4 A circular junction of a Hall element in which an electron exhibits skipping motion (the dashed line).

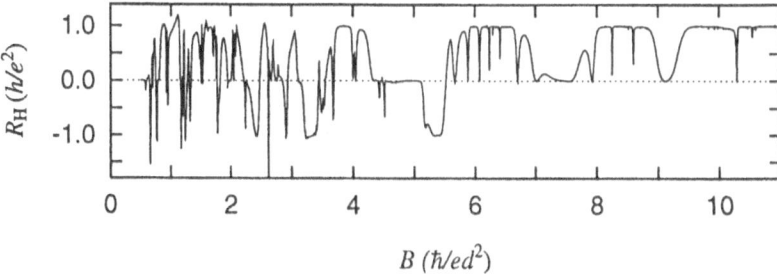

Fig. 5 The Hall resistance of an element shown in Fig. 1 exhibits suppression in wide range of magnetic induction.

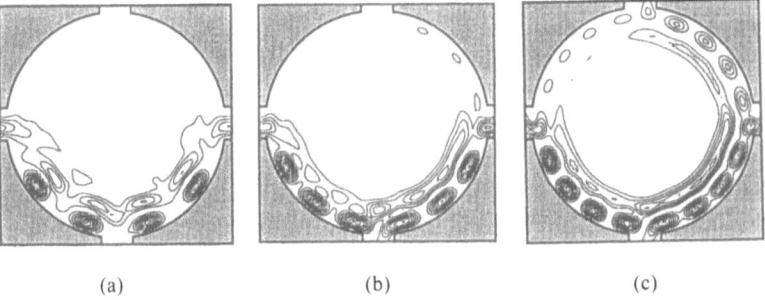

(a)   (b)   (c)

Fig. 6 The probability densities when the Hall resistance is suppressed in the wide range of magnetic induction.

elements with antidots of the size $w_a = 0.4d$ and $0.6d$ respectively. The Hall resistance is enhanced in both cases in agreement with the experiment of Ford *et al*. Even when a magnetic field is weak, electrons are forced to turn in the right direction by an antidot, as shown by the probability distribution for $\tilde{B} = 2.1$ in Fig. 2(c). The contours of probability distribution in such a Hall element for the case of $\tilde{B} = 10.0$ are not influenced by the antidot, as shown in Fig. 1(f).

For further studies of the Hall resistance, we assumed a circular device with large size, as shown in Fig. 4. We choose a circular structure instead of cubic one since such a quantum dot of higher symmetry is more fundamental.

The computed Hall resistance is plotted in Fig. 5. Behavior in low magnetic fields is chaotic so that we consider only the region of higher magnetic fields. It is seen that the Hall resistance exhibits suppression in rather broad regions of magnetic field as well as the sharp dips of the same nature as the curves in Fig.

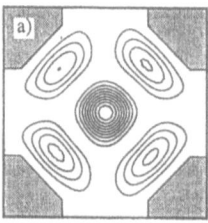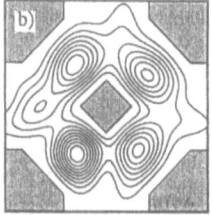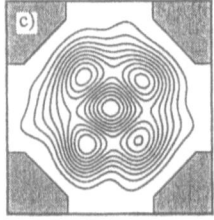

Fig. 7 The probability densities when the Hall resistance exhibits sharp dips as shown in Fig. 1 (b).

1(b).

In order to study such suppression of the Hall resistance in broad regions, the contours of probability densities of electrons are computed and represented in Figs. 6(a), (b) and (c). These figures imply that maxima of the probability density are arranged periodically along the edge of the circle. Let us suppose that these maxima of probability density correspond to the points where classical electrons collide with the wall. Then classical orbits which correspond to the probability densities in all three figures pass over the exit to the wave guide II but they terminate at the exit to the wave guide III. Consequently, the Hall resistance is suppressed as the quenching of the Hall resistance in Fig. 1 (b). This picture is confirmed from the fact that the distance from the entrance I to III along the edge is 2.4, 3.6 and 4.5 times the cyclotron diameter in each magnetic field.

Similarly, it is expected that when the distance along the edge from the entrance I to IV is an integer multiplied by the cyclotron diameter, electrons cannot go out from the exits II and III and they terminate at the exit to the wave guide IV. If this is the case the Hall resistance is expected to be negative valued when $\tilde{B} = 10.0$. This is confirmed by Fig. 5(a).

In this way, comparing the wave functions in a quantum dot with classical orbits, the suppression of the Hall resistance in broad ranges is explained. However, theory should invoke the concepts of the edge states to analyze these behaviors by genuine quantum mechanics. This has not been done yet.

3. Bound states within Hall elements [2]

In order to study the dips in the high magnetic region of the curves in Fig. 1(b), the wave functions are computed for values of the magnetic field for which the Hall resistance is sharply quenched. The probability densities represented in Fig. 7 exhibit almost circular structures. Contours of probability densities continue into wave guides but they are not seen because amplitude within the quantum dot is much larger than that within wave guides. This behavior of electrons is similar to acoustic resonance in musical instruments. The weak electronic flux which enters the quantum dot excites a stationary wave of large amplitude within the quantum dot.

Since those probability densities have fourfold rotational symmetry, these resonance states are combined almost equally to electronic states within the four wave guides. Consequently, the relation

$$T_{\mathrm{L}} = T_{\mathrm{R}} = T_{\mathrm{F}} \tag{2}$$

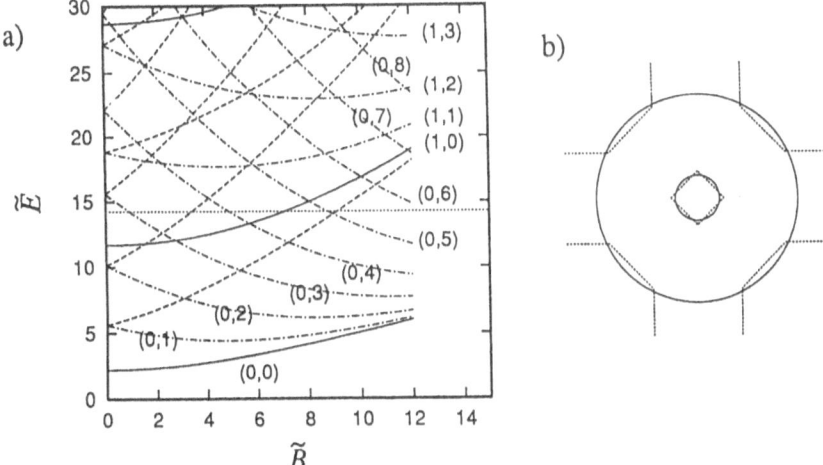

Fig. 8 (a) Energy levels of electrons in the presence of a magnetic field. (b) Circular potential barriers model the Hall junctions. Electrons whose energies are plotted in (a) are confined within the outer circle.

holds. The substitution of this relation in (1), yields $R_H = 0$.

Resonance takes place when oscillation in an external system weakly couples with eigenmodes within the resonator. In the present problem, the resonator is the octahedral junctions. We replace these octahedra with circles for the reason mentioned above. We consider the eigenstates of electrons confined within a disk in the presence of a magnetic field whose direction is perpendicular to the disk.

The Schrödinger equation

$$\frac{1}{2m}(p - eA)^2 \psi = E\psi \tag{3}$$

is separated into two ordinary differential equations, that is, the one in an angular variable and the other in a radial variable. The latter equation is transformed into a hypergeometric differential equation.

This equation was solved under the boundary condition that ψ should vanish on the outer boundary. This implies that the circle is a potential barrier of infinite height. The radius of the circle is determined so that the circle approximates the octahedron in the way indicated in Fig. 8(b). The antidot at the center of a certain Hall junction is modeled by another infinite circular potential barrier.

The eigenvalues are plotted in Fig. 8(a) as functions of applied magnetic fields for electrons confined only by a single circular potential barrier. Curves are labeled by (n, l) where n is a quantum number concerning the radial wave function and l is an eigenvalue of the angular momentum. Curves corresponding to negative valued l are not labeled but their labels can be easily assigned because the states (n, l) and $(n, -l)$ are degenerate in the absence of a magnetic field.

When B is small, the radial function changes slowly and is sensitive against the boundary condition. When B increases, the cyclotron radius becomes much smaller than the size of the confined region and almost all of the quantum states are those which correspond to cyclotron orbits in a bulk conductor of two dimensions. In this limit, the energy eigenvalues are expressed as

Table 1

Without Antidot		Radii of Antidots	
		0.233d	0.156d
7.07	9.11		
(n = 1, l = 0)	(n = 0, l = −1	7.58	8.34
$\tilde{B}_A = 6.72$	$\tilde{B}_D = 9.16$	$\tilde{B}_B = 7.68$	$\tilde{B}_C = 8.34$

$$E_{n,l} \rightarrow \omega_c \hbar \left(n + \frac{|l| - l}{2} + \frac{1}{2} \right) \tag{4}$$

In fact the curves in Fig. 8 asymptote to this limit. Since electrons which contribute to the Hall resistance have the Fermi energy E_F. electronic states which resonate lie on the energy level determined by

$$E_{n,l} = E_F . \tag{5}$$

Since $E_{n,l}$ takes discrete values, this condition is satisfied by some specific values of B.

The values of $\tilde{B}$ which satisfy (5) are computed for three disks and are listed in the upper column in Table 1. The radius of outer circles of all of the three disks is 1.14d. One disk has no inner circular potential barrier. The other two disks involve circular antidots at their center and their size is indicated in the table. In the lower column, the position in B of dips A through D in Fig. 1(b) are presented. The good agreement implies that the dips seen in Fig. 1(b) are caused by resonance with electron states confined within the junctions.

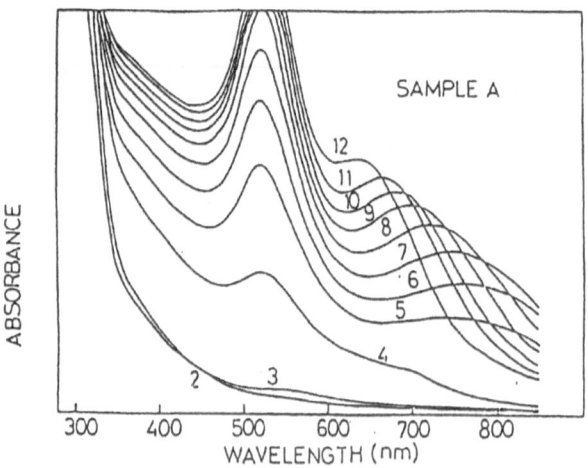

Fig. 9   Photo absorption of silicon particles coated by gold dispersed in a certain dielectric medium.
(After Ref. i)

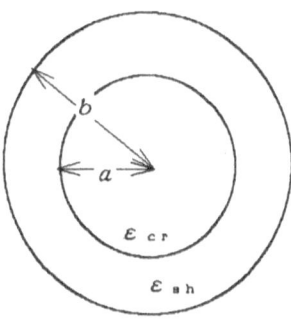

Fig. 10 Model of a spherical particle of a dielectric medium coated by a metallic shell within a dielectric solution.

4. Hartree approximation for electrons confined in metallic shell [3]

The implication of the line width of absorbance of metallic small particles was explained by Kawabata and Kubo [h] in the sixties.

Recently, absorbance of insulating particles coated by metal has been measured [i] and it has been observed that the line shape changes with the change of the metallic coat (see Fig. 9). The absorption spectrum consists of two peaks which mutually close with increase in the thickness of the metallic coat.

We have analyzed this behavior of the absorption spectrum to investigate the validity of the Hartree approximation for the study of mesoscopic conductors.

A model of the particle is shown in Fig. 10. An insulating sphere of the radius a and the dielectric constant ε_{cr} is surrounded by the spherical shell of a conductor of electric susceptibility $\chi(\omega)$. The outer radius is denoted by b. These coated spheres are suspended in a dielectric medium with the dielectric constant ε_m. The diameter of the particle is assumed much less than the wave length of visible lights.

The absorbance is proportional to $\Pi(\omega)$, which is expressed by

$$\Pi(\omega) = \frac{\omega}{2\pi} \Im\left[\int P(r)^* \cdot E_{ex}(r) dr\right]. \tag{6}$$

Here, $P(r)$ is a polarization vector which is proportional to the internal local electric field $E(r)$:

$$P(r) = \chi(\omega) \, E(r). \tag{7}$$

The relation between the internal field E and the external field E_{ex} is determined by the boundary condition of the dielectric medium. Consequently, we obtain

$$\Pi(\omega) = \frac{\omega}{2\pi} \, \Omega \Im[A(\omega)]|E_{ex}|^2 \tag{8}$$

where we have introduced

$$A(\omega) = \frac{3\varepsilon_m (\varepsilon_{sh} + 2\varepsilon_m)}{(2\varepsilon_{sh} + \varepsilon_{cr})(\varepsilon_{sh} + 2\varepsilon_m)f + 3\varepsilon_{sh}(\varepsilon_{cr} + 2\varepsilon_m)(1 - f)} \tag{9}$$

The dielectric function of the metallic shell is denoted by $\varepsilon_{sh}(\omega)$, which is expressed in terms of the electric susceptibility of the metal as $\varepsilon_{sh}(\omega) = \varepsilon_0 + \chi(\omega)$. In the above formula, $f = (b^3 - a^3)/b^3$ represent the volume fraction of the metallic part in the coated particle.

When the denominator is equated to zero, we obtain a quadratic equation for ε_{sh}. Its solutions, which are denoted by C_1 and C_2, are independent of ω. In terms of these two solutions, (8) is rewritten as

$$\Pi(\omega) = \frac{\omega\Omega}{2\pi} \Im\left[\frac{D_1}{\varepsilon_{sh}(\omega) - C_1} + \frac{D_2}{\varepsilon_{sh}(\omega) - C_2}\right] E_{ex}^2 . \tag{10}$$

The quantities D_1 and D_2 in this formula are independent of ω.

The quantity ε_{sh} on the right-hand side is expressed in term of the polarizability $\alpha(\omega)$ as

$$\varepsilon_{sh}(\omega) = \varepsilon_0 + \frac{\alpha(\omega)}{\Omega}. \tag{11}$$

We calculated $\alpha(\omega)$ from the eigenenergies and eigenfunctions of electrons confined within the spherical shell. Then we find

$$\alpha(\omega) = -\left(\frac{\omega_p}{\omega}\right)\left[2\varepsilon_0 + \frac{8e^2ab}{3\pi^2\Omega}\, g\left(\frac{\hbar\omega}{E_F}\right)\right] \tag{12}$$

where $g(x)$ is a complex function whose magnitude is almost unity and very weakly depends on x.

We note that the polarizability $\alpha(\omega)$ which appears in (12) is that of an independent electron system confined in the spherical shell and that the Coulomb interaction is not included in its calculation. The Coulomb interaction is taken into account in the depolarization field due to polarization charge.

The relation between the polarization vector and the electric field which we find in textbooks on electromagnetism is local relation. In a mesoscopic system, this relation should be nonlocal. Therefore, in a more rigorous theory the

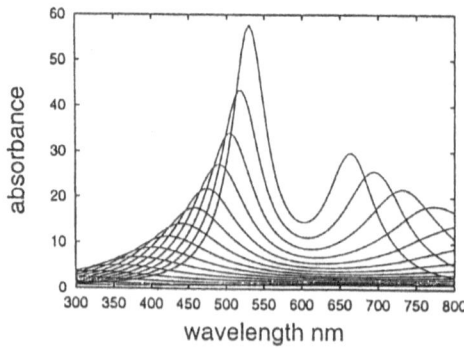

Fig. 11 Photo absorption spectra of a metallic shell computed from (10).

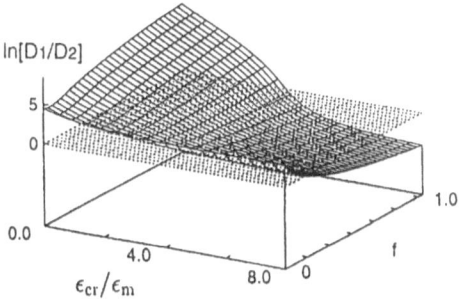

Fig. 12 The ratio of the height of the right peak to that of the left peak as functions of dielectric constants of the core material and dielectric medium.

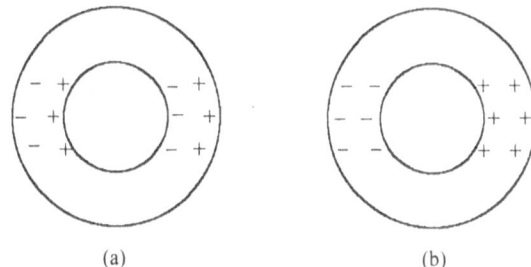

(a)   (b)

Fig. 13 The high frequency mode (a) and the low frequency mode (b) within a metallic shell.

polarization vector due to the external field should be represented by using the Kubo formula. When the effects of the Coulomb interaction is taken into account in RPA, we can take it in the same level as the electromagnetism.

We have computed the absorption spectrum by using (10), (11), and (12). The results are shown in Fig. 11. The agreement between the experiment and the present theory is quite good.

The size effect is represented by the prefactor to g on the right-hand side of(12). Since the volume Ω of the spherical shell is proportional to abd, in which d $(= a - b)$ represents the thickness of the shell, this prefactor is proportional to $(1/d)$. In particular, the line width, which is proportional to the imaginary part of g, is proportional to $(1/d)$.

It is also found that the two peaks approach each other mutually with increase in thickness. This does not come from the prefactor to g but from the size effect of C_1, C_2 on the right-hand side of (10).

In our theory, the relative height of the two peaks is determined by the ratio (D_1/D_2) whose expression is also known. Here, D_1 and D_2 represent the height of the peak on the right- and left-hand sides respectively. The ratio is plotted in Fig. 12. Note that the vertical axis represents the logarithm of (D_1/D_2). This suggests that the relative height of the two peaks changes appreciably according to the combination of a dielectric spherical core and a medium in which particles are suspended. This will be easily confirmed by experiment. At moment, optical absorption has not been measured with the exception Ref. i, but we hope experimental studies will be conducted for various combinations of dielectric media.

We can find eigenmodes of oscillation of electrons within a conducting shell in the literature.[1] The peak on the higher frequency side (on the left-hand side) is due to excitation of the mode in Fig. 13(a), whereas the peak on the lower frequency side is due to excitation of the mode in Fig. 13(b).

5. First principle theory of photo absorption of conducting shell [4]

The theory in the last section invokes the electromagnetism of dielectric media, which is not valid for a system of coherent electrons in fine particles. In a rigorous theory, P in (6) should be expressed by using the Kubo formula as a function of E_{ex}.

According to the Kubo formula, its expression is

$$P(r) = \int \langle\langle P(r); P(r') \rangle\rangle_\omega \cdot E_{ex}(r)dr' \tag{13}$$

where $\langle\langle P(r) ; P(r') \rangle\rangle_\omega$ is the Fourier transform of the retarded Green function. Substitution of this in (13) yields the following integral equation

$$\int E_{ex}(r) \cdot \langle\langle P(r); P(r') \rangle\rangle_\omega \cdot E_{ex}(r')drdr' = \sum_{nm} (P_z)_{nm} (P_z)_{sr} \langle\langle a_n^\dagger a_m ; a_s^\dagger a_r \rangle\rangle_\omega \tag{14}$$

where bases are appropriate one-electron states. The equation of motion for the Green function $\langle\langle a_n^\dagger a_m ; a_s^\dagger a_r \rangle\rangle$ is generally complicated, but it is written in RPA as follows.

$$\langle\langle a_n^\dagger a_m ; a_s^\dagger a_r \rangle\rangle_\omega = G_{nm}^{(0)}(\omega)\delta_{nr}\delta_{ms} + \sum_{pq} G_{nm}^{(0)}(\omega)(V_{pmqn} - V_{pmnq})\langle\langle a_p^\dagger a_q ; a_s^\dagger a_r \rangle\rangle_\omega \tag{15}$$

where we have introduced

$$G_{nm}^{(0)}(\omega) = \frac{f_m - f_n}{\hbar\omega - (E_n - E_m) - i\delta}. \tag{16}$$

The one-electron states which were employed as bases are the wave function in

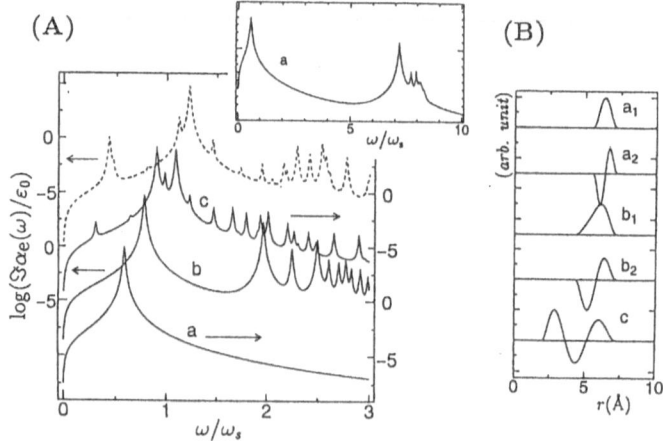

Fig. 14 (A) Photo absorption of a metallic shell by using the Kubo formula. (B) Induced electric charges within a metallic shell.

the Hartree-Fock approximation when RPA is employed to calculate the retarded Green function.

The absorption spectrum, which is obtained by solving (15), is plotted in Fig. 14(A). The outer radius is fixed to 7.2 angstroms and the electron numbers are a:50, b:72, c:90. The parameter r_s, which represents the electron density, is fixed at three. In the spectrum labeled by "a", the second peak is out of scale so that the same spectrum is shown in the inset using a reduced horizontal coordinate. The broken line represents the absorption spectrum of a metallic sphere whose radius is the same as the outer radius of the shell.

Each spectrum consists of two peaks, both of which are shifted to the surface plasma frequency with increase in the thickness d of the spherical shell.

To identify the eigenmodes which are responsible for these peaks, induced electron density was computed and plotted in Fig. 14(B). The spectra in a_1 and a_2 are induced electronic charges in the peaks on the lower and higher sides, respectively. The spectra in b_1 and b_2 have the same meaning. At first sight, the spectrum (c) also has two peaks, but it is suggested from the plot of the induced electron density that they are parts of a single peak. In fact electron density exhibits Friedel oscillation in the vicinity of the peak in (c).

The absorption spectra calculated by using such quantum mechanics is decomposed into line spectra, instead of a broad line calculated by using the classical electromagnetism. The low peaks are interpreted as satellite lines belonging to two main peaks. The rang of distribution of these satellite lines is expected to be the line width in the classical theory. The line width defined in this way is almost (v_F/d) where $d = b - a$ represents the thickness of the sphere and this has been confirmed in this study.

6. Electronic states in a quantum dot in the Hartree-Fock approximation [5]

Since the validity of the Hartree-Fock approximation has been verified in mesoscopic systems, we apply it to the study of electrons in a quantum dot.

When we talk about a quantum dot, we imagine the device which is illustrated in Fig. 15(a). Electrons can pass the narrowed regions only through tunneling

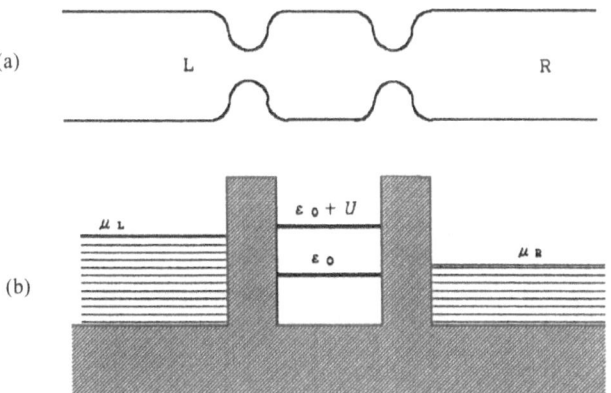

Fig. 15 An illustration of a quantum dot connected to electric leads (a) and the energy configuration when the Coulomb repulsion is strong enough within the quantum dot (b).

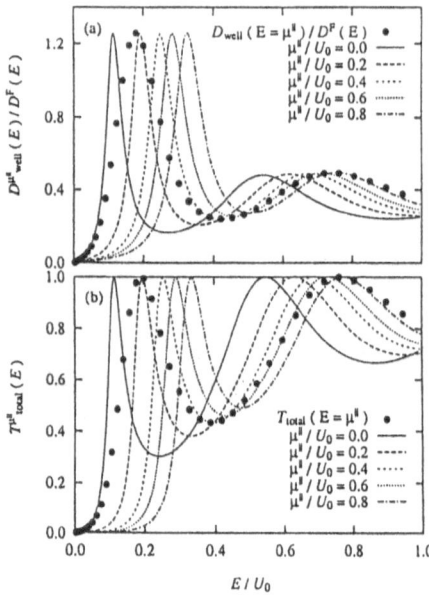

Fig. 16 (a) The local density of states as a function of the energy of an incident electron for various values of Fermi energy within the electrodes. (b) Similar plots of the transmission probability.

processes. Consequently, a quantum dot is equivalent to a space between two potential barriers, as shown in Fig. 15(b).

According to Landauer, conductance of a quantum dot in the linear approximation is proportional to the transmission probability from the electric lead on the left-hand side to the one on the right-hand side.

Calculation of the transmission probability in the one-electron approximation is nothing but textbook physics and is performed by solving the Schrödinger equation, which involves a double barrier potential.

We have studied the transmission probability assuming that electrons mutually interact when and only when they stay within an quantum dot. This is because inhomogeneous electron distribution due to a transfer between an electric lead and the quantum dot may be restricted, within an electric lead, to the region near potential barriers. On the other hand, the area of the quantum dot is so small that Coulomb interaction cannot be screened by other electrons.

The Green function formalism is useful to study manybody effects. However, we should notice that the unperturbed state is in an inhomogeneous potential field with double barriers. Because of this situation, the problem has an aspect which is different from the ordinary manybody problem in a bulk system.

We have begun looking for an appropriate from of the unperturbed Green function. The unperturbed Green function depends on two coordinates since it represents propagation of electrons within an inhomogeneous space. The Green function in the interacting system is written in perturbation series with respect to the Coulomb interaction, which works only when electrons stay within the quantum dot. The manybody effect is taken in the self-energy part in the Hartree-Fock approximation.

The transmission probability includes a factor which represents the density of states within the quantum dot. When electrons are mutually independent, electronic states within the dot are independent of electron number within the dot. When electrons interact each other, the density of states depends on the Fermi energy, since the Coulomb potential includes occupation number of other electrons.

Curves in Fig. 16(a) represent the density of states within the quantum dot for various values of the Fermi energy. The horizontal axis represents the electron energy scaled by the height of a potential barrier. This height is appropriately assumed so that two resonance levels are located in the dot in the absence of the Coulomb interaction.

The peak position of each curve is called the resonance level and the width of the peak is inversely proportional to the life time of the resonance. The density of states at the Fermi level is indicated by dots. A curve which combines these dots represents the density of states as a function of the Fermi energy.

The transition probability through a quantum dot is plotted in Fig. 16(b) with a thin line as a function of the Fermi level. The dots represents the transition probability at the Fermi level. Among the two peaks of the density of states, the peak on the higher energy side is lower since the energy of the peak position is near the barrier. On the other-hand, the transmission probability at the peak is just unity, as shown in Fig. 16(b).

7. Strong correlation in a quantum dot [6]

As mentioned above, the Coulomb correlation within a quantum dot causes a phenomenon called the Coulomb blockade. The Anderson model is useful to discuss this problem.

It was shown in the previous section that resonance levels rise within a quantum dot defined by double potential barriers. When the quantum dot was separated from electric leads with infinitely high potential barriers, bound states would be formed within the dot. We represent the energy of the lowest bound state by ε_0, the Coulomb energy stored in the dot by occupation of two electrons as U and electron energies in the two electric leads as E_{Lk}, and E_{Rk}. In terms of these quantities, the Hamiltonian of the electron system is written as follows:

$$\mathcal{H}_0 = \sum_{ik\sigma} E_{ik\sigma}\, a^\dagger_{ik\sigma} a_{ik\sigma} + \varepsilon_0 c^\dagger_\sigma c_\sigma + U\, n_\uparrow n_\downarrow \tag{17}$$

where i stands for L or R.

When the height of the barriers is finite, electronic states within the electric leads and the quantum dots are mixed, and the resonance level has finite width. This mixing is modeled by

$$\mathcal{H}_T = \sum_{ik\sigma} \left(V_{ik}\, a^\dagger_{ik\sigma} c_\sigma + V_{ik}^* c^\dagger_\sigma a_{ik\sigma}\right). \tag{18}$$

The combination of both Hamiltonians is nothing but the one-dimensional Anderson Hamiltonian.

The operator of a current which flows through the barrier on the left-hand side is

$$I_L = -\frac{ie}{\hbar} \sum_{k\sigma} [V_{Lk} a_{Lk\sigma}^\dagger c_\sigma + V_{Lk}^* c_\sigma^\dagger a_{Lk\sigma}]. \tag{19}$$

The average of this operator up to the first order of $\mathcal{H}_T$ gives a second-order expression for the current. The expression for the current obtained in this way is e multiplied by the transition probability from propagating electron states in the electric lead to a localized state in the quantum dot.

The power spectrum of current fluctuation is given by

$$S_L(\omega) = \int_{-\infty}^{\infty} d(t - t')[\langle I_L(t) I_L(t') \rangle + \langle I_L(t') I_L(t) \rangle] e^{i\omega(t-t')} \tag{20}$$

The lowest order term in $\mathcal{H}_T$ is its second order, since I_L is proportional to $\mathcal{H}_T$. Therefore, the power spectrum can be expressed in terms of the eigenvalues and eigenfunctions of the unperturbed Hamiltonian. The resultant expression includes the Fermi functions, which define the electron distribution in both the electric leads, and the average electron number $\langle n_\sigma \rangle$ in the quantum dot. We can determine $\langle n_\sigma \rangle$ as a function of μ_L, and μ_R from the continuity of the electric current $I_L = I_R$. The electrochemical potential in the right electric lead, that is, μ_R, is fixed at 0 as an example so that the applied voltage is given by $V = \mu_L/e$.

Computed current and $S(0)$ as a function of V are plotted in Fig. 17. In the present case, there is no electron in the electric lead on the right-hand side, since $\mu_R = 0$. When $\mu_L < \varepsilon_0$, there is no occupied state in the electric lead on the left-hand side, whose energy level is the same as that in the quantum dot, and transmission from the electric lead to the quantum dot cannot take place. This implies that a current cannot flow when $V < \varepsilon_0/e$.

When $(\varepsilon_0/e) < V < (\varepsilon_0 + U)/e$, an electron with energy $E_{Lk} = \varepsilon_0$ in the electric lead on the left-hand side can make a transition to the right lead by using the lower level within the quantum dot. When $(\varepsilon_0 + U) < \mu_L/e$, electrons with energies $E_{Lk} = \varepsilon_0$ and $\varepsilon_0 + U$ can flow to the right lead, the current is twice that in the case of $(\varepsilon_0/e) < V < (\varepsilon_0 + U)/e$. These arguments give the interpretation of the stepwise structure of I–V curve in Fig. 17(a).

Figure 17(b) represents the component which is proportional to the temperature T, that is, the Johnson-Nyquist noise. This noise appears when the current changes abruptly. The difference from the total noise is plotted in Fig. 17(c). This portion of the noise is called the shot noise. The ordinary shot noise is proportional to the current, but the present shot noise vanishes when V takes a value near $(\varepsilon_0 + U)/e$. More precisely, the shot noise vanishes when only one electron is accommodated in the quantum dot and the condition $\Sigma_\sigma \langle n_\sigma \rangle = 1$ holds.

8. Kondo effect due to a quantum dot [7]

When the electrochemical potential of both electric leads is located between ε_0 and $\varepsilon_0 + U$ there is no current which is proportional to $(\mathcal{H}_T)^2$. This is true, so far as we consider only a current which is accompanied by real transition processes from an electric lead to the quantum dot.

However, when we allow higher order terms, there is a current which is due to the transition from the left electric lead to the right electric lead over various intermediate states. In this transition process, an electron tunnels two barriers

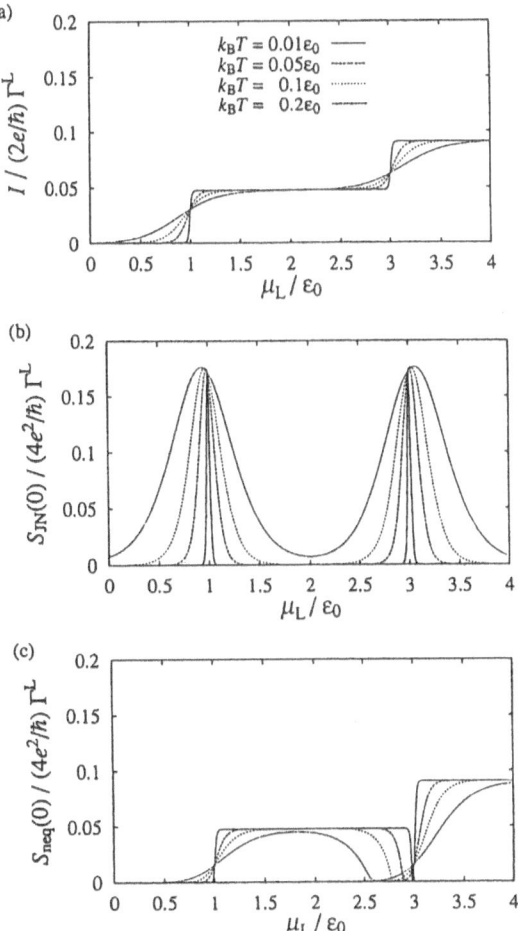

Fig. 17 Current through a quantum dot (a), its Johnson-Nyquist noise (b), and excess noise as functions of μ_L when μ_R is fixed to zero (the bottom of the conduction band).

and the current is proportional to fourth and higher orders in $\mathcal{H}_T$. In this process, both levels ε_0 and $\varepsilon_0 + U$ are virtually used as intermediate states.

When the chemical potential is located in the center of the levels within the quantum dot, the Kondo Hamiltonian

$$\mathcal{H}_0 = \sum_{ik} E_{ik}\, a_{ik}^{\dagger} a_{ik}\quad \mathcal{H}_T = (J/N) \sum_{ipq} a_{ip}^{\dagger}(\mathbf{S} \cdot \boldsymbol{\sigma}) a_{ip} \tag{21}$$

is more useful than the Anderson Hamiltonian. Here, a_{ik} is a column vector with two elements $a_{ik\uparrow}$ and $a_{ik\downarrow}$ and $a_{ik\uparrow}^{\dagger}$ is its conjugate vector. The quantity J is represented as

$$J = V_L \left(\frac{1}{\varepsilon_0} + \frac{1}{\varepsilon_0 + U} \right) V_R \tag{22}$$

where the dependence of V_R and V_L on the momentum has been ignored.

Perturbation calculation of the transition probability from the left to the right electric leads yields

$$\langle I \rangle = \frac{3\pi e}{\hbar} eV \left(\log \frac{eV}{kT_K} \right)^2 \tag{23}$$

$$\langle I^2 \rangle = \frac{6\pi e^2}{\hbar} eV \left(\log \frac{eV}{kT_K} \right)^{-2}. \tag{24}$$

Here, T_K represents the Kondo temperature which is defined by $\log(T_K/D) = 2JN(E_F)$, where $N(E_F)$ is the density of states at the Fermi level. Since these two expressions have been derived using the perturbation theory, they are valid only when $eV > kT_K$. The expression for the case of $eV < kT_K$ is calculated by using the Nozieres theory on the Kondo problem, and we obtain

$$\langle I \rangle = \frac{3\pi e}{\hbar} eV \left\{ 1 - \left(\frac{eV}{kT} \right)^2 \right\} \tag{25}$$

$$\langle I^2 \rangle = \frac{6\pi e^2}{\hbar} eV \left\{ 1 - \left(\frac{eV}{kT} \right)^2 \right\}. \tag{26}$$

9. Friedel sum rule and the quantum dot [8]

The Kondo effect concerning the quantum dot is a phenomenon in which electrons whose energies are near the Fermi level are trapped within and in the vicinity of the quantum dot. It is caused by degeneracy with respect to spin orientation of a single electron trapped in the quantum dot. This bound state arises when $\varepsilon_0 < E_F < \varepsilon_0 + U$ is fulfilled in the model of the last section.

When a bound state is formed, phases of conduction electrons are shifted and the shift is related to the number of bound electrons by the Friedel sum rule. The Friedel sum rule for the symmetric double barrier potential is discussed by Kawabata.[n] Since realistic double barriers are more or less asymmetric, we have studied a similar relation between the phase shift of a conduction electron and the number of electrons between asymmetric double barriers.

When double barriers are symmetrically located at $x = -l$ and l, the wave function is either symmetric or antisymmetric and have the form

$$\psi_s = \cos((k|x| - l) + \eta_s(k)) \tag{27}$$

$$\psi_a = \text{sign}(x) \sin((k|x| - l) + \eta_a(k)). \tag{28}$$

These expressions define the two phase shifts η_s and η_a.

In the case of asymmetric double barrier potential, the reflection amplitude r for an incident wave from the left is not equal to r' for an incident wave from the right. In such a case, we can define the phase shifts η_e and η_o. defined by

$$\psi_e = \cos \left[k \left(\left| x + \frac{\theta - \theta'}{4k} \right| - l \right) + \eta_e(k) \right] \tag{29}$$

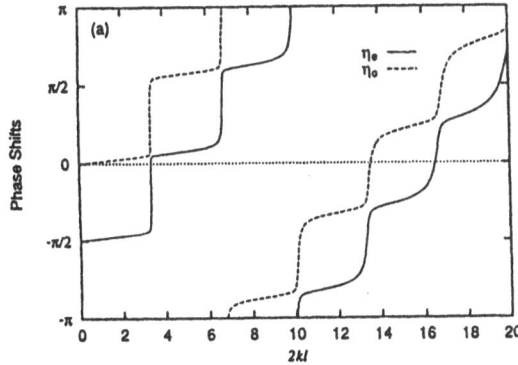

Fig. 18 η_s and η_a as a function of the wave number of the incident wave.

$$\psi_0 = \text{sign}(x)\sin\left[\left(\left|x + \frac{\theta - \theta'}{4k}\right| - l\right) + \eta_0(k)\right]$$ (30)

where $\theta = \arg(r)$ and $\theta' = \arg(r')$.

We have found that these phase shifts are related to the number of electrons within the quantum dot through the relation

$$\Delta N = \frac{2}{\pi}[\eta_s(k_F) + \eta_a(k_F)].$$ (31)

In addition, it has been found that the transmission probability through the double barrier potential is represented as

$$T = \cos^2[\eta_s(k_F) - \eta_a(k_F)].$$ (32)

and these relations imply that these phase shifts have physical meaning.

In Fig. 18, plots of η_s and η_a are presented for certain double barrier potential. An important fact is that the two curves do not cross anywhere. Consequently, the transmission probability in (32) is always less than unity. The well-known fact that the transition probability through an asymmetrmc double barrier potential is always less than unity is described in terms of the nonequality of two phase shifts.

When the energy passes over the value for which T takes a maximum, η_s and η_a simultaneously change by $\pi/2$ and ΔN changes by two. This behavior is the same as the symmetric potential barriers.

The extension of the Fmiedel sum rule to the ease of interacting electrons remains for a future study.

PUBLICATIONS

1. K. Amemiya and K. Kawamura: JRDC Intnl. Symp. Nanostructures & Quantum Effects. November 1993. K. Amemiya, T. Ueta, and K. Kawamura, The 48th Annual Meeting of Physical Society of Japan (March 1993). K. Amemiya and K. Kawamura, *J. Phys. Soc. Jpn.*, **bf 63**, 3087 (1994).
2. Katsuki Amemiya and Kiyoshi Kawamura, *J. Phys. Soc. Jpn.*, **64**, 1245 (1995).
3. Kiyoshi Kawamura, Kazunori Urata and Mikio Eto, N. Urata, M. Eto and K. Kawamura, The 48th Annual Meeting of the Physical Society of Japan (March 1993). N. Urata, M. Eto and K. Kawamura, Divisional Meeting of Physical Society of Japan 1993 (October 1993). M. Eto and

K. Kawamura, Divisional Meeting of Physical Society of Japan 1994 (September 1994).

4. M. Eto and K. Kawamura, *Phys. Rev.*, **B51**, 10119 (1995).
5. F. Yamaguchi, M. Eto and K. Kawamura, *J. Phys. Soc. Jpn.*, **63**, 209–222 (1994). F. Yamaguchi, M. Eto, and K. Kawamura, 48th Annual Meeting of Physical Society of Japan (March 1993).
6. F. Yamaguchi and K. Kawamura, *J. Phys. Soc. Jpn.*, **63**, 1258–1262 (1994). F. Yamaguchi and K. Kawamura, JRDC Intnl. Symp. Nanostructures & Quantum Effects November 1993. F. Yamaguchi and K. Kawamura, Divisional Meeting of Physical Society of Japan 1993 (October 1993).
7. F. Yamaguchi and K. Kawamura preprint. F. Yamaguchi and K. Kawamura, Divisional Meeting of Physical Society of Japan 1994 (September 1994).
8. T. Mizuno, M. Eto and K. Kawamura, *J. Phys. Soc. Jpn.*, **63**, 2658–2667 (1994). T. Mizuno and K. Kawamura, JRDC Intnl. Symp. Nanostructures & Quantum Effects. November 1993. T. Mizuno, M. Eto and K. Kawamura, 48th Annual Meeting of Physical Society of Japan (March 1994). T. Mizuno, M. Eto and K.Kawamura, Divisional Meeting of Physical Society of Japan (October 1993).

REFERENCES

a. R. Kubo, *J. Phys. Soc. Jpn.*, **17**, 975 (1962).
b. M. Büttiker, *Phys. Rev. Lett.*, **57**, 1761 (1986).
c. M. Loukes, A. Sherer, S. J. Allen,jr., H. J. Graighead, R. M. Ruthen, E. D. Beeke and J. P. Harbison, *Phys. Rev. Lett.*, **59**, 3011 (1987).
d. Y. Takagaki, K. Gamo, S. Namba, S. Ishida, S. Takaoka, K. Murase, K. Ishibashi and Y. Aoyagi, *Solid State Commun.*, **68**, 1051 (1988).
e. C. J. B. Ford, S. Washburn, R. Buttiker, C. M. Knoedler, and J. M. Hong, *Phys. Rev. Lett.*, **62**, 2724 (1989).
f. C. J. B. Ford, S. Washburn, R. Newbury, C. M. Knoedler and J. M. Hong, *Phys. Rev.*, **B43**, 7339 (1991).
g. R. B. Dingle, *Proc. Roy. Soc.*, **A211**, 500 (1952).
h. A. Kawabata and R. Kubo, *J. Phys. Soc. Jpn.*, **21**, 1765 (1966).
i. H. S. Zhou, I. Honma and H. Komiyama, *Phys. Rev.*, **50** (1994) to appear.
j. A. A. Lushnov, V. V. Maksimenko and A. J. Simonov, *Z. Physik,* **B27** (1977) 321.
k. R. Rupin in *Electromagnetic Surface Modes*, (A. D. Boadman ed.) John Wiley & Sons Ltd. (1982) Ch 9.
l. T. Inaoka, *J. Phys. Soc. Jpn.*, **62**, 1692 (1993).
m. R. Landauer, *IBM J. Res. & Der.*, **1**, 223 (1957).
n. A. Kawabata, *J. Phys. Soc. Jpn.*, **60**, 3222 (1991).

18

Synthesis and Optical Properties of Coated Composite Nanoparticles

Hao-Shen Zhou [a], Itaru Honma [b], Hiroshi Komiyama [c] and Joseph W. Haus [d]

[a] Frontier Research Program, Institute of Physical and Chemical Research (RIKEN)
Wako-shi, Saitama 351-0106, Japan
[b] Electrotechnical Laboratory (ETL)
Tsukuba-shi, Ibaraki 305-0045, Japan
[c] Department of Chemical System Engineering, Faculty of Engineering, University of Tokyo
Bunkyo-ku, Tokyo 113-0033, Japan
[d] The Department of Physics, Rensselaer Polytechnic Institute
Troy, NY 12180-3590, USA

1. Introduction

During the last decade, zero dimension materials (nanosize particles) have become very popular in various fields. Nanoparticles are clusters of atoms or molecules of metals and semiconductors, ranging in size from 1 nm to almost several 10 nm. The first observation of strong third nonlinearity and the blue shift of the bandgap in glass sample containing semiconductor, CdS or CdSe nanoparticles occurred in the beginning of the 1980's.[a] Brus[b] investigated ZnSe, ZnS, CdSe, and PbS nanoparticles by colloidal methods. Many optical properties were investigated and molecular orbits dealing with excited energies (HOMO-LUMO)[b] in nanoparticles were also studied. From 1985, Henglein and coworkers[c] have synthesized metal nanoparticles as catalyst materials and investigated the photoluminescence of semiconductor nanoparticles, as well as electron transfer between nanoparticles and the quench material.

The optical properties of semiconductor nanoparticles (SN) are very sensitive to the surface of the SN. Brus and his colleagues found that, in an equivalent CdSe system, the surface surfactant ligand was labile and could be transiently displaced by sequential additions of Se^{2-} and Cd^{2+}, which grew on the seed to form large crystals.[d] They synthesized ZnS/CdSe-coated semiconductor nanoparticles (CSN) by sequential addition of S^{2-} and Zn^{2+} in an equivalent CdSe colloidal system[e] in 1990. Henglein and his colleagues also synthesized the composite particles (sandwich structure) CdS-TiO_2, CdS-ZnO, CdS-Ag_2S, Cd_3P_2-ZnO, and AgI-Ag_2S^3 by surface modification. Some PL phenomena in AgI-Ag_2S are similar to our results in CdS/PbS;[1] this is simulated by our theoretical model. Bard and colleagues investigated the core-shell structure of CdSe/ZnSe SCN by X-ray photoelectron and Auger spectroscopies[f] because they thought that the structure was very difficult to determine by TEM. Eychmuller and his colleagues synthesized the CdS/HgS semiconductor coated nanoparticles (SCN) in the same way (ion exchange) as ours and reported the PL peak's shift as the core-shell ratio[g] in 1993. Recently, Eychmuller and his colleagues also synthesized a three-layer semiconductor coated nanoparticle CdS/HgS/CdS.[h] In the theoretical field of SCN structure, beside our work,[2] Hanamura and his colleagues calculated the bandgap change as the core-shell ratio by an effective mass model including coulomb interaction.[i] Kayanuma also calculated the exciton interaction in the shell layer of the SCN structure.[j]

On the other hand, research on metal particles began almost one hundred years ago.[k] Faraday synthesized gold colloidal particles by $NaAuCl_4$.[k,l] Recently, the enhancement of optical nonlinearity of the metal nanoparticles is also attracting much attention of researchers.[m,3] With technical advances and success in controlling the nanosize of metal particles or layers, the quantum size effect of metal layers and particles appears again, and has become a heated topic, even though it was pointed out about thirty years ago.[n] Silberberg et al. discussed the quantum size effect in a few monolayers of metal theoretically.[o] Kreibig et al. pointed out that the surface plasma peak's shift and the width's broadening of metal nanoparticles (MN) resulted from the quantum size effect in MN.[p] A composite material of semiconductor particles and metal, as one kind of enhancement catalysis material has been investigated.[q] Recently, the bistability, enhancement surface plasma resonance, and enhancement of the third nonlinearity in composite materials, have been theoretically predicted [r] and investigated in CdS-Ag [4,5] composite materials. And the enhancement of the conjugate reflectivity also has been observed on composite Ag- and Au-coated nanoparticles.[s]

In this article, we report synthesis of CdS/PbS SCN by ion exchange methods and investigate the optical properties.[1] The optical absorption spectra, PL spectra and theoretical results show that the excited energy (HUMO-LOMO) of the SCN system depends strongly not only on the size of the SCN but also the core-shell ratio, which agree with the experimental data.

We synthesize the Au_2S/Au metal-coated nanoparticles by the colloidal method [6] in order to obtain high enhancement surface plasma resonance according to our previous calculations.[r,5] We also discuss the optical absorption cross section. The theoretical results show that there are two surface plasma peaks belonging to the two surfaces of the metal shell layer. The optical absorption results show that there are two surface plasma resonant peaks in Au_2S/Au MCN. And the position of the long wavelength peak goes red shift at first, then goes blue shift, until at last two peaks become one peak (= Gold's surface plasma peak). This phenomenon can be explained by the quantum model.

2. Semiconductor-semiconductor coated nano-composite particles (SCN)

In coated particles, we have shown[2] that the carriers can spatially separate in the materials due to the band offset and mass differences among the carriers. We henceforth call these particles "coated nanoparticles" because they have two different materials in their constituent semiconductor, with an interface between them. The physical parameters of each semiconductor material will have great influence on the optical properties, such as luminescence, of coated nanoparticles. In fact, superlattice type-II-like coated nanoparticles can be achieved, where the electrons and holes are spatially separated. Charging and the band-bending effects can also be studied and modeled within the context of the coated particles; these effects may also be treated by a method [2] developed in our recent study.

We have also developed a theoretical treatment of coated semiconductor nanoparticles based on the single band effective-mass approximation.[2] It offers a simple analytical method to investigate the quantum confinement of the carriers. The coated semiconductor nanoparticle InP/InAs system has been well studied as quantum-well structures for carriers' spatially separating.[2] Coulomb effects between the electron and hole, and incomplete confinement of the carries can also be significant.[1] In the strong confinement regime, where the particle is smaller than the bulk exciton radius, the dominant contribution to the carrier's energy is the kinetic energy imposed by the boundary conditions. Here, we discuss only the synthesis and experimental optical properties of these SCN.

2.1 Synthesis of CdS/PbS SCN

8 ml 2.0 mM S^{2-} ion solution was injected to the mixture solution of 10 ml 2.0 mM Cd^{2+} ion solution and 5% PVP 10 ml to form CdS nanoparticles of 60–80 Å diameter. These nanoparticles stay stable in the solution because of the stabilizing agent PVP on the CdS surface which prevents coalescence. There were some excess Cd^{2+} ions in CdS colloidal solution because the quantity of Cd^{2+} ions was more than that of the S^{2-} ions. So the excess Cd^{2+} ion was presumably attached to the surface of CdS nanoparticles. Then 8 ml 2.0 mM Pb^{2+} ion aqueous solution was poured into the above solution. Because of the absence of excess S^{2-} ion in the solution, separated PbS single nanoparticles could not be formed directly. At first, Pb^{2+} ion was attached to the surface of CdS nanoparticles. Then the Pb^{2+} ion displaced Cd in the Cd-S bond to form Pb-S bond on the surface of CdS nanoparticles. The color of the SCN solution changed within several minutes or several hours according to the concentration of Pb^{2+} ion in the solution. These CdS nanoparticles with a shell of PbS on the surface could be called CdS/PbS semiconductor coated nanoparticles (SCN). All of the CdS nanoparticles will transform into PbS if the lead ion concentration is high enough. The procedure of ion displacing can be shown by:

$$Pb^{2+} + CdS \longrightarrow Cd^{2+} + PbS + G.$$

The reaction Gibbs energy G is identical to 4.618 kJ/mol, which guarantees that Pb^{2+} ion replaces Cd in the Cd-S bond of the CdS nanoparticles. We could control the size of the PbS shell in CdS/PbS CSN by changing the Pb^{2+} ion concentration.

The detailed procedure can be written as:

$$a\ Pb^{2+} + (CdS)_n \longrightarrow a\ Cd^{2+} + (PbS)_a(CdS)_{n-a}$$

$$m\ Pb^{2+} + (PbS)_a(CdS)_{n-a} \longleftrightarrow m\ Cd^{2+} + (PbS)_{m+a}(CdS)_{n-m-a}$$

$$n\ Pb^{2+} + n\ CdS \longrightarrow n\ Cd^{2+} + n\ PbS.$$

According to the phase diagram, the possibility of a CdS-PbS solid solution is very low[u] at room temperature. The change in the bandgap of CdS/PbS CSN takes about 2 hours from CdS to PbS.

We estimate the diffusion coefficient D to be of the order of 10^{-21} m²/s by the equation $r = (Dt)^{1/2}$ (r: radius of the particle, D: diffusion coefficient, t: diffusion time) when r of 60 A and t of 2 hours are assumed. This value is very close to the diffusion coefficient of Cd^{2+} in CdS solid (about 10^{-21} m²/s) and much smaller than that in the liquid (about 10^{-9} m²/s). The small value of D is realistic when the CdS/PbS CSN structure is assumed and there is diffusion resistance in the shell PbS layer.

2.2 Photoluminescence (PL) of the CdS/PbS SCN

The PL characteristics of the AgI/Ag_2S[3], CdS/HgS[8] and $CdS/HgS/CdS$[9] systems have been described. But in those papers, the results are correlated with the quantized energy structure of the coated semiconductor particles. However, in this report, we explore both the shift and the intensity change of PL spectra of the CdS/PbS coated nanoparticles, and present a theoretical treatment quantitatively.

2.2.1 Pb cation modified photoluminescence spectra of the CdS

When the Pb^{2+} solution is injected into the above CdS colloidal solution, the surface of the CdS colloidal is modified by attracted Pb^{2+} ions. Figs. 1 (a) and (b) show the optical absorption and PL spectra of the CdS colloidal nanoparticles modified by the Pb^{2+} ion as the quantity of the Pb^{2+} solution change. The absorption results show that the excitons of the CdS colloidal solution disappear after adding a little Pb^{2+} solution, although the absorption edge keeps same. The photoluminescence spectra of CdS modified by Pb^{2+}, in Fig. 1 (b) show that the S vacancy electron-hole recombination at 650 nm is quenched by the surface modification. The PL spectra of the CdS can be quenched by adding the Pb caption only as a hundredth (or the fiftieth) of Cd, and we can see that the S vacancy luminescence can be quenched mostly after adding the Pb cation at only about the tenth of the Cd caption at molar ratio.

Some researchers also reported quenching of the PL spectra of the semiconductor nanoparticles [c] by carriers transferring from the semiconductor nanoparticles to the surface.[v] So, there are two possibilities: one for forming PbS islands at the CdS surface,[w] the other for forming Pb- V_S. complex defect

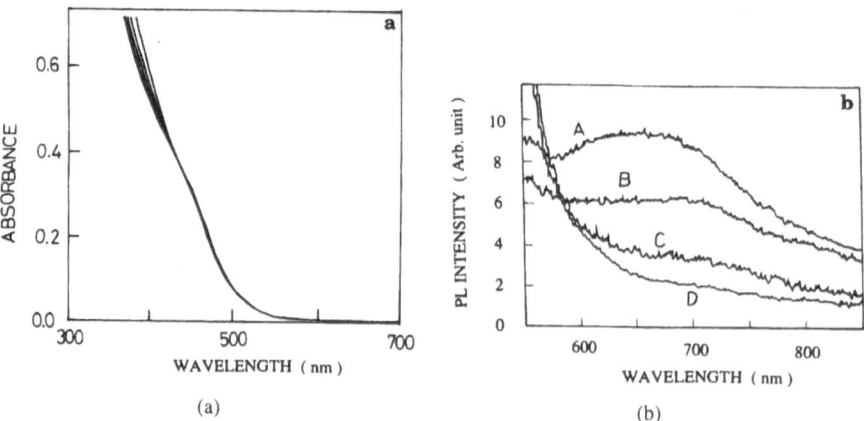

(a)   (b)

Fig. 1
(From H. S. Zhou and I. Honma *et al.*, *Chem. Mater.*, **6**, 1537 (1994)).

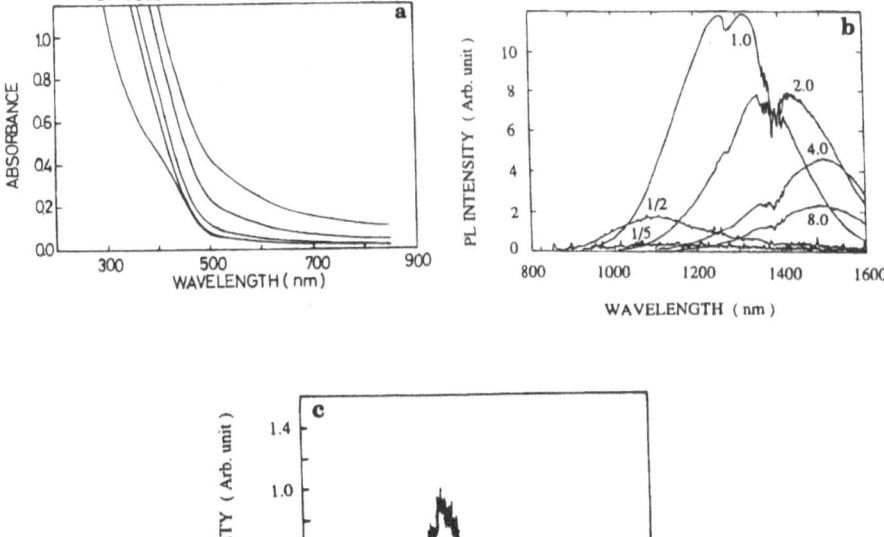

Fig. 2
(From H. S. Zhou and I. Honma *et al.*, *Chem. Mater.*, **6**, 1537 (1994)).

center to decrease the V_S PL center. Some researchers [v] have reported that the $V_S^{\&+}$ — $MV^{\{(\&+)-2\}}$ complex defect center quenched the electron-hole recombination with the sulphur vacancy. So there could be a similar complex defect center in the surface of the CdS colloidal nanoparticles.

2.2.2 The PL spectra of CdS/PbS SCN

As the quantity of Pb cations increase further, not only does the luminescence peak at 650 nm disappear, but also concurrently, the V_{Cd-S}^{PL} peak, which appears in the near infrared range, shifts to longer wavelengths. For larger Pb/Cd molar ratios, we consider the particles to be coated. The optical absorbance spectrum for larger cation ratios strongly shifts to longer wavelengths as shown in Fig. 2(a). The absorption edge shifts toward longer wavelengths as the molar ratio of Pb/Cd increases, indicating that the bandgap of the CdS/PbS coated nanoparticles decreases as the thickness of the PbS shell layer increases.

We find that the V_{Cd-S} of the photoluminescence peak at 1100 nm does not shift for the ratio in the range from 0.05 to 0.5, but the intensity of the peak slowly increases when the Pb/Cd ratio is very small; for these molar ratios, the absorption edge also does not change. For comparison, the PL spectrum for CdS particles is shown in Fig. 2(c); it is very weak and the signal-to-noise is small. However, the PL spectra peak changes from 1100 nm to 1500 nm as the molar

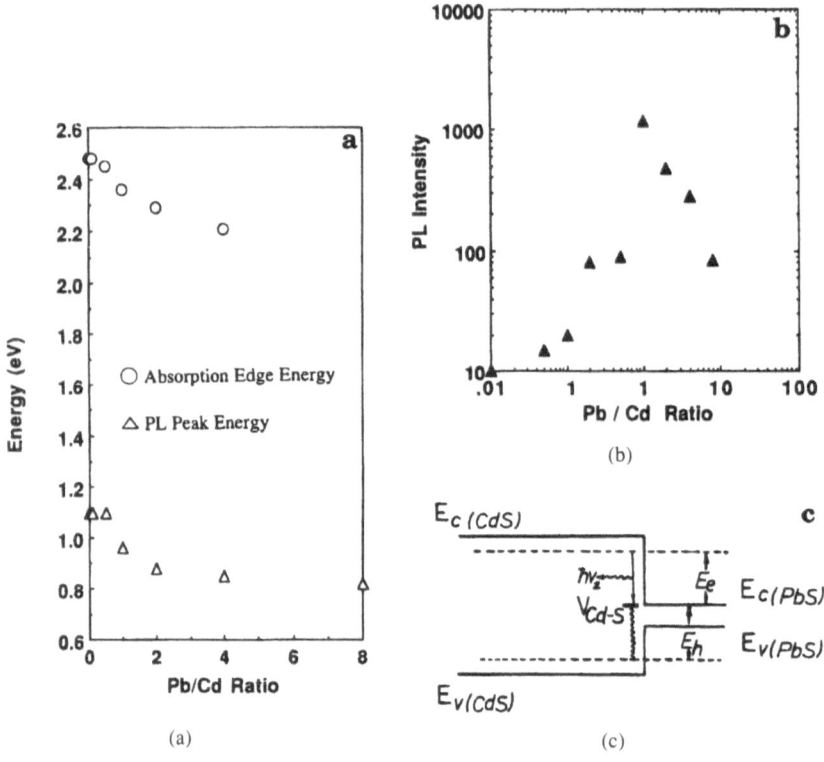

Fig. 3
(From H. S. Zhou and I. Honma *et al.*, *Chem. Mater.*, **6**, 1537 (1994)).

ratio Pb/Cd is increased from 0.05 to 4.0, as shown in Fig. 2(b). The large shift of the PL peak is correlated with the band edge shift of the absorbance in Fig. 2(a). The data for the PL peak photon energies and the bandgap energy deduced from the absorption data are plotted in Fig. 3(a); it displays the correlation between the two quantities, and the variation of the peak intensities in Fig. 3(b). as the Pb/Cd ratio is varied in the solutions.

2.2.3 Mechanism of PL in the CdS/PbS SCN

When the Cd cation in the CdS is exchanged by a Pb cation, the shell grows as PbS and the Cd cation migrates across the PbS shell layer into the solvent. Defects and vacancies can also be created at the interface by the kinetic exchange process and strains induced by the lattice mismatch between the two materials. The defects become new PL centers at the interface in the CdS/PbS system. Cd-S composite vacancies (V_{Cd-S}) is an acceptor vacancy PL center. The energy level for the ionization is located about 1.2 eV above the valence band.[x] This is the source of the infrared PL peak photon energies plotted in Fig. 3(a) and the variation of the peak intensities in Fig. 3(b) as the Pb/Cd ratio.

The photoluminescence center for the red fluorescence is identified as the S vacancies and the infrared fluorescence is the Cd-S composite vacancies. At first, the PL is observed in the red, then the S vacancy is reduced as the surface changes to a PbS layer and at the same time, Cd-S composite vacancy increases due to the ion exchange process. For example, the Pb atom does not enter the Cd atom's position in the neighbor of a V_S vacancy after the ion exchange process, because the radius of Pb ion is bigger than that of Cd ion, or the crystalline of PbS is different from that of CdS. So a V_S vacancy becomes a V_{Cd-S} composite vacancy in this way. The experimental results can be explained by the model. This has the effect of quenching the red luminescence and increasing the infrared luminescence.

Based on previously cited reports, we assume that the infrared PL of the CdS/PbS results from the V_{Cd-S} composite vacancy located at the interface between the two semiconductor materials. Fig. 3(c) is a schematic of the band structure scheme for the coated CdS/PbS nanoparticles including the recombination defect center. The bandgap of the CdS core is larger than the bandgap of the PbS shell layer. The sub-band states [1, 2] (excited energy levels) E1S and H1S, shown as dashed lines, can lie below the conduction band edge, and above the valence band of the CdS. The process from carrier excitation to recombination is

$$\text{CdS/PbS} + h\nu \longrightarrow \text{CdS/PbS} \ (h_{VB}^+ + e_{CB}^-)$$

$$e_{CB}^- + V_{Cd-S} \longrightarrow V_{Cd-S}^- + h\nu_2$$

$$V_{Cd-S}^- + h_{VB}^+ \longrightarrow V_{Cd-S}.$$

The electron is excited into an available sub-band state and has a radiative transition into the V_{Cd-S} defect state. The subsequent recombination of the V_{Cd-S} defect state and hole is a nonradiative process. The electrons and holes were excited by an argon laser, then the holes trapped into the V_{Cd-S} composite vacancy

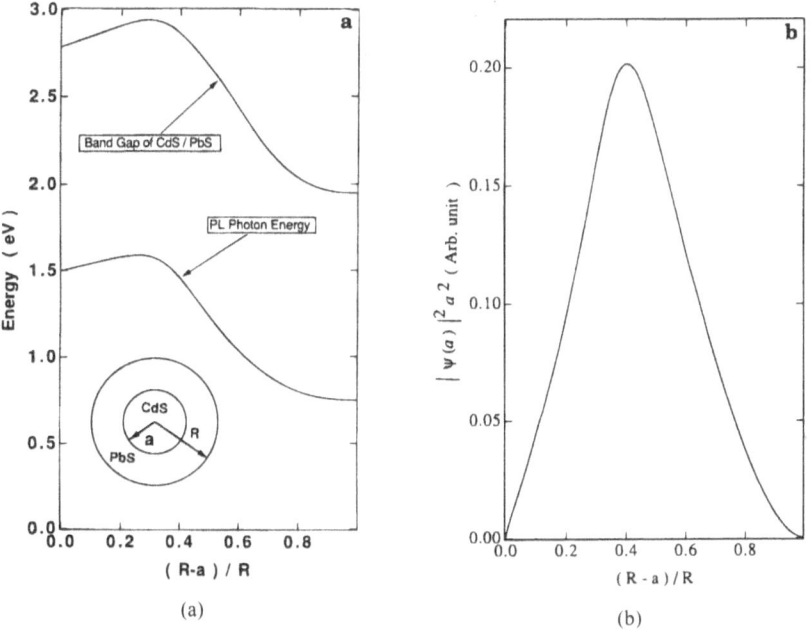

Fig. 4
(From H. S. Zhou and I. Honma et al., Chem. Mater., 6, 1537 (1994)).

acceptor, and recombined with the electrons in the excited electron level LUMO (1S level denoted by E_e) to give the infrared PL hv_2. E_e is measured with respect to the bottom of the PbS conduction band. The bandgap of CdS is taken to be 2.4 eV, while the conduction band offset between CdS and PbS is 1.2 eV Therefore, the valence band offset is 0.8 eV. Because the E_e also decreases as the core/shell ratio increases, the gap between the E_e and V_{Cd-S} vacancy level also decreases (red shift) as the shell/core ratio increases. The energy of the PL photon is

$$hv_2 = E_e + V_{Cd-S}$$

$$V_{Cd-S} = 0.0 \text{ eV}.$$

Expression of E_e has been discussed in our previous papers.[2] Fig. 4(a) is a plot of the excited energy bandgap (HOMO-LUMO), ($E = E_e + E_h$) , and the V_{Cd-S} PL energy shift, (E_e+V_{Cd-S}), versus the ratio of the shell thickness to the radius $(R - a)/R$. The theoretical results for the V_{Cd-S} [PL] energy shift, which corresponds to the infrared fluorescence peak, hv_2, is also shown in Fig. 4(a). It also decreases in energy as the shell layer increases, but by a smaller amount.

Comparison between Fig. 4(a) and Fig. 3(a) reveals that the trend of the energies for the model corresponds to the absorption band edge energy and the PL peak position. There are, however, quantitative differences, especially for the CdS-rich particles; the variation of the energies over the range of core/shell ratios is larger than those observed in the experiment. This is to be expected, considering the simplicity of the model; there are two improvements that are necessary in the model, which will also improve the agreement, the Coulombic effects, which include the interaction between the particles and medium polarization, and the finite height of the barrier between the shell and the host material. Both tend to lower the energy in the CdS particle, but have little effect on the PbS particle. Coulombic effects depend upon the dielectric constant of the material, which is 11 for CdS and 185 for PbS. This effect is expected to lower the sub-band energies by about 10% for CdS particles and is negligible for the PbS particles; the finite barrier height also affects the CdS particles more than the PbS particles because the bandgap is much larger for the CdS than for PbS; the barrier height depends on the host material in which the particles are embedded.

Also, the intensity of the PL peak can be estimated by our model. The defects are at the interface according to the transfer model from V_S vacancies to V_{Cd-S} composite vacancies in the ion exchange process, and the PL intensity depends on the number of defects and the electron density at the core/shell interface. We assume that the density of defects at the interface is constant with the core/shell interface area; then the PL peak intensity is proportional to

$$I_{PL} \sim |\psi_e(a)|^2 \, a^2.$$

The PL vanishes as a approaches zero; however, it also becomes small as a approaches R, since the wavefunction vanishes at the shell boundary. We find the maximum of the PL intensity in this model occurring at around $a \approx 0.6 \, R$.

The plot of the peak intensity versus the shell thickness to shell radius ratio is shown in Fig. 4(b). The experimental PL peak intensities in Fig. 3(b) can be correlated with those in Fig. 4(b) to infer a value for the core/shell ratio; for instance, when the maximum PL peak intensity is assumed to correspond to the curve maximum in Fig. 4(b), we find the ratio of the shell thickness to shell radius to be 0.4. From this information the PL photon energy from Fig. 3(a) can be compared with that in Fig. 4(a).

2.3 Summary

We developed an ion exchange colloidal method to synthesize CdS/PbS SCN. We also investigated the nanostructures and optical properties of the CdS/PbS SCN by TEM, electronic diffraction (ED), optical absorption and photoluminescence (PL).

The bandgap of the CdS/PbS SCN, which has been calculated by the effective mass model,[2] can be changed from the bandgap of the CdS nanoparticles to that of the PbS nanoparticles of the same size if the CdS nanoparticles become PbS nanoparticles across the CdS/PbS SCN.

The synthesis of CdS/PbS SCN of about 60 Å in diameter has been clearly demonstrated by free Pb^{2+} ion displaced Cd in the Cd-S bond on the surface of CdS colloidal solution. The PL spectra from dried PVP film containing coated CdS/PbS nanoparticles have been investigated at room temperature. The PL spectra of CdS/PbS coated nanoparticles appeared in the wavelength range from 1100 nm to 1500 nm and shifted to longer wavelengths as the quantity of Pb cations increased. We discussed the role of Cd-S composite vacancy PL center in the CdS/PbS system and calculated the energy gap from the electron excited level (LUMO) to the Cd vacancy level by the effective-mass model. The value, just as the excited energy gap in the CdS/PbS system, is not only dependent on the size but also strongly depends on the core/shell ratio. The theoretical results are comparable with the PL results in general, which show that the red shift and the intensity maximum of the PL spectra with shell layer growth can be explained by the recombination between the electron in the 1S electron quantum state and the V_{Cd-S} vacancy state on the core/shell interface of the coated CdS/PbS system.

Heterostructure particles can be used to obtain new information about the material properties. The defects at the interface can be probed by luminescence spectroscopy. New physical phenomena can also be expected in these particles; for instance, since the electron and hole wave functions no longer overlap, there is a static electric field generated within the particles due to the charge separation. Because of the small size of the particles, the static field can be quite large and Stark shifts of spectral features can be expected.

3. Metal-coated semiconductor nano-composite particles

The electromagnetic resonance effects can, in principle, greatly enhance the optical nonlinearity of dielectric particles. A large enhancement of the conjugate reflectivity was observed on the composite nanoparticle materials of Ag and Au.[m] Both have a strong surface plasma resonance effect in the visible regime, and therefore, metal-coated nanoparticles (MCN) have been suggested as a scheme for achieving very large effects.[m,3] The coated nanoparticle consists of a core and a shell, each with different material properties; those considered include metal or semiconductor as the core and shell, and the optical nonlinearity can be large for both case. Intrinsic optical bistability has been predicted for these coated particles. However, the major obstacle remains the fabrication of coated particles on a nanometer size scale. Several methods have been investigated with limited success thus far.

3.1 Au₂S/Au MCN Synthesis

Now, we will describe the method of synthesizing MCN and its optical properties in experiments to prove our calculation results and the possibility of a new photonic device for application.

We report the first synthesis of gold-coated nanoparticles by a colloidal method, and the extraordinary optical absorption spectra of the coated samples. The results are consistent with a theoretical approach that includes electromagnetic resonance effects[y] and the quantum confinement of the carriers in the thin gold shell layer. Interest in metal-coated particles comes from the expectation of large

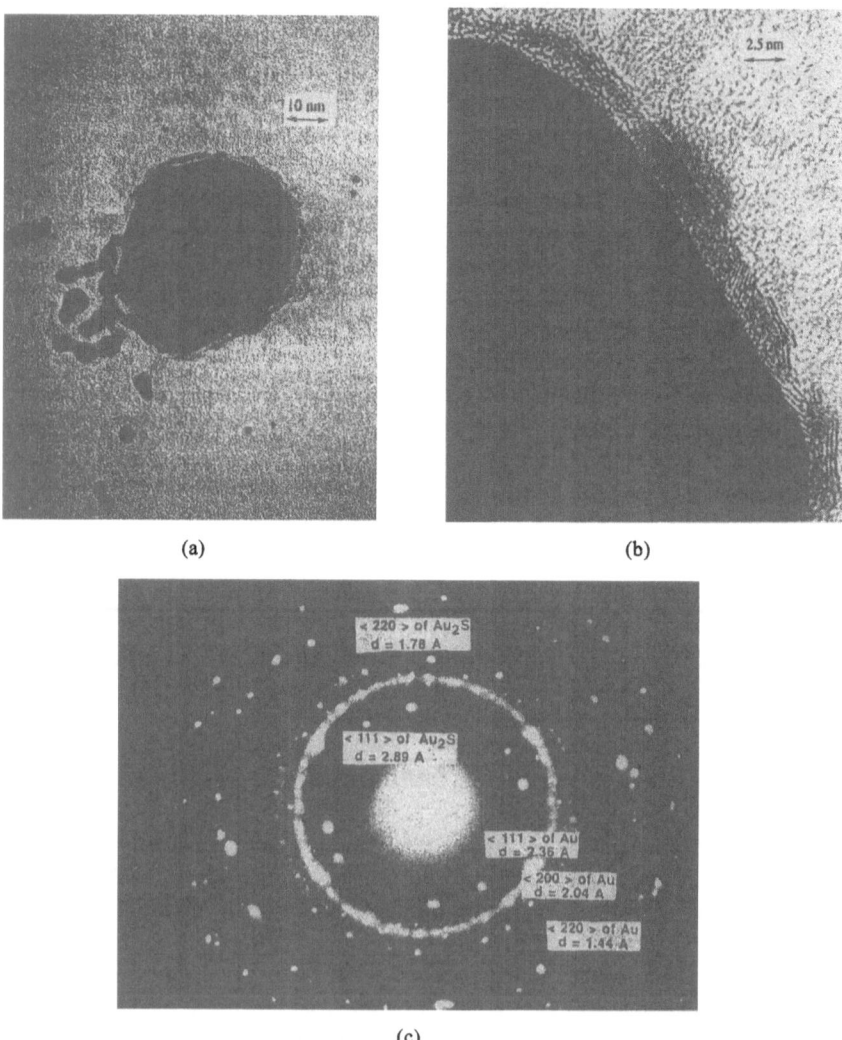

(a)   (b)

(c)

Fig. 5

(From H. S. Zhou and I. Honma *et al.*, *Phys. Rev. B*, **50**, 12053 (1994)).

nonlinear optical response due to the surface plasmon resonance.[r,5]

The particles were prepared by a two-step process. First, we dissolved chloroauric acid ($HAuCl_4 \cdot 4H_2O$) and sodium sulphide ($Na_2S \cdot 9H_2O$) into superpure water at room temperature to obtain a $HAuCl_4$ solution and a Na_2S solution. The chloroauric acid ($HAuCl_4 \cdot 4H_2O$) and sodium sulphide ($Na_2S \cdot 9H_2O$) are made by Wako Pure Chemical Industries, Ltd. We mixed the controlled amounts of the two solutions together to obtain the unstable gold sulphide Au_2S (if the quantity of Na_2S is more than the quantity of $HAuCl_4$, we get the stable Au_2S_3). The color of the Au_2S solution was light brown. In the second step, we injected a little Na_2S solution into the above Au_2S solution. The color changed from light brown to gray, light red, and dark red.

In the reaction, the gold atom in the Au-S bond on the surface of Au_2S was reduced by S^{2-} ion [z] and the surface became a gold-coated shell layer. The gold surface layer grew with time. This process is supported by TEM and electron diffraction; TEM images show small spherical particles average diameter about 4 or 5 nm, coexisting with large particles, of diameter of about 40 nm. The average particle diameter is determined by the concentration of the solution and the time interval before the addition of more Na_2S solution in the second step. Fig. 5(a), (b) and (c) displays a TEM image of the sample and an electron diffraction pattern. The lattice of the approximately 20 Å shell layer in Fig. 5(b) is about 2.30 Å, similar to the Au(111) lattice, 2.35 Å. So Fig. 5 shows a Au_2S particle of about 40 nm coated with a Au shell layer of about 20 Å. The electron diffraction patterns of the particles after gold precipitation include the pattern of gold mixed with that of Au_2S. [a'] We conjecture that the small particles are pure gold because the sulphur has been rapidly removed; the larger particles are Au_2S/Au MCN which naturally possess a longer diffusion and reducing time.

3.2 Optical properties of Au₂S/Au MCN

Our absorption experiments followed the reaction as it proceeded. The spectra were taken once every minute; the reaction time was slow and took about thirty minutes to complete. A selected data set on three samples (samples A, B and C) taken during the gold overcoat growth process are shown in Fig. 6. Curve 1 is the absorption of the $HAuCl_4$ solution and curve 2 is from the Au_2S precipitated solution in the absorption spectra of sample A, B and C, and all the following sequences in the three sample were taken after the addition of the Na_2S solution again in the second step. The relationships between the measuring time and longer wavelength peak's position of the all samples are shown in Fig. 7. They represent the characteristic absorption data for the gold-coated nanoparticles.

There are two peaks in the spectra of the three samples. The one at about 530 nm is from surface plasma resonance of the pure gold particles whose size is about 4 or 5 nm in the TEM image.

The other, which comes from the coated particles, shifts across the visible region and shows two time regimes for all samples; the peak has a red shift in the initial time regime, and then a blue shift in the second time regime. The later time regime is described by the electromagnetic theory of coated nanoparticles. [5] The resonance shifts from the particle resonance and depends on the ratio of the core and shell radii. The thicker the coating, the closer the resonance is to that of a homogeneous gold particle.[5,b'] The theoretical results are shown in Fig. 8(a),

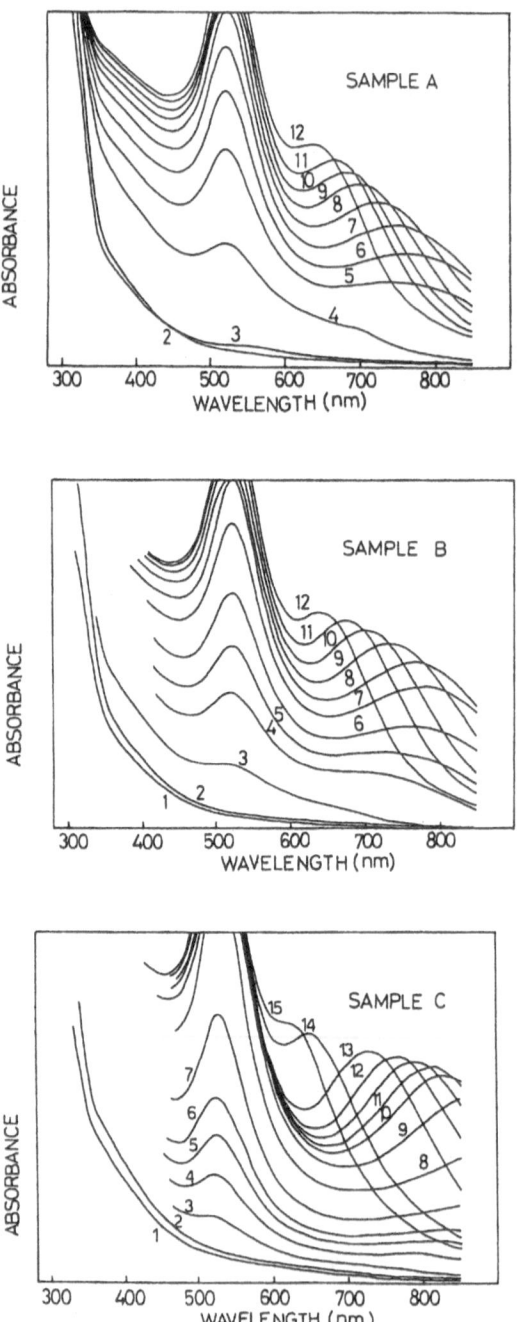

Fig. 6
(From H. S. Zhou and I. Honma *et al.*, *Phys. Rev B*, **50**, 12054 (1994)).

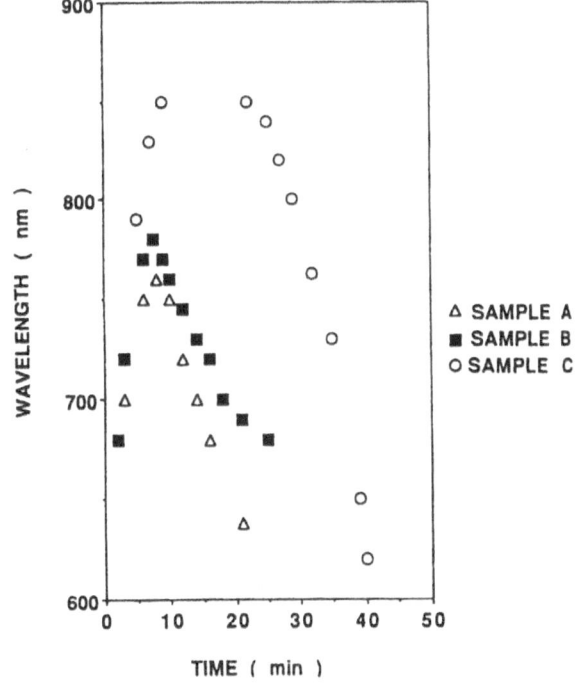

Fig. 7
(From H. S. Zhou and I. Honma *et al.*, *Phys. Rev B*, **50**, 12054 (1994)).

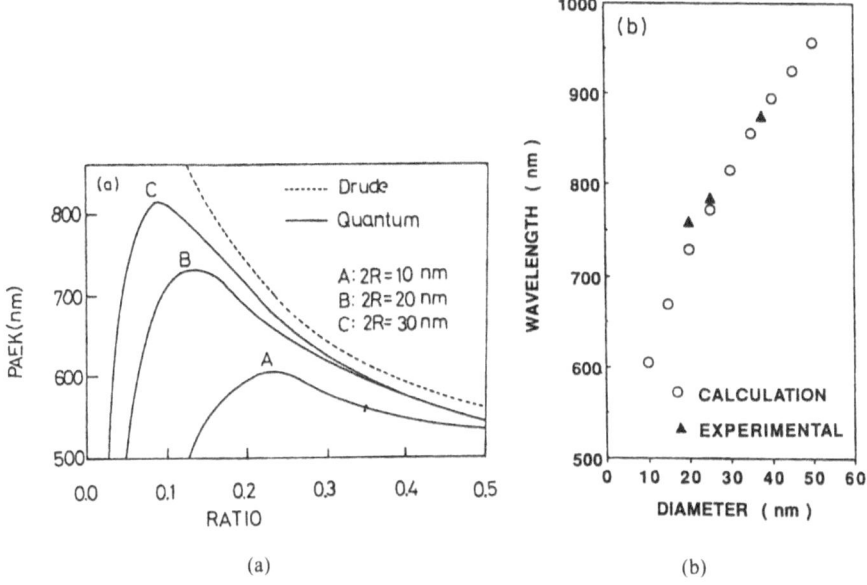

(a)   (b)

Fig. 8
(From H. S. Zhou and I. Honma *et al.*, *Phys. Rev B*, **50**, 12055 (1994)).

according to the classical electromagnetic model. There is an unexpected feature in the initial time regime of the absorption spectra, namely, the shift is initially toward longer wavelengths, which is attributed to quantum confinement of the electrons in the thin surface metal layer. The difference among the three samples is the turning point wavelength separating the first regime from the second regime; variation of the turning point wavelength among the samples is correlated with the different average particle size in each sample. The peak positions of the absorption spectra of the three samples are plotted in Fig. 7 as a function of time.

3.3 Quantum size effect in the MCN

A model description of the experimental results require the dielectric function for the metal and the electromagnetic response of the coated particles. The later function for coated nanoparticles can be analyzed within the Rayleigh regime, *i.e.* the coated particle size is much smaller than a wavelength in the material.[b'] The polarizability of coated nanoparticles dispersed in a host dielectric medium is given by[b']

$$p = \frac{(\varepsilon_s - \varepsilon_h)(\varepsilon_c + 2\varepsilon_s) + \Gamma(\varepsilon_c - \varepsilon_s)(\varepsilon_h + 2\varepsilon_s)}{(\varepsilon_c + 2\varepsilon_s)(2\varepsilon_h + \varepsilon_s) + 2\Gamma(\varepsilon_s - \varepsilon_h)(\varepsilon_c - \varepsilon_s)} a_s^3,$$

where $\Gamma = (a_c + a_s)^3$. The core, shell and host dielectric constants are denoted by ε_c, ε_s and ε_h, respectively; the radii of the core and shell are a_c and a_s respectively, the p equation is used in computing the absorption; the volume fraction is small (about 10^{-5}), so that the particles may be independently treated.[b']

Generally, the dielectric constant of the metal is described by the Drude model. However while the Drude model works well in the infrared regime, it is not directly applicable to the visible regime. Moreover, when the shell layer is as thin as several monolayers of gold, there is the very strong quantum confinement effect in the shell layer. To simplify the analysis, the quantum properties of the free electrons are modeled by a modified form of the Drude model,[p] in which individual electron resonances are replaced by an average resonant frequency ω_0,

$$\varepsilon = 1 + \frac{\omega_p^2}{\omega_0^2 - \omega^2 - i\gamma\omega} + \varepsilon_b.$$

To obtain an estimate of the effect of quantum confinement on the energy shift, we use the expression for the electron energy in the strong confinement regime.[p]

$$\omega_0 = \frac{1.3 V_F}{d}$$

where V_F is the Fermi velocity, which for gold is $V_F = 1.4 \times 10^6$ m/s, and d is the thickness of the metal shell. Since the width of the resonance is also strongly dependent on the width of the layer, d, we use the expression,[p]

$$\gamma = \frac{V_F}{d} + \gamma_0$$

with $\gamma_0 = 4.1 \times 10^{13}$ s^{-1}, the intrinsic width.[c'] The value of the fixed parameter is $\omega_p^2 = 1.9 \times 10^{32}$ s^{-2} for the plasma frequency [c'] and the contribution from the interband transition electrons[p] is taken as $\varepsilon_b = 9$. The remaining parameters, γ and ω_0^2, depend on the layer thickness.

Silberberg and Sands [o] have done work to calculate the dielectric constant of a thin metal layer on the quantum size effect. Here, we use a simple way described by Kreibig.[p]

When the shell layer is thin enough, the resonance frequency, ω_0, occurs in the visible regime and becomes small as the shell thickness increases. This strongly modifies the dielectric constant so that the absorption resonance peak moves to longer wavelengths as the shell layer d becomes thicker. This is the reason that the first regime has a red shift. Also, since the polarizability depends on the volume, and the coated particles are large, they contribute a strong peak to the absorption spectrum. The model approaches Drude's model when d is large enough. In Drude's model regime, the thicker the coating, the closer the resonance is to that of a homogeneous gold particle.[5]

The results of the theoretical analysis are displayed in Fig. 8. Fig. 8(a) shows the calculation results of the Au$_2$S/Au MCN using only the Drude model for the dielectric constant in Au shell layer. Fig. 8(b) displays the relationship between the peak position and the core-shell ratio for different size Au$_2$S/Au MCN, the modified model on the quantum size effect. We find that the results on the quantum size model is different from the classical Drude model and show a red shift as the ratio in the beginning. We compare the theoretical and experimental results on the peak position versus the size of the particles in Fig. 8(c), and find that the theoretical results agree with the experimental one.

The quantum analysis includes the effect of the resonant frequency, ω_0. Without this term the frequency shift of the surface plasmon resonance would be a monotonic decreasing function as the shell is made thicker; this curve is the dashed one labeled Drude plotted in Fig. 8(b). The initial shift of the peak toward the infrared and then toward the ultraviolet region agrees well with the experimental data. The return position of the shift peak goes to a longer wavelength when the size of the particles becomes large. The TEM images of samples B and C show that the average radius of samples B and C is about 12.5 nm and 17.5 nm respectively.

3.4 Summary

We have provided a recursive method to calculate the surface plasma resonance effects and the third optical nonlinearity in metal-coated nanoparticles (MCN) by a classical electromagnetic model and developed a two-step colloidal method to synthesize Au$_2$S/Au MCN. We also investigated the nanostructures and optical properties of the Au$_2$S/Au MCN by TEM, electronic diffraction (ED), and optical absorption.

Our theoretical results show that there are two surface plasma resonance peaks in the metal-coated nanoparticles, one from the surface of core and metal shell layer, the other from the surface of metal shell layer and the host material. The position of these two peaks strongly depends on the core-shell ratio, the long wavelength peak goes blue shift, and the short wavelength peak goes red shift when the core-shell ratio decreases, until at last, the two surface resonant peaks become one peak as the core-shell ratio becomes zero.

Two-step processes to synthesize Au_2S/Au metal-coated colloidal particles provide a unique method for extracting information on the quantum properties of metals. The materials can also be of use in studying the surface plasmon resonance enhancement of the optical nonlinear response of the solution.

The experimental phenomenon is different from the classical theoretical results which are obtained by the Drude-model for the dielectric constant of the metal layer. But this phenomenon can be explained by the quantum model which has been discussed above. When the shell layer is thin enough, the resonance frequency, ω_0, occurs in the visible region. The calculation position and the width of the surface plasma resonant peak agree generally with the experimental results in the Au_2S/Au MCN. And the changing of the resonant surface plasma is strongly dependent on the size of the Au_2S/Au MCN.

We stress that we have synthesized a particle with a nonconducting core and a metal shell. The quantum effects which we observe are possible for these particles because the initial size is large, and thus, has a large dipole moment. Moreover, because the core is not conducting, the carriers are confined to a thin shell region. This is not the case with solid metal particles, which would exhibit quantum effects when the particles are small, but unfortunately in that regime the dipole moment is also small and such particles do not significantly contribute to the absorption. When the particles grow to sizes at which the dipole moment is large, then the quantum nature of the particles is no longer important.

This research work on metal-coated nanoparticles (MCN) is of broader interest than the field of nanometer-size particles, although that field has been of fundamental interest to the scientific community. In our case, a new method is given to study the quantum nature of carrier confinement in metals. It is more sensitive because of the large dipole moment and this can translate to new applications to optoelectronics and sensitive assay tests. In the future this scheme can be used for other particles and the synthesis of coated nanometer-size particles, in general, can find new impetus.

4. Conclusion

The transition from the bulk to the molecular state is an important range where completely novel properties can be expected. Nanosize composite materials offer ideal systems for the investigation of various fundamental effects include quantum confinement effect of excited carriers, charge or energy transfer in the nanosize scale, which can be used for new electronic or photonic devices. We have entered a new field in which control of structure of nanophase materials, including size, shape, defects, impurities and dimensionality, can result in new properties.

PUBLICATIONS

a. R. K.Jain and R. C.Lind, *J. Opt. Soc. Am.*, **73**, 647–653 (1983); A. I. Ekimov, Al. L. Efros and A. A. Onushchenko, *Solid State Commun.* **56**, 921 (1985).
b. M. L. Steigerwald and L. E. Brus, *Acc. Chem. Res.*, **23**, 183 (1990).
c. A. Henglein, *Chem. Rev.*, **89**, 1861 (1989); L.Spanhel, H.Weller and A.Henglein, *J. Am. Chem. Soc.*, **109**, 6632–6635 (1987); A.Henglein, M.Gutierrez, H.Weller, A.Fojtik and J.Jirkovsky; *Ber. Bunsen-Ges. Phys. Chem.*, **93**, 5578 (1989); L.Spanhel, H.Weller, A.Fojtik and A.Henglein, *Ber. Bunsenges. Phys. Chem.*, **91**, 88 (1987); L.Spanhel, M.Haase, H.Weller and A.J.Henglein, *Am. Chem. Soc.*, **109**, 5649 (1987).

d. M. L.Steigerwald, A. P.Alivisatos, J. M.Gibson, T D.Harris, R.Kortan, A. J.Muller, A. M.Thayer, T.M.Duncan, D. C.Douglass and L. E. J.Brus, *Am. Chem. Soc.*, **110**, 3046 (1988).
e. A. R. Kortan, R. Hull, R. L. Opila, M. G. Bawendi, M. L. Steigerwald, P. J. Carroll and L. E. Brus, *J. Am. Chem. Soc.*, **112**, 1327 (1990).
f. C. F. Hoener, K. A. Allan, A. J. Bard, A. Campion, M. A. Fox, T. E. Mallouk, S. E. Webber and J. M. White, *J. Phys. Chem.*, **96**, 3812 (1992).
g. A. Hasselbarth, A. Eychmuller, R. Eichberger, M. Giersig, A. Mews and H. Weller, *J. Phys. Chem.*, **97**, 5333 (1993); Eychmuller, A., Mews, A. and Weller, H.; *Chem. Phys. Lett.*, **208**, 59–62 (1993); A.Eychmuller, A.Hasselbarth and H.Weller, *J. Lumin.*, **53**, 113–115 (1992).
h. A. Mews, A. Eychmuller, M. Giersig, D. Schoob and H. Weller, *J. Phys. Chem.*, **98**, 934, (1994); T. Vossmeyer, L. Katsikas, M. Giersig, I. G. Popovic, K. Diesner, A. Chemseddine, A. Eychmuller and H. Weller, *J. Phys. Chem.*, **98**, 7665 (1994); D. Schooss, A. Mews, A. Eychmuller,and H. Weller, *Phys. Rev.*, **B49**, 17072, (1994).
i. V. A. Shakin and E. Hanamura, private communication.
j. Y. Kayanuma and N. Saito, *Solid State Comm.* **84**, 771 (1992).
k. M. Faraday, *Philos. Trans. Roy. Soc. London*, **147**, 145, (1857); G. Mie, *Ann. Phys.* (Leipz.) **25**, 377 (1908).
l. M. Kerker, *J. Colloid Interface Sci.*, **112**, 302 (1986).
m. D. Richard, P. Roussignol and C. Flytzanis, *Opt. Lett.*, **10**, 511 (1985); F. Hache, D. Ricard and C. Flytzanis, *J. Opt. Soc. Am.*, **B3**, 1647, (1986); P. Roussignol, D. Ricard, J. Lukasik and C. Flytzanis, *J. Opt. Soc. Am.*, **B4**, 5, (1987); F. Hache, D. Ricard, C. Flytzanis and U. Kreibig, *Appl. Phys.*, **A47**, 347 (1988); C. Flytzanis, F. Hache, M. C. Klein, D. Ricard and P. Roussignol, *Prog. Opt.*, **29**, 323 (1991)
n. R. Kubo, *J. Phys. Soc. Japan*, **17**, 975, (1962).
o. Y. Silberberg and T. Sands, *IEEE J. Quantum Electronics,* **28**, 1663 (1992).
p. U.Kreibig and L.Genzel, *Surf. Sci.*, **156**, 678 (1985); L.Genzel, T. P.Martin and U.Kreibig, *Z. Phys.*, **B21**, 339 (1975).
q. K. Sayama and H. Arakawa, *Chem. Lett.*, **253**, (1992); *J. Chem. Soc., Chem. Commun.*, **150** (1992).
r. A. E. Neeves and M. H. Birnboim, *Opt. Lett.*, **13**, 1087 (1988); *J. Opt. Soc. Am.* **B6**, 787 (1989); N. Kalyaniwalla, J. W. Haus, R. Inguva and M. H. Birnboim, *Phys. Rev.*, **A42**, 5613 (1990).
s. P. Mulvaney, M. Giesig and A. Henglein; *J. Phys. Chem.*, **97**, 7061 (1993).
t. Y. Kayamuma and H. Momoji, *Phys. Rev.*, **B41**, 10261 (1990).
u. P. M.Bethke and P. B.Barton, *The American Mineralogist*, **56**, 2034–2039 (1971).
v. J. J.Ramsden and M.Gratzel, *J. Chem. Soc., Faraday Trans.* 1, **80**, 919–933 (1984); P. V.Kamat, M. N.Dimitrijevic and R. W.Fessenden, *J. Phys. Chem.*, **91**, 396–401 (1987).
w. R.Rossetti and L.Brus, *J. Phys. Chem.*, **86**, 4470–4472 (1982).
x. M. K.Sheinkman, I. B.Ermolovich and G. L.Belen'kii, *Soviet Phys. -Solid State*, **10**, 2069–2076 (1969); N.Susa, H.Watanabe and M.Wada, *Japan. J. Appl. Phys.*, **15**, 2365–2370 (1976).
y. A.Wokaun, *Solid State Physics,* **38**, 223 (1984).
z. E. H.Swift and W. P.Schaefer, Quantitative Elemental Analysis, p.280, W. H. Freeman, San Francisco (1962).
a'. H.Hirsh, A.deCugnac, M. C.Cadet and J. Pouradier, *Comp. Rend.*, **263B**, 1328 (1966).
b'. C. F.Bohren and D. R.Huffman, Absorption and Scattering of Light by Small Particles, John Wiley and sons, New York (1983).
c'. H. E.Bennett and J. M.Bennett, Optical Properties and Electronic Structure of Metal and Alloys, (F. Abeles ed.), p.175ff, North-Holland, Amsterdam (1966).

REFERENCES

1. H. S. Zhou, I. Honma, H. Komiyama and J. W. Haus, *J. Phys. Chem.*, **97**, 895 (1993); H. S. Zhou, I. Honma, H. Komiyama and Joseph W. Haus, *Chem. Mater.*, **6**, 1534, (1994).
2. J. W. Haus, H. S. Zhou, I. Honma and H. Komiyama, *Phys. Rev. B.*, **47**, 1357 (1993).
3. J. W. Haus, N. Kalyaniwalla, R. Inguva and C. M. Bowdeb, *J. Appl. Phys.*, **65**, 1420 (1989); J. W. Haus, R. Inguva and C. M. Bowdeb, *Phys. Rev.*, **A40**, 5729 (1989); M. J. Bloemer and J. W. Haus, *Opt. Lett.*, **17**, 598 (1992); *Appl. Phys. Lett.*, **61**, 1619 (1992); N. Kalyaniwalla, J. W. Haus, R. Inguva and M. H. Birnboim, *Phys. Rev.*, **A42**, 5613 (1990).
4. I. Honma, T. Sano and H. Komiyama, *J. Phys. Chem.*, **97**, 6692 (1993).
5. J. W. Haus, H. S. Zhou, S. Takami, M. Hirasawa, I. Honma and H. Komiyama *J. Appl. Phys.*, **73**, 1043 (1993).
6. H. S. Zhou, I. Honma, H. Komiyama and J. W. Haus, *Phys. Rev.*, **B50**, 12052, (1994).

19

Structure and Physical Properties of Chalcogen Microclusters

Hirohisa Endo [a], Makoto Yao [b] and Itsuro Yamamoto [c]

[a] Faculty of Engineering, Fukui Institute of Technology
Gakuen, Fukui 910-0028, Japan
[b] Department of Physics, Graduate School of Science, Kyoto University
Sakyo-ku, Kyoto 606-8224, Japan
[c] Faculty of Education, Hirosaki University
Hirosaki-shi, Aomori 036-8224, Japan

Purpose of the present study

Chalcogens show a wide variety of structural, optical and electrical properties in crystalline, amorphous and liquid states. Studies of microclusters, which are the building blocks of these condensed phases, may give valuable information to promote our understanding of the interesting and peculiar properties of chalcogens.

Content of the present study

1. Introduction

Chalcogens such as sulfur, selenium and tellurium are composed of polymeric chain or ring molecules connected by twofold coordinated covalent bonds. The chalcogens have much crystalline polymorphism and they show semiconducting behavior. The upper part of the valence bands are formed by lone pair (LP) orbitals, which play an important role in stabilizing the chain or ring structure, and the lower part of the conduction bands are formed by antibonding σ^* orbitals. When the chalcogen molecules are excited by photo illumination or thermal agitation, new defect states, which are hybridized with the LP orbitals, are expected to be created. In fact, various photo-induced phenomena have been observed in amorphous chalcogenides. Furthermore, when a sufficient number of defect states are generated in liquid Se under high temperature and pressure near the liquid-gas critical point, the semiconductor is transformed to a metal. Clustering effects due to intermolecular interaction are essential.[1-3] From these points of view studies of chalcogen microclusters will provide useful information to promote our understanding of the physics of disordered matter. [4-7]

In this report we discuss the structure and photo-induced phenomena of chalcogen microclusters confined in micropores of zeolites, the stability of chalcogen microclusters produced in free space by the supersonic jet expansion method, and the metal-semiconductor transition in tellurium nano-droplets.

2. Chalcogen microclusters confined in the micropores of zeolites

First we produced isolated Se chains in one-dimensional channels of mordenite, a kind of zeolite, and conducted structural analysis using EXAFS (see Fig. 1). We found that the covalent bond is contracted by 0.4 Å and the spiral period is changed into about 3.5 atoms, instead of 3, when the interchain

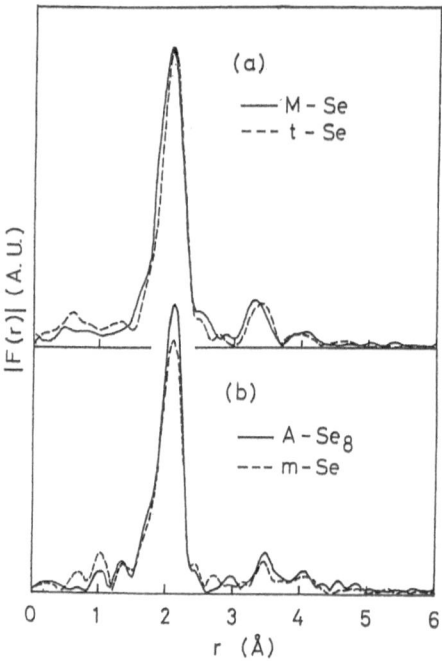

Fig. 1   Radial distribution functions determined from EXAFS measurements on the Se k-edge. (a) Se chains confined in the channels of mordenite (M-Se) and trigonal Se (t-Se), (b) Se_8-membered rings in zeolite 4 A (A-Se_8) and monoclinic Se (m-Se).

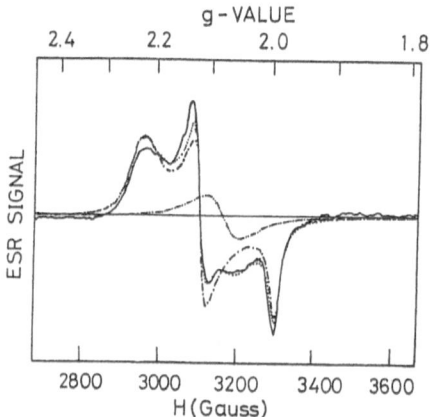

Fig. 2   Photo induced ESR signal in M-Se. The solid line represents the experimental result and the dotted line the fitting curve. The latter is composed of an anisotropic component (dash-dot line) and anisotropic component (dash-dot-dot line).
(From Y. Katayama, K. Maruyama, M. Yao. and H. Endo, *J. Phys. Soc. Jpn.*, **60**, 2237 (1991)).

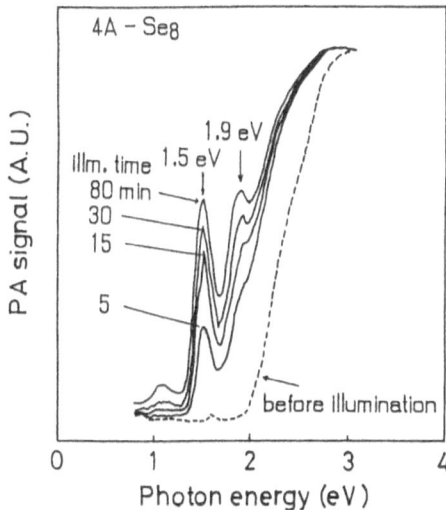

Fig. 3   Photoacoustic spectrum of A-Se$_8$ at 70 K. The dotted line represents the spectrum before photoillumination and the solid line after illumination with light of 2.8 eV. The illumination time is indicated in the figure.
(From K. Maruyama, T. Tsuzuki, M. Yao and H. Endo, *Surf. Rev. Lett.*, **3**, 713 (1996)).

interaction is removed. [a,b,c] Next we measured the Mössbauer effect of ^{125}Te nucleus for Se-Te mixed chains, and concluded that the anisotropic character of bonding is enhanced by isolating the chains.[d] The isolation widens the optical gap by 0.5 eV for the Se chains and by 1.2 eV for the Te chains.[c] Furthermore when the isolated Se chains are photoilluminated at low temperatures, new absorption peaks grow at 1.8 and 2.2 eV, accompanied by rather anisotropic ESR signals (see Fig. 2).[a,c] These phenomena are assigned to the creation of neutral dangling bonds.[e]

On the other hand, chalcogen 8-membered rings are produced within a micropore of zeolite 4A with the diameter of 11 Å.[f,8] These samples also exhibit photo darkening phenomena.[f,9] In particular, a very sharp peak appears at 1.5 eV (see Fig. 3),[f,9] which is not observed for the isolated Se chains. These photo-induced absorption bands can be erased by annealing at near room temperature and photo-bleaching effects are also found.

We have also studied photo-induced phenomena in chalcogenides such as CdS and AsS confined in zeolite Y.[g,h]

3. Production of chalcogen microclusters by the supersonic jet expansion method

We have constructed an apparatus with which chalcogen microclusters are produced in free space by the supersonic jet expansion method. The clusters produced are ionized by electron impact and their mass is measured by using time-of-flight mass spectrometer.[10] Sulfur is relatively stable in the forms of S_7 and S_8. When the electron impact energy is increased, S_2 dimers are evaporated from the sulfur clusters, leading to fragmentation (see Fig. 4). Eight-membered clusters such as $Se_{8-x}S_x$ are dominant species in Se-S mixtures. For pure Se, however, the most abundant species are 5-membered clusters.

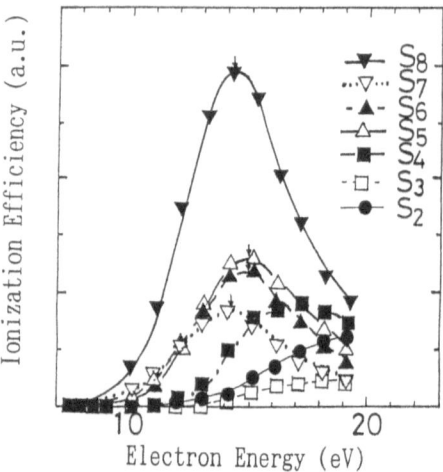

Fig. 4 Ionization efficiency of various sulfur clusters as a function of the electron impact energy.

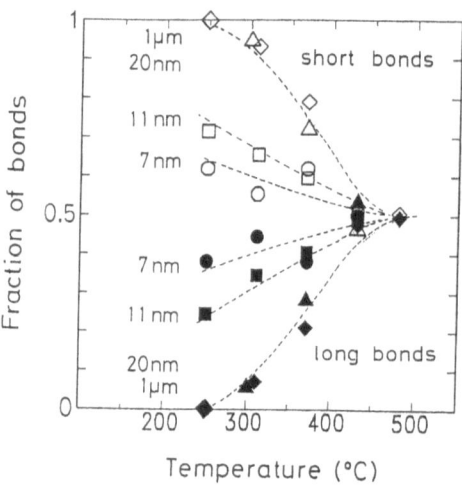

Fig. 5 Fractions of the long and short covalent bonds determined by EXAFS measurements plotted
as a function of temperature for supercooled liquid Te droplets. The numbers in the figure
denote the average diameter of the droplets.
(From T. Tsuzuki, M. Yao and H. Endo, *J. Phys. Soc. Jpn.*, **64**, 493 (1995)).

4. Metal-Semiconductor transition in chalcogen nano-droplets

Unlike other chalcogen elements tellurium is transformed to a metal on
melting. We have found that Te nano-droplets dispersed in a NaCl matrix are
maintained in the liquid state down to 250°C, which is lower than the melting
point by 200°C. The static structure of supercooled liquid Te was measured by
EXAFS and neutron diffraction. [11,12] The results indicate that liquid Te has a
twofold coordinated chain structure with long and short covalent bonds. The
fraction of the longer bond decreases with decreasing temperature and vanishes
in the strongly supercooled state where the metallic Te is changed to a
semiconductor, suggesting that the existence of the longer bond is closely related

to the metallicity of liquid Te (see Fig. 5). We have also investigated dynamic structures by means of quasielastic and inelastic neutron scatterings. The vibrational density of states changes dramatically in the supercooled state.

PUBLICATIONS

1. "The structure and properties of liquids" : H. Endo and M. Yao, Frontiers in Physics, **31**, Kyoritu, (1993) [in Japanese].
2. The optical properties of liquid chalcogens: H. Ikemoto, I. Yamamoto and H. Endo, *J. Non-Cryst. Solids*, **156–158**, 732–735 (1993).
3. The optical properties of liquid chalcogens at High Temperature and Pressure: H. Ikemoto, I. Yamamoto, M. Yao and H. Endo, *J. Phys. Sac. Jpn.*, **63**, 1611–1622 (1994).
4. Japanese translation of *The Physics of Structurally Disordered Matter* (by N. E. Cusack) Part I: H. Endo and M. Yao (Yoshioka, 1994).
5. Japanese translation of *The Physics of Structurally Disordered Matter* (by N. E. Cusack) Part II: H. Endo and M. Yao (Yoshioka, 1994).
6. Cluster formation in expanded fluid mercury and dilute amalgams near the liquid-gas critical point: M. Yao, K. Takehana and H. Endo, *J. Non-Cryst. Solids*, **156–158**, 807–811 (1993).
7. Liquids near the Critical Point: H. Endo and M. Yao, *Elementary Processes in Dense Plasmas*, (S. Ichimaru and S. Ogata, eds.) pp307–316 (1995).
8. EXAFS Study of Chalcogen Microclusters: H. Endo, K. Maruyama, T. Tsuzuki and M. Yao, *Jpn. J. Appl. Phys. Suppl.*, **32–2**, 773–775 (1993).
9. Photo Induced Phenomena of Chalcogen Microclusters Confined in the Zeolite Cages: K. Maruyama, T. Tsuzuki, M. Yao and H. Endo, *Surface Review and Letters* , 711–715(1996).
10. Production of Sulfur-Selenium Mixed Clusters by the Supersonic Jet Expansion Method: M. Yao, K. Nagaya, T. Hayakawa and H. Endo, *Surface Review and Letters*, 201–204 (1996).
11. Structure and electronic properties of liquid tellurium: H. Endo, *J. Non-Cryst. Solids*, **156–158**, 667–674 (1993).
12. Structure of chalcogen nano-droplets: T. Tsuzuki, A. Sano, Y. Kawakita, Y. Ohmasa, M. Yao, H. Endo, M. Inui and M. Misawa, *J. Non-Cryst. Solids*, **156–158**, 695–699 (1993).
13. Quasielastic and Inelastic Neutron Scatterings of Liquid Tellurium: H. Endo,T. Tsuzuki, M. Yao, Y. Kawakita, K. Shibata, T. Kamiyama, M. Misawa and K. Suzuki, *J. Phys. Soc. Jpn.*, **63**, 3200–3203 (1994) (letter).
14. Static and Dynamic Structures of Liquid Tellurium: T. Tsuzuki, M. Yao and H. Endo, *J. Plys. Soc. Jpn.*, **64** 485–503 (1995).

REFERENCES

a. The Isolated Se Chains in the Channels of Mordenite Crystal: K. Tamura, S. Hosokawa, H. Endo, S. Yamasaki, and H. Oyanagi, *J. Phys. Soc. Jpn.*, **55**, 528 (1986).
b. EXAFS Study on Selenium-Tellurium Mixed Chains: M. Inui, M. Yao, and H. Endo, *J. Phys. Soc. Jpn.*, **57**, 553(1988).
c. Photo-Induced Phenomena in Isolated Selenium Chains: Y. Katayama, M. Yao, Y. Ajiro, M. Inui, and H. Endo, *J. Phys. Soc. Jpn.*, **58**, 1811 (1989).
d. Mössbauer Studies on the Tellurium-Selenium Mixed Chains: H. Sakai, M. Yao, M. Inui, K. Maruyama, K. Tamura, K. Takimoto, and H. Endo, *J. Phys. Soc., Jpn.* **57**, 3587 (1988).
e. Spatial Correlations and Defects in Isolated Selenium-Sulfur Mixed Chains: Y. Katayama, K. Maruyama, M. Yao and H. Endo, *J. Phys. Soc. Jpn.*, **60**, 2229–2240 (1991).
f. Chalcogen 8-membered rings in the micropores of zeolite: K. Maruyama, T. Tsuzuki, M. Yao and H. Endo, *New Ceramics*, **10**, 41–44 (1992) [in Japanese].
g. Microclusters Confined in Zeolite Cage: Y. Katayama, K. Maruyama, and H. Endo, *J. Non-Cryst. Solids*, **117 & 118**, 485 (1990).
h. Photodarkening and photobleaching of CdS microclusters grown in zeolites: T. Moyo, K. Maruyama and H. Endo, *J. Phys.; Condens. Matter*, **4**, 5653–5664 (1992).

20

Synthesis and Photophysical Properties of III–V Semiconductor Nanocrystals

Hiroyuki Uchida*

Laboratory of Electrochemical Energy Conversion, Yamanashi University
4-3 Takeda, Kofu 400-0511, Japan

Purpose of the present study

Recently, great interest has been devoted to the photochemical and photophysical properties of quantized semiconductor nanocrystals. Much attention has focused on the optical nonlinearity of semiconductor nanocrystals, *i.e.* resonant third-order nonlinear optical processes for II–VII (CdS_xSe_{1-x}) or I–VII (CuCl) semiconductor-doped glasses or polymers. Since III–V semiconductor nanocrystals must exhibit large size-quantization effects due to very small effective masses of electrons and holes, they must show remarkable optical nonlinearity as expected from results obtained on GaAs quantum wells. However, it is very difficult to prepare monodispersed quantized III- V semiconductor particles that can be well characterized. The purpose of the present study is to prepare stable and monodispersed III–V nanocrystals by a wet process and to examine their photophysical properties.

Contents of the present study

1. Synthesis and photophysical properties of GaAs nanocrystals

Few reports are available regarding the preparation of size-quantized III–V compounds because of the difficulty involved. Recently, a synthesis of GaAs particles based on the reaction of $GaCl_3$ with $[(CH_3)_3Si]_3As$ in quinoline was reported.[a] Although this system produced quantized colloidal GaAs (diameter = 3–5 nm), the optical properties of the colloidal nanocrystals in quinoline were found[1] to be masked by the presence of interfering molecular species, which were believed to be either quinoline oligomers or Ga-quinoline complexes.

1.1 Synthesis and characterization of GaAs nanocrystals

We have succeeded in preparing GaAs nanocrystals without accompanying molecular interfering species.[2] This wet synthesis involves the reaction of gallium acetylacetonate, $Ga(acac)_3$, and $[(CH_3)_3Si]_3$ As at reflux (216°C) in triglyme for 70 h. Transparent orange GaAs colloids were obtained by filtering the resulting brown slurries through a 0.2 μm-pore size filter. We confirmed the formation of

---

* This research was performed at the Faculty of Engineering, Osaka University.

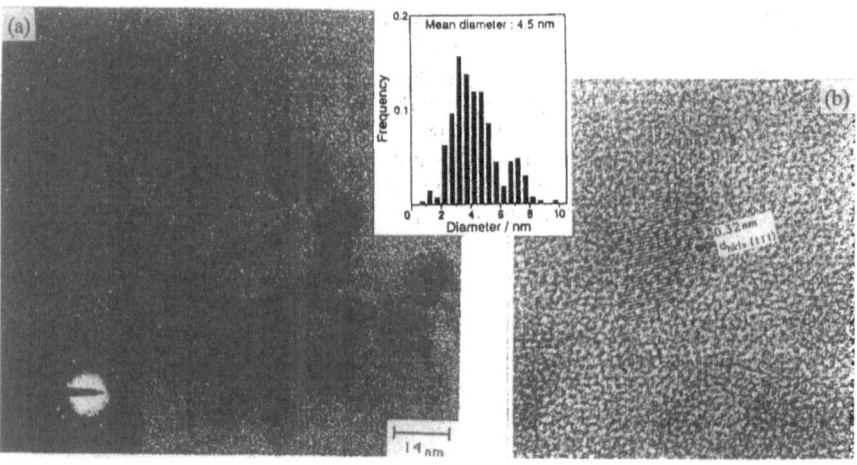

Fig. 1 (a) TEM image and the electron diffraction patterns of GaAs nanocrystals deposited onto amorphous carbon overlayers on a Cu grid. (b) High-resolution TEM image of GaAs nanocrystal.
(From H.Uchida, *J.Phys.Chem.*, **96**, 1158 (1992)).

GaAs nanocrystals in these colloids by TEM and electron diffraction data. Figure 1(a) shows GaAs particles in a colloid. The individual particle diameter ranges from 1.5 to 9 nm (mean diameter = 4.5 nm), and electron diffraction gives clear *hkl* zinc blende patterns of (111), (220), (311) and weak (422) for GaAs. In the high resolution picture (Fig. 1(b)), lattice planes can be seen and their spacing of 0.32 nm corresponds to d(111) of GaAs.

Figure 2 shows the optical absorption spectra of the GaAs colloid after a series of ultrafiltrations. The absorption onset of the GaAs colloid filtered with 70 nm-filter is about 600 nm, being blue-shifted from that of bulk material (874 nm) due to a quantization effect. The spectrum for the 1.5 nm-filtrate is different from that of 70 nm-filtrate and is similar to that of the blank solution containing no GaAs particles, which was prepared by heating only Ga(acac)$_3$ in triglyrme under the same conditions. These results clearly indicate that quantized GaAs

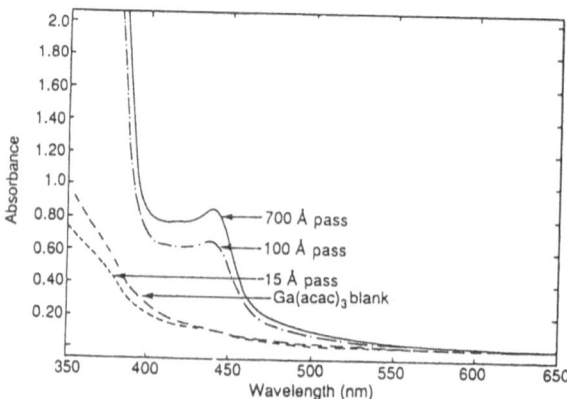

Fig. 2 Optical absorption spectra of the GaAs colloid after a series of ultrafiltrations.
(From H.Uchida, *J.Phys.Chem.*, **96**, 1158 (1992)).

Table 1 Content (mmol·dm⁻³) of Ga and As determined by fluorescent X-ray analysis

No.	Sample	Ga	As
1	loaded*	20	20
2	0.2 μm-filtrate	7.20	1.32
3	1.2 nm-filtrate	3.04	0.48
4	Δ (2–3)**	4.16	0.84

* : Original concentration of the starting materials.
**: Difference between runs No. 2 and 3, corresponding to the composition of GaAs nanocrystals greater than 1.2 nm.

(From H.Matsumoto and H.Uchida, *Res.Chem.Intermed.*, **20**, 727 (1994)).

nanocrystals (Q-GaAs) are prepared by this synthesis and are not accompanied by molecular interfering species.

The Q-GaAs colloids stored under nitrogen in the dark were very stable and did not aggregate for at least one year. We determined the composition of the Q-GaAs as follows. The Q-GaAs colloids were subjected to ultrafiltration using a 1.2 nm-pore size filter; the Q-GaAs was eliminated from the filtrate as judged from the size distribution (Fig. 1). The amount of Ga and As contained in the resulting filtrate was determined by fluorescent X-ray spectrometry. The results are summarized in Table 1. Results obtained prior to the filtration are included in this table. The 0.2 μm-filtrate (No. 2 in Table 1) contained 36 % of Ga and 6.6 % As of the loaded amount used in this synthesis of the GaAs. The difference in the concentration between No. 2 and No. 3, which is shown in No. 4, must be related to the GaAs particles (diameter = 1–9 nm) left on the 1.2 nm-filter. The amount of Ga given in No.4 was about four times that of As, suggesting that some Ga-species other than GaAs existed on the filter. The species were not detected by TEM or electron diffraction. Since triglyme is a good complexing agent for many metal ions, the excess Ga is probably a Ga-(acac)-triglyme complex, which may act as an effective stabilizer against the aggregation for the nanocrystals. If it is assumed that all the As contained in the 0.2 μm-filtrate were involved in the formation of GaAs particles, 0.84 mmol·dm⁻³ of GaAs was obtained. This is close to the upper concentration without aggregation reported for other Q-particles such as CdS or ZnO in colloidal solution with an appropriate stabilizing agent.

1.2 Quantum-size effect in GaAs nanocrystals

Using the simple effective mass approximation for the relation between the energy gap, E^*, and the radius, R, of the size-quantized semiconductor,

$$E^* = E_g + \frac{\hbar^2 \pi^2}{2R^2}\left(\frac{1}{m_e} + \frac{1}{m_h}\right) - \frac{1.8e^2}{\varepsilon R} \tag{1}$$

the absorption onset at 600 nm would correspond to a maximum particle size of about 6.5 nm. This is consistent with the TEM results (Fig. 1).

Next, the absorption spectra was analyzed on the model assuming a Gaussian distribution of particle sizes. Although the absorption spectrum for an ideal quantum dot is expected to consist of a series of discrete lines, it has been shown that for real quantum dots the discrete lines are broadened by various effects into

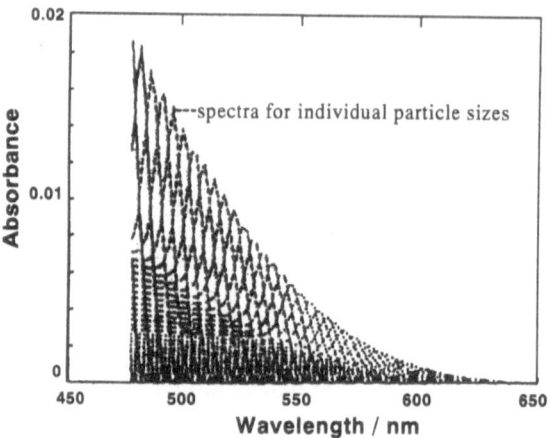

Fig. 3 Calculated spectra for individual particle sizes in the Gaussian distribution used for the fitting.
(From A.J.Nozik and H.Uchida, *Isr.J.Chem.*, **33**, 19 (1993)).

a complicated and very broad absorption band that onsets at the lowest hole to electron transition. Therefore, we have approximated the absorption spectrum of a quantum dot to be comprised of an exciton peak at the lowest hole to electron transition superimposed on a band-like absorbance with the same functional dependence on energy as the bulk semiconductor. The absorption coefficient (α, cm^{-1}) for the exciton peak is assumed to depend cubically on the ratio of the Bohr radius (a_B) for the bulk exciton to the particle radius (R). This follows from the fact that for quantum dots α is proportional to the oscillator strength per unit volume,[b] and the latter scales as $(a_B/R)^3$. The exciton is assumed to have a Gaussian shape with an α at the peak for the bulk exciton at 300 K of 4×10^3 cm^{-1} and a width of 20 meV. For the absorption continuum beyond the exciton peak, the dependence of α on $(1/R)$ is weaker [c]; we assume the absorption coefficient in this spectral region scales to be $(a_B/R)^2$, since we found that the mean particle size and its variance determined from the model fit was not very sensitive to the scaling factor $(a_B/R)^n$ for $n = 0$ to 3.

In our model, we first calculate the distribution of effective GaAs thickness as a function of particle size from a given particle size distribution. The bandgap distribution is calculated using Eq. (1). Then the absorption coefficient as a function of particle size and photon energy is determined. From this information we can calculate the absorbance as a function of particle size; summation of calculated spectra (Fig. 3) over the particle size distribution yields the overall absorption spectrum for the colloid. Application of a nonlinear least-squares fitting procedure to the experimental spectra produces an estimation of the particle size distribution. The concentration of GaAs in the colloid is also determined from the fit.

The absorption tail region from 650 to 470 nm in Fig. 2 was fitted to our model to estimate both the particle size distribution and the concentration of GaAs nanocrystals. Before fitting, the spectrum of the 1.5 nm-pass filtrate was subtracted from that of the 70 nm-pass filtrate. In Fig. 3 showing the calculated absorption spectra for individual particles, it is clearly displayed that the dominance of the lowest exciton peak increases with decreasing particle size.

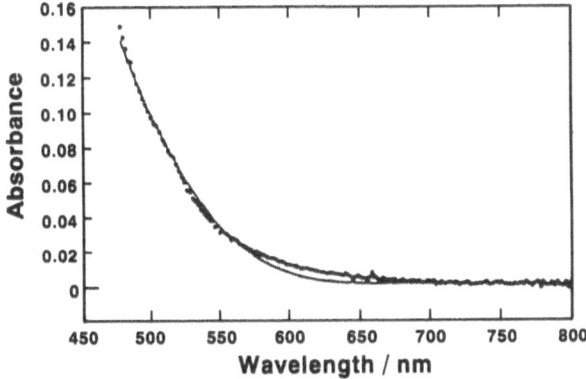

Fig. 4 Fit of model for absorption of a Gaussian distribution of particles sizes (solid line) to experimental spectrum (●).
(From A.J.Nozik and H.Uchida, *Isr.J.Chem.*, **33**, 19 (1993)).

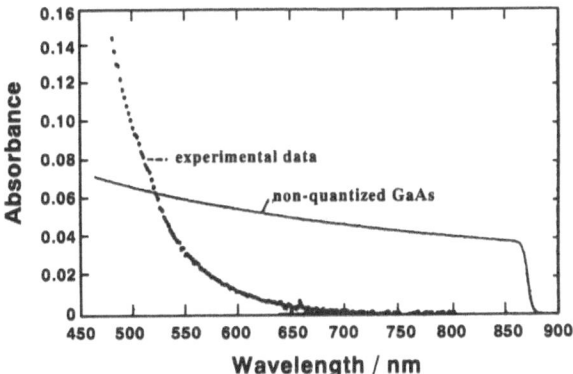

Fig. 5 Comparison of experimental data with expected spectrum of nonquantized GaAs of the same concentration in the colloid.
(From A.J.Nozik and H.Uchida, *Isr.J.Chem.*, **33**, 19 (1993)).

The summation over the particle size distribution yielded the absorption spectrum for the colloid, as shown by a solid line in Fig. 4. Comparison of the calculated spectrum with the experimental data (plotted by ●) indicated that the fit was very good over this region. From the calculation, we obtained the mean particle diameter of 4.0 nm with a standard deviation of 0.97 nm, being consistent with the TEM results (Fig. 1). Also, the calculation gave the concentration of the GaAs of 0.71 mM, being close to that determined by a fluorescent X-ray spectrometry (0.84 mM). For comparison, the calculated spectrum of non-quantized GaAs with the same concentration as determined above is also shown by the dotted line in Fig. 5. The size-quantization effect in the present GaAs colloid is clearly demonstrated with respect to the shape and the onset of the spectrum.

Below 470 nm the absorption increased more sharply and peaked at 440 nm while the 1.5 nm filtrate did not show a peak. Application of our model over the whole spectral region resulted in a poor fit (not shown) to the experimental data. Also, the peak at 440 nm showed widely varying intensities with repeated syntheses, and in some cases appeared as a small shoulder. At the present time, the origin of the 440 nm peak is not understood. However, the absorption above

470 nm can definitely be attributed to quantized GaAs nanocrystals.

1.3 Photobleaching of Q-GaAs

When semiconductor nanocrystals are excited by an intense laser beam, photobleaching caused by state-filling or surface-trapped electrons can be observed. However, so far, little is reported for visible-light-excited photobleaching, which is very important for optical switch applications.

Transient photobleaching of the GaAs colloid was observed during picosecond laser pump-probe experiments. A mode-locked Nd-YAG laser generating 532 nm pulses with a width of 18 ps was employed as the exciting (pump) beam. Figure 6 shows the changes in absorbance observed for Q-GaAs colloids taken 100 ps after the pump pulse. It is seen that with excitation at 532 nm the 70- and 10-nm filtrates show appreciable photobleaching beginning near 600 nm and peaking near the pump wavelength, whereas the blank sample of Ga(acac)$_3$ in triglyme shows no significant change. In the wider wavelength range for the probe spectrum shown in Fig. 7, the Q-GaAs colloid shows photobleaching in two regions; one centered at the pump energy and another at higher energies

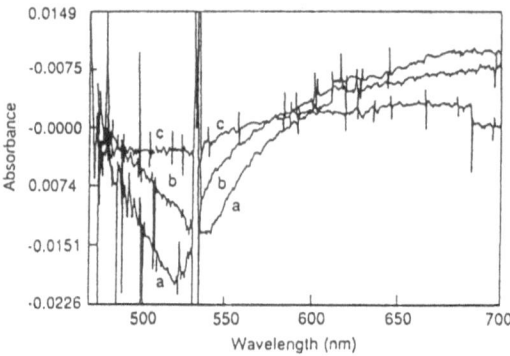

Fig. 6 Photobleaching in pump-probe experiment for GaAs colloid at 100 ps after 532 nm excitation, (a) 70 nm-filtrate, (b) 10 nm-filtrate, (c) blank sample of Ga(acac)$_3$ in triglyme. (From H.Uchida, *J.Phys.Chem.*, **96**, 1158 (1992)).

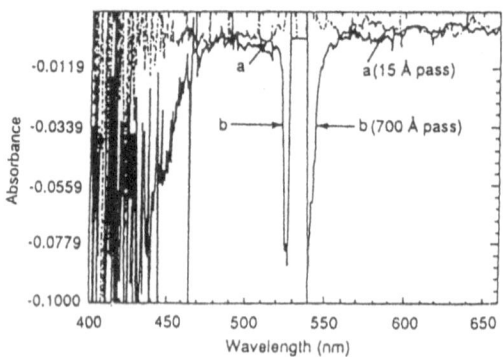

Fig. 7 Photobleaching of the GaAs colloids showing wider wavelength range at 0 ps after the pump pulse, (a) 70 nm-filtrate, (b) 1.5 nm-filtrate containing no GaAs particles. (From H.Uchida, *J.Phys.Chem.*, **96**, 1158 (1992)).

(below 475 nm). The photobleaching at high energy is quite unusual and may be caused by a two-photon absorption process in the Q-GaAs particles. The photobleaching centered around the pump energy is the usual process that is well established for semiconductor nanocrystals.

1.4 Optical nonlinearity of Q-GaAs in DFWM experiments

The third-order optical nonlinearity of GaAs colloids was examined using the configuration of the degenerative four-wave mixing (DFWM) with two forward incident beams at room temperature.

When a nonlinear optical material is excited with two forward laser beams in the directions denoted by k_1 and k_2 wave-vectors, diffracted signals due to the third-order nonlinearity must be observed in the directions of $2k_2 - k_1$ and $2k_1 - k_2$ on the opposite side of the sample. The intensity of the diffracted signal beam I_s is related both to the incident beam intensity I_0 and to the absolute value of the third-order susceptibility $|\chi^{(3)}|$ as follows,[d]

$$I_s = \frac{2^8 \pi^4 \omega^2 (1-T)^2 T |\chi^{(3)}|^2 I_0^3}{n^4 c^4 \alpha^2} \tag{2}$$

where α is the absorption coefficient, T the transmittance, n the refractive index, c the light velocity, and ω the angular frequency of the laser beam, respectively. As given by Eq.(2), the intensity of the diffracted signal I_s should depend cubically on that of the incident beam I_0.

We performed the DFWM experiments using the original and the concentrated GaAs colloids, a blank sample, and a commercial CdS_xSe_{1-x}-doped glass (HOYA, Y-52) as a reference sample whose $|\chi^{(3)}|$ at 532 nm is reported to be 1.3×10^{-9} esu. The sample was excited by two laser beams intersecting at an angle of 1.5° from a Q-switched Nd-YAG laser generating 532 nm pulses with a width of 7–11 ns. Distinct diffracted signals were observed for the GaAs colloids, and their intensities indeed increased cubically with increasing laser power, as

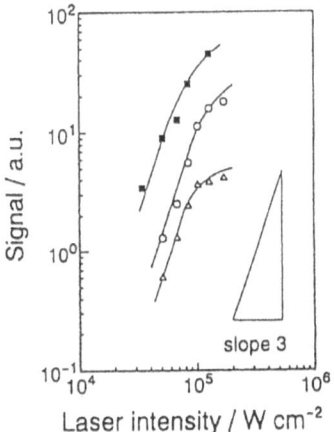

Fig. 8 Beam power dependence of diffracted DFWM signal. △: original Q-GaAs, ○:concentrated Q-GaAs, ■:Y-52 glass filter.
(From H.Matsumoto and H.Uchida., *Denki Kagaku*, **61**, 919 (1993)).

shown in Fig. 8. On the other hand, no signals were detected for the blank sample, indicating that the GaAs nanocrystals are responsible for the optical nonlinearity. The signal intensity is higher for the concentrated GaAs colloid than that for the original colloid. The value of $| \chi^{(3)} |$ for the concentrated GaAs colloids was determined to be 3.5×10^{-10} esu by applying a reference method to Eq.(2). The volume fraction of GaAs in the colloid, which was determined by a fluorescent X-ray spectrometry, was very small ($\sim 10^{-5}$), compared to the case of the Y-52 glass filter where $2-5 \times 10^{-3}$ volume fraction of CdS_xSe_{1-x} is contained. Consequently, $| \chi^{(3)} |$ for the GaAs nanocrystals will be on the order of 10^{-8} esu if the volume fraction of the semiconductor particles is adjusted to the same value as that of Y-52.

2. Synthesis and photophysical properties of InAs nanocrystals

InAs ($E_g = 0.35$ eV) is a narrow direct-gap III–V semiconductor and its optical bandgap must be tunable from near infrared to visible range, depending on the degree of size quantization. However, only two reports are available regarding the preparation of InAs nanocrystals for optical applications. Wang and Heron have reported an application of metalorganic chemical vapor deposition (MOCVD), *i.e.* reaction of $In(C_2H_5)_3$ with AsH_3, in porous glass supports,[e] although the optical bandgap of InAs-doped glass seemed to be still narrow as judged from its black color. The other report concerns a magnetron rf-sputtering of InAs, GaAs and SiO_2, targets, which resulted in $In_xGa_{1-x}As$ nanocrystals embedded in a SiO_2, glass substrate.[f]

We report a new synthesis of colloidal InAs nanocrystals by a wet process, and the resulting crystal structure and chemical composition are evaluated. Furthermore, it is demonstrated that the quantized InAs nanocrystals prepared in this study show a large third-order optical nonlinearity.

2.1 Synthesis and characterization of InAs nanocrystals

A mixture containing 0.5 mmol of $In(acac)_3$ and 0.5 mmol of $[(CH_3)_3Si]_3As$ in 25 cm³ of triglyme was heated at reflux (216°C) for 70 hours. Brown turbid slurries were obtained by this reaction. Transparent brown colloids were obtained

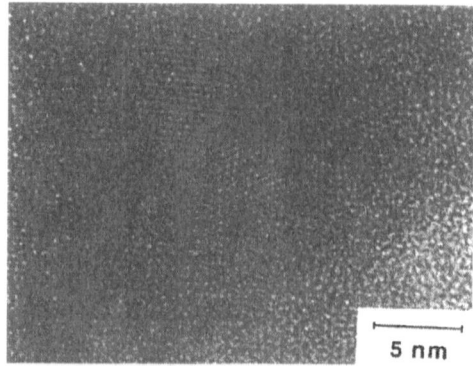

5 nm

Fig. 9 TEM image of InAs nanocrystals.
(From H.Uchida, *Chem.Mater.*, **5**, 717 (1993)).

Table 2 Lattice spacing, d, of InAs nanocrystals determined by electron diffractions

d/nm (JCPDS data)[21]	hkl	d/nm (obsd)[a]	error/%[b]
0.3498	111	0.3441	−1.7
0.3030	200	0.2964	−2.2
0.2142	220	0.2141	−0.1
0.1826	211	0.1835	+0.5

[a] Lattice spacing determined by reference to $d(200)$ of Au single crystal measured under the same conditions. [b] Experimental error for d, compared with JCPDS data.

(From H.Uchida, *Chem.Mater.*, **5**, 717 (1993)).

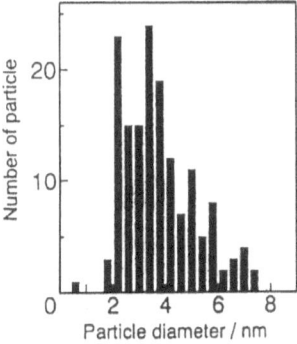

Fig. 10 Size distribution of InAs nanocrystals. (From H.Uchida, *Chem.Mater.*, **5**, 717 (1993)).

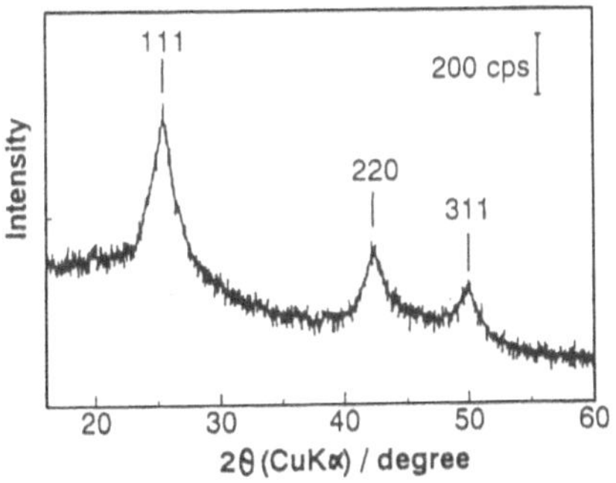

Fig. 11 X-ray diffractogram of the dark brown powder obtained by filtration of original slurries with 0.2 μm-filter. (From H.Uchida, *Chem.Mater.*, **5**, 718 (1993)).

from the slurries by filtering them through a 0.2 μm-filter. The colloids stored under nitrogen in the dark were very stable and did not aggregate for at least 6 months.

Figure 9 shows a typical TEM photograph of the InAs particles deposited onto the carbon overlayer of a Cu grid from the 0.2 μm-filtrate. Clear lattice planes are seen in Fig. 9. By reference to the lattice spacing of (200) plane of gold single crystal in the TEM picture obtained under the same conditions, the lattice spacing of the particles was determined to be 0.346 ±0.003 nm, which is assigned to $d(111) = 0.3498$ nm of InAs. The electron diffraction of the particles gave clear zinc-blende type patterns of (111), (220), (311) and weak (200) of InAs, as summarized in Table 2. Figure 10 shows the size distribution of the InAs particles, which was obtained for 154 particles with clear lattice fringes on several TEM photographs. The particle size ranges from 1 to 8 nm, and the average diameter is 3.76 nm with a standard deviation of 1.35 nm.

Dark brown powder produced in the original slurries were collected on the 0.2 μm-filter. After drying under vacuum, 0.193 g of dark brown powder was obtained. As shown in Fig. 11, the X-ray diffractogram of the dark brown powder shows very broad peaks which can be assigned to zinc-blende type InAs. The particle size estimated by Scherrer's equation is about 4 nm, being consistent with that observed by TEM for nanocrystals in the colloids. This result suggests that the dark brown powder contains either InAs nanocrystals which were not

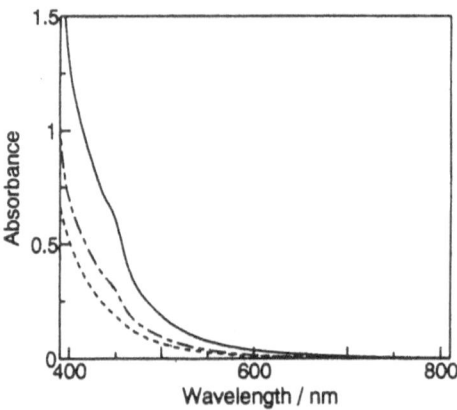

Fig. 12 Absorption spectra of the InAs colloid filtered through 0.2 μm(———) and 1.2 nm-
(— · —) pore size filters (- - - - -): blank sample, In(acac)₃ in triglyme.
(From H.Uchida, *Chem.Mater.*, **5**, 718 (1993)).

dispersed in the colloids or poorly crystalline InAs bulk powder.

2.2 Optical absorption spectra and quantum size effect of InAs nanocrystals

Figure 12 shows the absorption spectra of the 0.2 μm- and 1.2 nm-filtrates
of the InAs colloids. For the 0.2 μm-filtrate, the absorption onset is about 730 nm
(1.7 eV), being blue-shifted greatly from that of the bulk material (3543 nm).
Since the InAs colloid which was concentrated to one-half the original volume
exhibited almost the same absorption onset, the absorption onset at about 730 nm
seems to be a characteristic of the InAs nanocrystals. Using Eq. (1) for the relation
between the bandgap and the particle size, the diameter of the InAs particle is
estimated to be 6.6 nm. This is consistent with the TEM results (Fig. 10), since the
optical absorption threshold of the colloids must be determined by the largest
particles in the colloids. The spectrum for the 1.2 nm-filtrate is quite different from
that of the 0.2 μm-filtrate and is rather similar to that of an In(acac)₃ triglyme
blank solution containing no InAs particles. The weak absorption from ca. 600 nm
for the blank solution is probably attributed to an In-(acac)-triglyme complex,

Table 3 Content of In and As in the filtrates of InAs colloid and in the dark brown powder determined
by fluorescent X-ray analysis

no.	sample	In/mmol[a]	As/mmol[a]
		Solutions	
1	loaded[b]	0.500	0.500
2	0.2-μm filtrate	0.187	0.033
3	1.2-nm filtrate	0.123	0.019
4	Δ(2–3)[c]	0.064	0.014
	Dark Brown Powder		
		0.360 (21.4 wt%)	0.304 (11.8 wt%)

[a] In 25 cm³ of each solution or in 0.193 g of the dark brown powder. [b] Original amount of the starting
material. [c] Difference between the run nos. 2 and 3, corresponding to the composition of InAs
nanocrystals greater than 1.2nm- particles.

(From H.Uchida, *Chem.Mater.*, **5**, 718 (1993)).

since triglyme is a good complexing agent for many metal ions.

Table 3 shows analytical result for the content of In and As in the InAs colloids. The content of In is much higher than that of As for both 0.2 μm- and 1.2 nm-filtrates. The composition of the InAs particles in the colloids was estimated by subtracting the amount of In and As obtained for the 1.2 nm-filtrate from that for the 0.2 μm-filtrate, since, as mentioned above, almost all InAs particles were removed by the 1.2 nm pore-size filter. The composition of the InAs nanocrystals obtained in this way contained excess In, whose source may be the In-(acac)-triglyme complex, which may serve as a stabilizing agent to prevent the aggregation of the InAs nanocrystals. The amount of InAs produced in this synthesis is estimated to be 0.014 mmol (0.55 mM), with the assumption that the amount of As determined for the residue on the 1.2 nm filters is that from the InAs nanocrystals. This means that 2.8 % of $[(CH_3)_3Si]_3$ As in the original solution was used in the chemical synthesis of InAs. A large portion of the loaded materials was found to be included in the dark brown powder as shown in Table 3. Judging from the content of In and As, the dark brown powder contains an appreciable amount of organic moieties besides InAs crystals, as confirmed by XRD.

2.3 Optical nonlinearity of InAs nanocrystals

The optical nonlinearity of the InAs nanocrystals was examined by DFWM experiments in the same manner as described in the previous section. Distinct diffracted signals were observed for the InAs colloids, and their intensities indeed increased cubically with increasing laser power, as shown in Fig. 13. On the other hand, no signals were detected for the 1.2 nm-filtrate and the blank sample,indicating that it is the InAs nanocrystals that are responsible for the optical nonlinearity. The signal intensity is higher for the concentrated InAs colloid than for the original colloid. The value of $|\chi^{(3)}|$ for the concentrated InAs colloids was determined to be 3×10^{-10} esu, being similar to that of Q-GaAs colloid. Since the colloid solution contained only a small volume fraction of InAs (~10^{-5}), compared to the case of the Y-52 colored glass where $2–5 \times 10^{-3}$ volume

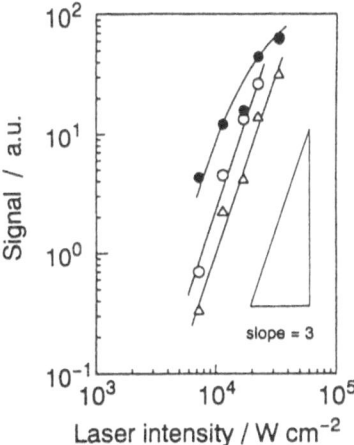

Fig. 13 Beam power dependence of diffracted DFWM signal; ○: concentrated InAs colloid , △: original InAs colloid, ■: Y-52 glass filter.
(From H.Uchida, *Chem.Mater.*, **5**, 719 (1993)).

fraction of CdS_xSe_{1-x} is contained, $| \chi^{(3)} |$ for the InAs nanocrystals will be on the order of 10^{-8} esu if the volume fraction of the semiconductor particles is adjusted to the same value as that of Y-52. Furthermore, the InAs particles that can respond to monochromatic light of 532 nm must be limited to those having a diameter larger than 5.4 nm if the bandgap vs. the particle size relation is given by Eq. (1). Large particles occupy only 12 % of the prepared InAs nanocrystals as judged from the particle size distribution given in Fig. 10. Since $| \chi^{(3)} |$ depends on the wavelength and volume fraction, much higher $| \chi^{(3)} |$ may be obtained at shorter wavelength if the InAs nanocrystals are doped in a solid matrix at high concentration.

PUBLICATIONS

1. H. Uchida, C. J. Curtis and A. J. Nozik, *J. Phys. Chem.*, **95**, 5382 (1991).
2. H. Uchida, C. J. Curtis, P. V. Kamat, K. M. Jones and A. J. Nozik, *J Phys. Chem.*, **96**, 1156 (1992).
3. H. Matsumoto, H. Uchida, T. Sakata, H. Mori, T. Sasaki and H. Yoneyama, *Denki Kagaku*, **61**, 918 (1993).
4. A. J. Nozik, H. Uchida, P. V. Kamat and C. J. Curtis, *Israel J. Chem.*, **3**, 15 (1993).
5. H. Matsumoto, H. Uchida, H. Yoneyama, T. Sakata and H. Mori, *Res. Chem. Intermed.*, **20**, 723 (1994).
6. H. Uchida, T. Matsunaga, H. Yoneyama, T. Sakata, H. Mori and T. Sasaki, *Chem. Mater.*, **5**, 716 (1993).

REFERENCES

a. M. A. Olshavsky, A. N. Goldstein and A. P. Alivasatos, *J. Am. Chem. Soc.*, **112**, 9438 (1990).
b. S. Schmitt-Rink, D. A. B. Miller and D. S. Chemla, *Phys. Rev.*, **B35**, 8113 (1987); Y. Kayanuma,. *ibid*. **B38**, 9797 (1988).
c. Y. Masumoto, M. Matsuura, S. Tarucha, H. Okamoto, *Phys. Rev.*, **B32**, 4275 (1985).
d. A. Nakamura, T. Tokizaki, H. Akiyama and T. Kataoka, *J. Lumines.*, **53**, 105 (1992).
e. Y. Wang and N. Herron, *Res. Chem. Intermed.*, **15**, 17 (1991).
f. K. Tsunetomo, M. Yamamoto and Y. Osaka, *Jpn. J. Appl. Phys.*, **30**, L 521 (1991).

21

Absorption and Photoemission Studies of Au-55 Metal Clusters

Hiroshi Miyauchi, Kuniyasu Taketomi, Akihiro Egami, Sinya Hosoi and Hirohito Fukutani
Institute of Physics, University of Tsukuba
Tsukuba-shi, Ibaraki 305-0006, Japan

Absorption spectra of chemically synthesized Au-55 metal clusters were measured in the photon energy range from 0.5 to 40 eV. Apart from the well-known plasma resonance at 2.4 eV, small but distinct features were found at 4.5 and 5.0 eV.

The valence band photoemission spectra were also investigated. The Au 5d-band maxima were found at -4.0 and -6.5 eV below the Fermi level. We attributed the two features at 4.5 and 5.0 eV to transitions from the d-band maximum at -4.0 eV to the unoccupied discrete states which arise from the level quantization of Au 6sp electrons confined in the small cluster with a radius of 7 Å.

1. Introduction

The fundamental question on the electronic states of metal clusters containing small number of atoms, is still open, irrespective of many experiments which have been made so far.[a] The electronic structure of clusters is known to depend on their size and atomic numbers. Many properties such as optical, electrical and thermal properties, are also size-dependent.

At present, two methods are known to obtain size-controlled metal clusters; one is the method of selecting clusters with TOF (time of flight) technique[b] and the other is the chemical synthesis method.[c] The latter has several advantages over the former; it is easier to obtain sufficient amounts of sample and the chemically synthesized clusters are more stable and protective against forming aggregates by coalescence because they usually have ligand shells, as in the case of Au-55, $Au_{55}(P(C_6H_5)_3)_{12}Cl_6$.[d]

In this paper, optical absorption spectra and valence band photoemission spectrum of Au-55 are reported. Previous reports on absorption published so far are limited to energy regions within the near ultraviolet and infrared (0.5 to 5 eV).[e] This is the first investigation extended to the vacuum ultraviolet region up to 40 eV.

We have found two well-distinguished optical structures in the ultraviolet region at about 5 eV, which can be attributed to a cluster containing 55 gold atoms. These structures have been assigned as optical transitions from the d-band state density maximum to the level-quantized sp-levels.

2. Experimental

The Au-55 metal cluster was synthesized according to the method of Schmid *et al.*,[f] reducing $(C_6H_5)_3PAuCl$ in benzene with BH_3 resolved in tetrahydrofuran. The ^{31}P-NMR spectrum of the sample resolved in pyridin showed only one sharp signal at the position of NMR shift of 30 ppm in accordance with the reported value,[f] indicating that the twelve phosphor atoms are located at the chemically equivalent sites as anticipated from the atomic geometry of $Au_{55}(P(C_6H_5)_3)_{12}Cl_6$.[d] It has been reported that another signal at 25 ppm emerges when the Au-55 clusters are disintegrated.[f] That signal was not detected in the present sample. The stretching vibration mode between the chlorine and peripheral gold atom was detected in IR spectrum at 280 cm^{-1} and it was distinctly located at a different position from that of the starting material, $(C_6H_5)_3PAuCl$ of 330 cm^{-1}. Results by NMR and IR thus confirm good quality of the present sample. The sample was fairly stable in air at normal condition but disintegrated at temperatures higher than about 150°C, changing in color from black to dark brown, which suggests development of bulky gold.

The Au-55 clusters were dispersed in dichloromethane and sprayed on transparent LiF, colodion film or quartz plates which were used for absorption measurements from 0.5 to 40 eV. Samples for photoemission measurement were made by embedding the Au-55 powder in indium foil or pressing the powder to a disc. The two kinds of samples were confirmed to be electrically conductive.[g]

The photoemission and vacuum ultraviolet absorption (between 4 and 40 eV) were measured at the Photon Factory, National Laboratory for High Energy Physics, using Beam Line BL-11D and BL-11C. From 0.5 to 6 eV, a conventional IR-UV monochromator, with W-lamp or deuterium lamp, was used. The photoemission spectrum was measured using an angle-integrated cylindrical mirror analyzer. The energy resolution was 0.3 eV, estimated from Fermi edge broadening of bulk gold.

3. Results

The optical ,absorption spectrum of Au-55 in the photon energy region from

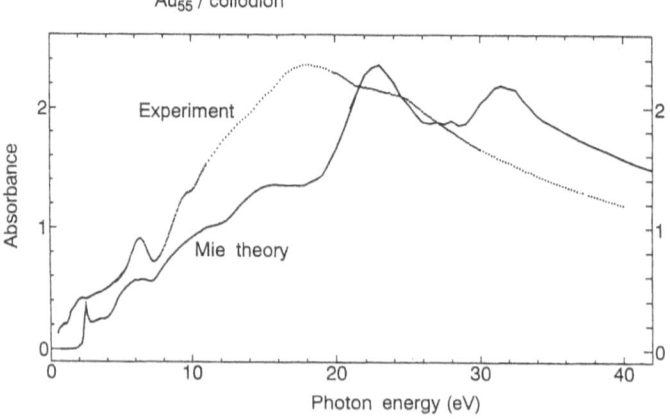

Fig. 1 Absorbance of Au-55 in arbitrary scale.

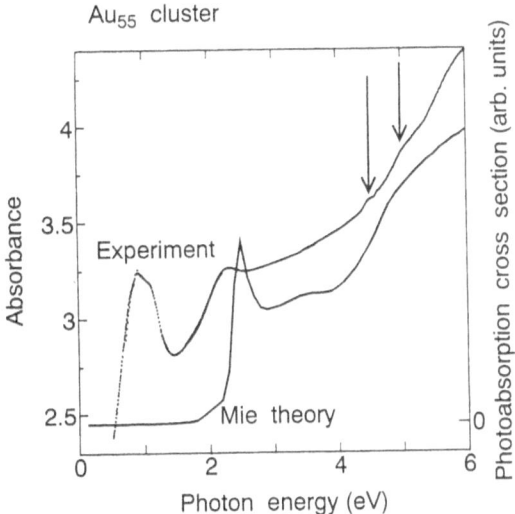

Fig. 2 Absorbance of Au-55 from 0.5 to 6 eV.

0.5 to 40 eV, is shown in Fig. 1 (solid dots). In the spectrum, the absorption due to the substrate (in this case, collodion) is subtracted. Also in the figure, the absorption expected by Mie scattering theory for a spherical particle, using the following expression,[h]

$$\alpha(\omega) = \frac{4\pi\omega}{c} \, \text{Im} \, \frac{\varepsilon - 1}{\varepsilon + 2}$$

is shown for comparison. In the above expression, ε is the complex dielectric constant of gold and the bulk value is used.[i] The general features in the two spectra are in good accord, suggesting that the effective dielectric constant of Au-55 cluster is not so different from that of the bulk Au.

The absorption spectrum in the low energy region from 0.5 to 6 eV shown in Fig. 2, however, reveals clearly some features which can not be explained by the Mie expression (thin solid lines). First, two small but distinct structures are found at 4.5 and 5.0 eV, overlapping a rather smooth background. The two structures were always observed, irrespective of different kinds of substrate *i.e.* quartz, LiF, or collodion. Second, a peak is located around 1 eV. The intensity and also the energy location of this peak was sample-dependent, whereas the 4.5 and 5.0 eV features were sample-independent.

When Au-55 clusters are heated above 150°C, they are observed to disintegrate and coalesce to form larger particles or bulk gold material. The two distinct structures at 4.5 and 5.0 eV in Fig. 2 disappear completely and instead a broad absorption band appears at around 4.8 eV, which can be ascribed to the absorption of bulk gold.[i] The 1 eV peak also disappears and the well-known plasmon peak at 2.4 eV increases in intensity, suggesting that the 1 eV peak is ascribable to ellipsoidal-shaped particles,[j] which are made in the course of sample preparation and remain partly in the present sample. The 4.5 and 5.0 eV structures, on the other hand, can be assigned to intrinsic features of isolated Au-55 spherical clusters, because they completely disappear after heat treatment and convert to bulk-like features.

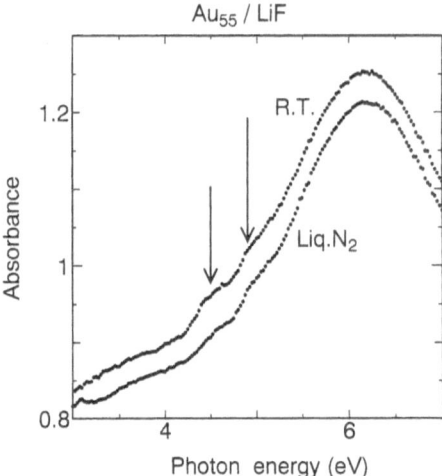

Fig. 3 Temperature dependence of absorbance of Au-55 around 5 eV.

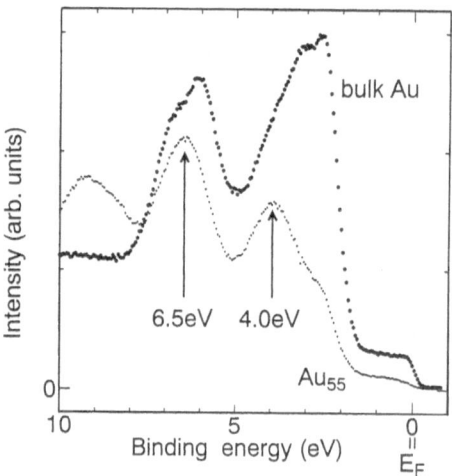

Fig. 4 Valence-band photoemission of Au-55 compared with that of bulk Au.

Figure 3 shows the temperature dependence of the 4.5 and 5.0 eV structures, at room temperature, and liquid nitrogen temperature. The two structures do not change intensity or energy, indicating that they are not vibration-related but electronic-related structures, contrary to the assignments by Benfield et al.[e]

The valence-band photoemission spectrum of Au-55 measured at room temperature is given in Fig. 4. Compared to that of bulk Au, the d-band maxima shift to higher binding energies. Separation of the two maxima decreases from 3.6 (bulk) to 2.5 eV (Au-55) and this is the well-known d-band narrowing due to decrease in the number of neighboring atoms.[k] The d-band maxima are located at −4.0 and −6.5 eV below the Fermi level. These observations are in good agreement with previous reports.[k] The sp-band photoemission is seen in the low binding energy from Fermi level to −1.5 eV. Two interesting points are noted; Fermi edge broadening of Au-55 is about 0.44 eV, which is about 1.5 times larger than that of bulk Au, and the photoemission intensity of Au-55 decreases gradually towards the Fermi edge, while remaining almost constant in the case of bulk gold. This finding suggests that the density of the sp-band states of Au-55 is different from that of bulk Au.

4. Discussion

In this section, we give a tentative assignment for the 4.5 and 5.0 eV structures of Au-55 cluster in teams of transitions from the d-band maximum to unoccupied discrete levels. As mentioned in the previous section, the two structures are intrinsic to isolated Au-55 clusters and their origins are electronic rather than vibronic, because they disappear completely when the sample is heated so that the clusters disintegrate and their energy and intensity do not depend on temperatures (Fig. 3).

If we assume that 55 6sp electrons of an Au-55 cluster behave like free electrons, confined in a spherical potential well with radius 7.2 Å, the 55 6sp electrons will occupy up to the 1g subshell. [l] As the 1g subshell is partially-filled, this level will be the Fermi level of Au-55. High lying unoccupied 2d and (3h,1h) subshell are expected at about 1.0 and 1.5 eV above the Fermi level, respectively.

The dominant optical transitions relevant to bulk Au are known to be interband transitions from the 5d-band to unoccupied 6sp band. [m] Therefore, it is reasonable to assume that the same transitions are relevant to the 4.5 and 5.0 structures of Au-55. The d-band maximum was found at −4.0 eV by the photoemission experiment (Fig. 4). Thus, we can expect transitions from the d-band maximum to Fermi level or to 2d sublevel, at about 4.0 and 5.0 eV, respectively, in good accord with the present experiment. The small difference between the observed and the calculated values is allowable if we consider rather simple assumptions in this calculations.

A tight-binding calculation has been recently reported in an attempt to interpret the optical absorption of Au-55. [n] Density of states (DOS) of the Au-55 particles was computed for different values of the external potential, which take into account interactions between bare Au-55 and surrounding ligands. It was found that DOS of the d-band is not affected by the external potential, and the separation between 1g and 2d is about 1.3 eV. These results can explain well the present experimental findings. In Fauth et al.'s reports, [e] in which well isolated Au-55 clusters embedded in polystyrene were used as samples, the plasmon resonance at 2.4 eV was found only as a faint structure. As indicated by Bifore's calculation, their sample may be influenced by the external potential caused by surrounding polystyrene. In the present experiments, Au-55 was sprayed on substrates and less affected by external perturbation.

If the above assignments are correct, we can expect to observe the quantization effect in the sp-band photoemission spectrum. The sp-band photoemission spectrum shown in Fig. 5 dose not show any discrete feature, instead gradually decreasing its intensity toward the Fermi level. The sp-band of bulk Au, on the other hand, shows a nearly constant DOS. As occupied sublevels, we expect the 2p sublevel at about −0.5 eV and the 1f sublevel at about −1.3 eV, respectively. The large degeneracy of the 1f level than the 2p level will cause gradual decrease of the photoemission intensity if level broadening is sufficiently large to make an apparant continuous decrease of DOS for the sp-electrons. This assumption may be reasonable because the sp-electrons confined in a small sphere

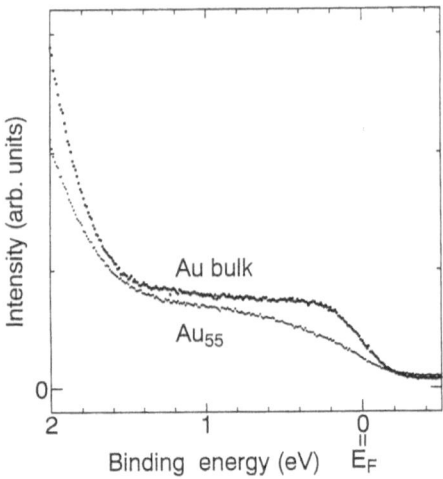

Fig. 5 sp-Band photoemission of Au-55 compared with that of bulk Au.

will suffer scattering by the external ligands to cause large level broadening. In fact, the broadenig of the Fermi level of the Au-55 cluster is 1.5 times larger than that of bulk Au (Fig. 5).

In conclusion, we have found two distinct structures in the optical absorption of Au-55 cluster and assigned them to transitions from the d-level to the size-quantized sp-levels in a 7.2 Å spherical cluster. This is also consistent to our observation of the photoemission spectra.

We thank the staff of the Photon Factory for their helpful support during the experiment.

PUBLICATIONS

1. H. Miyauchi, K. Taketomi, A. Egami, S. Hosoi and H. Fukutani, Proc. Int.Conf. Optical Properties of Nanostructures, Sendai, 1994, *Jpn. J. Appl. Phys.*, **34**, Suppl. 34-1, 19 (1995).

REFERENCES

a. L. L. Chang, Highlight in Condensed Matter Physics and Future Prospects, (L. Esaki, ed.) p83, Plenum Press (1991).
    W. Halperin, *Rev. of Mod. Phys.*, **58**, 533 (1986).
b. W. Eberhardt, P. Fayet, F. M. Cox, A. Kolder, R. Sherwood and D. Sondericker, *Phys. Rev. Lett.*, **64**, 780 (1990).
c. K. P. Hall and D. M. P. Mingos, *Progress in Inorganic Chemistry*, **32**, 237 (1984).
d. G. Shmid, *Structure and Bonding*, **62**, 51 (1985).
e. K. Fauth, U. Kreibig and G. Shmid, *Z. Phys. D-Atoms, Molecules and Clusters*, **12**, 515 (1989).
    R. E. Benfield, J. A. Creighton, D. G. Eadon and G. Shmid, *ibid*, **12**, 533 (1989).
    M. A. Marcus, M. P. Andrews and J. Zegenhagen,*Phys. Rev.*, **B42**, 3312 (1990).
f. G. Shmid, R. Pfeil, R. Boese, F. Bandermann, S. Meyer, G. H. Calis and J. W. A. van der Velden, *Chem. Ber.*, **114**, 3634 (1981).
g. M. P. J. van Staveren, H. B. Brom and L. J. de Jongh, *Solid State Commun.*, **60**, 319 (1986).
h. G. Mie, *Ann. Phys.*, **25**, 377 (1908).
i. D. W. Jynch and W. R. Hunter: *Handbook of Optical Constants of Solids*, (E. D. Palik, ed.), p286, Academic Press (1985).
j. U. Kreibig and L. Genzel, *Surface Science*, **156**, 678 (1985).
k. S. B. DiCenzo, S. D. Berry and E. H. Hartford, Jr., *Phys. Rev.*, **B38**, 8465 (1988).
l. C. Kittel, *Introduction to Solid State Physics*, p.155, John Wiley and sons (1986).
m. R. Lasser, N. V. Smith and R. L. Benbow, *Phys.Rev.*,**B24**, 1895 (1981).
n. A. Bifore and F. Bassani, *Z. Phys.*, **D29**, 73 (1994).

22

Stability of Spin-glass Ordering in the Nanoparticle System

Tetsuya Sato[a] and Tomoyasu Taniyama[b,c]

[a] Department of Applied Physics and Physico-Informatics, Faculty of Science and Technology, Keio University
Kohoku-ku, Yokohama 223-8522, Japan
[b] Department of Materials Science, Faculty of Science and Technology, Keio University
Kohoku-ku, Yokohama 223-8522, Japan
[c] National Research Institute for Metals
Tsukuba-shi, Ibaraki 305-0047, Japan

Purpose of the present study

Two dimensionality becomes remarkable with decreasing size of particle ascribed to an enhancement of the surface-to-volume ratio in the itinerant electron nanoparticle system. This may result in the localization of itinerant electrons in the surface region and stabilize the spin-glass ordering. In contrast, stability of the spin-glass ordering is influenced by mean free path of conduction electrons since the metallic spin-glass ordering is determined through the strength of the indirect interaction between the localized spins via conduction electrons. As a result, SG ordering becomes unstable according to the localization of itinerant electrons in the particle. Thus, the competition between the two conflicting ingredients determines the stability of spin-glass ordering in the nanoparticle system. In the present work, we prepared AuFe nanoparticles by two different techniques and performed magnetization measurements in order to investigate the stability of spin-glass ordering in the nanoparticle system. We expect this work to provide some information about the size effect of spin-glass ordering.

Contents of the present study

1. Introduction

$Au_{1-x}Fe_x$ alloy has some interesting magnetic properties. A system whose concentration is below 15 at% Fe exhibits spin-glass or cluster-glass behavior. Above this critical concentration, the system undergoes a transition from paramagnetism to ferromagnetism at Curie temperature followed by a transition which is characterized by the freezing of the transverse spin component.[a-c] Further, a strong irreversible behavior is observed in the magnetization process at much lower temperatures.

In recent years, studies on the size effect and three-dimensional (3D) to 2D crossover behavior of spin-glasses (SG)[c,d] have been extensively performed. Vloeberghs et al.[c] and Hoines et al.[d] found that the typical SG characteristics persist in AuFe thin film even in the SG layer with thickness down to 1 nm. In such cases, the characteristic relaxation time (t_M) of the system exceeds the experimental time scale so that the temperature-dependent susceptibility shows a cusp at a finite temperature. On the other hand, in the particle system, two-dimensionality becomes remarkable with decreasing size of particle ascribed to an enhancement of the surface-to-volume ratio, which may result in the localization of itinerant electrons in the surface region. This localization works as an ingredient to stabilize the SG ordering. On the contrary, stability of the spin-glass ordering is influenced by the mean free path of conduction electrons since the metallic spin-glass ordering is determined by the strength of the indirect interaction between the localized spins via conduction electrons. Consequently, SG ordering becomes unstable according to the localization of itinerant electrons in the particle. Thus, the competition between the two conflicting ingredients determines the stability of SG ordering in the nanoparticle system.

Generally speaking, the stability of the SG ordering is determined experimentally by a change in the spin freezing temperature (T_f) in the confined SG system such as thin film or particle. Therefore, we can obtain some information regarding the stability of SG ordering by observing the change in T_f with decreasing size of the system. Further, the relaxation time (t_M) of the system is an important factor for judging the stability of SG system. The bulk SG system exhibits slow dynamics which originates from the complex multi-valley structure of the free energy and possesses a very long characteristic relaxation time beyond 10^5 s. The t_M is determined by the stability of the SG ordering so that the long characteristic relaxation time suggests that the system is a very stable SG one. Thus, we can probe the stability of the SG system through the change in t_M in the particle system.

In the present work, we prepared AuFe nanoparticles by two different techniques and performed magnetization measurements in order to investigate the stability of the SG nanoparticle system. We discuss the stability of the SG ordering through the size dependence of T_f and the relaxation process of the magnetization. This work is expected to provide convincing information on the stability of the AuFe nanoparticle system.

2. Experimental procedure

2.1 The sandwiches of $Au_{85.5}Fe_{14.5}$ nanoparticles and LiF films

The sandwiches of $Au_{85.5}Fe_{14.5}$ nanoparticles and LiF films were prepared by

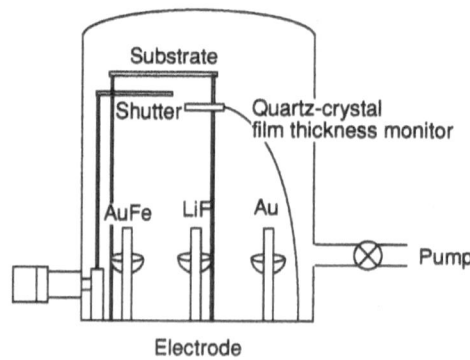

Fig. 1 The apparatus for preparing the sandwiches of $Au_{85.5}Fe_{14.5}$ nanoparticles and LiF films.

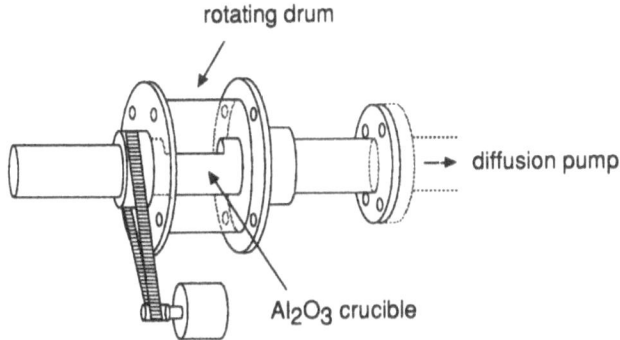

rotating drum

→ diffusion pump

Al_2O_3 crucible

Fig. 2 The apparatus for preparing the $Au_{81.7}Fe_{18.3}$ nanoparticles dipersed in oil.

evaporating the $Au_{71}Fe_{29}$ alloy and LiF alternately on quartz substrate using the vacuum evaporation apparatus shown in Fig. 1. The size of nanoparticles and the thickness of the interlayer were controlled by the quartz crystal film thickness monitor. The particle size was measured using transmission electron microscopy, which reveals a narrow distribution of particle size. The magnetization measurements were performed in the temperature range of 2.0–40 K under varying magnetic fields using a Quantum Design MPMS superconducting quantum interference device (SQUID) magnetometer.

2.2 $Au_{81.7}Fe_{18.3}$ nanoparticles dipersed in toluene

The $Au_{81.7}Fe_{18.3}$ nanoparticle specimen was prepared by evaporating the homogenized $Au_{72}Fe_{28}$ alloy on the liquid hydrocarbon oil substrate using the apparatus shown in Fig. 2 [f]. The evaporation was started after evacuation to a pressure of 8×10^{-6} Torr. The composition of the specimen was determined by means of electron probe microanalysis. The particle size was measured using transmission electron microscopy, which reveals a narrow distribution around 3 nm in diameter with a standard deviation of 0.9 nm. The magnetization measurements were performed in the temperature range of 2.0–20 K under a magnetic field of 100 Oe using a Quantum Design MPMS SQUID magnetometer after the hydrocarbon oil had been removed from the as-made colloidal metal by mixing with ethyl alcohol and then the sample was mixed with toluene.

3. Results and discussion

3.1 Size dependence of the spin freezing temperature of the sandwiches of $Au_{85.5}Fe_{14.5}$ nanoparticles and LiF films

Figure 3 shows the size dependence of T_f of various AuFe thin films and the present sandwiches of nanoparticles. The size dependence of nanoparticles is different from that of thin film below 100 Å, which may be ascribed to the difference in the geometrical confinement, $i.e.$, the one-dimensional size effect in thin films and three-dimensional size effect in nanoparticles. Further, the rate of decrease in T_f slows down in the region of less than 70 Å in size.

Now, we classify the size of particle in the following four regions: (a) $d > 500$

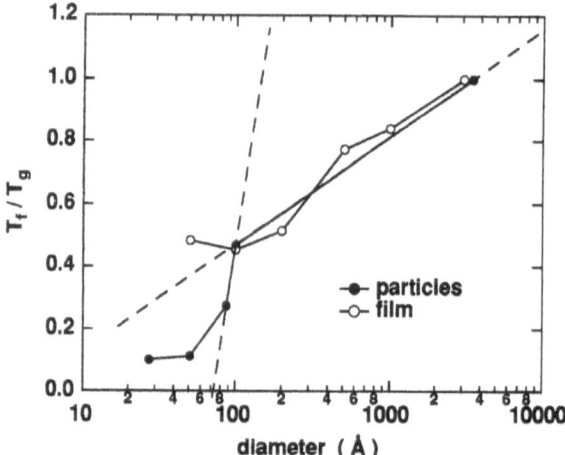

Fig. 3 The size dependence of spin freezing temperature of $Au_{85.5}Fe_{14.5}$ nanoparticles.

Å, (b) $d = 500$ Å–100 A, (c) $d = 100$ Å–70 Å, (d) $d < 70$ Å. In region (a), since the size dependence of T_f of particle coincides with that of thin film, we can employ the usual finite size scaling analysis [g] in order to interpret the size dependence of T_f. The finite size scaling has succeeded in representing the size dependence of T_f the SG thin film system. In this theory, the size dependence of T_f is followed as equation (1):

$$\frac{T_g - T_f}{T_g} \propto d^{-\lambda} \tag{1}$$

where, T_f is spin freezing temperature, T_g is spin freezing temperature of bulk sample, d is the size of the system and λ is the shift parameter. We obtained the shift parameter λ as 0.69 by fitting the data in terms of equation (1). However, the finite size scaling failed in interpreting the size dependence in the SG thin film in regions (b), (c) and (d) in which the droplet model developed by Fisher and Huse should be applied.[h] By the droplet model, the size dependence of T_f is represented in terms of equation (2):

$$\frac{T_f}{T_g} \propto \frac{d^K}{(\ln t_m)^c} \tag{2}$$

where, t_m is the relaxation time of the system and c is the arbitrary parameter. By fitting the data in terms of equation (2), we obtained the parameters for each region and summarized them in Table 1, in which the parameters of some different materials reported by the other authors are also shown.

The obtained shift parameter λ of $Au_{85.5}Fe_{14.5}$ nanoparticles is much smaller compared with those of the other metal spin-glasses; this may originate from the difference in the mechanism of spin freezing. That is, the $Au_{85.5}Fe_{14.5}$ alloy is classified as not spin-glass but cluster-glass. In cluster-glass, it is expected that the decrement in T_f with decreasing size is reduced according to the existence of local anisotropy. The parameter K is also much reduced and the rate of decrease in T_f slows down with decreasing size. We can speculate that $Au_{85.5}Fe_{14.5}$ particles are superparamagnetic rather than cluster-glass in region (d), in which the rate of

Table 1 Shift parameter λ and parameter K of various materials

			AgMn	CuMn	NiMn
	AuFe				
λ	0.69		0.83	0.90	1.25
K	500–100 Å	0.21	0.65	0.5	0.62
	100–70 Å	0.39			
	70 Å–	0.15			

decrease in T_f is moderated, because size of particle is comparable to that of the magnetic domain. Further comparison with the droplet model is not made here since the present result is beyond the theory.

Next, we measured the nonlinear susceptibilities of two samples of $Au_{85.5}$ $Fe_{14.5}$ particle sandwiches whose mean sizes are 25 Å and 87 Å in diameter, respectively. We analyzed the nonlinear susceptibilities in terms of the following scaling function of equations (3)–(5) and plotted them in Figs. 4 (a) and (b).

$$\chi_{nl}(H, T) \propto H^{2/\delta} f(t/H^{2/\phi}) \tag{3}$$

$$t = (T - T_g)/T_g \tag{4}$$

$$f(x) = \begin{cases} const, & x \to 0, \\ x^{-\gamma}, & x \to \infty \end{cases} \tag{5}$$

$$\gamma = \phi(1 - 1/\delta) \tag{6}$$

The derived critical exponents are $\delta = 10.6$ and $\phi = 20$ for the particle of 25 Å and $\delta = 7$ and $\phi = 12$ for the particle of 87 Å. The critical exponents are summarized in Table 2, in which those of the Au_{91} Fe_9 thin films and the bulk are also shown. Although the critical exponents of the thin film and the bulk satisfy the scaling relation of equation (6), the scaling relation fails in the particle samples. This result suggests that the system of particles is in the nonequilibrium state in which the scaling relation is not satisfied and the equlibrium cluster-glass nature of

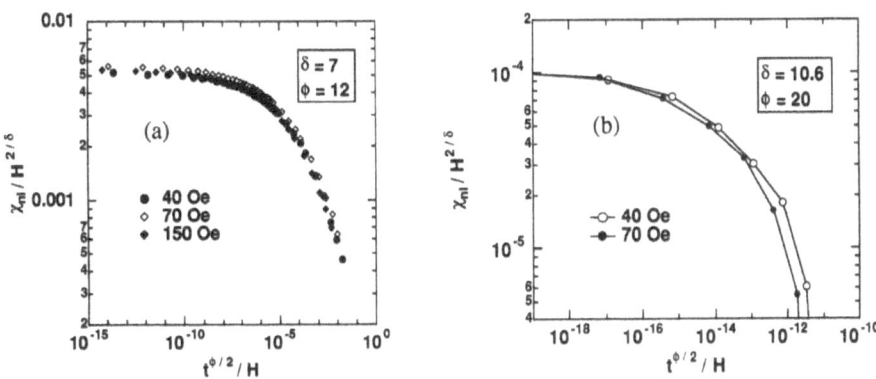

Fig. 4 Scaling of nonlinear susceptibility of the particle of (a) 87 Å and (b) 25 Å.

Table 2 Critical exponents of AuFe material of various sizes

	25 Å	87 Å	700 Å	bulk
δ	10.5	7	8.2	5.1
ϕ	20	12	8.0	4.5

$Au_{85.5}Fe_{14.5}$ particle disappears in the size region of less than 100 Å.

3.2 Time-dependent magnetization of $Au_{81.7}Fe_{18.3}$ nanoparticles dispersed in toluene

Figures 5 (a) and (b) show the time-dependent magnetization of the $Au_{81.7}$ $Fe_{18.3}$ nanoparticles under the field-cooled (FC) and zero-field-cooled (ZFC) conditions in a field of 100 Oe, respectively. In the ZFC measurement the sample is cooled under zero field from a temperature well above the spin freezing temperature to the measurement temperature T_m. Thereafter the magnetic field is applied and the magnetization is recorded as a function of time. In the FC

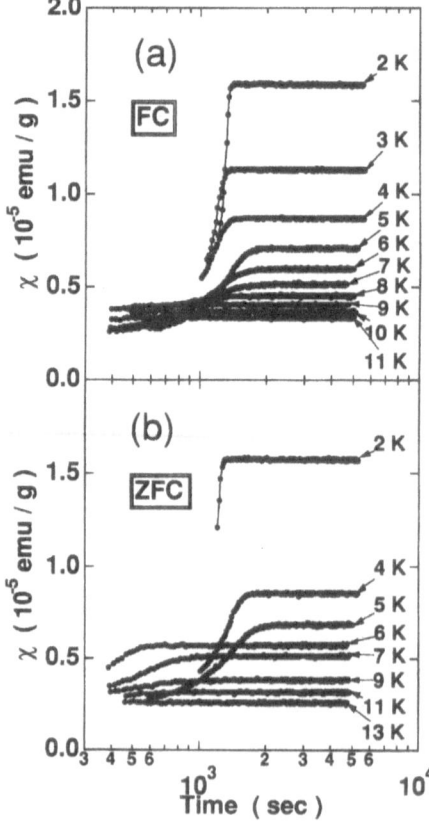

Fig. 5   Time-dependent magnetization of $Au_{81.7}Fe_{18.3}$ nanoparticles at the various temperatures under the (a) FC and (b) ZFC conditions.
(From T.Taniyama *et al.*, *Mater.Sci.Eng.A.*, **217/218**, 319-321 (1996)).

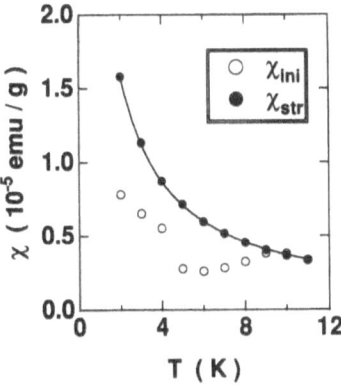

Fig. 6   Temperature dependence of initial magnetization χ_{ini} and saturated magnetization χ_{str} under the FC and ZFC conditions. The solid line represents the fitted curve in terms of the Curie-Weiss law.
(From T. Taniyama *et al.*, *Mater. Sci. Eng. A.*, **217/218**, 319-321 (1996)).

measurement the sample is cooled in the field, then the magnetization is recorded in the same way as ZFC measurements. The time $t = 0$ was set as the moment at which the temperature goes down through 10 K below which the relaxation of the magnetization was observed. The present magnetization increases with time and reached the upper limit values within a period of time shorter than 1500 s. It is important to note that the peculiar time-dependent magnetization was observed below 10 K even under the FC condition, which is in contrast to the time-independent behavior of FC magnetization in bulk SG. This relaxation process is not expressed by the standard relaxation formula such as the stretched exponential used for the SG system and beyond the reach of the current SG theory. We note that the time-dependent magnetization under FC condition was also observed in the 2D CuMn thin film [i)] just below the freezing temperature, whereas the magnetization increases monotonically with time beyond the time scale of 10^4 s.

Figure 6 shows the temperature-dependent initial and saturated magnetization, where the initial magnetization and the saturated magnetization are defined as those at the first sampling point and $t = 4000$ s, respectively. The temperature-dependent initial magnetization has a peak which suggests an onset of the spin freezing process around 10 K in the nanoparticles. The saturated magnetization follows well the Curie-Weiss law over the measuring temperature range, as shown by the solid curve in Fig. 6. The estimated Curie temperature is nearly zero, that is, the present nanoparticle system exhibits paramagnetic or superparamagnetic behavior after the relaxation process is completed. Thus, we claim this to be the first observation of the entire relaxation process from the non-equilibrium state to the equilibrium state in the SG-like system.

What does the peculiar relaxation process imply? We are able to give the following two possible origins for the short relaxation time: (1) decrease in the number of the energy barriers to relax, and (2) a lowered energy barrier to relax. In the bulk SG, the free energy is much degenerated owing to the frustration of the interaction among a great deal of interacting magnetic moments and the degeneracy results in a multi-valley structure of free energy having many metastable states. The SG system takes a very long time to relax from the non-equilibrium state to the equilibrium one compared with the case of the ferromagnet or anti-ferromagnet because of the great number of the energy barriers to relax, *e.g.*, the typical relaxation time of the bulk SG materials is beyond a time scale of 10^5 s. Such long relaxation behavior has been observed even in 2D-like geometry. On the other hand, the structure of free energy would be much simplified in the nanoparticle system since the interacting magnetic moments must be much reduced compared with the bulk and 2D-like systems. Thus, it becomes easier to relax to the equilibrium state from the non-equilibrium state in $Au_{81.7}Fe_{18.3}$ nanoparticle, which must result in this short relaxation process, as shown in Fig. 5. Next, we note the second factor of the reduced relaxation time in the $Au_{81.7}Fe_{18.3}$ nanoparticle, *i.e.*, the origin of the lowered energy barrier. The surface-to-volume ratio of the nanoparticle is much larger than that of bulk so that the surface effect becomes remarkable. Generally speaking, the conduction electrons are likely to localize according to the enhancement of the randomness in the surface region (Anderson localization). Such a change in the itineracy of the conduction electrons should change the indirect interaction between the Fe spins via the conduction electrons (RKKY interaction) in the surface region. Thus, it is expected that the strength of the frustration would be much weaker in the surface region in the SG system, and the weakened frustration must result in a lower energy barrier to relax to the equilibrium state or the short relaxation time.

By way of comment on the droplet model in the previous section, we mention the non-equilibrium behavior in SG system. In the study of CuMn 2D thin film, the unique relaxation process of the FC magnetization was interpreted in terms of the domain growth picture based on the droplet model. Although this picture may be utilized for the interpretation of the relaxation process in the $Au_{81.7}Fe_{18.3}$ nanoparticle, it is not clear whether or not this domain growth picture is appropriate for the 3D confined system. This is the reason we refrained from utilizing the droplet model.

Finally, we discuss the equilibrium magnetic phase in $Au_{81.7}Fe_{18.3}$ nanoparticle system. Since the equilibrium state should be the paramagnetic or superparamagnetic phase as mentioned above, we can now raise some speculations for the equilibrium magnetic phase. The temperature-dependent saturated magnetization which was followed by the Curie-Weiss law evidently suggests the emergence of magnetic moments. However, this may not necessarily conclude the ferromagnetic spin arrangement in the nanoparticle, since the difference between the χ_{ini} and the χ_{str} is too small. Although not described in detail, we speculate some magnetically ordered state consisting of spins insufficiently aligned in each particle under a magnetic field. We need some microscopic magnetic information using Mössbauer spectroscopy and detailed field dependent-magnetization measurements in order to elucidate the magnetic spin structure of the equilibrium state. This is now in progress.

4. Concluding remarks

Conclusively, we can state that the three-dimensional size effect peculiar to the nanoparticle system is remarkable in the region of less than 100 Å in size in the sandwiches of $Au_{85.5}Fe_{14.5}$ nanoparticles and LiF. Moreover, the decrement in T_f with decreasing size slows down below 70 Å in size, suggesting that the system exhibits superparamagnetic behavior rather than the nature of cluster-glass.

In experiments on the $Au_{81.7}Fe_{18.3}$ nanoparticles dispersed in toluene, we first observed the entire relaxation process from the non-equilibrium state to the equilibrium state in an SG-like system. In the $Au_{81.7}Fe_{18.3}$ nanoparticle an SG-like phase appears at temperatures below 10 K, and the relaxation process to the equilibrium state is completed within the experimental time scale. This peculiar relaxation behavior may be interpreted by the decrease in the relaxation time which may be ascribed to (1) the simplification of the multi-valley structure of the free energy with decrease in the size and/or (2) the lowered energy barrier originating from the weakened frustration by the reduction in the RKKY interaction in the surface region.

PUBLICATIONS

1. T. Taniyama, R. Ogawa, T. Sato and E. Ohta, International Conference on Nano-Clusters and Granular materials (Sendai, 1995), *Materials Science and Engineering A* 217/218, 319–321 (1966).

REFERENCES

a. S. Mitsuda, H. Yoshizawa, T. Watanabe, S. Itoh, Y. Endoh and I. Mirebeau, *J. Phys Soc. Jpn.*, **60**, 1721 (1991).
b. W. Marsschmann, J. Lauer and W. Keune, *J. Magn. Magn. Mater.*, **31–34**, 1345 (1983).
c. I. A. Campbell, S. Senoussi, F. Varret, J. Teillet and A. Hamzic, *Phys. Rev. Lett.*, **50**, 1615 (1983).
d. H. Vloeberghs, J. Vranken, C. Van Haesendonck and Y. Bruynseraede, *Europhys. Lett.*, **12** (6), 557 (1990).
e. L. Hoines, J. A. Cowen and J. Bass, *Physica*, B, **194–196**, 309 (1994).
f. I. Nakatani, T. Furubayashi and H. Hanaoka, *J. Magn. Magn. Mater.*, **65**, 261 (1987).
g. G. G. Kenning, J. Bass, W. P. Pratt, Jr., D. Leslie-Pelecky, L. Hoines, W. Leeach, M. L. Wilson, R. Stubi and J. A. Cowen, *Phys. Rev.*, B, **42**, 2393 (1990).
h. See e.g. K. H. Fischer and J. A. Hertz, *Spin Glasses*, Cambridge (1991).
i. C. Djurberg, J. Matsson, P. Nordbald, T. Stubi and J. A. Cowen, *Physica*, B, **194–196**, 303 (1994).

23

Screening Effects in Metal Clusters

Kohji Sonoda[a], Fuyuki Shimojo[b], Kozo Hoshino[b] and Mitsuo Watabe[b]

[a] Venture Business Laboratory, Hiroshima University
Higashihiroshima-shi, Hiroshima 739-0046, Japan
[b] Faculty of Integrated Arts and Science, Hiroshima University
Higashihiroshima-shi, Hiroshima 739-0046, Japan

Purpose of the present study

Screening is one of the most fundamental and important many-body effects in bulk metals and has been studied for many years. However, little study has been done on the screening effects in metal clusters and fine particles. In metal clusters and fine particles, owing to the finiteness of the system, the screening effects are expected to be different from those in bulk metals. The purpose of the present work is to investigate the size- and the density-dependence of the screening effects in neutral metal clusters by introducing an extra positive point charge. In this work, we employ a jellium sphere as a model for a metal cluster and calculate the electronic state using some approximations based on the density functional theory and the Hartree-Fock approximation. We discuss the above problems based on the results.

Contents of the present study

1. Introduction

Screening effect is one of the most important many-body effects in metals. The effect has been studied by calculating the displaced charge caused by an extra positive charge (a proton) or a hydrogen atom embedded in jellium.[a-d] The screening charge around such an impurity in an infinite jellium is characterized by the formation of a bound state at the impurity, which gives rise to a screened short-range interaction, and by the long-range Friedel oscillation due to the effect of the Fermi surface.

On the other hand, it is interesting to investigate how the screening effect in metal clusters differs, due to the finiteness of the system, from that in bulk metals. For this purpose, a spherical jellium model is useful, because it is the simplest model successfully applied to alkali-metal clusters and it has been extensively used to study the size dependence of physical properties such as the ionization potential,[e,f] work function,[g] and electron charge density.[g] As far as we know there are two studies on the size dependence of the change in the electronic density due to the hydrogen atom embedded in the Jellium sphere. Hintermann and Manninen[h] investigated the applicability of the cluster method to describe the electronic states of a hydrogen (an impurity) in bulk simple metals. Ekardt[i] also studied the size dependence of the static and the dynamic polarizability of the same model.

The purpose of the present work is to study the size- and the density-dependence of the screening effects in metal clusters in detail.[1,2] For this purpose, we calculated the electronic states of a jellium sphere, the radius of which is determined by the number of ions in the cluster and the electron sphere radius. The electronic states are calculated by solving the Kohn-Sham (KS) equation[j,k] in the local density approximation (LDA) and in the local spin-density approximation (LSDA) with the self-interaction correction (SIC)[l] and by solving the spherical Hartree-Fock (HF) equation.[m]

In the present study, we introduce an extra positive point charge, instead of a hydrogen atom, at the center of the jellium sphere. To extract the screening charge density, we need to subtract the surface charge density, because the system is not neutral but positively charged. We calculate the screening charge density by the above three methods, in which the self-interaction is treated in a different way, as a function of the cluster size, and compare these results, paying special attention to the effect of the self-interaction. The effect of the self-interaction on the polarizability of the small metal clusters was studied by Stampfli and Bennemann.[n]

2. Method

2.1 Model

We use the jellium sphere of radius R as a model for metal clusters. In this sphere positive charge of the ions is smeared out into a uniform positive spherical background with the density $n_0 = (4 \pi r_s^3/3)^{-1}$, where r_s is the electron sphere radius. The radius R is related to the number of ions N contained in the cluster as $R = (ZN)^{1/3}r_s$, Z being the valence of the ion. In the following, we take $Z = 1$ since we are concerned with alkali metal clusters.

2.2 Method of calculation

We employ the following three approximate methods to calculate the electronic states of the jellium sphere: (i) the Kohn-Sham equation in the LDA, where the expression for the exchange-correlation potential due to Perdew and Zunger[1,o] is used. (ii) the Kohn-Sham equation in the LSDA with the SIC(LSDA+SIC)[l]; (iii) the Hartree-Fock equation in the spherical approximation. The reason for employing these approximations and comparing the results is as follows: In the LDA, the exchange interaction is not treated correctly, because the original nonlocal exchange potential is replaced by the local potential, which is written in terms of the electron number density, so the self-interaction is included. We can correct this error of the LDA by taking account of the SIC to the LDA, though the SIC is itself an approximation. Alternatively, we can employ the HF approximation (HFA), in which the exchange interaction is treated exactly, although the correlation is not included. Therefore, we can investigate the effect of the self-interaction on the screening charge density by comparing the results of these approximations.

Since spherical symmetry is imposed in these methods, we only need to solve self-consistently the equation for the radial part $R_{nl}(r)$ of the wavefunction $\psi_{nlm}(r) = R_{nl}(r) Y_{lm}(\theta, \phi)$, where $Y_{lm}(\theta, \phi)$ is the spherical harmonics. The $R_{nl}(r)$ and the corresponding eigenvalue ε_{nl} is labeled with number n (n: positive integers), where $n - 1$ is called the radial quantum number and it is equal to the number of nodes of $R_{nl}(r)$, and the orbital-angular-momentum quantum number l ($l = 0, 1, 2, ...$, for which we use the usual symbol s, p, d,..., respectively). The electron number density $n(r)$ is given by $n(r) = \sum_{nl}^{occ} n_{nl}(r) = \sum_{nl}^{occ} W_{nl} \mid R_{nl}(r) \mid^2 / 4\pi$, where $n_{nl}(r)$ and W_{nl} are the electron number density and the occupation number of state (n, l), respectively.

2.3 Method for estimating screening charge in jellium sphere

In order to investigate the screening effect in metal clusters, we consider the change in the electronic charge density caused by the introduction of an extra positive unit point charge at the center of the neutral jellium sphere. To extract the screening charge density in the jellium sphere, we calculate the electron number density of the jellium sphere, $n_I(r)$, $n_{II}(r)$ and $n_{III}(r)$, for the following three cases, respectively:

(I) The neutral jellium sphere with N electrons.
(II) The singly-ionized jellium sphere with $N - 1$ electrons.
(III) An extra positive unit point charge is introduced at the center of the jellium sphere.

By using these densities, we define the following two quantities:

$$\Delta n(r) = n_{III}(r) - n_I(r), \tag{1}$$

$$n_{surface}(r) = n_{II}(r) - n_I(r). \tag{2}$$

Here $\Delta n(r)$ describes the change in the electron number density caused by introducing the point charge at the center in the neutral jellium sphere, while $n_{surface}(r)$, which is the difference between the electron number density in Case I and that in Case II, is considered as the surface charge. Using $\Delta n(r)$ and $n_{surface}(r)$ thus obtained, we define the screening charge density $n_{scr}(r)$ in the metal cluster as follows:

$$n_{scr}(r) = \Delta n(r) - n_{surface}(r) = n_{III}(r) - n_{II}(r). \tag{3}$$

It should be noted that, although the procedure of extracting the screening charge density is well justified when the cluster size is large enough, it is not always the case when we consider small clusters, because the screening charge density and the surface charge density overlap each other and they are not unambiguously separable. For this reason we consider the above definition of $n_{scr}(r)$ as a natural and convenient way for estimating the screening change density in metal clusters.

3. Results and discussion

3.1 Screening charge distribution in jellium sphere

In this section, we show the process of deriving the screening charge in the LDA. Figures 1(a), 1(b) and 1(c) show $n(r)$ and $v_{eff}(r)$ for Cases I, II and III in section 2.3, respectively, for $N = 20$. The contribution of each orbital's $n_{nl}(r)$ to $n(r)$ is also shown in these figures. Horizontal lines drawn above the curve describing $v_{eff}(r)$ show the occupied energy levels ε_{nl}; thick lines show the energy

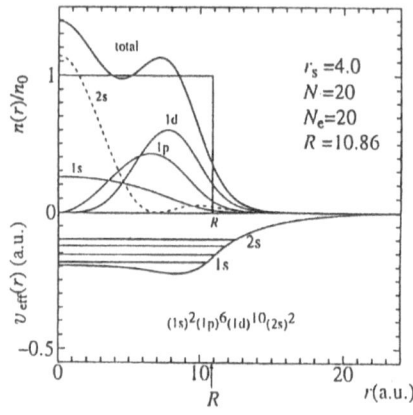

(a) The electron number density $n(r)$ and its components for each orbital $n_{nl}(r)$, the effective potential $v_{eff}(r)$ and the occupied energy levels for the neutral jellium sphere calculated in the LDA with $N = 20$ and $r_s = 4.0(Na_{20})$.

(b) The same as Fig. 1(a), but $(Na_{20})^+$.

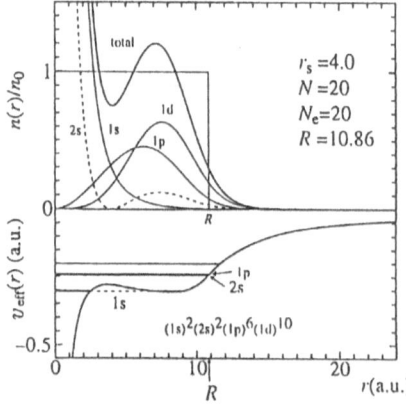

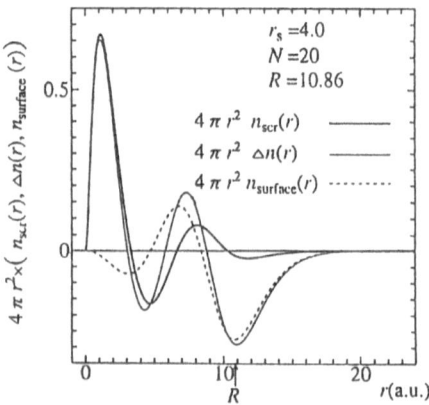

(c) The same as Fig. 1(a), but Na_{20} with an extra positive unit point charge introduced at its center.

(d) The screening charge density $n_{scr}(r)$, the surface charge density $n_{surface}(r)$ and the change density induced by the introduction of the extra charge $\Delta n(r)$ for the jellium sphere.

Fig. 1

levels of s orbitals ($l = 0$) and thin lines those with $l \geq 1$. In Fig. 1(d) the thick line shows $n_{scr}(r)$, which is derived by subtracting $n_{surface}(r)$ shown by the dotted line from $\Delta n(r)$ shown by the thin line. As is seen from Figs. 1(a) and 1(b), the effective potential $v_{eff}(r)$ and all energy levels in Na_{20}^+ shift downwards in parallel from those in Na_{20}.[g] The highest energy level, which is 2s orbital level both in Na_{20} and in Na_{20}^+, is doubly occupied in Na_{20} and singly in Na_{20}^+. The 2s orbital distributes substantially near the center of the sphere, and the lack of one electron in the 2s state in Na_{20}^+ causes the shift of all orbitals towards the center, resulting in the decrease of the electron charge near the surface. Moreover, this shift of all orbitals towards the center produces the oscillation of $n_{surface}(r)$ which extends to the center, as shown in Fig. 1(d). Figure 1(c) clearly indicates how the electron states change with the introduction of an extra positive point charge at the center of the neutral Na_{20}; electrons gather around the excess charge and screen it. By comparing the curves for $n_{nl}(r)$ in Fig. 1(c) with those in Fig. 1(a), it can be seen that the shift of both 1s and 2s orbitals towards the center gives rise to the screening charge. As is seen from the curve for the effective potential $v_{eff}(r)$ in Fig. 1(c), the potential due to the extra charge, which is originally $1/r$, becomes short-ranged. While the energy levels of the orbitals except 1s and 2s are almost unchanged from those un Fig. 1(b), the 1s and 2s levels shift downwards considerably owing to the attractive force from the positive point charge at the center. The 1s electrons are almost bound by the point charge, although the electron number density of the 1s orbital spreads beyond the broad hump of the effective potential around 4 a.u. by the tunnel effect.

Near the surface of the jellium sphere, the electron number densities $n_{II}(r)$ and $n_{III}(r)$ nearly coincide with each other. Also the effective potential $v_{eff}(r)$ for Case III has almost the same value as that for Case II near and beyond the surface.

From these results we can estimate the screening charge density $n_{scr}(r)$ following the procedure explained in section 2.3, as shown in Fig. 1(d). The orbitals which mainly contribute to the screening charge are the s orbitals.

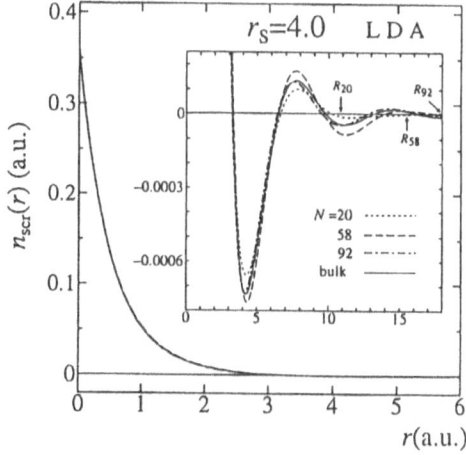

Fig. 2　The screening charge densities $n_{scr}(r)$ in the jellium spheres calculated in the LDA for $N = 20$, 58, 92 and ∞ (a bulk jellium) with $r_s = 4.0$. The corresponding Friedel oscillations are magnified in the inset. R_{20}, R_{58}, and R_{92} are the jellium sphere radii for each cluster size.

3.2 Size-dependence of screening charge

We show in Fig. 2 the screening charge densities $n_{scr}(r)$ in the jellium spheres calculated in the LDA for $N = 20$, 58 and 92 with $r_s = 4.0$. The screening charge density induced around the positive point charge embedded in the infinite (bulk) jellium with $r_s = 4.0$ is also shown in the figure by the solid line. It is seen from this figure that the screening charge density, even for small metallic clusters considered here, is very close to that in the bulk jellium, in spite of the discreteness of the electronic energy levels due to the finiteness of the cluster size. This result is quite understandable because the main contribution to the screening charge density comes from the s orbitals and in particular the 1s orbital is almost bound by the extra positive charge, which suggests the local nature of the screening effect. The formation of the bound state, by embedding an extra positive charge or a hydrogen atom, has already been pointed out by several authors [b,d,i] for jellium spheres as well as for infinite jellium and it is considered to be a H$^-$ -like state.

3.3 Density-dependence of screening charge

In Fig. 3 we show the screening charge densities $n_{scr}(r)$ in the jellium spheres calculated in the LDA for $r_s = 3.0$, 4.0 and 5.0 with $N = 58$, where r is scaled by r_s. From this figure we can see that the period of the oscillation of the $n_{scr}(r)$ can be almost scaled by r_s and it is very close to the period of the Friedel oscillation in the infinite jellium. The amplitude of the oscillation, on the other hand, depends on the density. These features of the screening charge density are also obtained for other sizes of jellium sphere we have investigated.

3.4 Comparison of approximations

In order to investigate the effect of the self-interaction on the screening

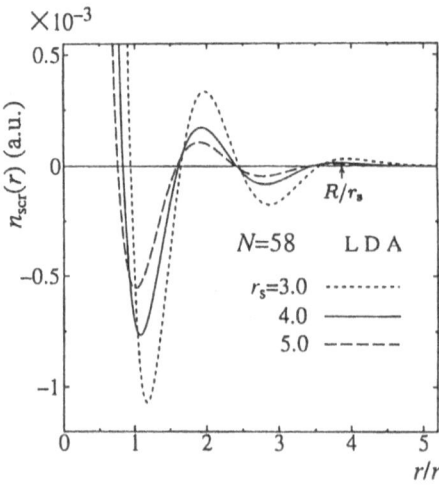

Fig. 3   The screening charge densities $n_{scr}(r)$ in the jellium spheres calculated in the LDA for $r_s = 3.0$, 4.0 and 5.0 with $N = 58$, where r is scaled by r_s. The radius of the jellium sphere scaled by r_s, $R/r_s = N^{1/3}$, is shown by the arrow.

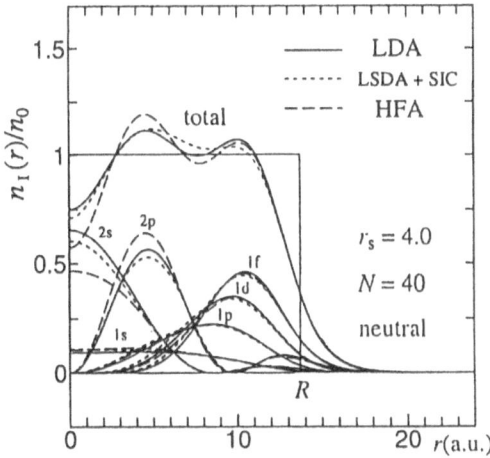

Fig. 4   The electron number densities $n_l(r)$ in the neutral jellium spheres for $N = 40$ and $r_s = 4.0$, calculated in the LDA(solid), the LSDA with the SIC(dotted) and the HFA(broken).

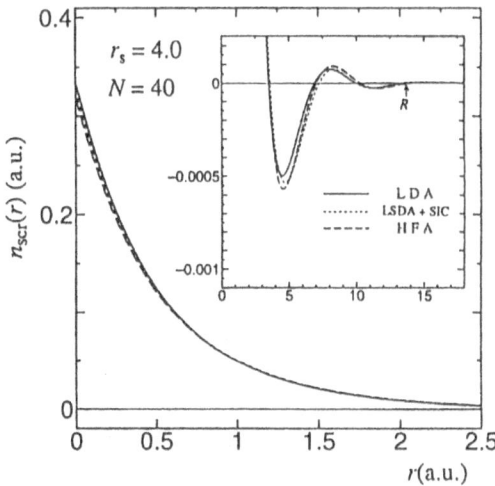

Fig. 5   The screening charge densities $n_{scr}(r)$ for $N = 40$ and $r_s = 4.0$ calculated in the LDA(solid), the LSDA with the SIC(dotted) and the HFA(broken). The inset shows the magnified Friedel oscillations, where the radius of the jellium sphere is shown by the arrow.

charge density $n_{scr}(r)$, we have also carried out calculations of the electronic states of the jellium sphere for the three cases mentioned above, based on two methods, i.e. the LSDA+SIC and the HFA, in addition to the LDA. In Fig. 4 we show the electron number density $n_l(r)$ in the neutral jellium sphere for $N = 40$ and $r_s = 4.0$. In the figure we also show the contribution of each orbital to $n_l(r)$. We have also carried out similar calculations for other sizes of cluster. The characteristic features of $n_l(r)$ thus obtained are as follows: As for the total charge density at the center of the jellium sphere, the result of the LSDA+SIC is smaller than that of the LDA irrespective of cluster size. The result of the HFA, on the other hand, becomes the smallest when $n_l(r)/n_0 < 1$ (for $N = 40$ and 58) and the biggest when $n_l(r)/n_0 > 1$ (for 20 and 92), in other words, the amplitude of the oscillation in $n(r)$

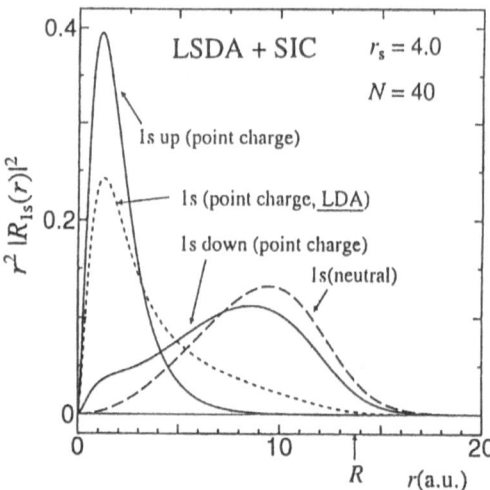

Fig. 6   The wavefunctions of 1s orbital with an up and a down spin, calculated in the LSDA with the
SIC for the case III where $N = 40$ and $r_s = 4.0$ (solid lines). For comparison, we also show the
1s wavefunctions calculated for the case I in the LSDA with the SIC and for the case III in
the LDA by broken and dotted lines, respectively. See section 2.3.

obtained in the HFA is always the biggest among the three results. Very recently
Lipparini et al.[p] made a comparison of the HFA with the LDA and obtained the
same result as we.

In Fig. 5 we show the screening charge densities $n_{scr}(r)$ calculated by the
three methods for $N = 40$ and $r_s = 4.0$. As is seen from this figure, the $n_{scr}(r)$'s of
the HFA as well as the LSDA+SIC spread out more than that of the LDA. In the
LDA, since exchange interaction is not treated correctly, i.e. the self-interaction
is included, the local exchange potential falls faster than $1/r$ outside the jellium
sphere. Therefore $n(r)$'s in the LSDA+SIC and in the HFA spread farther than that
in the LDA. In the same way, $n_{scr}(r)$'s, which are derived as the difference of
$n(r)$'s, in both approximations spread farther than that in the LDA. As a result,
$n_{scr}(r)$'s in the LSDA with the SIC and in the HFA are smaller than that in the
LDA in the center of jellium sphere. In the case of the HFA, however, the $n_{scr}(r)$
at the center of the jellium sphere is smaller than those in the LDA for $N = 40$ and
58 and is larger than those in the LDA for $N = 20$ and 92, which reflects the
feature of $n_1(r)$ mentioned above. As for the exchange-correlation effects on the
electronic states of spherical jellium clusters, see Ref. 3.

Finally, to see the characteristic features of the electronic states of the jellium
sphere calculated in the LSDA with the SIC, we show in Fig. 6 the wavefunctions
of 1s orbital with an up and a down spin, calculated for the case III with $N = 40$
and $r_s = 4.0$, by the solid lines. For comparison, we also show the 1s
wavefunctions calculated for the case I in the LSDA with the SIC and for the case
III in the LDA by broken and dotted lines, respectively. From this figure we can
see the followings: The 1s orbital obtained in the LDA is a rather weakly bound
state, which is occupied by two electrons. In contrast to this, in the LSDA with the
SIC, a 1s orbital with an up (or a down) spin is strongly bound by the positive
point charge, while the other 1s orbital with the opposite spin is not bound.
Therefore the 1s wavefunctions obtained in the LSDA with the SIC are

qualitatively different from that in the LDA, although the $n_{\text{scr}}(r)$'s obtained in these approximations as the differencein charge densities are quantitatively slightly different. The above features of 1s orbitals are also obtained for $N = 92$, although it is not obtained for $N = 18, 20, 34$. Note that the unrestricted HFA gives a similar result to the one in the LDA for all sizes of cluster investigated, *i.e.* the 1s orbital is bound and doubly occupied.

4. Summary

We have presented the results of our investigation on screening effects in metal clusters. In this investigation we have employed a jellium sphere model and calculated the change in the electronic states when an extra positive point charge is introduced at the center of the neutral jellium sphere, by solving the Kohn-Sham equation in the LDA. The screening charge density in the jellium sphere was estimated by subtracting the charge induced near the cluster surface from the total induced charge density thus calculated. In order to investigate how the self-interaction of the electrons affects the screening effects, we studied the change in the screening charge by solving the Kohn-Sham equation in the LSDA with the SIC and by solving the HF equation. The calculation of the screening charge densities was carried out systematically for jellium spheres of sizes $N = 20, 40, 58$, 92 and ∞ (a bulk jellium) with density parameters $r_s = 3.0, 4.0$ and 5.0.

The results of this paper are summarized as follows: (i) The behavior of the screening charge in metal clusters is very similar to the bulk behavior, though it is truncated at a finite distance by the finiteness of the cluster size. (ii) The tail part of the screening charge distribution in clusters shows an oscillatory behavior whose period, even for clusters with smaller sizes $N = 20, 40$ and 58, is close to the value of the Friedel oscillation in the bulk, although its amplitude differs slightly from the bulk value. The screening charge distributions for $N = 92$ and 198 almost coincide with those in the bulk. (iii) The orbitals mainly contributing to the screening charge are s orbitals. In particular, 1s orbital plays an important role in the screening, which is nearly or perfectly bound by the extra positive unit point charge introduced in the cluster. (iv) As for the density dependence, we have shown by carrying out the calculations for various densities that the period ofthe oscillation of the $n_{\text{scr}}(r)$ can be almost scaled by r_s and it is very close to the period of the Friedel oscillation of the screening charge density in the infinite jellium, though the amplitude of the oscillation depends on the density. (v) To investigate the effect of the self-interaction on the screening charge density $n_{\text{scr}}(r)$, we compared the electronic states of the jellium sphere calculated by three methods, *i.e.* the LDA, the LSDA with the SIC and the HFA. Since, in the LDA, the exchange interaction is not treated correctly, *i.e.* the self-interaction is included, the local exchange potential falls faster than $1/r$ outside the jellium sphere. On the other hand, the self-interaction is approximately eliminated in the LSDA with the SIC and exactly canceled out in the HFA. As a result of this fact, the $n_{\text{scr}}(r)$'s in the LSDA with the SIC and in the HFA spread out of the jellium sphere more than that in the LDA. As for the screening charge density near the center of the jellium sphere, the LSDA with the SIC always gives a smaller value than the LDA. (vi) Although the 1s orbital is always bound by the extra positive charge in the LDA, the situation is somewhat different in the case of the LSDA with the SIC. In the latter approximation, for $N = 40$ and 92, the 1s orbital with an up (or a down) spin is strongly localized by the extra positive charge, while the 1s orbital with the opposite spin is not bound. For a smaller jellium sphere such as $N = 18$, 20 and 34, both approximations give rise to the bound 1s orbitals. In this respect, the unrestricted HFA gives results similar to those of the LDA.

PUBLICATIONS

1. K. Sonoda, K. Hoshino and M. Watabe, *J. Phys. Soc. Jpn.*, **64**, 540 (1995).
2. K. Sonoda, F. Shimojo, K. Hoshino and M. Watabe, *Surf. Rev. and Lett.*, **3**, 329 (1996).
3. K. Sonoda, F. Shimojo, K. Hoshino and M. Watabe, in *Structures and Dynamics of Clusters*, (T. Kondou, K. Kaya and A. Terasaki, eds.), p.541, and references therein, Universal Academy Press, Tokyo, Japan (1996).
4. K. Sonoda, F. Shimojo, K. Hoshino and M. Watabe, in *Science and Technology of Atomically Engineered Materials*, (P. Jena, S. N. Khanna, and B. K. Rao, eds .), p.305, World Scientific (1996).

REFERENCES

a. Z. D. Popovic and M. J. Stott, *Phys. Rev. Lett.*, **33**, 1164 (1974).
b. C. O. Almbladh, U. von Barth, Z. D. Popovic and M. J. Stott, *Phys. Rev.*, **B 14**, 2250 (1976).
c. M. Manninen, P. Hautojärvi and R. Nieminen, *Solid State Commun.*, **23**, 795 (1977).
d. P. Jena and K. S. Singwi, *Phys. Rev.*, **B 17**, 3518 (1978).
e. J. L. Martins, R. Car and J. Buttet, *Surface Sci.*, **106**, 265 (1981).
f. Y. Ishii, S. Ohnishi and S. Sugano, *Phys. Rev.*, **B 33**, 5271 (1986).
g. W. Ekardt, *Phys. Rev.*, **B 29**, 1558 (1984).
h. A. Hintermann and M. Manninen, *Phys. Rev.*, **B 27**, 7762 (1983).
i. W. Ekardt, *Phys. Rev.*, **B 37**, 9993 (1988).
j. P. Hohenberg and W. Kohn, *Phys. Rev.*, **136**, B864 (1964).
k. W. Kohn and L. J. Sham, *Phys. Rev.*, **140**, A1133 (1965).
l. J. P. Perdew and A. Zunger, *Phys. Rev.*, **B 23**, 5048 (1981).
m. For example, S. E. Koonin and D. C. Meredith, *Computational Physics*, Chap. 3, Addison-Wesley (1990).
n. P. Stampfli and K. H. Bennemann, *Phys. Rev.*, **A 39**, 1007 (1989).
o. D. M. Ceperley and B. J. Alder, *Phys. Rev. Lett.*, **45**, 566 (1980).
p. E. Lipparini, Ll. Serra and K. Takayanagi, *Phys. Rev.*, **B 49**, 16733 (1994).

24

Electronic States of Transition Metal Clusters

Nobuhisa Fujima[a] and Tsuyoshi Yamaguchi[b]

[a] Faculty of Engineering, Shizuoka University
Hamamatsu-shi, Shizuoka 432-8561, Japan
[b] Faculty of Engineering, Shizuoka University
Shizuoka-shi, Shizuoka 422-8529, Japan

Purpose of the present study

We calculated the electronic states of the 3d transition-metal clusters by using the local density functional method, and clarified their extraordinary characteristics, which are mainly caused by the chemical bonding of 3d electrons in the clusters. Here we first clarify the structure of chemical bonding in Mn clusters and its size-dependence, and discuss the origin of the magic number which has been observed in the cationic Mn clusters. Second, we calculate the electronic states of binary Co-V clusters, and discuss the extraordinary stability to the hydrogen-adsorption of $Co_{12}V$ cluster.

Contents of the present study

1. Chemical bonding in Mn clusters

1.1 Introduction

Manganese bulk is known to show strange characteristics among the 3d-transition metal elements. It gives the lowest modulus and the lowest cohesive energy among the 3d-elements. Both Mn-metal and Mn-compounds have various crystal structures, where numerous inequivalent sites exist. For example, there are 58 atoms in a unit cell of the α Mn phase, where the interatomic distances range from 2.25 Å to 2.95 Å.

Manganese clusters also show some interesting characteristics different from other 3d-transition metal clusters. For example, it is suggested from the following two experimental results that Mn atoms in small clusters behave like rare-gas atoms. First, it is observed that the binding energy of the neutral dimer is less than 10^{-1} eV. [a] This value is one or more order weaker than those of other 3d-metal dimers. It is also observed that the interatomic distance of the dimer, 3.4 Å is 130% of that in the bulk crystal. [b] These facts suggest that metallic 4s-4s and 3d–4s interactions in the Mn dimer are weak. Second, magic numbers 5, 14, 16, 29 ... are observed in the cationic Mn clusters. [c,d] These magic numbers are similar to those observed in cationic rare-gas clusters. [e–g] That is, except for the first magic number 5, the differences between the magic numbers of cationic Mn clusters and those of cationic rare-gas clusters are at most 1.

These strange characteristics originate mainly from the electronic configuration of the isolated Mn atom $3d^5 4s^2$. The 3d (4s) orbital is half-(fully-) filled and has a spherical symmetry. It is an interesting configuration, as well as or more than that of divalent metals, from the point of view of how the fully- or half-filled orbitals are broadened to form the band structure.

Considering the electronic configuration of the Mn atom, one can forecast an outline of the electronic features in the condensed matter that Mn atoms constitute. When the cohesion of atoms is smaller than a certain value, the interactions between atoms are weak, and the electronic configuration maintains its spherical character. In this situation, the Mn atom behaves like a divalent atom or a rare-gas atom. However, the electronic configuration of the isolated Mn atom is frailer than that of rare-gas atoms and is easily changed. Once the configuration is changed, there appears a strong interaction between 3d electrons and the electronic feature changes drastically as compared to divalent atoms in which only s-p hybridizations occur. Thus, manganese has both divalent and 3d transition metallic characteristics. It is interesting to compare the electronic structure of the Mn clusters and its size dependence with those of divalent clusters and other 3d-transition metal clusters.

In the present study, we calculate the electronic states of small neutral and cationic Mn_N clusters ($N = 2$–7) by using the local spin-density functional method, discuss the electronic feature of Mn clusters, and clarify its size-dependence, especially the change from non-metallic to metallic bonding.[1,2]

1.2 Calculations

The electronic states of Mn clusters are calculated by the spin-polarized discrete variational Xα (DV-Xα) method with the linear combination of atomic orbitals approximation (LCAO). We employ numerical 1s-4s, 2p-4p and 3d orbitals of Mn atom as the LCAO bases. The bases are symmetrized according to the irreducible representations of the point group of the cluster.

The equilibrium interatomic distance is found by minimizing the total energy of the cluster E,

$$E = \sum_{i,\sigma} n_{i\sigma}\varepsilon_{i\sigma} - \frac{1}{2}\iint \rho(r)\frac{2}{|r-r'|}\rho(r')drdr'$$

$$+ \sum_{\sigma}\left[\frac{3}{2}\alpha\int\left\{\frac{3}{4\pi}\rho_\sigma(r)\right\}^{4/3}dr\right] + \sum_{i>j}\frac{Z_i Z_j}{|R_i - R_j|},$$

where $\varepsilon_{i\sigma}$ denotes the eigenvalue of the i-th one-electron orbital with σ-spin, $n_{i\sigma}$

Table 1 Atomic structure and symmetry of bases

N	Atomic structure	Symmetry of bases
2		$C_{\infty v}'$
3	linear chain	$D_{\infty h}$
4	square	D_{2h}
5	square pyramid	C_{2v}
6	octahedron	D_{4h}
7	pentagonal dipyramid	D_{5h}

the occupation number, $\rho_\sigma(r)$ the spin density at r with σ-spin ($\rho(r) = \rho_\uparrow(r) + \rho_\downarrow(r)$), and Z_i and R_i the charge and the position of the i-th nucleus. The parameter α of the exchange potential is equal to 0.7.

The atomic symmetries are fixed as shown in Table 1. We evaluate the spin order of the clusters within the restriction of atomic symmetries.

1.3 Chemical bonding in neutral Mn clusters

Figure 1 shows the energy level diagrams for the neutral Mn$_N$ clusters at the equilibrium interatomic distance. Clusters are (a) the dimer, (b) the linear chain trimer, (c) the square, (d) the square pyramid, (e) the octahedron and (f) the pentagonal dipyramid for N = 2,3,4,5,6 and 7, respectively. The left(right)-hand part of each diagram indicates the energy levels of up(down)-spin electron. The solid (dashed) line indicates the occupied (unoccupied) level. The thin line crossing the energy levels shows the 3d electron components estimated by the Mulliken charge analysis.

All the clusters show the antiparallel spin order within the symmetrical restrictions. Note that the antiparallel spin order has been observed in the Mn dimer.[b]

For the neutral Mn dimer in Fig. 1(a), it is shown that the 3d-derived and 4s-derived levels are separate from each other. All the 4s-derived levels exist under the highest occupied level, that is, the 4s orbitals are fully occupied. They are shown as two levels in each spin. The lower side levels correspond to the bonding part and the upper side levels the antibonding part. This means that the neutral dimer is not bound by the 4s electronic bonding. The 3d-derived levels are divided into two parts, fully occupied levels shown as the lowest levels and unoccupied levels just above the highest occupied level. The levels of each group are almost 5-fold degenerate respectively. The 3d orbitals are thus nearly half-filled and have characteristics similar to those of the isolated Mn atom.

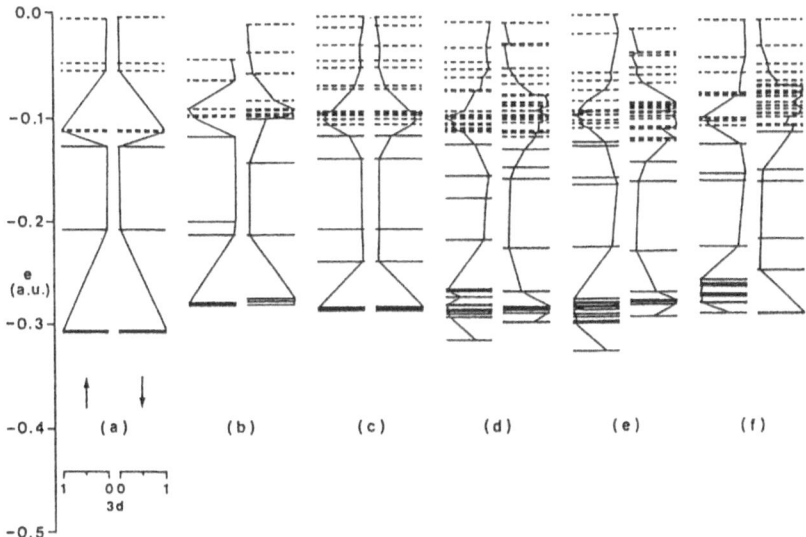

Fig. 1 Energy level diagram of neutral Mn clusters, Mn$_N$ (N = 2–7).

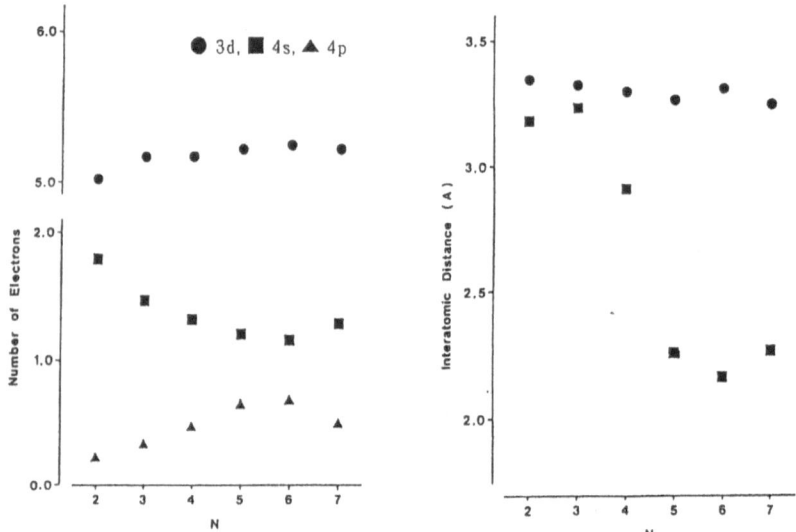

Fig. 2 Average electronic configurations for neutral Mn clusters.

Fig. 3 Equilibrium interatomic distances of Mn clusters, Mn_N ($\bullet$) and Mn_N^+ ($\blacksquare$).

The averaged electronic configurations of the Mn clusters at their equilibrium atomic distances are shown in Fig. 2. The electronic configuration of the neutral dimer is $3d^{5.02} 4s^{1.77} 4p^{0.21}$. This is close to that of the isolated atom, $3d^5 4s^2$.

The size dependence of the equilibrium interatomic distance of the Mn clusters is indicated in Fig. 3 ($\bullet$). The equilibrium interatomic distance of the neutral dimer is 3.4 Å. This value corresponds to about 130% of that of the bulk crystal and agrees well with the experimental value. [b]

These results suggest that the neutral Mn dimer is non-metallic and atoms in the dimer have features close to those of an isolated atom.

For the larger neutral clusters in Figs. 1(b)–1(f), the gross electronic structures seem to be similar to that of the dimer although the 3d-4s mixing around the highest occupied level becomes gradually larger with cluster size. This tendency is shown in the average electronic configurations in Fig. 2. With increasing cluster size, the 3d component with the minority spin increases and the 4s component decreases. However, comparing the population of the s electron in the bulk crystal ~0.6, the 4s population in the Mn clusters is significantly large.

The equilibrium interatomic distance in Fig. 3 tends to be small with cluster size, but these values are still larger than 120% that of the bulk crystal.

Thus, it is concluded that the neutral Mn_N clusters ($N = 2–7$) have non-metallic features. These features should be compared with those of the Fe clusters, which, even for the dimer, have an atomic distance smaller than that of the bulk crystal and have metallic features. [h]

1.4 Chemical bonding in cationic Mn clusters

Figures 4(a)–4(f) show the same energy levels as Fig. 1 for the cationic Mn_N^+ clusters. As well as the neutral clusters, all cationic clusters show the antiparallel spin order within the symmetrical restrictions.

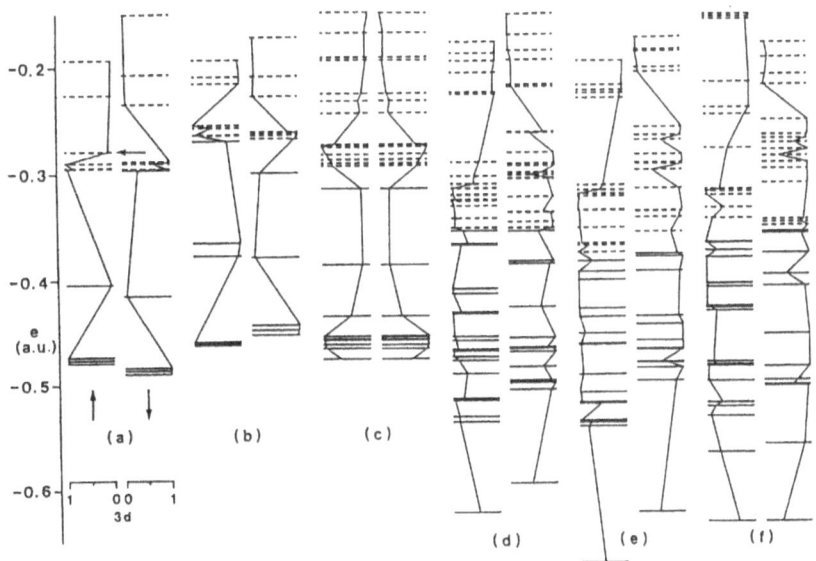

Fig. 4   Energy level diagram of cationic Mn clusters, Mn_N^+ (N = 2–7).

For the cationic dimer in Fig. 4(a), the gross feature of the electronic structure is similar to that of the neutral dimer in Fig. 1(a) although the s-d mixing near the highest occupied level becomes larger. The level designated by an arrow corresponds to the unoccupied level, which has been caused by removing an electron from an occupied level in the neutral dimmer. The removing an electron from the antibonding states causes an increase in the 4s bonding, $i.e.$ metallic bonding. Nevertheless, the equilibrium interatomic distance keeps its value of 3.1 Å. This value is about 10% smaller than that of the neutral dimer, but still 20% larger than that of the bulk crystal. Thus, the cationic dimer maintains the non-metallic characteristics.

These characteristics remain for the cationic trimer and tetramer in Figs. 4(b) and 4(c). The average electronic configurations of these clusters shown in Fig. 5 are similar to that of the cationic dimer. The equilibrium interatomic distances are also similar to that of the dimer (~120% of that of the bulk crystal). The cationic trimer and tetramer still maintains non-metallic characteristics.

For cationic clusters larger than tetramers, in Figs. 4(d)–4(f), the features of the energy level are quite different from those of the smaller clusters. The 3d levels broaden in energy and the s-d mixing is so large that we cannot identify the 3d-derived or the 4s-derived level. It is shown in Fig. 5 that a stepwise change in the size-dependence of electronic configurations occurs between the tetramer and the pentamer. The 3d component increases dramatically and the 4s component becomes close to the value in the bulk crystal.

The equilibrium interatomic distances shown in Fig. 3 also change in a stepwise manner and become about 2.2 Å in the cationic pentamer (■). The value is much smaller than those of smaller clusters, and corresponds to about 90% that of the bulk crystal.

The binding energies per atom are one order of magnitude larger than that of the neutral dimer. The strong binding energy is due to the 4s bonding, which is caused by the charge transfer from the antibonding 4s levels to the unoccupied 3d

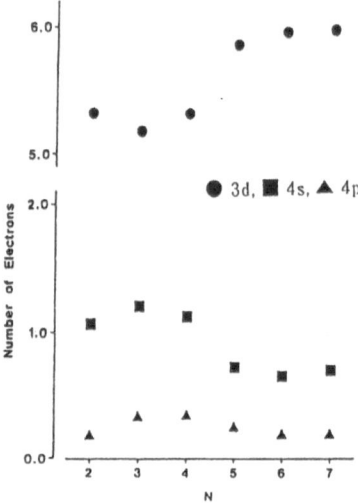

Fig. 5 Average electronic configurations for cationic Mn clusters.

levels around the highest occupied level. Thus, we conclude that the cationic Mn_N^+ clusters ($N = 5$–7) have metallic features.

The fact that the neutral clusters keep their non-metallic features from $N = 2$ to 7, but the cationic clusters change features at $N = 5$ corresponds to the magic number "5," which is observed for cationic Mn clusters, but has not been observed for neutral Mn clusters.[c,d] That is, the magic number 5 originates in the transition from non-metallic to metallic. It should be mentioned that the magic number 5 has not been observed in divalent metal clusters, rare-gas clusters or other transition metal clusters. The magic number is due to the electronic features specific to the Mn atom.

2. H-adsorption of Co-V clusters

2.1 Introduction

Recently reaction studies of various metal clusters have been done and have revealed that reactivities of metal clusters with molecules have anomalous size dependencies.[i–o] For example, some particular Co clusters are extraordinarily stable against H-adsorption, although Co clusters in general and Co metal surfaces are unstable. Nonose et al. have observed, in bi-metal cluster beam experiments using the laser vaporization method, that a $Co_{12}V$ cluster is stable against H-adsorption while Co_{13}, $Co_{11}V_2$ and Co_NV ($N < 12$) clusters are unstable. [o] Furthermore, Co_NV ($N = 13$–18) clusters show similar stability, but it was not so clear as the $Co_{12}V$ cluster. These results suggest that the clusters which have a V atom and consist of more than 13 atoms are stable against H-adsorption. In addition, they suggest that the V atom is located at an interior site of the cluster since clusters consisting of more than 13 atoms are likely to have interior sites. Thus, the interior atom seems to play an important role in the reaction with the H molecule.

In the present study, we calculate the electronic states of icosahedral Co-V

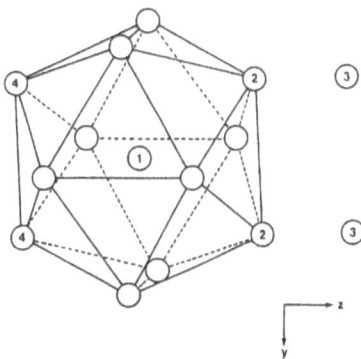

Fig. 6 Atomic structure of $Co_{12}MH_2$ cluster.

clusters, Co_{13}, $Co_{12}V$, and those adsorbed with hydrogen atoms using the local density functional method. We discuss the extraordinary stability of the $Co_{12}V$ cluster against H-adsorption.[3,4]

2.2 Calculations

All calculations are performed using the self-consistent-charge DV-Xα-LCAO method with the 1s–4s, 2p–4p and 3d orbitals of transition-metal atoms and the 1s orbital of H atom. Calculations are performed with a basis set of C_{2v}-symmetrized block.

Figure 6 shows the atomic structure of the clusters, that is, the icosahedral (I_h)13-atom cluster where the center is a Co atom for Co_{13} and a V atom for $Co_{12}V$, and those with two H atoms, $Co_{13}H_2$ and $Co_{12}VH_2$ clusters where two H atoms are located directly on the exterior Co atoms with the distance of a sum of atomic radii, 3.00 a.u. We assume the interatomic distances in the icosahedral cluster to be the bulk bond-length.

Note that, in the C_{2v} symmetry of the H-adsorbed cluster, only the levels of the A_1 and B_1 representations have the 1s component of H atom, and that the B_1 representation to which the $3d_{yz}$ orbital of the center atom belongs plays an important role in the stability against H-adsorption, as discussed below.

2.3 Electronic states of $Co_{12}V$ and Co_{13} clusters

Figures 7(a) and 7(b) show the energy level diagrams and the partial density-of-states (PDOS) of the 3d orbitals of I_h-$Co_{12}V$ and I_h-Co_{13} clusters without H-adsorption. The straight lines show the energy levels. The dotted lines point to the highest occupied level. The solid curves indicate the 3d-PDOS of the whole cluster which is multiplied by a factor of 0.1. The broken curves indicate the 3d-PDOS of the center atoms, V for (a) and Co for (b). The PDOS of the 3d orbitals of the whole cluster has two large peaks at about –0.26 and –0.19 a.u. in both clusters. These peaks are mainly composed of the 3d orbitals of the exterior Co atoms.

The PDOS near the highest occupied level of the $Co_{12}V$ cluster is 15% smaller than that of the Co_{13} cluster. The shape of the PDOS is broader for the $Co_{12}V$ than for the Co_{13}.

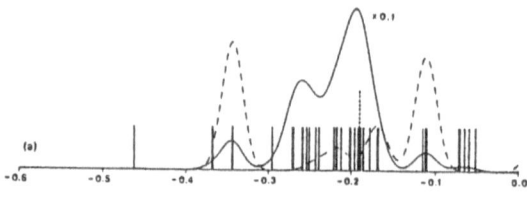

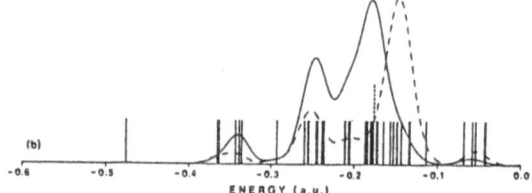

Fig. 7 Energy level diagram of (a) $Co_{12}V$ and (b) Co_{13} clusters.

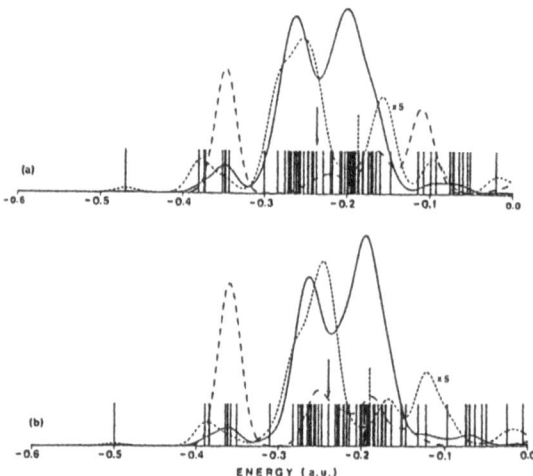

Fig. 8 Energy level diagram of (a) $Co_{12}VH_2$ and (b) $Co_{13}H_2$ clusters.

The 3d-PDOS of the center V atom in the $Co_{12}V$ cluster has two large peaks between which the highest occupied level exists. On the other hand, the 3d-PDOS of the center Co atom in the Co_{13} cluster has a large peak at −0.35 a.u.

These facts suggest that the center V atom of the $Co_{12}V$ cluster interacts with the exterior Co atoms more strongly than the center Co atom of the Co_{13} cluster.

2.4 Electronic states of $Co_{12}VH_2$ and $Co_{13}H_2$ clusters

Figures 8(a) and 8(b) show the energy level diagrams of $Co_{12}VH_2$ and $Co_{13}H_2$ clusters. The solid, broken and dotted curves indicate the 3d-PDOS of Co atoms adjoining the H atoms, the 3d-PDOS of the center atom and the 1s-PDOS of H atoms, respectively. The H 1s-PDOS is multiplied by a factor of 5.

The 1s-PDOS of H atom of both clusters has a large peak at the energy below the highest occupied level and small peaks above the highest occupied level. The peak below (above) the highest occupied level has a bonding (anti-bonding) orbital component between the H-1s electron and the 3d electron of the adjacent Co atom. Comparing Figs. 8(a) for $Co_{12}VH_2$ cluster with Fig. 8(b) for $Co_{13}H_2$ cluster, it is shown that the peak corresponding the anti-bonding orbital shifts to the high energy side and the energy difference between the two peaks becomes larger when the center atom changes from a V atom to a Co atom. This means that the interaction between the exterior Co and H atoms is larger in the $Co_{13}H_2$ cluster than in the $Co_{12}VH_2$ cluster.

Figure 9 shows the contour map of electron density of the B_1 level, which contains the 1s orbital of H atom and the $3d_{yz}$ orbital of the center atom in the central region of clusters. The section plane is the y–z plane including the center atom site, four exterior sites and two H sites. The calculated levels are designated by arrows in Fig. 8. In Fig. 9(a), the charge density is small in the region between the exterior Co and H atoms. Thus this orbital is the non-bonding orbital. In Fig. 9(b), the charge density is larger and the orbital seems to be a bonding orbital. These facts are consistent with the adsorption experiments [o] that a H molecule hardly adsorbs to the $Co_{12}V$ cluster and easily adsorbs to the $Co_{13}V$ cluster.

Figure 10 shows the electron densities of the B_1 level at the middle point between the Co and H atoms and the average of energy separation between the peaks of the 1s-PDOS of H atom above and below the highest occupied level, where the center atom of the cluster changes from Ti to Co. Both the electron density and the energy separation become smaller (larger) for the center atom of the earlier (later) transition metals.

Therefore, the interaction between the exterior Co and adsorbed H atoms depends on the type of the center atom, *i.e.* it increases as the center atom varies from the V to the Co atom.

The magnitude of the 3d-3d interaction is expected to be derived from the length of the radius of the atomic orbital relative to the inter atomic distance. We estimate the ζ value of the 3d orbital in the single ζ-STO ($r^2 e^{-\zeta r}$) from the numerical orbital of the DV-Xα calculation: it varies from 2.80 for the V atom to

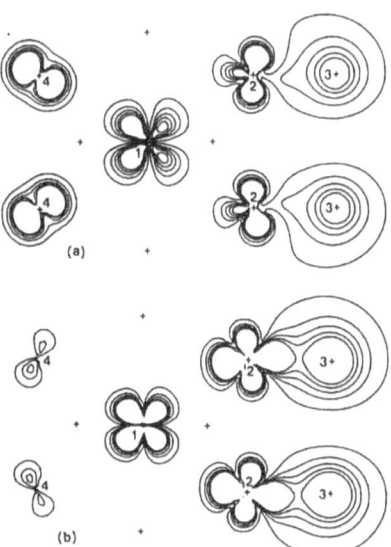

Fig. 9 Contour map of electron density of the B_1 level on the y–z plane (in arbitrary units). (a) $Co_{12}VH_2$, (b) $Co_{13}H_2$.

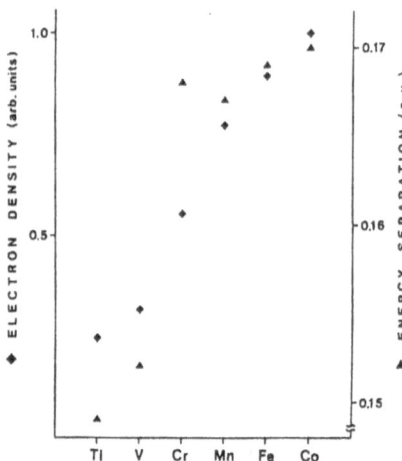

Fig. 10 Electron density at the middle point between the Co and H atoms in the $Co_{12}MH_2$ clusters ♦, and average of the energy separation between the two peaks of the 1s-PDOS of H atom ▲.

3.93 for the Co atom. Note that the observed atomic radius is 2.53 a.u. for V crystal and 2.42 a.u. for Co crystal. Thus we expect the interaction between the center and exterior atoms to be large for the $Co_{12}V$ cluster and small for the Co_{13} cluster.

In conclusion, the extraordinary stability against H-adsorption of the $Co_{12}V$ cluster results from the strong interaction between the center V atom and the exterior atom, resulting in a weak interaction between the exterior Co atom and the adsorbed H atom.

PUBLICATIONS

1. N. Fujima and T. Yamaguchi, *Z. Phys.*, **D26**, S150 (1993).
2. N. Fujima and T. Yamaguchi, *J. Phys. Soc. Jpn.*, **64**, 1251 (1995).
3. N. Fujima and T. Yamaguchi, *NATO ASI*, **C34**, 1095 (1995).
4. N. Fujima and T. Yamaguchi, *J. Phys. Soc. Jpn.*, **61**, 1724 (1992).

REFERENCES

a. A. Kant, S. Lin and B. Strauss, *J. Chem. Phys.*, **49**, 1983 (1968).
b. C. A. Baumann, R. J. Van Zee, S. V. Bhat and W. Weltner Jr, *J. Chem. Phys.*, **78**, 190 (1983).
c. Y. Saito, H. Ito and I. Katakuse, *Z. Phys.*, **D19**, 189 (1991).
d. Y. Sone, K. Hoshino, T. Naganuma, A. Nakajima and K. Kaya, *J. Phys. Chem.*, **95**, 6830 (1991).
e. A. Ding and J. Hesslich, *J. Chem. Phys. Lett.*, **94**, 64 (1983).
f. I. A. Harris, R. R. Kidwell and J. A. Northby, *Phys. Rev. Lett.*, **53**, 2390 (1984).
g. O. Echt, K. Sattler and E. Recknagel, *Phys. Rev. Lett.*, **47**, 1121 (1981).
h. M. Castro and D. R. Salahub, *Phys. Rev.*, **B49**, 11842 (1994).
i. J. L. Elkind, F. D. Weiss, J. M. Alford, R. T. Laaksonen and R. E. Smalley, *J. Chem.Phys.*, **88**, 5215 (1988).
j. E. K. Parks, G. C. Nieman, L. G. Pobo and S. J. Riley, *J. Chem. Phys.*, **88**, 6260 (1988).
k. M. R. Zakin, D. M. Cox, R. O. Brickman and A. Kaldor, *J. Phys. Chem.*, **93**, 6823 (1989).
l. K. Fuke, S. Nonose, N. Kikuchi and K. Kaya, *Chem. Phys. Lett.*, **147**, 479 (1988).
m. S. Nonose, Y. Sone, N. Kikuchi, K. Fuke and K. Kaya, *Chem. Phys. Lett.*, **158**, 152 (1989).
n. M. D. Geusic, S. C. Morse and R. E. Smalley, *J. Chem. Phys.*, **82**, 590 (1985).
o. S. Nonose, Y. Sone, K. Onodera, S. Sudo and K. Kaya, *J. Phys. Chem.*, **94**, 2744 (1990).

25

SIMS Experiment

Itsuo Katakuse[a], Toshio Ichihara[a], Hiroyuki Ito[b], Tohru Sakurai[b] and Takekiyo Matsuo[b]

[a] Department of Earth and Space Science, Graduate School of Science, Osaka University
Toyonaka-shi, Osaka 560-0043, Japan
[b] Department of Physics, Faculty of Science, Osaka University
Toyonaka-shi, Osaka 560-0043, Japan

Purpose of the present study

The sputtering technique, or secondary ion mass spectrometry (SIMS), is a useful method for investigating clusters. Positively and negatively charged clusters are produced from solid samples by bombardment of energetic Xe ions or other ions. As the clusters have high internal energies, it is suitable for the study of dissociation mechanism and stabilities. In this work, we describe the characteristics of clusters, MgO, Hg-Cs, rare gas, TaC, Zn-Cs, Cd-Cs, Te, Ag, Ta and Pb clusters, produced from a spattering ion source.

Experimental

Secondary ions were produced by bombarding a sample with 7 keV Xe ions from a discharge type primary ion gun.[a] The primary current was about 10 µA at the sample tip and the spot size was about 1 mm in diameter. In some cases, cesium ions produced from a thermal ion source (Antek Model No. 160-250B, Palo Alto, CA) was used for primary ions. The cluster ions were accelerated and focused by a lens system in the ion source and mass-analyzed by a single focusing mass spectrometer, CQH type double focusing mass spectrometer or QQHQC type mass spectrometer, where C, Q and H denote a cylindrical electric field, an electric quadrupole lens and homogeneous magnetic field, respectively. The single focusing mass spectrometer consists of a 100-cm magnet with 110-degree deflection angle. The magnet radius of the CQH type mass spectrometer is 50 cm[b] and that of the QQHQC is 140 cm.[c] The mass analyzed ions were amplified by a 16-stage Cu/Be conventional secondary electron multiplier (Hamamatsu Photonics Model No. R667) and a DC amplifier. The mass spectra were recorded by a UV recorder or data system (JEOL Model No. JMA-DA5000) by scanning the magnet current in a linear mode. As the ion current varied by several orders of magnitude between the low and high mass regions, it was difficult to obtain the mass spectra with the same multiplier potential in the whole mass range. The entire mass spectra was obtained from several partial mass spectra with overlapping regions, which were obtained using different multiplier potentials.

Results

1. Cluster ion abundance and geometrical structure of magnesium oxide clusters

Mass spectra were obtained using the CQH type mass spectrometer. The single focusing mass spectrometer was also used to compare the mass spectra obtained by the double focusing mass spectrometer. By bombarding magnesium metal with mixed ions of Xe and oxygen, two types of clusters, $(MgO)_n^+$ and $(MgO)_n Mg^+$, were observed. The ion intensity of the former series was considerably stronger than that of the latter.

Figure 1 shows the cluster ion abundance of $(MgO)_n^+$ up to $n = 50$, represented on a logarithmic scale. The ion intensity decreases pseudo-exponentially as the cluster size, n, increases. Superimposed on the decrease, enhanced peak intensities are observed at certain n-values. These n-values are listed in Table 1. Above a cluster size of $n = 50$, the abundances of all cluster ions, except the 52-, 54-, 56-, 60-, 70- and 75-mers are under the detection limit.

As an MgO crystal has an NaCl type structure, the stable cluster structure of $(MgO)_n^+$ is expected to be the cubic-like structure used to explain the stabilities of alkali halide clusters.

The high ion intensities of the clusters at $n = 4(2 \times 2 \times 2)$, $6(3 \times 2 \times 2)$, $9(3 \times 3 \times 2)$, $18(4 \times 3 \times 3)$, $32(4 \times 4 \times 4)$, $40(5 \times 4 \times 4)$, $50(5 \times 5 \times 4)$, $60(6 \times 5 \times 4)$ and $75(6 \times 5 \times 5)$ are attributed to the cubic-like structures. The stabilities of other clusters with high ion intensities are attributed to the chair-like structures, which can be obtained by removing or attaching MgO molecules along the side of the rectangular structure, $i \times j \times k$. For instance, the structure of the 34-mer is obtained by attaching two MgO units to the $4 \times 4 \times 4$ structure. This can be represented by a formula, $4 \times (4 \times 4 + 1)$. The proposed structures of clusters with high ion intensities are represented in Table 1.

The ion intensity of the $(MgO)_n Mg^+$ type is relatively weak, and above $n = 40$, only a peak at $n = 62$ is observed. Figure 2 shows the cluster ion abundance of $(MgO)_n Mg^+$. Table 2 shows the cluster sizes with high ion intensities, total number of atoms in the clusters and possible structures. The structure with high ion intensities can be explained in the same manner as for above $(MgO)_n^+$ clusters. The structures at $n = 23$ and 29 cannot be explained by the cubic- or chair-like structures.

The ion intensities of the $(MgO)_n Mg^+$ type clusters seem to be correlated to those of $(MgO)_n^+$ type clusters. At the cluster size, where the ion intensities of the $(MgO)_n Mg^+$ type clusters are strong, those of the $(MgO)_n^+$ type are weak and vice versa. This behavior can be understood as follows.

It is supposed that the yields of both types of cluster ions at the moment of ejection from the bulk, decrease monotonical with increasing n, as is the case for CsI clusters.[d] The discontinuous variation arises from fragmentation of clusters during the flight time in the mass spectrometer. If an n-mer is stable, then clusters larger than n-mers will decay mainly to n-mers and the intensities of both types of neighboring sizes will become weak.

The high intensities of $(MgO)_n Mg^+$ are correlated to weak intensities of $(MgO)_n^+$ and those of $(MgO)_n^+$ are to $(MgO)_{(n-1)} Mg^+$. The strong intensities of

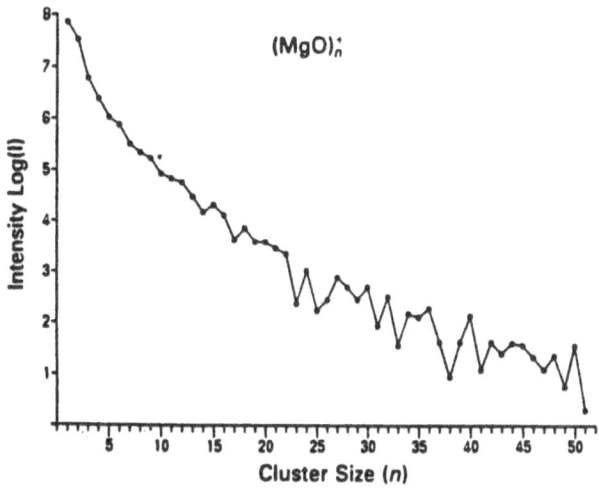

Fig. 1   Cluster-ion abundances of $(MgO)_n^+$ ions up to $n = 51$. Above $n = 50$, peaks of 52-, 54-, 56-, 60-, 70- and 75-mers are observed (not shown).

Table 1 Cluster sizes giving enhanced peak intensities, total number of atoms and proposed stable structures of $(MgO)_n^+$ clusters

Cluster sizes n	Number of atoms	Proposed structures
4	8	$2 \times 2 \times 2$
6	12	$3 \times 2 \times 2$
9	18	$3 \times 3 \times 2$
12	24	$3 \times (3 \times 3 - 1)$
15	30	$3 \times (3 \times 3 + 1)$
18	36	$4 \times 3 \times 3$
20	40	$4 \times (3 \times 3 + 1)$
22	44	$4 \times (4 \times 3 - 1)$
24	48	$4 \times 3 \times 4$
27	54	$3 \times (4 \times 4 + 2)$
30	60	$4 \times (4 \times 4 - 1)$
32	64	$4 \times 4 \times 4$
34	68	$4 \times (4 \times 4 + 1)$
36	72	$4 \times (4 \times 4 + 2)$
40	80	$4 \times 4 \times 5$
42	84	$4 \times (4 \times 5 + 1)$
44	88	$4 \times (4 \times 5 + 2)$
45	90	$5 \times (4 \times 4 + 2)$
48	96	$4 \times (5 \times 5 - 1)$
50	100	$4 \times 5 \times 5$
52	104	$4 \times (5 \times 5 + 1)$
54	108	$4 \times (5 \times 5 + 2)$
56	112	$4 \times (5 \times 5 + 3)$
60	120	$4 \times 5 \times 6$
70	140	$5 \times (5 \times 6 - 2)$
75	150	$5 \times 5 \times 6$

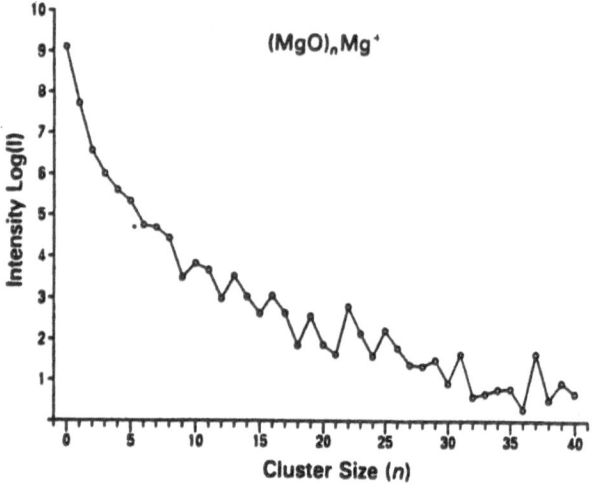

Fig. 2 Cluster ion abundances of $(MgO)_nMg^+$ ions up to $n = 40$.

Table 2 Cluster sizes giving enhanced peak intensities, total number of atoms and proposed stable structures of $(MgO)_nMg^+$ clusters

Cluster sizes n	Number of atoms	Proposed structures
10	21	$3 \times (3 \times 3 - 2)$
13	27	$3 \times 3 \times 3$
16	33	$3 \times (3 \times 3 + 2)$
19	39	$3 \times (3 \times 4 + 1)$
22	45	$3 \times 3 \times 5$
23	47	?
25	51	$3 \times (4 \times 4 + 1)$
29	59	?
31	63	$3 \times (4 \times 5 + 1)$
37	75	$3 \times 5 \times 5$
62	125	$5 \times 5 \times 5$

the $(MgO)_{23}Mg^+$ and $(MgO)_{29}Mg^+$ may be attributed to weak intensities of $(MgO)_{23}^+$ and $(MgO)_{29}$ clusters.

2. Mercury-cesium clusters —From van der Waals clusters to metallic clusters—

Cluster ions $(Hg)_nCs^+$ were produced by bombarding a mercury-silver amalgam with Cs^+ ions produced by the thermal ion gun. Primary ion energy and current on the sample tip were 15 keV and about 0.1 μA respectively. The secondary clusters of $(Hg)_nCs^+$ type were mainly produced and other types of clusters such as $(Hg)_nK^+$, $(Hg)_nAg^+$, $(Ag)_n^+$ and $(Hg)_n^+$ were less than 1%. The mass analysis was performed using the QQHQC type mass spectrometer.

2.1 Low mass region

Mass spectra were obtained under such conditions that individual cluster ion

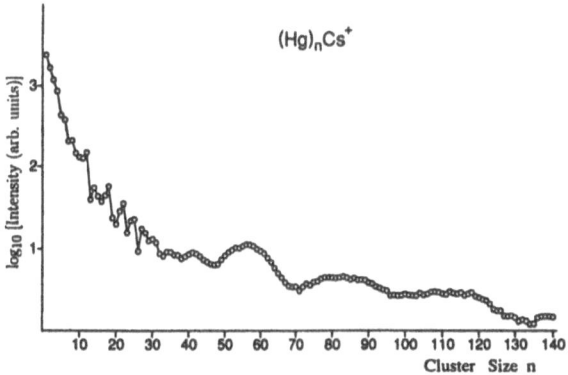

Fig. 3 Mass destribution of mercury-cesium cluster ions $(Hg)_nCs^+$ up to $n = 140$. The individual cluster ion peak were completely resolved.

peaks with the same n-value were completely resolved by adjusting the slit width, filter frequency and scan speed. The acceleration potential was 2 kV.

Figure 3 shows the size distribution in the low mass region. The raw mass spectra up to $n = 600$ are shown elsewhere.[e] Superimposed on a pseudo-exponential decrease, generally seen in sputtering mass spectra, two specific features are observed. First, with a cluster size less than 40, the distribution shows discontinuous variation at $n = 4, 6, 8, 12, 14, 18, 22, 25, 28$ and 31. Second, when $n > 40$, the distribution shows a consecutive hill and valley pattern. The cause of this pattern will be discussed in the section on high mass region.

Peaks that are somewhat more intense than their neighbors ($n = 4, 6, 8, 12, 14, 18, 22, 25, 28, 31$) are explained by icosahedral or double icosahedral structures (DIC).[f] In the case of $(Ar)_n$, typical van der Waals clusters, magic numbers are observed at $n = 13, 19, 23, 26, 29$ and 32. Total numbers of mercury-cesium clusters at magic numbers coincide completely with the magic numbers of argon clusters. The bond energy of mercury-mercury atoms in dimers is considerably weak compared with normal metallic bond. The mercury-cesium binary clusters are supposed to be van der Waals clusters. At the center of the cluster, a cesium ion exists with mercury atoms surrounding it.

2.2 High mass region

The mass spectrum $(Hg)_nCs^+$ in the high mass region is shown in Fig. 4. The mass spectra were obtained at 1 kV acceleration using wider slits and higher scan speed to clear ion intensity patterns and to obtain high ion transmission. Peaks of each cluster size were not resolved and only a global structure was observed. The maximum detectable mass to charge ratio was $m/z = 240000$.

The cluster sizes of the local maxima are 40, 56, 79, 114, 135, 153, 201, 255, 320, 400, 480, 585 and 700, while those of the local minima are 47, 70, 100, 132, 137, 170, 223, 280, 348, 430, 528, 648 and 764. The positions of the local maxima and minima were determined by compensating for the influence of the general decrease in the ion intensity pattern. The cluster sizes of local minima less than 280 were determined from mass spectra in which peaks of individual sizes were resolved completely.

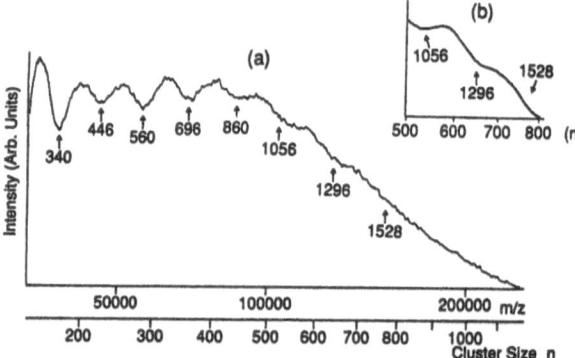

Fig. 4   A mass spectrum of mercury-cesium cluster ions $(Hg)_nCs^+$ up to $n = 1200$. The individual
cluster ion peaks were not completely resolved. The arrows shown below the spectrum
indicate the total number of electrons ($Ns = 2 \times n$) of shell-closing clusters $(Hg)_n^0$. The
narrow range between $n = 500$ and 800 is shown in the upper corner, which is determined to
precisely the shell-closing numbers in this region.

We can find special meaning in the cluster size of local minima, not local
maxima. The mercury atom has two valence electrons, so the clusters of $(Hg)_nCs^+$
have $2n$ electrons. The cluster sizes at local minima were roughly equal to half the
value of the shell closing numbers predicted by a shell model based on a Wood-
Saxon potential.[g] From this fact, we conclude that the valence electrons in the
large cluster of $(Hg)_nCs^+$ were delocalized and the clusters have metallic
characteristics.

The reason why magic numbers of $(Hg)_nCs^+$ appears at the local minima is as
follows. In the ionization region, two types of clusters are generated.

$$(Hg)_nCs^+ \quad \text{and} \quad (Hg)_n^0$$

The neutral clusters are mainly produced in the sputtering source and they may
influence the production yield of the charged clusters. At the size where clusters
of $(Hg)_n^0$ type are stable, the production rate of $(Hg)_n^0$ increases and that of
$(Hg)_nCs^+$ is suppressed. We can observe the stability of neutral clusters $(Hg)_n^0$
through the mass spectra of $(Hg)_nCs^+$. We have obtained other results which
support the above argument. The consecutive hill and valley pattern was
experimentally observed in mass spectra of $(CsI)_nCs^+$ cluster ion.[h] We could
interpret the size of the bottom of a valley by considering the stability of neutral
clusters $(CSI)_n^0$, which are cubic-like structures.

2.3   Supershell structure

The sudden change in ion intensity slope was observed at around $n = 500$. We
think this sudden decrease can be explained by the supershell effect. It is expected
from the theoretical calculation for large sodium clusters, that the shell part of the
total binding energy at shell closing number should have the beating pattern,
namely, supershell.[g] The shell energy should have local minima at around $Ns = 1000$ and 4000. This theory could be applied to mercury clusters, since mercury

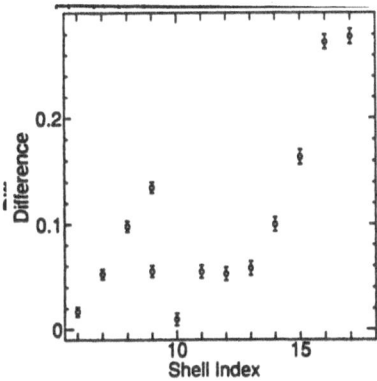

Fig. 5   The difference (subtraction) between the cube root of the shell-closing number of the mercury data set and a straight line ($Y = 0.61X + 0.87$). vs. the shell index. The straight line is explained in the text. The error bars shown in the figure are derived from the ambiguity in determining the position of cluster size at a local minimum.

clusters are also expected to have an electronic shell structure. Considering the influence of the stability of neutral clusters on the cluster ions $(Hg)_nCs^+$, we can expect to observe such a sudden change in the ion intensity slope in the real mass spectra at approximately $n = 500$ ($Ns = 1000$).

If the mercury clusters have electronic shell structure and supershell structure, the cube root of the number of valence electrons (magic number $\times$ 2) and shell index have a linear relationship and the phase shift occurs at around $Ns = 1000$. Figure 5 shows the difference (subtraction) between the cube root of the shell closing number of the mercury data set and a straight line ($Y = 0.61X + 0.87$) vs. the shell index. The straight line employed here is expressed by $Y = 0.61X + 0.87$, where X is the shell index and Y is the cube root of the shell closing number. We obtained this linear equation using least square fitting from the data of ref. 7. From Fig. 5, it is seen that the interval of the cube root of the shell closing number has changed at the region of shell index 15, which corresponds to cluster size $n = 500$. We can conclude that the supershell structure is observed in mercury clusters.

3. Rare gas clusters

We describe here a unique method to produce the rare gas clusters. By bombarding rare gas (Xe, Kr and Ar) on an aluminum metal, we can obtain rare gas cluster $(R)_n^+$ and binary clusters of rare gas and aluminum $(R)_nAl^+$, where R means a rare gas atom. Here we discuss mainly argon clusters.

The cluster ions were produced by bombarding an appropriate aluminum metal sheet with Ar ions from the discharge type primary gun. Five types of cluster ions, $(Al)_n^+$, $(Al_2O_3)_n(AlO)^+$, $(Al_2O_3)_n(AlO)_2^+$, $(Ar)_n^+$ and $(Ar)_nAl^+$ were observed. At the beginning of bombardment, aluminum oxide clusters were observed and no Ar and Ar-Al clusters were found. With decreasing intensity of aluminum oxide clusters, the Ar and Ar-Al clusters increased and when the aluminum oxide clusters disappeared, the Ar and Ar-Al clusters reached constant intensities. The intensity of the aluminum clusters was constant throughout the

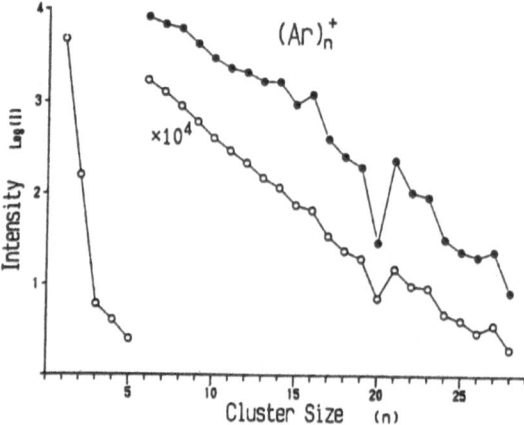

Fig. 6   Size distribution of $(Ar)_n{}^+$ clusters. The vertical scale is logarithmic. Open circles show the
present results, which are generated by argon ion bombardment from argon bubbles in
aluminum metal. For comparison, the size distribution obtained from the free jet expansion[i]
is represented by solid circles.

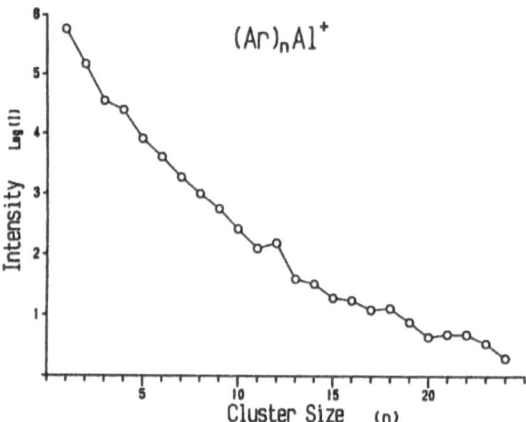

Fig. 7 Size distribution of $(Ar)_n Al^+$ clusters plotted on a logarithmic scale.

experiment. The size distribution of Ar clusters was similar to those obtained by
the supersonic expansion method.[i] Open circles in Fig. 6 show the size
distribution of $(Ar)_n{}^+$ produced by the sputtering. Discontinuity is seen at $n = 4$,
14, 16, 19, 21, 23 and 27. For comparison with those produced by another method,
the size distribution produced by the supersonic expansion method is shown by
solid circles.[i] Figure 7 shows the size distribution of Ar-Al clusters. The magic
numbers are 12 and 18, the total number of atoms in clusters, 13 and 19, which
can be explained by icosahedral structures.

Argon bubbles are produced easily by bombardment of excess argon ions in
an aluminum metal.[j] Argon cluster ions are supposed to be produced when the
bubbles in the aluminum metal are smashed into vacuum by argon ion
bombardment. The argon atoms are released from a high pressure state into
vacuum. This process is similar to those of the supersonic expansion method.

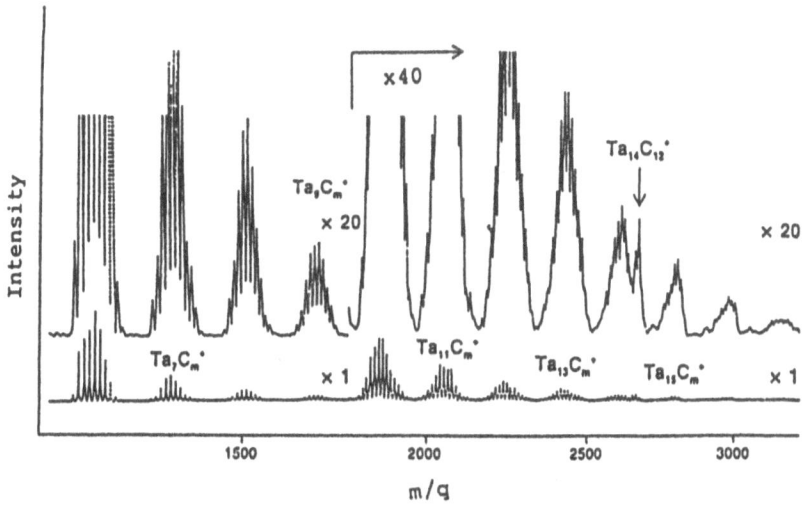

Fig. 8 Size distribution of TaC clusters. $Ta_{14}C_{12}^+$ has $3 \times 3 \times 3$ cubic-like structures having a vacancy at the center of the cluster (C atom site).

In the case of Xe ion bombardment, only Xe-Al binary clusters were produced.[k)]

4. TaC clusters —The smallest structure having point defect?—

Alkali halide, magnesium oxide and calcium oxide have a rock salt structure and their clusters are stable when they have the cubic-like structure. Many kinds of carbides such as TaC, TiC and WC also have rock salt structure; however, some positions of carbon atoms are vacant. Here we bombarded a TaC sample with Xe ions and observed clusters.

Figure 8 shows a mass spectrum of TaC clusters produced by spattering. The observed clusters were a series of $Ta_nC_m^+$. The cluster ion intensity as a function of m (n = constant) has a maximum at $n > m$ and a Gaussian-like distribution.

The only exception to this distribution is the $Ta_{14}C_{12}^+$ cluster. It has high ion intensity. Considering the cubic-like structure and the symmetry of the alignment of atoms, the most reasonable structure is the $3 \times 3 \times 3$ cube having a point defect. The carbon atom site at the center of the cluster is vacant. This is the smallest structure which can have a point defect.

Above $n = 14$, the distribution of the ion intensity seems to be asymmetric structures, ion intensity of the larger m side decreases suddenly. This is assumed to come from the dissociation to the $Ta_{14}C_{12}^+$ in the flight time in the mass spectrometer.

5. Size distributions of zinc-cesium and cadmium-cesium clusters and electronic shell structure

Abundance distributions were studied for Zn-Cs and Cd-Cs clusters. The cluster ions were produced by the Cs^+ ion bombardment on an investigated metal sheet and mass analysis was performed using the QQHQC type mass

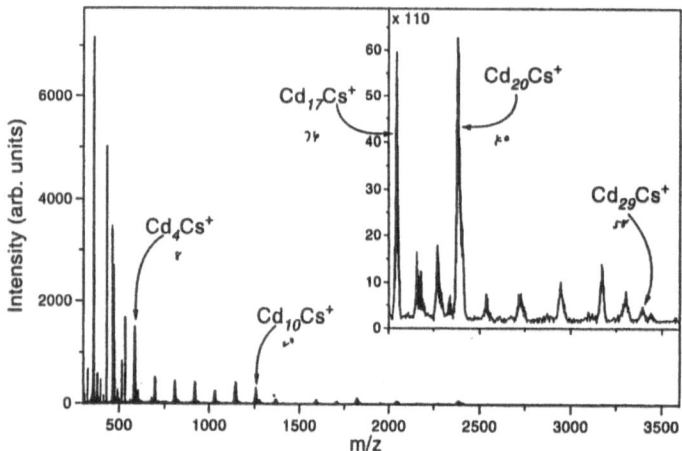

Fig. 9   A mass spectrum of cadmium-cesium cluster ions, m/z = 300–3,600. Peaks somewhat more
intense than their neighbors are observed at n = 4, 7, 10, 13, 15, 20, 23, 25, 27 and 29. A step-
like feature is observed at n = 4, 10, 20 and 29.

spectrometer.

5.1 Cadmium-cesium clusters

We obtained the mass spectra of cadmium-cesium positively charged cluster
ions of $(Cd)_nCs^+$ and $(Cd)_nCs^{2+}$ types. We can distinguish the single cesium group
$(Cd)_nCs^+$ from the double cesium group $(Cd)_nCs^{2+}$, because they have enough
mass difference. Figure 9 shows the mass spectrum of Cd-Cs clusters. In the
region $n < 30$, the peaks of the $(Cd)_nCs^+$ which are somewhat more intense than
their neighbors are observed at n = 4, 7, 10, 13, 15, 17, 20, 23, 25, 27 and 29. The
total number of electrons 8(n = 4), 20(10), 34(17), 40(20) and 58(29) agrees with
the theoretical electronic shell closing number based on the square well potential.
The other magic numbers, 7, 13, 15, 23, 25 cannot be explained by the electronic
shell structure.

5.2 Zinc-cesium clusters

Figure 10 shows a mass spectrum of Zn-Cs clusters. The most abundant ion
is the single cesium group $(Zn)_nCs^+$. Here it should be noted that there is a
considerable abundance of some double cesium ions $(Zn)_nCs^{2+}$ of some sizes.
The position of single and double cesium group peak overlaps in mass spectra and
minimum resolving power that resolves the single cesium group from the double
cesium group is 2×10^4. Fortunately, we can distinguish them in the mass spectra
of resolution of about 1000 by considering the isotope peak pattern. The magic
numbers of the singly cesium group are 10, 17, 20, 29 and those of double cesium
group are 9, 19, which coincide with the shell closing number of the square well
potential.

Comparing mass spectra of Zn-, Cd- and Hg-cesium clusters, different results
were obtained, even they are IIb metals. Zn- and Cd-Cs clusters have an electronic
shell structure, while Hg-Cs clusters have an atomic shell structure. It is significant

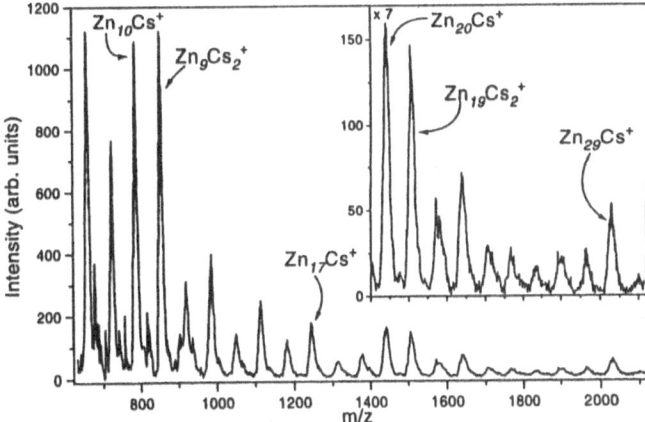

Fig. 10 A mass spectrum of zinc-cesium cluster ions, $m/z = 620\text{–}2{,}120$. Peaks of $[Zn_{10}Cs]^+$, $[Zn_{17}Cs]^+$, $[Zn_{20}Cs]^+$, $[Zn_{29}Cs]^+$, $[Zn_9Cs]^+$ and $[Zn_{19}Cs]^+$ are somewhat more intense than those of their neighbors.

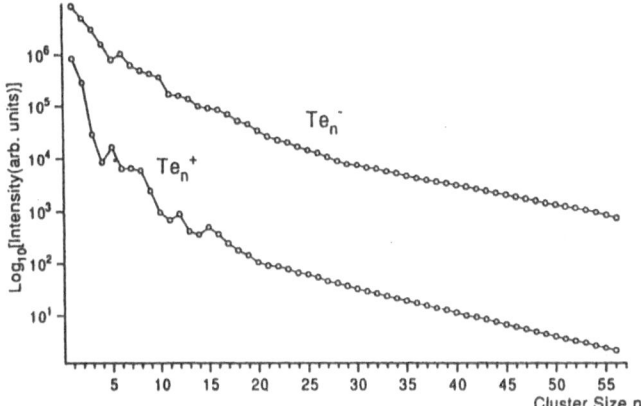

Fig. 11 Size distribution of positive and negative tellurium cluster ions (up to $n = 56$) obtained by SIMS. Peaks which are somewhat more intense than those of neighbors are observed at $n = 5, 8, 12, 15, 19$ and 23 for positive cluster ions. They are observed at $n = 6, 10, 13$ and 16 for negative cluster ions.

question, whey they have different shell structures. A determination of the exact origin will have to await detailed theoretical investigations.

6. Tellurium clusters

The size distribution of positive and negative tellurium clusters in the size range from 2 to 56 atoms was investigated. Figure 11 shows the size distribution of positive and negative cluster ions. A discontinuous variation of cluster ion intensity appeared at cluster sizes $n = 5, 8, 12, 15, 19$ and 23 for positive clusters and $n = 6, 10, 13$ and 16 for negative clusters.

Dissociation patterns were obtained for positive and negative cluster ions by scanning the acceleration voltage in the linear mode. In the case of the sputtering method, an internal energy is high enough to cause a spontaneous

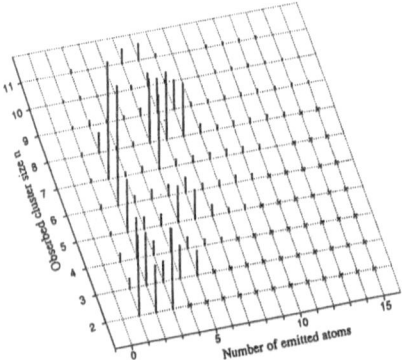
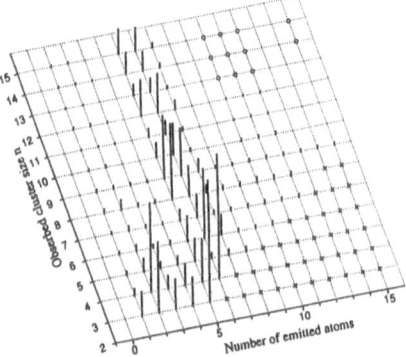
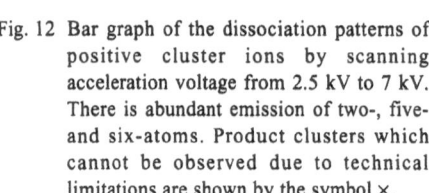

Fig. 12 Bar graph of the dissociation patterns of positive cluster ions by scanning acceleration voltage from 2.5 kV to 7 kV. There is abundant emission of two-, five- and six-atoms. Product clusters which cannot be observed due to technical limitations are shown by the symbol ×.

Fig. 13 Bar graph of the dissociation patterns of negative cluster ions by scanning acceleration voltage from 2.5 kV to 7 kV. There is abundant emission of two-, five and six-atoms. Product clusters which cannot be observed due to technical limitations are shown by the symbol ×. The symbol ○ shows intensity which is not strong enough to show by bars.

fragmentation during flight in a mass spectrometer. By scanning the acceleration potential, all n_1-mers dissociating from precursor ions of different sizes are observed. Figure 12 and 13 show the fragment patterns for positive and negative Te clusters. The height of the bars indicates the probabilities of the fragmentation and the symbol of open circles shows very small probability. Due to technical limitation of the upper limits of the acceleration potential, we cannot study some fragmentation pathways. These positions are shown by the symbol ×.

We note that almost all product cluster ions, Te_n^+ and Te_n^- have an enhanced relative abundance from the $n+2$, $n+5$ and $n+6$ precursor ions. Though we cannot neglect the possibility of sequential fragmentation, the observation of specific fragmentation patterns strongly suggests the existence of non-sequential fragment channels and of stable neutral clusters, Te_2, Te_5 and Te_6. Considering that Te_2, Te_5 and Te_6 are more stable than their neighbors, it is conceivable that the abundant emission of 10, 11 and 12 atoms can be interpreted as the result of the sequential fragmentation of the neutral Te_5 and Te_6.

7. Dissociation spectra of doubly charged clusters and metal-nonmetal transitions

Many kinds of clusters consisting of metal atoms have the electronic shell structure. However, stabilities of some kinds of metal clusters of small size come from the geometrical configuration. Magic numbers of mercury-cesium clusters, manganese[1] clusters are similar to those of van der Waals clusters. We tried to determine the electronic or geometrical structure of the clusters from the dissociation patterns of doubly charged clusters.

We now consider the following dissociation.

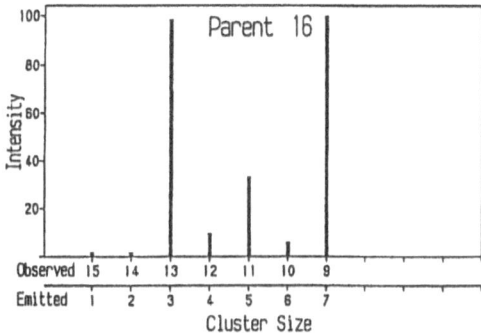

Fig. 14 Fragment pattern of Ag_{16}^{++}. Dominant pathway is the decomposition to Ag_3^+ and Ag_9^+.

$$n^{++} \rightarrow n_1^+ + n_2^+$$

For simplicity of expression, $(M)_n^+$ and $(M)_n^{++}$ are shown by n^+ and n^{++}.

The potential of the energy analyzer is set at E_0, and in this condition, ions having energy of V_0 can pass it. If the n^{++} clusters decompose in the flight between the source slit and the entrance of the energy analyzer of the CQH mass spectrometer, n_1^+ ions cannot pass the analyzer. However, bar adjusting the acceleration potential to $(n/2n_1)V_0$, ions of n_1^+ can pass the analyzer. If the $n_1 > n/2$, the acceleration potential to pass the n_1^+ is smaller than V_0. The fragment ions n_1^+ decomposing from the n^{++} can detect without the interference of the fragments from singe charged clusters. Fragment ions decomposing from single charged clusters are always detected at a higher acceleration potential than V_0. By scanning the acceleration potential from $V_0/2$ to V_0 and setting the magnet is to detect n_1-mers, all fragment ions n_1 can be detected. By setting the magnet strength stepwise, we can detect fragments of different sizes. Rearranging those fragment spectra, we obtain the fragment pattern from certain sizes of parent clusters shown in Fig. 14.

7.1 Ag clusters

Figure 14 shows the fragment pattern from the 16^{++} parent clusters. The main channel is decomposition to 3- and 9-mers. The 3- and 9-mers have shell closing structures and the 16^{++} is assumed to have electronic shell structure. About silver clusters, main channels of decomposition are those to 3- and 9 mers in small size and the symmetrical decomposition is dominant above $n = 20$.

7.2 Ta clusters

Figure 15 (a), (b), (c) shows the dissociation pattern of Ta clusters. The main dissociation channel of Ta_7^{++} is atomic ion evaporation. If the cluster has a geometrical structure, the atomic ion evaporation is energetically advantageous, because least numbers of bonds must be cut, while in a liquid drop model, or jellium model, symmetrical fragmentation is advantageous.

At $n = 11$, probability of an atomic ion evaporation and symmetrical fragmentation is comparable, while at $n = 14$, the symmetrical fragmentation

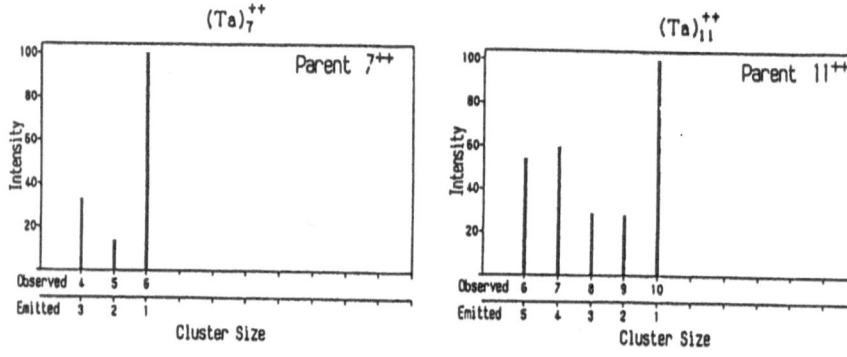

(a) Fragment pattern of double charged Ta   (b) Fragment pattern of double charged Ta
     clusters. $(Ta)_7^{++}$ clusters. $(Ta)_{11}^{++}$

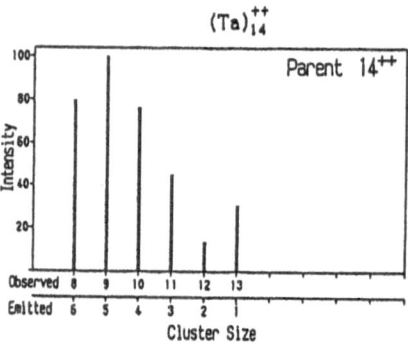

(c) Fragment pattern of double charged Ta clusters. $(Ta)_{14}^{++}$

Fig. 15

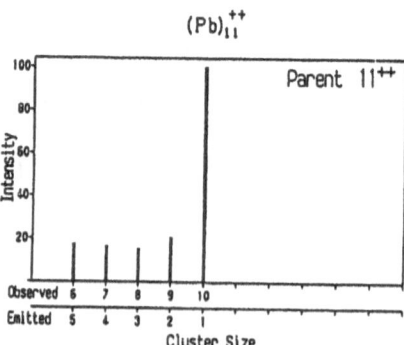

Fig. 16 Fragment pattern of $(Pb)_{11}^{++}$.

becomes dominant. From the dissociation patterns, we can assume that the nonmetal-metal transition occurs at around $n = 11$.

7.3 Pb clusters

Figure 16 shows the fragment pattern from 11-mers. We could not detect fragment patterns larger than 11-mers because of the weak intensity of the parent ions. The transition to the metallic seems to occur at cluster size larger than 11.

PUBLICATIONS

1. I. Katakuse, T. Ichihara, H. Ito and M. Hirai, Rapid Commun, *in Mass Spectrom.*, **4**, 16 (1990).
2. H. Ito, T. Sakurai, T. Matsuo, T. Ichihara and I. Katakuse, *Phys Rev.* **B48**, 4741 (1993).
   I. Katakuse, Proc. 7th Int. Symp. Small Particles and Inorganic Clusters, Kobe (1994), (to be published).
3. I. Katakuse, H. Ito and T. Ichihara, Int. J. Mass Spectrom. *Ion Processes*, **99**, 207 (1990).
4. H. Ito, T. Ichihara, T. Sakurai, T. Matsuo and I. Katakuse, *Mass Spectrom. Jpn*, to be published.
5. H. Ito, T. Sakurai, T. Matsuo, H. Matsuda, T. Ichihara and I. Katakuse, *Mass Spectrom. Jpn.*, submitted.
6. H. Ito, T. Sakurai, T. Matsuo, T. Ichihara and I. Katakuse, Proc. 7th Int. Symp. Small Particles and Inorganic Clusters, Kobe (1994), (to be published).

REFERENCES

a. I. Katakuse, T. Ichihara, H. Nakabushi, T. Matsuo and H. Matsuda, *Mass Spectrom. Jpn.*, **31**, 111 (1983).
b. H. Matsuda, *Atomic Masses and Fundamental Constants*, **5**, 185 (1976).
c. H. Matsuda, T. Matsuo, Y. Fujita, T. Sakurai and I. Katakuse, *Int. J. Mass Spectrom. Ion Processes*, **91**, 1 (1989).
d. W. End, R. Beeves and K. G. Standing, *Phys. Rev. Lett.*, **50**, 27 (1983).
e. T. Sakurai, H. Ito, T. Matsuo and I. Katakuse, Rapid Commun. *Mass Spectrom*, **5**, 437 (1991).
f. I. A. Harris, R. S. Kidwell, and J. A. Northby, *Phys. Rev. Lett.*, **53**, 2390 (1984).
g. H. Nishioka, K. Hansen and B. A. Mottelson, *Phys. Rev.* B42, 9377 (1990).
h. I. Katakuse, T. Ichihara, H. Ito, T. Matsuo, T. Sakurai and H. Matsuda, Rapid Commun. *Mass Spectrom.*, **2**, 191 (1988).
i. A. Ding and J. Hesslich, *Chem. Phys. Lett.*, **94**, 54 (1983).
j. A. vom Felde, J. Fink, Th. Muller-Heinzelring, J. Pfluger, B. Scheerer, G. Linker and D. Kaletta, *Phys. Rev. Lett.*, **53**, 922 (1984).
k. Y. Saito, I. Katakuse and H. Ito, *Chem. Phys. Lett.*, **161**, 332 (1989).
l. Y. Saito and I. Katakuse, *Z. Phys.*, **D19**, 189 (1991).

26

Preparation of Nanometer-sized Composite Particles by Flowing Gas Evaporation Technique

Saburo Iwama

Department of Applied Electronics, Daido Institute of Technology
Minami-ku, Nagoya 457-0811, Japan

Purpose of the present study

Particle size control is one of the most important requirements for a preparation technique, because the properties of fine particles are strongly dependent on their size and size distribution. The present author developed a new technique named FGE-Technique (Flowing Gas Evaporation Technique), where the vaporization and condensation of metals are carried out in a forced flowing gas with high velocity. This technique enables us to make ultrafine particles with a narrow size distribution. The purpose of the present study is to investigate the crystal structure of particles formed by FGE-Technique using a single and dual evaporation source, and to make clear the characteristic features of the technique in comparison with the ordinary GET (Gas Evaporation Technique).

Contents of the present study

1. Introduction

GET[a] has been widely used to prepare ultrafine particles of nanometer size on a laboratory scale. In the ordinary GET, metal particles are formed in a convection flow of inert gas rising from the hot evaporation source.[b] These particles usually show a broad size distribution (log-nornal distribution) caused by coalescence growth,[c] which takes place very frequently in a narrow region called the intermediate zone. Such a broad size distribution has not been entirely satisfactory for the research or applications of fine particles. This is one of the important reasons why improvements in preparation technique are eagerly anticipated. In order to avoid the convection flow of inert gas, GET has been tried under a micro-gravity or a non-gravity condition using a dropping capsule or a space station, and the following result concerning particle size has been reported; *i.e.* larger particles of micron-meter order tend to grow under convection-less conditions.[d,e] The method is unsuitable for preparing namometer-sized particles.

The characteristics of the FGE-Technique is that the formation of the vapor zone and particle growth zone along the flow of inert gas can be controlled by the inert gas species and the flow velocity.[1] The typical flow velocity is 5 m/s, which is faster by about one order of magnitude than that of convection flow[f,g] in GET.

The metal vapor zone extends to several tens of centimeters, which is wider by about one order of magnitude than that obtained by ordinary GET. Nucleation and particle growth by ad-atom process take place in the stage following the vapor zone. By traveling in the gas flow the particles grow by coalescence. The FGE apparatus is equipped with two Knudsen cells as the evaporator to make fine alloy particles. Various kinds of alloy particles have already been formed by mixing two kinds of smoke prepared by GET.[h–j] The present study is the first attempt to produce fine alloy particles by mixing two kinds of metal vapor under a flowing gas condition.

2. Experimental

A schematic illustration of the evaporation apparatus is shown in Fig. 1. The apparatus consists of mainly a cylindrical chamber with an inner diameter of 9 cm and 150 cm in length, two Knudsen cells as the evaporation source and the gas flowing system. A steady flow of inert gas was formed in the chamber by introducing the gas through the flow meter and removing it from the other side of the chamber by a mechanical booster pump. The gas pressure in the present experiment was 0.8–2.0 kPa. The gas velocity of 3–7.2 m/s is given by the following formula :

$$V = (P_0 \cdot Q)/(P \cdot A),\tag{1}$$

where P_0, P, Q and A are the atmospheric gas pressure (101 kPa), pressure in the chamber, gas flow rate and cross-sectional area of the chamber, respectively. Evaporation temperature was kept at a fixed level from 1400°C to 1500°C for Mn, from 700°C to 800°C for Sb and from 1200°C to 1550°C for In. Electron microscopic observation was carried out for specimens collected at sampling positions A, B and C, which are 35, 75 and 115 cm downstream from the evaporation source, respectively.

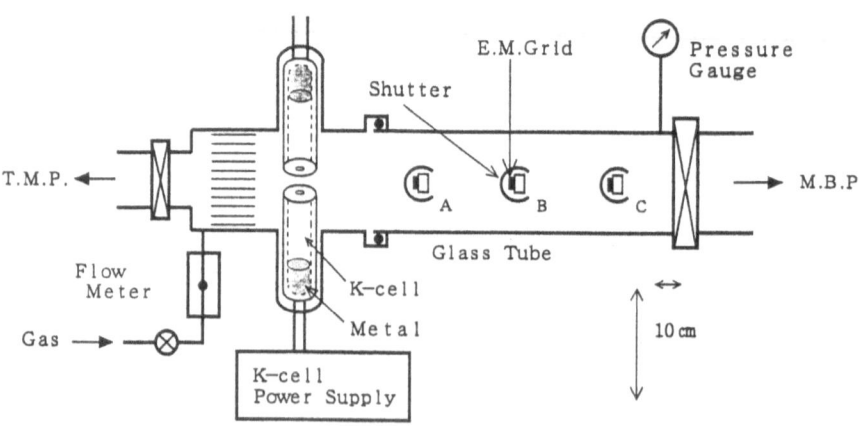

Fig. 1 A schematic illustration of the evaporation apparatus.

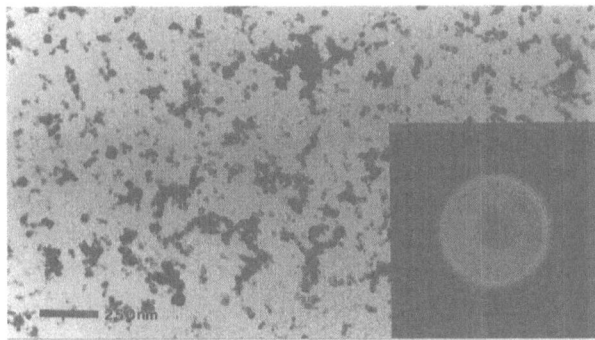

Fig. 2 β-Mn particles formed in an Ar gas flow.

3. Results and discussion

3.1 Mn Particles [2]

Manganese was evaporated at the source temperature of 1450°C in an Ar gas flow, the gas pressure and the velocity being 0.93 kPa and 5 m/s, respectively. Figure 2 shows an electron micrograph and the corresponding electron diffraction pattern of Mn particles collected at the sampling position C. Most particles are smaller than 30 nm. Particles are identified as β phase, which is stable in the temperature range from 725°C to 1095°C. In the ordinary GET, the both α- and β-Mn are known to form [k–m] and in general, the former grows in the intermediate zone and the latter in the inner zone.[k] It should be noted that the formation of a uniform growth zone such as the inner zone can be established, which is impossible to do with the ordinary GET.

Figure 3 shows Mn particles formed in the Ar flow of 2.0 kPa pressure and 6 m/s velocity at the elevated source temperature of 1500°C. Larger particles of size from 20 to 300 nm were observed. Many particles show the clear rhombic dodecahedron crystal shape which is characteristic to β-Mn.[l] The corresponding electron diffraction pattern indicates a β-Mn structure, although a trace of α-Mn is detected. At higher pressure of inert gas the FGE-Technique behaves like ordinary GET.

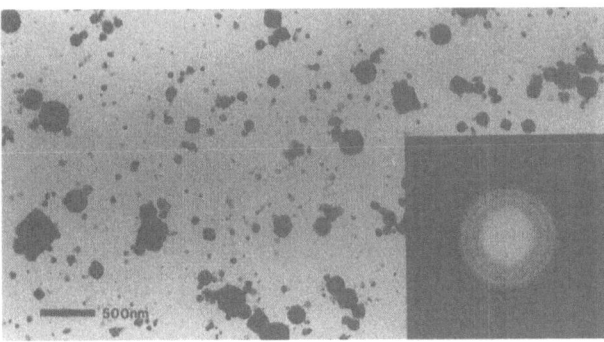

Fig. 3 Mn particles formed in a high pressure Ar flow, showing a broad size distribution.

3.2 In particles

Indium was vaporized in a He gas flow, the gas pressure and the velocity being 0.8 kPa and 4.2 m/s, respectively. Figure 4 shows an electron micrograph and the corresponding electron diffraction pattern of In particles collected at the sampling position C. Most particles have definite crystal habits of triangle with truncated corners as indicated by T in the photograph, and of square as indicated by S. In the diffraction pattern, the relative intensities of 101 and 220 are very strong, while 111 and 002 are very weak. This feature in the diffraction intensity can be explained by the fiber structure with [111] and [001] fiber axes of tetragonal index of In (a_0 = 4.5986 and c_0 = 4.9459). This means that the metal vapor zone extended to the collection position, and the nucleation and growth took place on a substrate of amorphous carbon. The same trend in morphology and the diffraction intensity was recognized in In islands grown by ordinary vacuum deposition on amorphous carbon, as shown in Fig. 5.

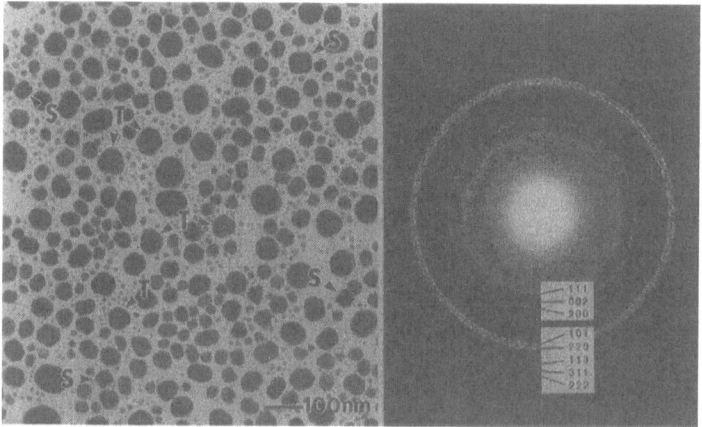

Fig. 4 In particles formed on amorphous carbon in a He gas flow.

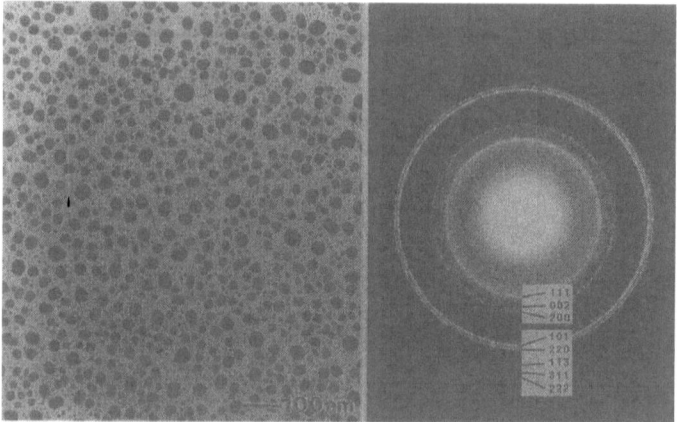

Fig. 5 In particles formed by vacuum deposition on amorphous carbon.

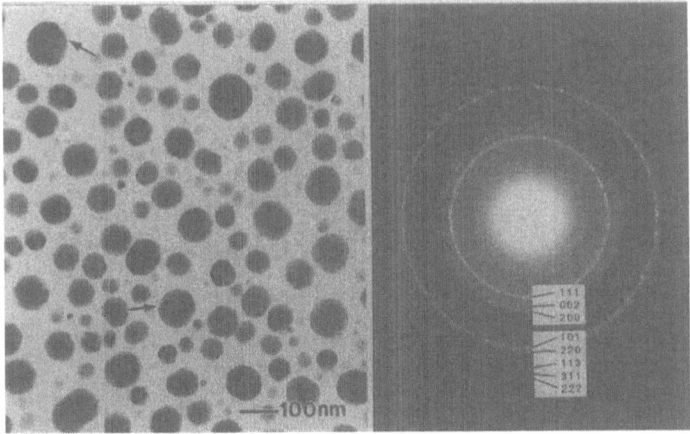

Fig. 6 In particles formed in a free space of a mixed gas flow composed of He and Ar.

Figure 6 shows In particles formed in a mixed gas flow composed of He and 27 vol% of Ar, the total gas pressure being 0.8 kPa. These particles, collected at the same sampling position C as In particles in Fig. 4, have a rounded shape and frequently show the typical contrast in them due to a lattice defect as indicated by arrows. In the diffraction pattern, the relative intensity becomes distinguishable from that of Fig. 4, *i.e.* 111 and 002 show moderate intensities in contrast to the weak 220. These results indicate quite clearly that the particles shown in Fig. 6 have grown in the free space before deposition on the substrate. The use of a mixed gas is practically effective to control the region of the metal vapor zone and the following particle growth zone in FGE-Technique.

3.3 Sb particles [2]

Figures 7(a) and (b) are the bright and the dark field images of Sb particles formed in Ar gas flow, the gas pressure and the velocity being 0.93 kPa and 7.2 m/s, respectively, and collected at sampling position C. Spherical morphology can be seen in most particles. From the diffuse halo in the diffraction pattern (Fig.7(c)) these particles of size smaller than 20 nm are found to be amorphous, in contrast to the ordinary GET where crystalline Sb particles are formed.[n] All Sb particles prepared under the other condition of gas flow in the present experiment show the amorphous structure, which may be attributed to the quenching effect enhanced in the forced flow of inert gas. Figures 7(d) and (e) show the bright and the dark field images of Sb particles transformed to a rhombohedral structure (Fig. 7(f)) by the elevation of electron beam brightness for the microscopic observation.

Crystallization of amorphous Sb particles has been found to depend on the annealing temperature,[2] annealing duration and particle size, and detailed experiments concerning these factors are now under way.

3.4 InSb particles [3]

In and Sb were evaporated simultaneously in a He gas flow, the pressure and the velocity being 1.33 kPa and 5 m/s, respectively. The evaporation

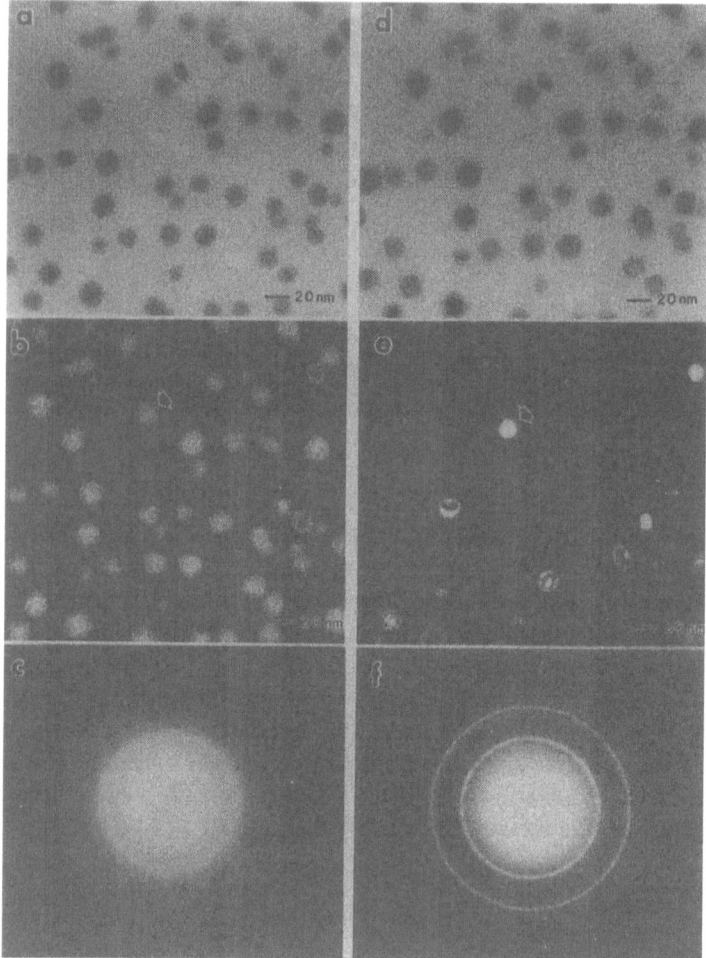

Fig. 7 Amorphous Sb particles; (a) bright field, (b) dark field images and (c) the diffraction pattern, and crystallized Sb particles by elevation of electron beam brightness; (d) bright field, (e) dark field images and (f) the diffraction pattern. Arrows indicate the same particle.

temperature was 1550°C for In and 750°C for Sb. Electron micrographs of three specimens are shown in Fig. 8, where (a), (b) and (c) are those of different sampling durations of 1 s, 3 s and 20 s, respectively, at the sampling position A. Island growth can be seen in the consecutive micrographs, suggesting that the nucleation and growth took place on the substrate in the metal vapor zone. The crystal growth in FGE-Technique, however, succeeds in a heavy shower of inert gas atoms which is roughly 10^7 times that in ordinary vacuum deposition. In such a dense gas, the coalescence may decrease and result in the granular characteristics shown in the mosaic structure in Fig. 8(c). In the diffraction pattern in Fig. 8(d), two structures can be assigned; one is a zinc blende type of the ordinary intermetallic compound InSb, and the other is a wurtzite type. A number of small areas in the dark field image show the alternative parallel contrast as indicated by arrows in Fig. 9. These characteristic contrasts may reflect the stacking fault on

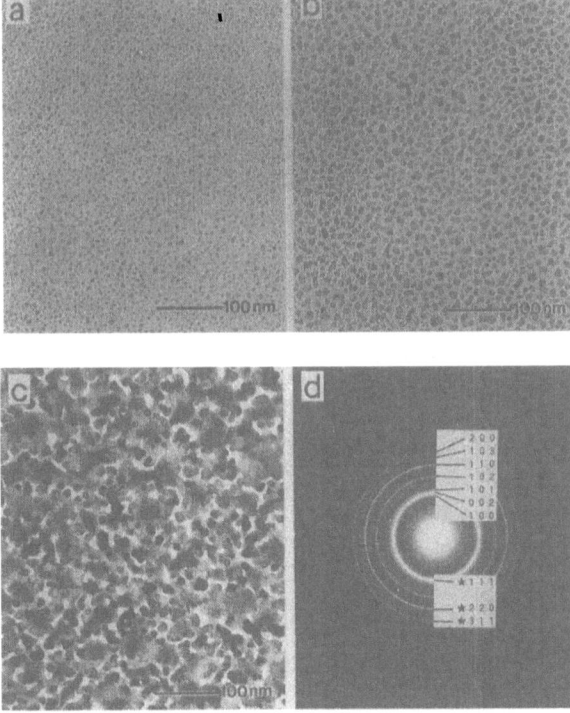

Fig. 8   InSb films formed by co-evaporation of In and Sb in a He gas flow and collected at sampling
position A for different sampling durations of (a) 1 s, (b) 3 s and (c) 20 s. (d) shows the
diffraction pattern of specimen (c). Miller indices with ★ represent zinc blende and others of
wurtzite type.

(111) lattice plane, and are consistent with the result of a mixed structure of zinc
blende and wurtzite type, because the stacking in the former is ABCABC......,
while that in the latter is ABAB...... .

Figures 10(a), (b) and (c) show the specimens of different sampling duration
of 5 s, 20 s and 1min, respectively, at the sampling position C. Particles in Fig.
10(a) show a broad size distribution in comparison with that of Fig. 8(a). In Fig.
10(b), a piling effect is observed with accompanying coalescence among particles.
The piling becomes more clear, as shown in Fig. 10(c). This suggests that the
particles formed initially in the gas stream and deposited on the substrate
accompanying the coalescence of particles on the substrate. The crystal structure
of particles grown in the gas stream (Fig. 10(d)) is the same as that grown on the
substrate shown in Fig. 8(d).

The composition of metal vapor can be controlled easily by adjusting the
evaporation rate individually. On the In-rich side from the stoichiometric
composition, a mixed phase composed of InSb and In was formed. On the Sb-rich
side, however, the amorphous Sb was dominant and the intermetallic compound
InSb was hardly recognized in the diffraction pattern.

In summary, we have developed a new gas evaporation technique for the
preparation of homogeneous intermetallic compound particles of nanometer size
by using two Knudsen cells as the evaporation source in a flowing inert gas with

Fig. 9 Dark field image of InSb fine crystals.

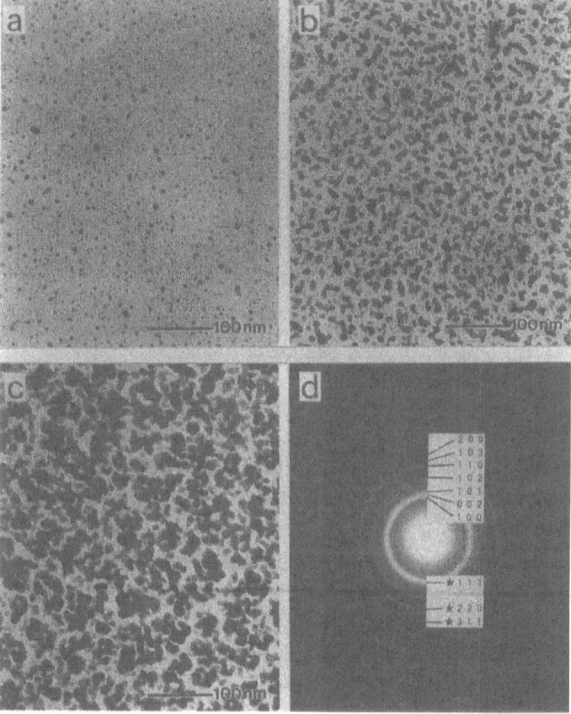

Fig. 10 InSb fine particles formed under the same evaporation condition as that of Fig. 8 and collected at sampling position C for different sampling durations of (a) 5 s, (b) 20 s and (c) 1min. (d) shows the diffraction pattern of specimen (c). Miller indices with ★ represent of zinc blende and others of wurtzite type.

high velocity. The granular film of InSb, grown on the amorphous carbon, and fine particles of InSb, grown in a free space, were found to include stacking faults, in general, which resulted in a mixed structure of zinc blende and wurtzite types. The FGE-Technique was found to have a higher quenching effect than the ordinary GET at the nucleation and subsequent growth stages, as demonstrated by the formation of amorphous Sb and β-Mn particles. We have successfully controlled the region of the vapor zone by using a mixed gas of He and Ar as mentioned in In experiments. This technique enables us to make size-controlled particles.

Part of this work was supported by a Grant-in-Aid for Scientific Research from the Ministry of Education, Science and Culture, and also by the Naito Research Grant.

PUBLICATIONS

1. S. Iwama and K. Hayakawa, *Nanostructured Materials*, **1**, 113 (1992).
2. S. Iwama and K. Mihama, *Nanostructured Materials*, **6**, 305 (1995).
3. S. Iwama and K. Mihama, *Surface Rev. and Lett.*, **3**, 49 (1996).

REFERENCES

a. K. Kimoto, Y. Kamiya, M. Nonoyama and R. Uyeda, *Jpn. J. Appl. Phys.*, **2**, 702 (1963).
b. S. Yatsuya, S. Kasukabe and R. Uyeda, *Jpn. J. Appl. Phys.*, **12**, 1675 (1973).
c. C. G. Granqvist and R. A. Buhrman, *J. Appl. Phys.*, **47**, 2200 (1976).
d. N. Wada, Proc. 14th Int. Symp. Space Tech. Sci., 1599 (1984).
e. N. Wada, M. Kato, M. Dohi, T. Sato and T. Goto, Proc. 15th Int. Symp. Space Tech. Sci., 2173 (1986).
f. S. Yatsuya, A. Yanagida, K. Yamauchi and K. Mihama, *J. Cryst. Growth*, **70**, 536 (1984).
g. S. Yatsuya, K. Yamauchi, T. Kamakura, A. Yanagida, H. Hayakawa and K. Mihama, *Surface Sci.*, **156**, 1011 (1985).
h. C. Kaito and M. Shiojiri, *Jpn. J. Appl. Phys.*, **21**, L421 (1982).
i. C. Kaito, *Jpn. J. Appl. Phys.*, **23**, 525 (1984).
j. Y. Saito, K. Mihama and H. S. Chen, *Phys. Rev.*, **B, 35**, 4085 (1987).
k. S. Kasukabe, S. Yatsuya and R. Uyeda, *Jpn. J. Appl. Phys.*, **13**, 1714 (1974).
l. I. Nishida, *J. Phys. Soc. Jpn.*, **26**, 1225 (1969).
m. T. Okazaki, *Jpn. J. Appl. Phys.*, **27**, 2037 (1988).
n. T. Okazaki, *J. Jpn. Assoc. Cryst. Growth*, **9**, 175 (1982).

27

Tungsten and Chromium Having the A15-Structure

Masashi Arita[a], Noriyoshi Suzuki[b] and Isao Nishida[c]

[a] Department of Electronics and Information Engineering, Faculty of Engineering, Hokkaido University
Kita-ku, Sapporo 060-0813, Japan
[b] Kawasaki Steel Co.
Kurashiki-shi, Okayama 712-8511, Japan
[c] Faculty of Science, Nagoya University
Chikusa-ku, Nagoya 464-0814, Japan

Purpose of the present study

Physical and chemical features of materials are known to change by the size reduction of crystals, and many studies have been performed on optical, magnetic and electric properties of small particles and clusters. Small particles of metal elements sometimes have high-temperature phases or new phases which are not known for bulk crystals while they usually have the stable phases. These characteristic crystal structures also influence the properties of materials. For example, A15-Cr shows paramagnetism while the stable bcc-Cr shows spin density wave. In the present work, W and Cr were selected as examples having the characteristic structure in small particles, and the crystallographic features such as morphological shapes and defects were studied.

Contents of the present study

1. Island particles of W and Cr having the A15-structure in thin films

Thin films of some metals such as W and Cr whose stable structures are bcc are composed of small island particles having a size of several nanometers, and the characteristic crystal structure named A15 (Fig. 1) is frequently observed at the early stage of film deposition [a-c]. Concerning the appearance of the A15-structure, there has been a polemic discussion concerning the role of oxygen.[c] In order to obtain information on the factors for producing the A15-structure, thin

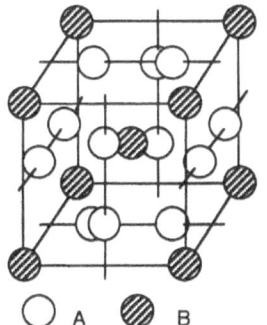

Fig. 1 A15-structure of A_3B.

films of W and Cr were deposited under different conditions, *e.g.* atmosphere, substrate temperature and substrate material. At the same time, the microstructure of films, *e.g.* morphology of islands and crystal defects, was observed and analyzed.

1.1 Experimental procedure

After evacuation of the vacuum chamber down to 10^{-6} Pa, some gases (O_2, N_2, Ar or He; 10^{-6}–10^{-8} Pa) were introduced substrates heated depending on the purpose. Sample films were prepared by the electron beam evaporation of W or Cr ingots, and they were analyzed by X-ray diffractometry and transmission electron microscopy (TEM). Films (30–200 nm thick) for X-ray diffractometry were deposited on glass with a deposition rate of 0.5 nm/s. Samples for TEM were 3–20 nm thick and deposited on carbon films or (001) of alkali-halides (AH) such as NaCl , KCl and KBr kept at room temperature (RT) to 300°C. The deposition rate in this case was 0.05 nm/s.

1.2 Influence of atmospheric gas on the production of A15-films [1,2]

Very thin films with the thickness of several nanometers prepared under a vacuum of 10^{-4} Pa gave electron diffraction patterns of halo or fcc. By the increment of the film thickness, A15- and bcc-phases were obtained. Under a better vacuum, the major phase was bcc. This tendency was the same as in earlier reports.[d] Based on these phenomena, it has been reported that oxygen stabilizes the A15-structure. However, for a detailed discussion, films should be prepared under a vacuum containing gases of which the kind and the amount are controlled.[e–f]. In Fig. 2, an example is presented for W-films, where films deposited under a vacuum of 10^{-4} Pa containing different gases. O_2- and N_2-gases gave the A15-structure of W, while films under a vacuum containing Ar showed

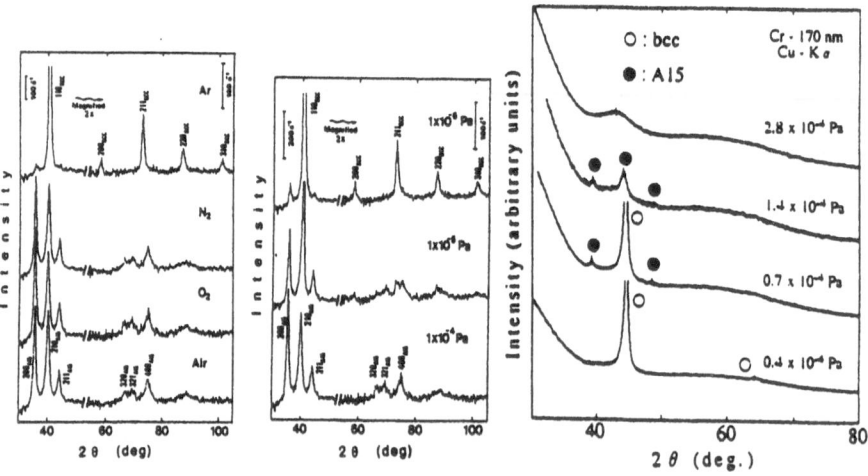

Fig. 2   Appearance of A15-W under different atmosphere. Total pressure: 10^{-4} Pa.

Fig. 3   Amount of O_2 and the appearance of A15-W.

Fig. 4   Amount of O_2 and the appearance of A15-Cr.

the bcc-pattern. He gas gave the same tendency as Ar.

In order to obtain information on the influence of oxygen, films were prepared under different amounts of O_2-gas. The result is presented in Fig. 3 for W-films (50 nm thick). Under the deposition rate (0.5 nm/s) used in this experiment, O_2-gas of 10^{-4} Pa is required to obtain A15-W films. Figure 4 is the result for Cr. In this work, we succeeded in the fabrication of A15-Cr films as thick as about 200 nm. The range of the amount of O_2-gas required to give the A15-Cr is narrower than that for A15-W. This must be related to the fact that A15-Cr films have only been in very thin films as thick as several nanometers. Under the condition containing much O_2, the halo pattern is recognized at the early deposition stage.

In thin W-films (of several nanometers thickness) deposited on (001) of alkali-halides kept at 300°C for TEM observations, another tendency was recognized. In this experiment, both Ar- and O_2-gases gave the A15-structure. Without any gases, A15-W was also obtained. This result could be caused by the influence of substrate surfaces. Electron diffraction patterns of W films on NaCl indicated bcc-W as the major phase, while films on KCl and KBr were composed almost only of A15-W.

From the present X-ray and TEM observations, the following summaries were obtained, although the effects by impurities and crystal size still remain obscure: (a) Grain growth was depressed by O_2-gas. Grain size of the films prepared under oxygen atmosphere was mainly less than 10 nm for both A15- and bcc-phases. Large bcc-particles seen in films without the introduction of O_2-gas were not recognized in films with oxygen. (b) The A15-crystal obtained was not W_3O ordered phase. Even if the O-atoms are within the A15-lattice, O-atoms randomly substitute for W-atoms or they are at interstitial positions.

1.3 Microstructure of A15-islands [2]

Increasing the temperature of alkali-halides (001) substrate during the deposition of W, the size of A15-islands became larger. The shapes of these

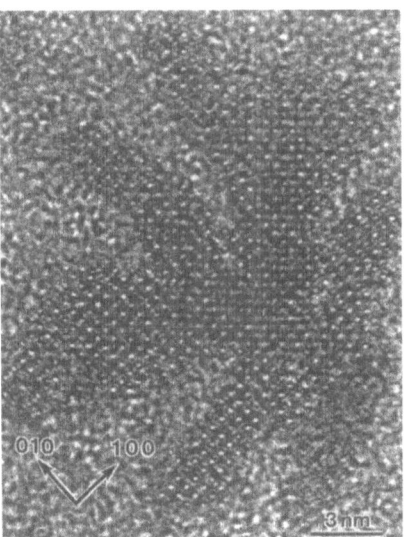

Fig. 5 A15-W single crystal particle having a dendrite-like shape.

islands were like the dendrite [g] (Fig. 5). Dendrite-like bcc-islands were also seen. Each island was a single crystal. Epitaxial grain growth of A15 on alkali-halides were seen. Main orientation relations observed were

$$(001)_{AH} \,/\!/\, (001)_{A15} \text{ and } [100]_{AH} \,/\!/\, [110]_{A15}$$

and

$$(001)_{AH} \,/\!/\, (\bar{1}10)_{A15} \text{ and } [100]_{AH} \,/\!/\, [110]_{A15}$$

where the subscripts denote the material or crystal structure.

1.4 Defect structure within A15-island particles [2]

High resolution TEM observations were performed on W island particles in very thin films of several nanometers deposited on $(001)_{KCl}$ kept at 100–300°C. As described above, each particle was an A15 single crystal some of which contained planar defects. Translation vector of these defects was [$\sqrt{3}$ /2, 1/2, 0], and the defect was composed of Zr_4Al_3-type [h] structure units. This type of defect was the same as the defects observed in V_3S [i] and A15-Cr [3] small particles prepared by the gas evaporation technique while it is different from the defect in Nb_3Ge A15-tapes. [i]

Another type of defect observed in A15-W thin films was the grain boundary

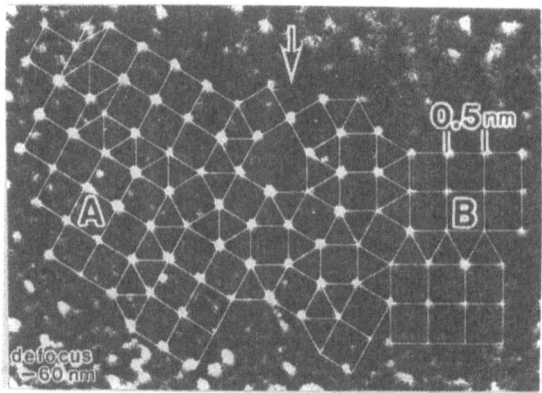

Fig. 6 Grain boundary within an A15-W particle.

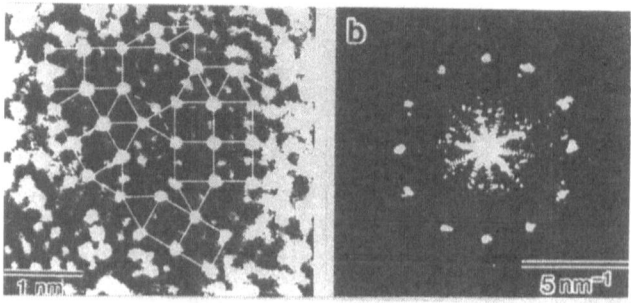

Fig. 7 Nanoparticle of W. (a) TEM image and (b) the optical diffraction pattern.

as presented in Fig. 6. Adjacent grains have [001] as a common axis and rotated by exactly 30° around the common axis. Also at this boundary region, triangular contrast denoting the Zr_4Al_3-type structure unit was recognized. This boundary structure is completely different from those in other A15 grain boundaries,[k] and must be characteristic of very small A15-particles. In order to understand the atomic structure of this defect, contrast calculations of A15-, Zr_4Al_3-type structure as well as related structures such as σ-phase [h] and H-phase [h] were performed using the program made by one of the authors. Assuming a thickness of 3 nm and defocus of –60 nm (underfocus), atom chains along [001] of all the structures examined were white dots as shown in Fig. 5. Therefore, the Fig. 5 was explained as a combination of A15 and Zr_4Al_3-type structure units.

Within the W-films prepared at 100°C and 200°C, small particles of several nanometers were observed, of which the image contrasts were those at the boundary in Fig. 6 (Fig. 7a). The optical diffraction pattern obtained from this image is presented in Fig. 7b. The pattern showed almost 12-fold symmetry. Structure images containing Zr_4Al_3-type triangular contrast have been observed only in these very small particles. The existence of a Zr_4Al_3-type structure unit may be due to the size reduction of the particle. At lower temperatures, image contrast like amorphous was observed. Small periodic areas (of a few nanometers) were also seen within the amorphous-like contrast.

2. Morphology and lattice defects in A15-Cr small particles prepared by gas evaporation method

Since the first discovery of the A15-Cr particles (*i.e.* δ-Cr) prepared by the gas evaporation method [l,m] A15 small particles of other IVa elements (*i.e.* W and Mo) were produced by a similar technique.[n,o] The small particles of VIa elements with the characteristic A15-structure had the morphological shapes of icositetrahedron, rhombic dodecahedron and cube. [a] In this section, a discussion of the morphological shape of the A15-Cr particle based on the surface energy will be presented. Crystal defects observed in the particles will also be discussed.

2.1 Experimental procedure

After evacuation of a metallic bell jar down to 4×10^{-4} Pa, purified Ar-gas was introduced to $7–13 \times 10^2$ Pa (in some cases 4×10^4 Pa), and Cr was evaporated by electrical heating. Generally by the gas evaporation method, collected particles were a mixture of bcc- and A15-particles. The proportion of A15-particles depended on the evaporation conditions. For example, higher evaporation temperature gave a lower proportion of A15-particles. Particles collected at the *inner zone* where the temperature is high were small and had the bcc-structure. On the other hand, those collected at the *intermediate zone* where the particle size and the number of particles are generally high were well faceted A15-particles. By selecting adequate evaporation conditions, 90% of the particles obtained at the *intermediate zone* were A15-particles.

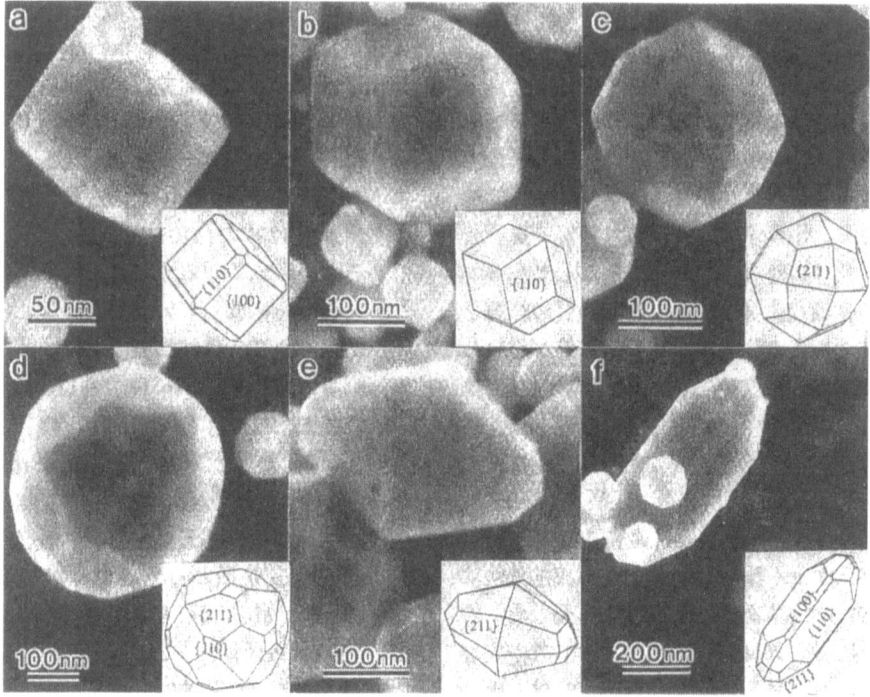

Fig. 8 Cr small particles by the gas evaporation method. (a) bcc and (b)–(f) A15.

2.2 Morphological shape of A15-Cr particle and discussion based on thermodynamics [3,4]

The morphological shape of the bcc-particles was basically the cube (Fig. 8(a)). By considering of surface energy, the stable shape of any bcc-crystal is the rhombic dodecahedron formed by twelve {110} planes.[p] The cube is therefore considered as the growth form. On the other hand, the particles of A15 single crystals had several shapes as shown in Figs. 8(b) - 8(f). They were mainly the icositetrahedron formed by twenty-four {211} planes (Fig. 8(c)) while the rhombic dodecahedron (Fig. 8(b)) was also be observed. In our present experiment, truncated icositetrahedra by {110} (Fig. 8(d)) and elongated icositetrahedra (Figs. 8(e) and 8(f)) were also frequently observed.

The morphological shapes of A15-Cr particles were strongly influenced by the Ar gas pressure. For example, rhombic dodecahedra were observed only under Ar pressure of 1.3×10^4 Pa or higher. It is believed that the higher gas pressure gives quicker cooling speed; therefore the rhombic dodecahedron must not be a stable shape. In contrast, under low gas pressure, morphological shapes based on the icositetrahedron were mainly observed. Not only for Cr, A15 small particles have been reported to have shapes based on the icositetrahedron.[n,o,q] Thus. it is expected that {211} planes of the A15-crystal minimize the surface energy of the particles. However, no detailed discussion of the shapes of A15-crystals has been mode. In this subsection, a discussion of the thermodynamically stable shape of A15-Cr will be presented taking into consideration surface energy and the Wulff polyhedron.[r]

The stable shape of a crystal particle formed by n planes satisfies the relation

$$\gamma_1/h_1 = \gamma_2/h_2 = \gamma_3/h_3 = \cdots = \gamma_n/h_n = (\text{const.}) \tag{1}$$

where γ_n is the surface energy per unit area of the n-th plane and h_n is the distance between the particle center and the n-th plane. The *Broken-Bond-Method*[p] was used for the evaluation of the surface energy. This method is based on the following assumptions: (a) by cutting a crystal at a crystal plane, interatomic bonds near this plane are broken, and (b) the surface energy can be obtained by summing up the half values of bond energy which are estimated as the interatomic two-body potentials. For the discussion of A15-Cr particles, bonds shorter than the third nearest distance were considered and the Morse potential[s] was used. The discussion is on the stable shape at 0 K. It is believed that the sharpness of the edges and corners of the particle is worse at higher temperature.

The stable shapes of bcc- and fcc-crystals have been discussed in detail by a method (named *Method-1* in this report) using the assumption that the number of broken bonds is influenced only by the direction of the cutting plane but not by its position.[p] As Mackenzie *et al.*[p] pointed out, some complicated structures such as the A15 structure, where atom-bond vectors are not symmetric, do not satisfy this assumption. For example, the surface energy of the (110)-plane of A15-Cr has two different values depending on the position of the cutting plane, *i.e.* 13.595 and 13.895 eV/nm². The surface energy evaluated by *Method-1* is the averaged value of these two values (*i.e.* 13.745 eV/nm²). For the discussion on particle shape giving the lowest surface energy, a method where only the lower energy values are considered is more reasonable than *Method-1*. This method will be called *Method-2* throughout this report. In the present work, Wulff polyhedra were obtained by these two methods (*i.e. Method-1* and *Method-2*) and compared.

Since the A15-structure is cubic, only the plane orientations within a stereo triangle formed by (100), (110) and (111) poles should be considered. The map of the surface energy evaluated by *Method-1* has continuous distribution within this stereo triangle. Thus, the surface energy is a function of the orientation angles (*i.e.* θ and ϕ of the polar coordinate) and only the planes giving the local minimum are required to make the Wulff polyhedron. They were (100), (110), (111), (210), (211), (311), (221), (421) and (432). On the other hand, the energy value by *Method-2* changes abruptly by orientation deviation of the cutting plane, and it cannot be treated as a simple function of θ and ϕ, as pointed out by Mackenzie *et al.*[p] In this method, therefore, surface energies of 205 planes having indices of {100} to {10 10 9} were calculated and the Wulff polyhedron was constructed following Eq. (1).

The obtained Wulff polyhedra are presented in Fig. 9(a) for *Method-1* and in Fig. 9(b) for *Method-2*. The polyhedron of Fig. 9(a) has a complicated shape formed by 9 types of planes. This is not the experimentally observed shape. On the other hand, the polyhedron by *Method-2* (more reasonable method than *Method-1*) is almost an icositetrahedron formed by {211} and very similar to the observation. Actually, *Method-2* is good for the consideration of A15-particles, and the icositetrahedron of {211} is almost a thermodynamically stable shape. Since only the discussion using the lower energy values explains the observed

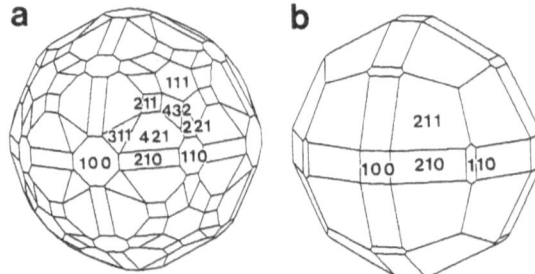

Fig. 9 Wulff polyhedra of A15-Cr obtained (a) by *Method-1* and (b) by *Method-2*.

particle shape, cutting positions of the crystal by {211} planes can be decided. This indicates that the atomic arrangement at the particle surface is derived as a model without atomic reconstruction. Interesting information will likely be obtained if scanning tunneling microscopy is applied to the surface of the A15-Cr small particle.

2.3 Multi-grained particle and grain boundary [3,5]

While all the particles mentioned above were single crystals, A15-Cr particles containing plural grains (multi-grained particles: MGP) were also frequently observed (Fig. 10). This is the first experiment discovering Cr-particles containing plural grins. MGP was usually observed among the particles under relatively low gas pressure (about 10 Torr or less). In this case, about 10% of the particles obtained at the *intermediate zone* was MGP. The number of grains was 3 to 12, and in many cases 10 or 11. The grains have [001] as the common axis. Figure 10(a) is a TEM image of MGP where the spoke-like contrast indicates the set of grain boundaries. This particle contains 11 grains; 9 thin grains having boundaries on {410} with a central angle of about 28° and two thick grains with a central angle of about 53°. The morphological shape is schematically drawn in Fig. 10(b), which was obtained from a set of TEM and SEM (scanning electron microscopy) images. The corresponding electron diffraction pattern (Fig. 10(c)) is a superposition of 11 patterns observed along [001]; thus the (100) ring contains 44 spots. In this figure, the arrowed spots are (100) reflections from the grains in the

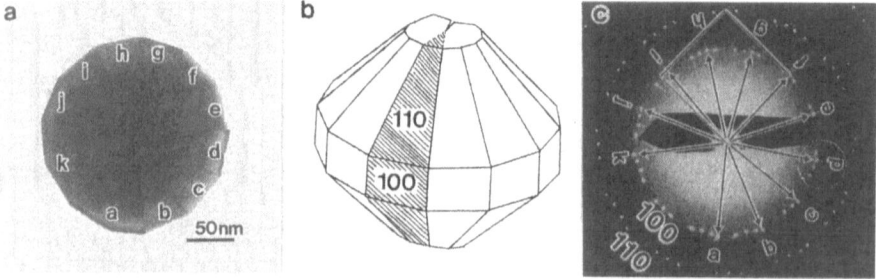

Fig. 10 Multi-grained particle. (a) TEM image, (b) electron diffraction pattern and (c) schematic drawing of the particle shape.

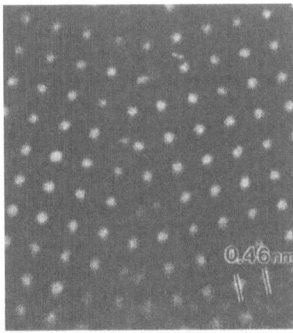

Fig. 11 High resolution TEM image of Σ17 tilt boundary.

particle, and the unit for the grain f is drawn by a square. The angle between adjacent (100) spots is about 28° or 78°. By comparing Figs. 10(a) and 10(c), the following result was obtained: 10 of these boundaries were Σ17/[001] tilt boundaries on {410}, and the 11-th boundary was Σ25/[001] on {430}. Ideal rotation angles of these two types of tilt boundaries are 28.1° and 73.7°, respectively. Since eleven 28.1° and one 73.7° gives a gap of 5.3°, lattice relaxation must occur as in multiple-twin particles of Au and Ag. For observed MGP composed of N grains. N-1 or N-2 boundaries had the Σ17 relation. The other relations recognized were Σ5, Σ13, Σ25. etc. While the Σ17 boundary was usually flat, some of the other boundaries were bent.

In order to elucidate the structure of the Σ17 boundary of A15-Cr, high resolution TEM observations were carried out. An example is shown in Fig. 11. The periodic square of white spots corresponds to the unit cell observed along [001]. Thus the angular and translational relations of two adjacent lattices are directly realized. Assuming the white spots correspond to atom chains along [001] of the A15-lattice, this image fits the Σ17 boundary found in a Nb_3Ge superconductor tape.[1] From the above results, it is expected that the Σ17 boundary is in favor among many other possible tilt boundaries around [001].

In addition, a region composed of a Zr_4Al_3-type structure unit was observed at the corner of MGP. As expected from the result of W thin films, very small Cr-particles composed of Zr_4Al_3-type structure units are expected to be found.

Acknowledgments

We are obliged to Dr. T. Ishimasa for helpful suggestions on the production of Cr small particles and to Mr. M. Suzuki (JEOL) for help on high resolution TEM observations. Messrs. E. Takeuchi, Y. Hayashi and T. Tagawa are acknowledged for technical assistance on the instrumentation of the vacuum system. Part of the work was performed using JEM 2010 microscopes at the Shizuoka Institute of Science and Technology and Jeol as well as a S-900 scanning microscope at the Instrument Center, Institute of Molecular Science. This work was supported by a Grant-in-Aid for Scientific Research from the Ministry of Education, Science and Culture.

PUBLICATIONS

1. M. Arita and I. Nishida, *Jpn. J. Appl. Phys.*, **32**, 1759 (1993).
2. M. Arita and I. Nishida, *Surface Rev. Lett.*, **3**, 1191 (1996).
3. M. Arita and I. Nishida, *J. Cryst. Soc. Jpn.*, **36**, 320 (1994) [in Japanese].
4. I. Nishida, N. Suzuki and M. Arita, Res. Bull. Col. Gen Educ. Nagoya Univ., **B38**, 25 (1994) [in Japanese].
5. M. Arita and I. Nishida, Proc. 38th Jpn. Cong. Mater. Res. 33, Kyoto (1995).
6. H. Ishibashi, M. Arita, I. Nishida, A. Yanase and K. Nakahigashi, *J. Phys. Cond. Matter,* **6**, 8681 (1994).
7. M. Arita, N. Suzuki and I. Nishida, *J. Cryst. Growth*, **132**, 71 (1993).
8. M. Arita, N. Suzuki and I. Nishida, Res. Bull. Col. Gen Educ. Nagoya Univ. **B37**, 39 (1993).
9. S. Matsuo, T. Matsuura, I. Nishida and N. Tanaka, *Jpn. J. Appl. Phys.*, **33**, 3907 (1994).
10. M. Arita, N. Suzuki and I. Nishida, Res. Bull. Col. Gen. Educ. Nagoya Univ. **B38**, 53 (1994).

REFERENCES

a. K. Kimoto, *J. Japan Assoc. Crystal Growth*, **6**, 122 (1979).
b. R. Uyeda, in, Morphology of Crystals, Part B, Ed. I. Sunagawa, Chap. 6, Reidel, Dordrecht (1987).
c. E. Koch-Bienemann, L. Berg, G. Czack and J. Wagner Gmelin Handbook of Inorganic Chemistry, Tungsten, Suppl. Vol. **A2**, Sec. 5.3, Springer-Verlag, Berlin (1987).
d. S. Basavaiah and S. R. Pollak, *J. Appl. Phys.* **39**, 5548 (1968).
e. W. L. Bond, A. S. Cooper, K. Andres, O. W. Hull, T. H. Geballe and B. T. Matthias, *Phys. Rev. Lett.*, **15**, 260 (1965).
f. F. M. Kilbane and P. S. Habig, *J. Vac. Sci. & Technol.*, **12**, 107 (1975).
g. T. Kizuka, T. Sakamoto and N. Tanaka, *J. Cryst. Growth*, **131**, 439 (1993).
h. A. K. Sinha, *Prog. Mater. Sci.*, **15** (1973) 79.
i. T. Ishimasa and Y. Fukano, *Jpn. J. Appl. Phys.*, **22**, 1092 (1983).
j. M. Arita, H.-U. Nissen, Y. Kitano and W. Schauer, *J. Solid State Chem.*, **107**, 76 (1993).
k. Y. Kitano, H.-U. Nissen, R. Wessicken, D. Yin and W. Schauer, *J. Appl. Phys.*, **58**, 1904 (1985).
l. K. Kimoto, Y. Kamiya, N. Nonoyama and R. Uyeda, *Jpn. J. Appl. Phys.*, **2**, 702 (1963).
m. K. Kimoto, I. Nishida, *Jpn. J. Appl. Phys.*, **6**, 1047 (1967).
n. Y. Saito, S. Yatsuya, K. Mihama and R. Uyeda, *J. Cryst. Growth*, **45**, 501 (1978).
o. S. Iwama and K. Hayakawa, *Surf. Sci.*, **156**, 85 (1985).
p. J. K. Mackenzie, A. J. W. Moore and J. F. Nicholas, *J. Phys. Chem. Solid,* **23**, 185 & 197 (1962).
q. T. Ishimasa and Y. Fukano, *Jpn. J. Appl. Phys.*, **22**, 6 (1983).
r. G. Wulff, *Z. Krist.*, **34**, 449 (1901).
s. L. A. Girifalco and V. G. Weizer, *Phys. Rev.*, **114**, 687 (1959).
t. H.-U. Nissen, Y. Kitano and R. Wessicken, *J. Microscopy*, **142**, 171 (1986).

28

Dynamic Properties of Cluster Ions in Relation to the Geometric and Electronic Structures

Tamotsu Kondow[a], Takashi Nagata, Shinji Nonose and Akira Terasaki[a]

Department of Chemistry, School of Science, University of Tokyo
Bunkyo-ku, Tokyo 113-0033, Japan

present address:
[a] Cluster Research Laboratory, Toyota Technological Institute: in East Tokyo Laboratory, Genesis Research Laboratory Inc., Ichikawa-shi, Chiba 272-0001, Japan

Purpose of the present study

The dynamics of clusters (or their ions) is characterized by the collective properties of the clusters and is closely related to their geometric and electronic structures, which depend characteristically on the number of their constituent particles (cluster size); the specificity of the dynamics originates mainly from the fact that the clusters consist of a limited number of constituent particles. In order to elucidate the specificity inherent to the cluster dynamics, we investigated (1) photo-induced dissociation of size-selected cluster ions, (2) collisional dissociation of size-selected cluster ions, (3) electron detachment processes and electronic structure, and (4) collective motions of clusters. Five topics are selected for detailed description.

Contents of the present study

1. Photodissociation of argon cluster ions

Photodissociation dynamics of a cluster ion is closely related to its geometric and electronic structure, particularly the distribution of the charge in the cluster ion. [a] In this section, we describe the study on the photodissociation of argon cluster ions in relation to their geometric and electronic structures. [1,17] The experimental results were analyzed by a diatomics-in-molecules (DIM) calculation. [18,23]

Figure 1 shows a schematic diagram of the apparatus employed in the present study. An ion beam was produced by electron impact on a supersonic expansion from a 500-μm orifice with a stagnation pressure of 4 atm. An argon cluster ion of a given size, Ar_n^+, was dissociated by laser irradiation, and the kinetic energy and angular distribution of neutral and ion fragments were measured.

Figure 2 shows the time-of-flight (TOF) mass spectrum of the photofragment ion, Ar^+, produced by the photodissociation of Ar_3^+ with the polarization vector of the excitation laser parallel ($\Theta = 0°$) and perpendicular ($\Theta = 90°$) to the beam axis of the parent cluster ion. The central and the outer peaks for $\Theta = 0°$ are assigned as the slow and the fast components of the photofragment ion, Ar^+, respectively. The angular and kinetic energy distributions (see also the spectrum

295

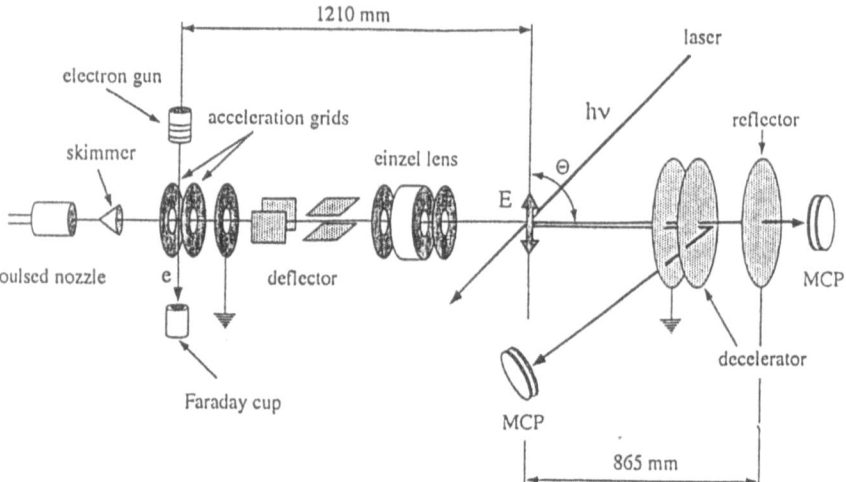

Fig. 1   Schematic diagram of the experimental apparatus based on a time-of-flight mass spectrometer. A cluster ion is photodissociated by a laser with its polarization vector parallel ($\Theta = 0°$) and perpendicular ($\Theta = 90°$) to the incident ion beam axis.

$$Ar_3^+ \longrightarrow Ar^+$$

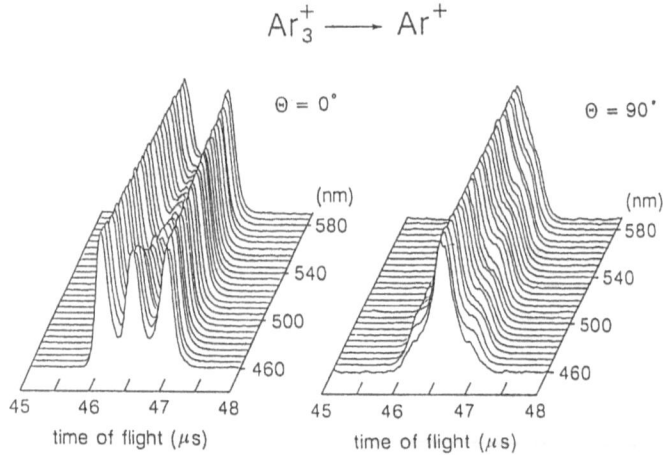

Fig. 2   Spectra of Ar^+ produced front Ar_3^+ by irradiation of a laser with its polarization vector parallel ($\Theta = 0°$) and perpendicular ($\Theta = 90°$) to the incident Ar_3^+ beam axis.

for $\Theta = 90°$) of the photofragment ion, together with the theoretical calculation lead us to conclude that the equilibrium structure in the ground state is linear ($^2\Sigma_u^+$) and the dissociation proceeds through the transition to the repulsive excited state ($^2\Sigma_g^+$). The positive charge is almost completely delocalized in the ground state, whereas it is localized in the outer argon atoms in the repulsive excited state, $^2\Sigma_g^+$. The slow component results from nonadiabatic transition during the photodissociation.

Similar measurements were performed for larger argon cluster ions, Ar_n^+ ($4 \leq n \leq 25$). In this experiment, the kinetic and angular distributions of charged and neutral photofragments were measured simultaneously. Figure 3 shows the spectra of the photofragment ions from parent cluster ions with different sizes. In the

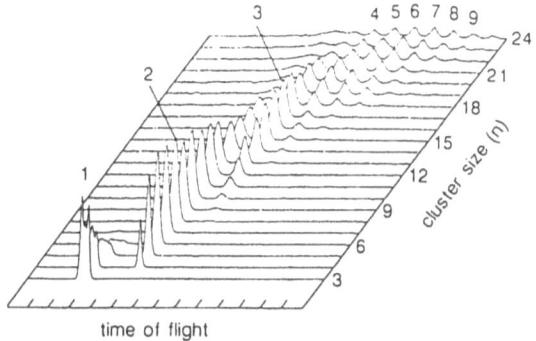

Fig. 3 Time-of-flight (TOF) spectra of photofragment ions from Ar_n^+ with different n. The digit on each peak shows the size of the fragment ion.

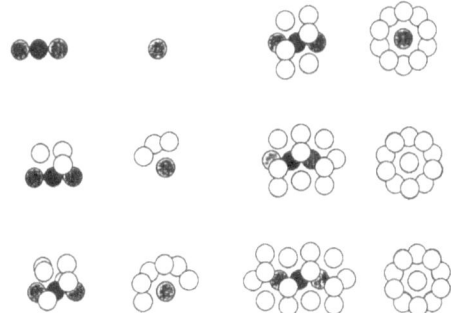

Fig. 4 The most stable structures and charge distributions of Ar_n^+ (n = 3, 6, 9, 13, 16, and 25) calculated by the DIM method. The charge is delocalized in the core ion, Ar_3^+.

photodissociation of the parent cluster ions with $4 \leq n \leq 14$, the photofragment ion, Ar^+, which corresponds to the fast component for the Ar_3^+ dissociation, is observed. Argon atoms associated with the photofragment ions were also detected. These results show that the core ion Ar_3^+ in Ar_n^+ is directly fragmented by absorbing the laser light. As the cluster size of the parent cluster ion increases, the photofragment ions produced directly from the core ion were no longer observed. The angular and the kinetic energy distributions of the photofragment ions support a statistical distribution of the excess energy among the internal degrees of freedom. Argon atoms evaporate after the energy distribution is completed.

The DIM method was employed to calculate the geometric and electronic structures.[11,18,23] Figure 4 shows the geometric structures and the charge distributions of Ar_n^+ with typical sizes. As shown in Fig. 4, the charge is localized at the central three atoms with a linear spatial arrangement, and is surrounded by the solvent argon atoms keeping a five-fold symmetry. The photodissociation process of Ar_3^+ was also calculated by hemiquantal dynamics and coupled surface hopping methods. These calculations predict the experimental results.

2. Collective motions of van der Waals clusters

Collective vibrational modes of a cluster have large density of states and play an important role in cluster dynamics. We developed a method to derive collective vibrations, such as breathing, quadrupole spheroidal and torsional ones from the complex cluster motion in terms of the vibrational eigen functions of a dense sphere.[4,7] The calculated spectrum of these collective modes for argon clusters, especially Ar_{55}, reproduces the experimental spectrum obtained by inelastic scattering of He by argon clusters.[b]

The molecular dynamics (MD) method was used to trace the time evolution of argon atoms in spherical argon clusters, such as Ar_{13} and Ar_{55}. The Lernnard-Jones (LJ) potential function with $\sigma = 3.40$ Å and $\varepsilon = 84$ cm^{-1} between an Ar-Ar pair was employed and the potential energy of the cluster system was made by overlapping the two body LJ potentials. The collective modes studied are breathing (S00), quadrupole spheroidal (S20-S22), and torsional (T21-T22) modes (see Fig. 5). Figure 6 shows the spectra of Ar13; panels (a), (b), and (c) represent the spectra for S00, S20-S22, and T21-T22 modes, respectively, where the internal energy of Ar_{13} is set to be 8.9 meV. These vibrations were found to be nonlinear

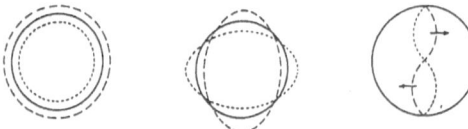

Fig. 5   Vibrational modes of a dense sphere. (a) Breathing, (b) quadrupole spheroidal, and (c) torsional modes.

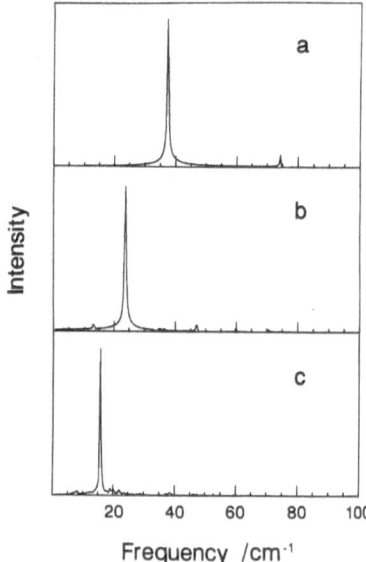

Fig. 6   Spectra of (a) breathing, (b) quadrupole spheroidal, and (c) torsional vibrations calculated by the expansion of complex cluster vibrations in terms of the vibrational eigen functions of a dense sphere.

because of the peak shift observed with increase in the internal energy of Ar_{13}. The change in the peak frequency and width with the internal energy is closely related to the "phase" change of Ar_{13}.

3. Collision of argon cluster ions with ^{36}Ar

The collision-induced reaction of Ar_n^+ (n = 2 - 23) with ^{36}Ar was studied at the collision energies of 0.2 and 2.0 eV by use of a tandem mass spectrometer equipped with octopole ion guides. The product ions were $Ar_{n'}^+$ and $^{36}ArArAr_{n'-1}^+$ ($n'< n$). The absolute cross sections for formation of the product ions were obtained as a function of the size of the parent cluster ion. Figure 7 shows the total absolute cross section at the collision energy of 0.2 eV as a function of the size of the parent cluster ion. The cross sections for the formation of $Ar_{n'}^+$ (evaporation process) and $^{36}ArAr_{n'-1}$ (fusion process) were also measured. In order to clarify how the evaporation and the fusion proceed, the molecular dynamics simulation based on the diatomics-in-molecules method was performed. The total (evaporation and fusion) cross sections calculated from the simulation reproduce the measured ones within an accuracy of 15% at n = 6, 13, and 19 at the collision energy of 0.2 eV (see Fig. 7). The experimental and simulation studies lead us to conclude that the dissociation proceeds via collisional excitation of the parent cluster ion, Ar_n^+, and subsequent unimolecular dissociation (evaporation), or formation of a collision complex, $^{36}ArAr^+$, followed by subsequent unimolecular dissociation (fusion). The molecular dynamics simulation for the 0.2-eV collision with Ar_{10}^+ also shows that the fusion proceeds via perfectly in elastic collision followed by unimolecular dissociation.

4. Collision of sodium cluster ions with rare gas atoms

Collisional dissociation of sodium cluster ions, Na_n^+, by a helium or a neon atom was investigated by a tandem mass spectrometer equipped with octopole ion

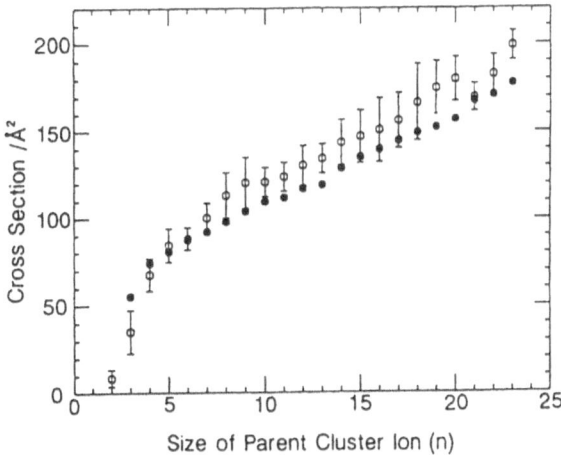

Fig. 7   Absolute cross sections (open circles) for the collisional dissociation of Ar_n^+ by ^{36}Ar at different n. The calculated cross sections (closed circles, see text) are also shown for comparison.

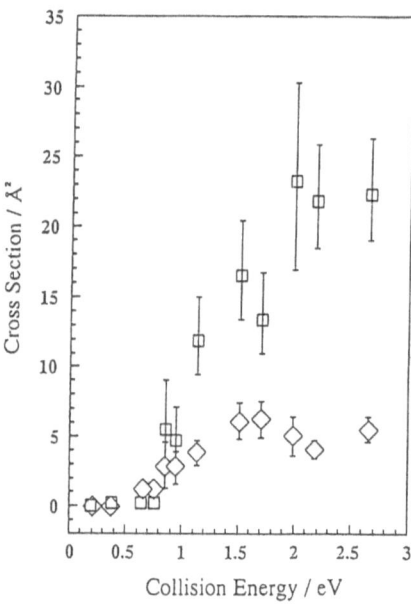

Fig. 8 Cross sections for release of Na (open diamonds) and Na_2 (open squares) from Na_9^+ in collision with a helium atom.

guides.[46] A smoke ion source was specially designed to prepare sodium cluster ions such as Na_9^+. A size-selected parent cluster ion was allowed to collide with a rare gas atom, He or Ne. The product ions show that release of Na and Na_2 from Na_9^+ are the dominant dissociation channels. The absolute cross sections for the Na and Na_2 release were measured at different collision energies. Figure 8 shows the absolute cross sections for the Na and Na_2 release from Na_9^+ as a typical example. As shown in Fig. 8, the energies for the Na and Na_2 release should be almost the same because both cross sections begin to rise at almost the same energy, ~ 0.7 eV. The cross section for the Na_2 release becomes much larger than that for the Na release as the collision energy increases. This tendency is explained in terms of the orbital correlation diagrams for $Na_8^+ + Na$ and $Na_7^+ + Na_2$ systems on the basis of collisional deformation of a spherical jellium of Na_9^+. Similar results were obtained for the collisional dissociation of sodium cluster ions smaller than $n = 9$.

5. Electronic structures of cobalt cluster anions studied by photoelectron spectroscopy

The photoelectron spectrum of a size-selected cobalt cluster anion, Co_n^- ($3 \leq n \leq 70$), was measured by a magnetic-bottle time-of-flight electron spectrometer; Co_n^- was irradiated by a XeCl excimer laser (4.025 eV) and the photoelectrons produced were energy-analyzed in the spectrometer with an energy resolution, $\Delta E/E$, of less than 0.05.

The geometric and electronic structures of Co_n^- with $n = 3$, 4, and 6 were obtained by comparing the measured spectra with those calculated by the spin polarized DV-$X\alpha$ method. The measured spectra are reproduced reasonably well

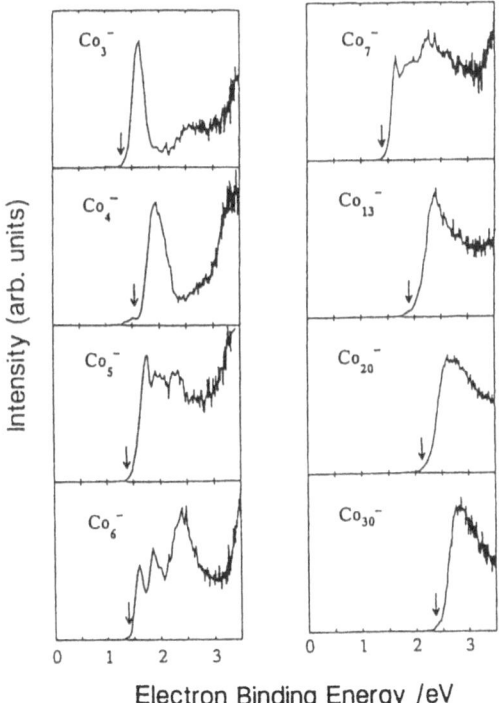

Fig. 9 Photoelectron spectra of Co_n^-. The onset energies indicated by arrows give the electron affinities.

by the calculation, postulating the most probable geometric structures. [c,d] The 3d energy levels with the majority spin are found to be separated by 1.0–2.8 eV from those with the minority spin; the latter levels are partly filled. A spin difference of about 2.0 per Co atom suggests ferromagnetic spin couplings in the cluster ions.

For Co_n^- with $n \leq 7$, the electron affinity measured depends linearly on the reciprocal of the cluster radius and approaches the work function of a cobalt metal with increase in n, whereas, for Co_n^- with $n \leq 6$, the electron affinity deviates from the linear relationship. The deviation from the linear dependence indicates a size-dependent transition at $n \approx 7$.

PUBLICATIONS

1. T. Nagata, J. Hirokawa and T. Kondow, *Chem. Phys. Lett.*, **176**, 526 (1991).
2. T. Nagata and T. Kondow, *Z. Phys. D-Atoms, Molecules and Clusters*, **20**, 153 (1991).
3. Y. Ozaki, M. Ichihashi and T. Kondow, *Z. Phys. D-Atoms, Molecules and Clusters*, **20**,161 (1991).
4. Y. Ozaki, M. Ichihashi and T. Kondow, *Chem. Phys. Lett.*, **182**, 57 (1991).
5. T. Tsukuda and T. Kondow, *J. Chem Phys.*, **95**, 6989 (1991).
6. T. Tsukuda and T. Kondow, *Chem. Phys. Lett.*, **185**, 511 (1991).
7. Y. Ozaki, M. Ichihashi and T. Kondow, *Chem. Phys. Lett.*, **188**, 555 (1992).
8. M. Ichihashi, Y. Ozaki and T. Kondow, *Chem. Phys. Lett.*, **189**, 353 (1992).
9. T. Tsukuda and T. Kondow, *J. Phys. Chem.*, **96**, 5671 (1992).
10. T. Tsukuda and T. Kondow, *Chem. Phys. Lett.*, **197**, 438 (1992).
11. T. Ikegami, T. Nagata and T. Kondow, in: *Physics and Chemistry of Finite Systems: From Clusters to Crystals*, **Vol**. I, (P. Jena, S. N. Khanna and B. K. Rao, eds.), p.417, Kluwer Academic Publishers, Dordrecht (1992).

12. S. Nonose, J. Hirokawa, M. Ichihashi, M. Sakamoto, T. Tahara and T. Kondow, in: *Physics and Chemistry of Finite Systems: From Clusters to Crystals*, **Vol. II**, (P.Jena, S. N. Khanna and B. K. Rao, eds.), p.1125, Kluwer Academic Publishers, Dordrecht (1992).

13. T. Tsukuda and T. Kondow, in: *Physics and Chemistry of Finite Systems: From Clusters to Crystals*, **Vol. II**, (P. Jena, S. N. Khanna and B. K. Rao, eds.), p.1147, Kluwer Academic Publishers, Dordrecht (1992).

14. T. Nagata, H. Yoshida and T. Kondow, *Chem. Phys. Lett.*, **199**, 205 (1992).

15. F. Mafuné, Y. Takeda, T. Nagata and T. Kondow, *Chem. Phys. Lett.*, **199**, 615 (1992).

16. T. Tsukuda, A. Terasaki, T. Kondow, M. G. Scarton, C. E. Dessent, G. A. Bishea and M. A. Johnson, *Chem. Phys. Lett.*, **201**, 351 (1993).

17. T. Nagata and T. Kondow, *J. Chem. Phys.*, **98**, 290 (1993).

18. T. Ikegami, T. Kondow and S. Iwata, *J. Chem. Phys.*, **98**, 3038 (1993).

19. F. Misaizu, P. L. Houston, N. Nishi, H. Shinohara, T. Kondow and M. Kinoshita, *J. Chem. Phys.*, **98**, 336 (1993).

20. M. Ichihashi, J. Hirokawa, S. Nonose, T. Nagata and T. Kondow, *Chem. Phys. Lett.*, **204**, 219 (1993).

21. S. Nonose, J. Hirokawa, M. Ichihashi, M. Sakamoto, H. Tanaka and T. Kondow, *Z. Phys. D-Atoms, Molecular and Clusters*, **26**, 223 (1993).

22. T. Nagata, H. Yoshida and T. Kondow, *Z. Phys. D-Atoms, Molecular and Clusters*, **26**, 367 (1993).

23. T. Ikegami, T. Kondow and S. Iwata. *J. Chem. Phys.*, 99, 3588 (1993).

24. T. Tsukuda, M. Nagamine, T. Tahara, T. Nagata and T. Kondow, *Chem. Phys. Lett.*, **218**, 1 (1994).

25. F. Mafuné, J. Kohno, T. Nagata and T. Kondow, *Chem. Phys. Lett.*, **218**, 7 (1994).

26. F. Mafuné, Y. Takeda, T. Nagata and T. Kondow, *Chem. Phys. Lett.*, **218**, 234 (1994).

27. M. Ichihashi, S. Nonose, T. Nagata and T. Kondow, *J. Chem. Phys.*, **100**, 6458 (1994).

28. S. Nonose, H. Tanaka, T. Nagata and T. Kondow, *J. Phys. Chem.*, **98**, 8866 (1994).

29. M. Nagamine, K. Someda and T. Kondow, *Chem. Phys. Lett.*, **229**, 8 (1994).

30. J. Kohno, F. Mafuné and T. Kondow, *J. Am. Chem. Soc.*, **116**, 9801 (1994).

31. J. Hirokawa, M. Ichihashi, S. Nonose and T. Kondow, *Z. Phys. D-Atoms, Molecules and Clusters*, **31**, 187 (1994).

32. T. Tsukuda and T. Kondow, *J. Am. Chem. Soc.*, **116**, 9555 (1994).

33. J. Hirokawa, M. Ichihashi, S. Nonose, T. Tahara, T. Nagata and T. Kondow, *J. Chem. Phys.*, **101**, 6625 (1994).

34. H. Matsumura, F. Mafuné and T. Kondow, *J. Phys. Chem.*, **99**, 5861 (1995).

35. M. Ichihashi, J. Hirokawa, T. Tsukuda, T. Kondow, C. E. H. Dessent, C. G. Bailey, M. G. Scarton and M. A. Johnson, *J. Phys. Chem.*, **99**, 1655 (1995).

36. F. Mafuné, J. Kohno and T. Kondow, *J. Chinese Chem.*, **42**, 449 (1995).

37. T. Tsukuda, H. Yasumatsu, T. Sugai, A. Terasaki, T. Nagata and T. Kondow, *J. Phys. Chem.*, **99**, 6367 (1995).

38. D. M. Cyr, M. G. Scarton, K. B. Wiberg, M. A. Johnson, S. Nonose, J. Hirokawa, H. Tanaka and T. Kondow, R. A. Morris and A. A. Viggiano, *J. Am. Chem. Soc.*, **117**, 1828 (1995).

39. H. Yoshida, A. Terasaki, K. Kobayashi, M. Tsukada and T. Kondow, *J. Chem. Phys.*, **102**, 5960 (1995).

40. Y. Fukuda, T. Tsukuda, A. Terasaki and T. Kondow, *Chem. Phys. Lett.*, **242**, 121(1995).

41. F. Mafuné, Y. Hashimoto, M. Hashimoto and T. Kondow, *J. Phys. Chem.*, **99**, 13814 (1995).

42. M. Ichihashi, T. Tsukuda, S. Nonose and T. Kondow, *J. Phys. Chem.*, **99**, 17354 (1995).

43. H. Yasumatsu, T. Kondow, H. Kitagawa, K. Tabayashi and K. Shobatake, *J. Chem. Phys.*, **104**, 899 (1996).

44. A. Terasaki, T. Tsukuda, H. Yasumatsu, T. Sugai and T. Kondow, *J. Chem. Phys.*, **104**, 1387 (1996).

45. H. Yasumatsu, A. Terasaki and T. Kondow, *Science*, **106**, 3806 (1997).

29

Structure and Reactivity of Binary Clusters

Koji Kaya [a,b] and Atsushi Nakajima [a,b]

[a]   Department of Chemistry, Keio University
      Kohoku-ku, Yokohama 223-8522, Japan
[b]   Institute of Physical and Chemical Research (RIKEN)
      Wako-shi, Saitama 351-0106, Japan

Purpose of the present study

Over the past few decades, there has been much active work on the properties (structure, reactivity, stability, etc.) of single component clusters. When one synthesizes binary clusters containing two different components, the binary clusters may exhibit unexpected superior properties induced by the interaction between two components. In the present study, we aim to clarify the properties of binary clusters containing metal atoms. Ionization energies, electronic spectra, electron affinities and reactivities of various binary clusters are discussed.

Contents of the present study

1. Experimental

Figure 1 is a schematic diagram of the laser vaporization apparatus for the generation and detection of binary clusters. We have modified the laser vaporization method for binary cluster formation. Two rods of different metal (or semiconductor) elements are positioned a few millimeters apart. Outputs of the second harmonic (532 nm) of two independent pulsed Nd^{3+} YAG lasers are

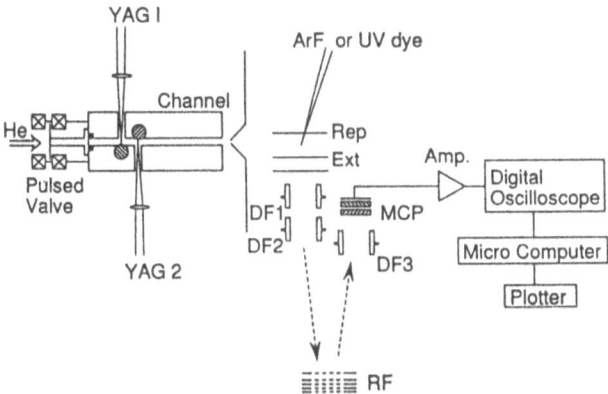

Fig. 1   Modified laser vaporization set-up for binary cluster generation (YAG: YAG laser, DF: deflection plate, RF: reflectron).

focused on the surfaces of the rods with a time delay of a few microseconds. Hot vaporized atoms and ions from respective rods are mixed together and cooled by high pressure He gas pulses to form binary clusters. The intensity distribution of binary clusters is determined by a TOF mass spectrometer equipped with a reflectron. Advantages of the modified laser vaporization method over others [a,b] are: (1) one is able to adjust the composition ratio of binary clusters by controlling the laser fluence of respective vaporization lasers, and (2) one can arbitrarily choose any combination of elements for the binary cluster formation.

In the present study, we have generated binary clusters in which an alkali metal atom such as sodium and cesium is a component. For the laser ablation of such soft elements, sodium or cesium atoms are deposited on the aluminum metal rod in advance. The details have been given in previous papers.[1,2]

We have also conducted photoelectron spectroscopic measurements of cluster anions. In order to generate cluster anions, an electron must be injected into a neutral cluster. The electrons for the injection were produced by focusing the second harmonic of Nd^{3+} YAG laser gently on the Y_2O_3 disk surface located 3 mm downstream from the nozzle exit. Because the work function of Y_2O_3 is 2.0 eV, the photoelectrons ionized by the 2.3 eV photons from the 532 nm laser has a kinetic energy of less than 0.3 eV. These slow electrons make it possible to generate cluster anions efficiently even though corresponding neutral clusters have low electron affinity. Figure 2 shows the experimental set-up of photoelectron spectroscopy constructed in our laboratory.[3,4]

The apparatus consists of a cluster anion source, a TOF mass spectrometer and a magnetic bottle-type photoelectron spectrometer. The cluster source part is almost the same as described in the modified laser vaporization. The cluster anions were extracted by applying a pulsed electric field of about - 1 kV. After a 1.5-m flight path, the target cluster anion was mass selected by a mass gate, and decelerated to the ion energy of 20 eV by the potential elevator. The decelerated cluster anion was infected into the magnetic bottle photoelectron spectrometer where the detachment laser (typically 2nd or 3rd harmonic of YAG laser) irradiates the cluster anion. The detached photoelectrons are collected efficiently

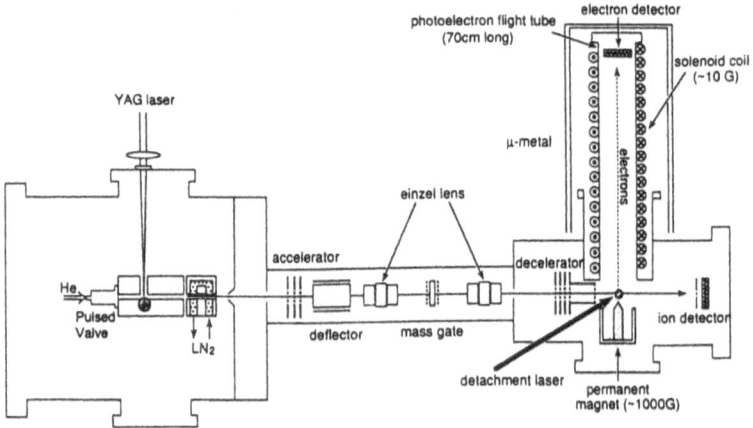

Fig. 2 Experimental scheme of photoelectron spectrometer for mass selected cluster ions.

by a magnetic bottle and the kinetic energies were determined by a TOF analyzer. The advantage of the magnetic bottle spectrometer is of high sensitivity because of the high efficiency of electron collection. Instead, the resolution of the electron energy must be sacrificed. In our case, the resolution is 60 meV.

We have also measured the electronic spectra of small clusters by the mass selected resonance enhanced multiphoton ionization method (REMPI) described in our previous paper.[5]

Ionization energy of a neutral cluster was determined by the measurement of photoionization efficiency against the photon energy of the tunable dye laser as the ionization light source.

2. Ionization energies and stability of Al$_n$ Cs$_m$ clusters [3]

Since the success in the explanation for the magic numbers in alkali metal clusters, the jellium shell model has been extensively applied to clusters of elements other than alkali elements. The aluminum atom, which has three valence electrons, was first predicted by Chou and Cohen to have shell structure when Al atoms form clusters.[c]

However, because of the odd number of electrons, complete closing of the electron shell cannot be attained in small aluminum clusters. Instead, aluminum cluster anions can close their electron shell at cluster size 13 and 23 corresponding to the closing of 2p and 3s shells, respectively. Leuchtner et al. have reported anomalous stabilities of Al$_{13}^-$ and Al$_{23}^-$ against the oxidation reaction.[d]

We have reported that the electron shell effect can be applied to the sodium-doped bimetallic clusters such as Al$_n$Na$_m$ [1] and In$_n$Na$_m$.[6]

Among sodium-doped Al clusters, Al$_{13}$Na, and Al$_{23}$Na were found to have anomalously high ionization energies because of the closing of electron shells. It was also concluded that the electron migration from the alkali metal atom to the aluminum clusters takes place in Al$_n$Na$_m$ clusters. Among the alkali metal elements, cesium atom has the lowest ionization energy. We have produced the binary clusters Al$_n$Cs$_m$ by a combination of the laser vaporization and high temperature nozzle method. Figure 3 depicts the mass distribution of cesium-doped aluminum clusters Al$_n$Cs$_m$ ionized by an ArF excimer laser (6.42 eV). The spectrum reflects the stability or respective clusters because the energy of the

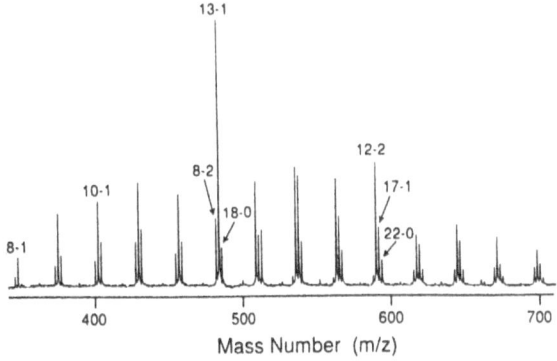

Fig. 3   TOF mass spectrum of Al$_n$Cs$_m$ clusters. n-m in the figure corresponds to the number of Al and Cs atoms, respectively.

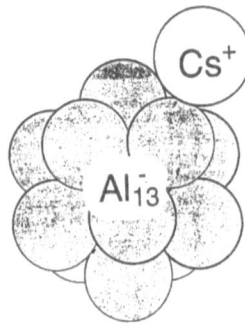

Fig. 4 The most probable structure of $Al_{13}Cs$.

ionization laser photon (6.42 eV) is well above the ionization threshold of these clusters. One easily finds high stability of $Al_{13}Cs$ among neighboring clusters. The ionization energy was found to exhibit an abrupt drop in going from $Al_{13}Cs$ to $Al_{13}Cs_2$. However, it was also observed that the $Al_{12}Cs_{14}$, which also has 40 electrons as valence electrons, does not exhibit stability.

From these results. it was concluded that the stability of $Al_{13}Cs$ comes from two factors: the closing of the electron shell and the stable icosahedral structure of Al_{13} moiety. Moreover, by the electron donation to the Al cluster, the cesium atom also closes the electron shell. Finally, $Al_{13}Cs$ behaves like alkali halide molecules $Al_{13}^-Cs^+$, as seen in Figure 4.

3. Photoionization electronic spectroscopy of AlNa [5]

As a basic unit of the binary clusters Al_nNa_m, we have studied the electronic structure of the diatomic molecule AlNa by means of mass selected REMPI method under supersonic molecular beam condition. Figure 5 shows the electronic spectrum of AlNa in the 635–650 nm region. The origin band is located at 15424 cm^{-1} and a vibrational progression with the 115 cm^{-1} spacing is identified. From the sequence bands, the vibrational frequency 186 cm^{-1} of the ground state is also identified. *Ab initio* calculation was conducted for AlNa. Combined with

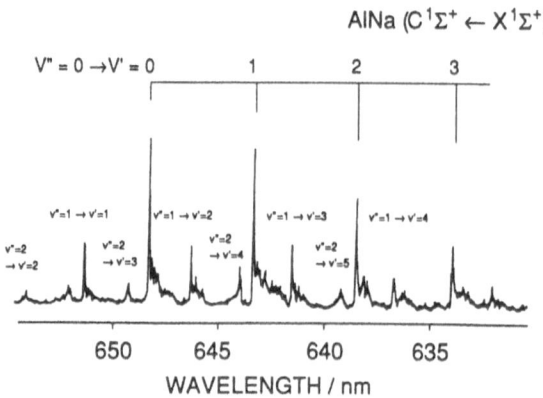

Fig. 5 REMPI spectrum of AlNa in C–X transition region.

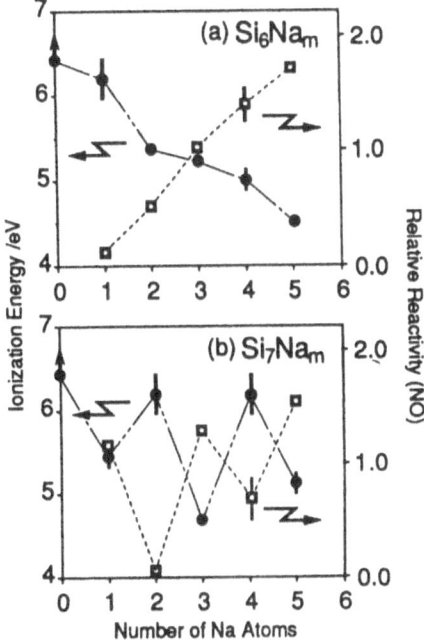

Fig. 6 Ionization Energy and Reactivity toward NO adsorption of Si_6Na_m and Si_7Na_m ($m = 0$–5).

the theoretical result, it was concluded that the observed spectrum in Figure 5 corresponds to the electronic transition from the ground state ($X^1\Sigma^+$) to the upper C state with $^1\Sigma^+$ symmetry. We also observed another electronic transition $^1\Pi$–X $^1\Sigma^+$ at the 590 to 615 nm region.

The experimental results were analyzed by the *ab initio* calculation and it is suggested that electron migration from Na to Al is negligibly small in the ground state. The conclusion drawn here is in contrast to the case of larger Al_nNa_m clusters where an electron migrates from the Na atom to Al clusters.

4. Structure and reactivity of Si_nNa_m clusters [7,8]

There is currently intense interest in the physical and chemical properties of small clusters of semiconductor materials, especially silicon (Si). Much experimental effort has been expended on the survey of silicon cluster ions because the neutral silicon clusters have high ionization energy which cannot be reached by the one photon excitation of commercially available lasers. We have developed a method to produce sodium containing metal or semiconductor clusters. Because of the low ionization energy of the Na atom the ionization energy of a sodium doped cluster decreases compared with the genuine cluster. In this study, we produced the binary clusters Si_nNa_m by modified laser vaporization method. The ionization energy and surface reactivity toward NO adsorption of these clusters were examined. Figure 6 shows the ionization energies and surface reactivities of Si_6Na_m and Si_7Na_m ($m = 0$–5) clusters plotted against the number of doped Na atoms. In Si_6Na_m series, the ionization energy decreases as the number m of doped Na atoms is increased. At the same time, the reactivity toward NO adsorption is increased as the number m is increased. Si_7 is known to have an electronic shell closing structure and the large HOMO-LUMO gap causes even-odd alternation of ionization energy due to pairing and unpairing of electrons in HOMO of respective clusters.[e]

Common to Si_6Na_m and Si_7Na_m, the reactivity toward NO adsorption has beautiful anticorrelation with the ionization energy. That is to say, a Si_nNa_m cluster which has high ionization energy is inactive to NO adsorption and a cluster with low ionization energy has high reactivity. The anticorrelation between reactivity and ionization energy is interpreted in terms of the electronic factor of the reaction. In the other words, if the rate-determining step of the reaction is the electron migration from the HOMO of the cluster to the antibonding orbital of adsorbed NO molecule, the reaction rate is determined by the ionization energy of the cluster. Hydrogen adsorption on the Fe clusters [f] was reported to be determined by such an electronic factor. There still exists some controversy on this subject because many examples also suggest the importance of a geometric factor to the reaction rate.

In the Si clusters, Si atoms are bound by strong covalent bonds. Therefore, sodium doping to silicon clusters may not perturb the geometric structure or genuine silicon clusters. Instead, due to the electron migration from the doped Na atom to Si clusters, the electronic factor may be strongly perturbed. This assumption well explains the experimental result in Figure 6.

In order to confirm this assumption, theoretical calculation was conducted on the geometric and electronic structures of Si_nNa clusters.[8] It was concluded from the calculation that the structure of the stable Si_nNa keeps the frame of the corresponding Si cluster unchanged and that the electronic structure of Si_nNa is similar to that of the corresponding genuine cluster anion Si_n^-.

5. Photoelectron spectroscopy of Al_nS^- cluster anions [9,10]

Recently, photoelectron spectroscopy of anions has become a well-established technique for studying the size dependent geometric and electronic structures of metal and semiconductor cluster anions, because it is possible to select the cluster size by mass spectrometric means. The photodetachment process is the transition

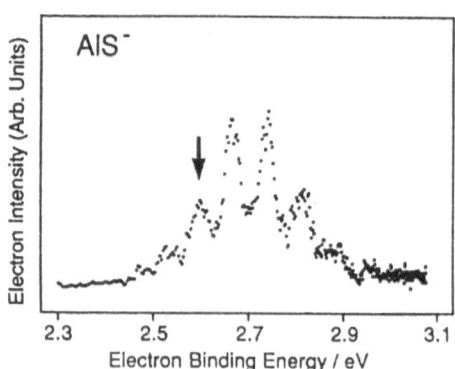

Fig. 7   Photoelectron spectrum of AlS^- at 2.5 eV region. Downward arrow indicates the origin band of the neutral ground state.

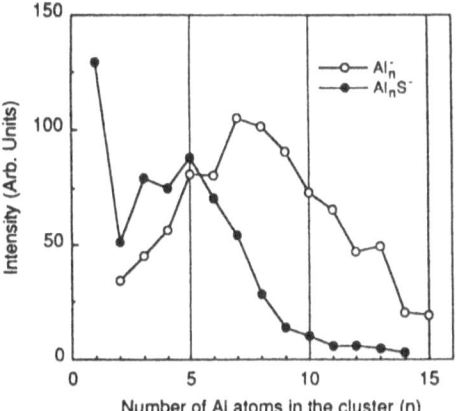

Fig. 8 Mass abundance of Al_nS^- and Al_n^-.

from the ground state of the anions to an eigenstate of the corresponding neutral. Thus, the photoelectron spectra give information on the neutral states of the clusters. As described in the experimental section, we have constructed a magnetic bottle-type photoelectron spectrometer for size-selected cluster anions. In this section, we discuss the results of AlS^- and Al_nS^-.

Figure 7 shows the photoelectron spectrum of AlS^- detached by 355 nm (3.49 eV) photon. A vibrational progression corresponding to the neutral ground state (640 cm^{-1}) is clearly observed. At the same time, we conducted REMPI measurement of neutral AlS and determined the vibrational temperature to be 500 K. Also combined with *ab initio* calculation, the origin band of the neutral AlS was determined to be located at 2.6 eV, corresponding to the electron affinity of AlS. Moreover, by using the 266 nm photon as the detachment light, the low lying excited electronic state which is located 0.4 eV above the ground state of the neutral AlS was discovered. The existence of this excited state was well confirmed by *ab initio* calculation.

Figure 8 depicts the abundance of the Al_nS^- and Al_n^- cluster anions obtained by intensities of respective mass peaks. As is clear from the figure, the abundance of Al_nS^- has a maximum at $n = 5$ and decreases above $n = 6$, and at $n > 10$ the abundance becomes very small. Theoretical calculation and electron affinity measurement suggest that the S atom, which is known to be a hypervalent atom, can bridge between 5 Al atoms at most. Then by increasing the number of Al atoms in the cluster by one from 5, a sudden structural change takes place. The stable structure of the cluster for $Al_nS(n > 5)$ is the structure where the S atom bridges the Al_5 cluster and n-5 Al atoms. This reduces the electron affinity. The cluster size-dependence of the electron affinity for Al_nS clusters behaves almost similarly with true mass abundance pattern in the figure, which also confirms the above conclusion.

PUBLICATIONS

1. A. Nakajima, K. Hoshino, T. Naganuma, Y. Sone and K. Kaya, *J. Chem. Phys.*, **95**, 7061 (1991).
2. K. Hoshino, K. Watanabe, Y. Konishi, T. Taguwa, A. Nakajima and K. Kaya, *Chem. Phys. Lett.*, **231**, 499 (1994).

3. A. Nakajima, T. Taguwa, K. Hoshino, T. Sugioka, T. Naganuma, F. Ono, K. Watanabe, K. Nakao, Y. Konishi, R. Kishi and K. Kaya, *Chem. Phys. Lett.*, **214**, 22 (1994).
4. A. Nakajima, T. Taguwa and K. Kaya, *Chem. Phys. Lett.*, **221**, 436 (1994).
5. A. Nakajima, K. Hoshino, K. Watanabe, Y. Konishi, T. Kurikawa, S. Iwata and K. Kaya, *Chem. Phys. Lett.*, **222**, 353 (1994).
6. A. Nakajima, K. Hoshino, T. Sugioka, T. Naganuma, T. Taguwa, Y. Yamada, K. Watanabe and K. Kaya, *J. Phys. Chem.*, **97**, 86 (1993).
7. K. Kaya, T. Sugioka, T. Taguwa, K. Hoshino and A. Nakajima, *Z. Phys. D*, **26**, 201 (1993).
8. R. Kishi, A. Nakajima, S. Iwata and K. Kaya, *Chem. Phys. Lett.*, **224**, 200 (1994).
9. A. Nakajima, T. Taguwa, K. Nakao, K. Hoshino, S. Iwata and K. Kaya, *Chem. Lett.*, 1525 (1994).
10. A. Nakajima, T. Taguwa, K. Nakao, K. Hoshino, S. Iwata and K. Kaya, *J. Chem. Phys.*, **102**, 660 (1995).

REFERENCES

a. V. E. Bondebey and J. H. English, *J. Chem. Phys.*, **74**, 6978 (1981).
b. D. E. Powers, S. G. Hansen, M. E. Geusac, A. C. Pulu, J. B. Hopkins, T. G. Dietz, M. A. Duncan, P. R. P. Langridge-Smith and R. E. Smalley, *J. Phys. Chem.*, **86**, 2556 (1982).
c. M. Y. Chou and M. L. Cohen, *Phys. Lett.*, **A 113**, 420 (1986).
d. R. E. Leuchtner, A. C. Harms and A. W. Castleman, Jr., *J. Chem. Phys.*, **94**, 1093 (1991).
e. O. Chesnovsky, S. H. Yang, C. L. Pettiette, M. J. Craycraft, Y. Liu and R. E. Smalley, *Chem. Phys. Lett.*, **138**, 119 (1987).
f. R. L. Whetten, D. M. Cox, D. J. Travor and A. Kaldor, *Phys. Rev. Lett.*, **54**, 1494 (1985).

30

Formation and Application of High Intensity Cluster Beams
—Study of reactions of cluster ions with O₂ molecules—

Nobuo Kobayashi

Department of Physics, Tokyo Metropolitan University
Hachioji-shi, Tokyo 192-0364, Japan

Purpose of the present study

We have aimed to study formation mechanism and physical and chemical properties of various clusters by producing high intensity cluster beams in vacuum. In this work, we have conducted three different experiments: (1) reactions of size-selected cluster ions with molecules by using an rf ion trap combined with TOF mass spectrometer, (2) magnetic property of small carbon clusters by employing hexapole magnets [1] and (3) stability of helium cluster ions with an injected-ion drift tube cooled by liquid helium.[2-4] In this report, the studies of chemical reactions of cluster ions with O_2 molecules are described.

Contents of the present study

1. Introduction

Ion-molecule reactions of microclusters are widely studied because we can obtain not only fundamental quantities of reactivity of clusters but also useful information about their geometric or electronic structure. Fourier-transform ion cyclotron-resonance mass spectrometry (FTMS) is a useful technique and widely used for the study of chemical reactivities of cluster ions. However, the apparatus is not easy to construct in a laboratory because it uses a strong magnetic field.

On the other hand, an rf ion trap is easily constructed and has the great advantage of storing ions for so long a time that we can determine a rate constant of a very slow reaction. Furthermore, the rf trap has the capability of confining ions with a specific mass selectively. Therefore, we have tried to apply an rf ion trap to the study of ion-molecule reactions of size-selected clusters.

In this work, we have measured the rate constants for the reactions of positively and negatively charged carbon clusters and carbon-metal and carbon-silicon binary clusters with O_2 molecules. The presence of isomers and the geometrical structures have been discussed.

2. Principle of the rf ion trap

Paul[a] developed an rf trap which consists of three electrodes in the shape an axially symmetric ellipsoid. The electrodes are schematically shown in Fig. 1. By

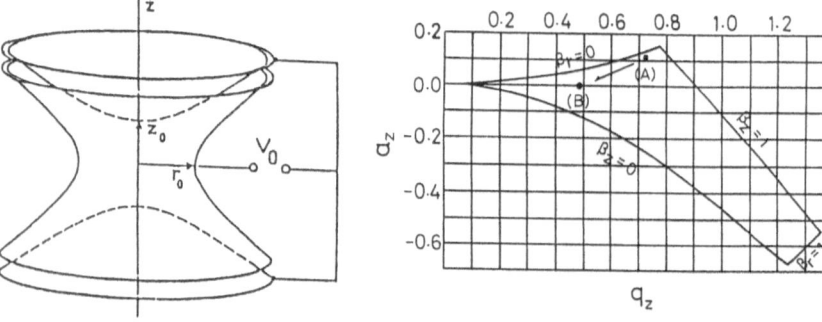

Fig. 1 Electrodes of an rf ion trap.   Fig. 2 Stability diagram of an rf ion trap.

applying an alternating current voltage ($V\cos \Omega t$) and a direct current voltage (U) to the electrodes, ions are stored at the center of the trap. The electrodes in the middle and up and down are called ring electrode and end caps, respectively. When the parameters given by

$$a_z = -16qU/m\Omega^2 (r_0^2 + 2z_0^2),$$ (1)

$$q_z = 8gV/m\Omega^2 (r_0^2 + 2z_0^2),$$ (2)

where q and m are the charge state and the mass of the ion, respectively, are the values inside the stability region shown in Fig. 2, ions can be confined in the trap. The $2z_0$ is 16 mm and $r_0 = 11.3$ mm, that is, $r_0^2 = 2z_0^2$. If the parameters a_z and q_z are tuned to the value located at the corner of the stability diagram like position (A), only the ions with specific mass are stored in the trap whereas others are ejected from the trap. Therefore, we can selectively trap cluster ions of a particular size among those of various sizes by tuning the values of U, V and Ω. However, if we keep these values constant, the product ions by the reactions with molecules are also ejected from the trap. So, we must tune again the values of U, V and Ω just after the size selection is performed to shift the parameters a_z and q_z to the values at the point like (B). After that, both primary and product ions can be stored in the trap.

3. Experimental procedure

A schematic diagram of the apparatus is shown in Fig. 3. The trap electrodes are settled in an ultrahigh vacuum chamber evacuated with a 300 l/s TMP. Ultimate pressure is 1×10^{-9} Torr. Cluster ions are produced by laser ablation of a sample disk settled outside of the trap electrodes and drift toward the center of the trap through a hole at the ring electrode. By applying rf and dc fields to the ring electrode at the time cluster ions arrive at the center of the trap, the cluster ions with a specific size are selectively confined. Then, the rf and dc fields are modified to a reaction mode. Reactant gas is filled in the vacuum chamber at a pressure of 5×10^{-7}–2×10^{-5} Torr according to the reactivity of the selected ions and the reactant gas. One of the end caps is made of gold-plated tungsten mesh. After a specific reaction time, the all ions stored are ejected from the trap by

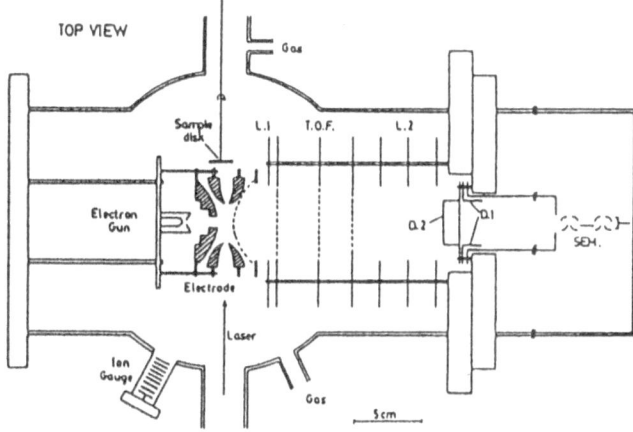

Fig. 3 Schematic diagram of the apparatus.

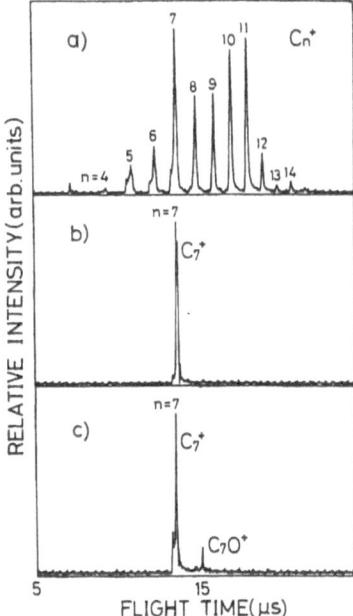

Fig. 4 TOF spectra of carbon cluster ions (a) without size-selection, (b) with size-selection and (c) after the reaction of C_7^+ with O_2.

applying a repelling pulse to the other end cap and then accelerated to measure mass spectra by TOF method.

Figure 4 shows the TOF spectra obtained for carbon cluster ions produced with graphite disk without size selection (a), with size selection (b) and after the reaction of C_7^+ with O_2 (c). Various TOF spectra are measured by varying the reaction time, and the attenuation curve of the parent ions, intensity ratios of the parent ions to the total ions, is plotted versus the reaction time (T_r). If the parent cluster ions consist of a single isomer, the number of the ions in the trap is given by

$$N = N_0 \exp(-knt),\qquad\qquad(3)$$

when N_0 is the number of ions at $t = 0$, k is the rate constant of the reaction and n is the density of the reactant gas. Therefore, we can determine the rate constant from the measured attenuation curve of the parent ions. On the other hand, if the parent ions consist of two isomers with the fractions of f and $(1 - f)$ and the rate constants for both species are k_1 and k_2, the attenuation curve is given by

$$N/N_0 = f \exp(-k_1 nt) + (1 - f) \exp(-k_2 nt).\qquad\qquad(4)$$

By fitting the formula to the measured attenuation curve, we are able to obtain the rate constants together with the relative populations of each component of the isomers.

4. Reaction of C_n^+ ($n = 4, 5$) and C_n^- ($n = 4$–8) with O_2

An ion trap method has another advantage in that in both positively and negatively charged ions can be stored. Since the cluster beam produced by laser ablation contains positive and negative cluster ions, both are confined in the trap.[5] If the densities of both species are high, it is possible that the existence of different charge states affects the measurements of the reaction rate constants. However, in the present case, the densities are so low that the experiments are not influenced.

First, we measured the rate constants for the reaction of C_n^+ ($n = 4$ and 5) with O_2 to test the ability of this new technique. The measured values were compared with those previously obtained by the other technique. The rate constants obtained for C_4^+ and C_5^+ are 1.8×10^{-10} and 2.1×10^{-10} (cm^3/s),[6] which agree well with those determined by McElvany et al.[b] with the FTMS.

Since the mass analysis of the product ions in this experiment is made by the TOF method after ejection of stored ions, the reaction can be studied under relatively higher pressure (10^{-6}–10^{-5} Torr) compared to the case of FTMS (typically $\sim 10^{-7}$ Torr). Thus, the rf trap method is suitable for the study of reactions with lower reactivity. We have tried to the rate constants of reactions of C_n^- with O_2 which are very slow.

The rate constants obtained for C_n^- are 4.3×10^{-12}, 1.2×10^{-12} and 8.8×10^{-13} (cm^3/s) for $n = 4$, 5 and 6, respectively.[6] For $n = 7$ and 8, the rate constants were too small to measure in this experiment. Characteristic feature seen for $n = 4$, 5 and 6 is that only the addition reaction producing $C_n O^-$ was observed for even numbers, $n = 4$ and 6, while for an odd number, $n = 5$, the abstraction reaction forming C_4^- was observed. From this result, it has been suggested that the cluster ions of even numbers are more stable compared with those of odd numbers.

5. Reaction of $C_n Sc^+$ ($n = 2$–6) with O_2

$C_n Sc$ clusters have attracted great interest since scandium atoms were first observed to be encapsulated in fullerenes.[c,d] Although the carbon-scandium binary clusters are considered to be absolutely necessary in producing the Sc-a)

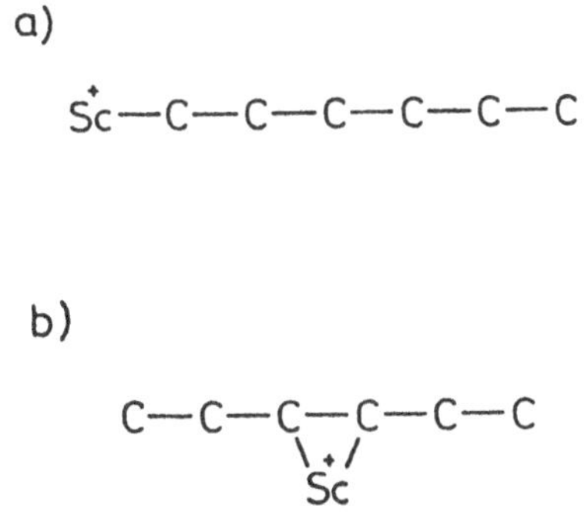

$$\overset{+}{Sc}-C-C-C-C-C-C$$

b)

$$C-C-\underset{\underset{Sc}{\backslash\,_+\,/}}{C}-C-C-C$$

Fig. 5 Possible geometric structures.

Table 1 Rate constants and branching ratios for the reaction of $C_nSc^+ + O_2$

| | | | n | | |
	2	3	4	5	6
k^*	0.12	1.00	0.81×10^{-2}	0.47	0.024
branching ratios	ScO^+: 1.00	ScO^+: 0.13	ScO^+: 1.00	ScO^+: 0.58	ScO^+: 0.09
		ScO_2^+: 0.87		ScO_2^+: 0.42	
					C_4Sc^+: 0.09

*The rate constants are normalized by that for $C_3Sc^+ + O_2$. The absolute value for C_3Sc^+ is 4.7×10^{-10} cm³/s.

encapsulated fullerene, C_nSc clusters have not been studied very well. Therefore, we studied the reaction of C_nSc^+ ($n = 2$–6) with O_2 molecules.[7]

The product ions observed are ScO^+ for all sizes of cluster ions and ScO_2^+ only for odd n. In case of the reaction of C_6Sc^+ ions, the C_4Sc^+ product was observed in small amounts. The product ions C_nScO^+ could not be observed. It must be noted that all product ions result from Sc abstraction in all reactions except for C_6Sc^+, in which C_3Sc^+ were also observed.

The reaction rate constants are obtained from the slopes of attenuation curves of the parent cluster ions. Estimated reaction rate constants of $C_nSc^+ + O_2$ for $n = $ 2–6 and branching ratios of ScO^+ and ScO_2^+ are listed in Table 1. Alternation in reactivities with even-odd n is clearly seen. Odd species have much larger rate constants than even ones.

Although the geometric structure of C_nSc^+ clusters has not yet been determined, the linear chain structure with a Sc atom combined with the terminal carbon atom shown in Fig. 5(a) seems to be an important candidate. The other type of structure, Sc attached onto the side of a C-C bond of a carbon linear chain, shown in Fig. 5(b), is also feasible. It is likely that the alternation of the rate constants with even-odd n can be explained by the existence of different geometric structure (end or side) for n even and odd as described above.

The possible presence of isomer ions is suggested for C_6Sc^+. The rate constants listed in Table 1 are estimated assuming a single component of the attenuation curve of the parent ions. However, in the case of C_6Sc^+, a better fit can be attained with the assumption of two components.

6. Reaction of $Si_mC_n^+$ with O_2

Silicon-carbon binary clusters have been widely studied in experimental and theoretical works, since they were found in interstellar clouds and have the possibility of chemical application. In addition, the similarities and disparities among neat carbon and silicon clusters and binary ones seem quite important in connection with the possibility of forming hollow-caged binary clusters. That is, if the structure and reactivity of binary clusters is the same as carbon clusters, the production of binary fullerenes may be realized, since small carbon clusters must play an important role in forming the fullerenes. The structures of carbon clusters are classified into a chain, a ring, an aggregate of rings and a fullerene. On the other hand, silicon clusters form close packed structures. If one or two silicon atom(s) is combined with carbon cluster, what is expected for the reactivity and structures of such the binary clusters?

The reactions of the small Si-C binary cluster ions have been studied using FTMS method. Parent has made detailed study on the reactions of SiC_n^+ ($n \leq 8$) and $Si_2C_n^+$ ($n \leq 6$) with acetylene.[e] From the similarity in the size dependence of the rate constants for SiC_n^+ to C_{n+1}^+, he has suggested that the structure of SiC_n^+ is a chain form. Greenwood et al.[f] have also studied various types of reactions of Si-C binary cluster ions, $Si_mC_n^{+,-}$ ($m \leq 4$, $n > 2$). However, the reactions of these ions with O_2 molecules, which is quite important for the discussion on the air-stability of the clusters, have not been studied, probably due to small reactivity. Therefore, we have applied the rf ion trap with TOF mass spectrometer to the studies of the reactions of SiC_n^+ ($n = 1-6$),[8] SiC_n^- ($n = 2-4$)[8] and $Si_2C_n^+$ ($n = 1-2$)[9] with O_2 molecules.

6.1 SiC_n^+ ($n = 1-6$)

The mass spectra obtained after different reaction times of SiC_3^+ with O_2 are shown in Fig. 6. As seen in the spectra, a small amount of product ions of SiC_3O^+ was observed while the growth of this signal was stopped after ~40 ms reaction time. The attenuation curve of the primary ions could only be fitted by a two-component exponential decay arising from reactive and unreactive ions. The rate constant obtained for the reactive component was determined to be 3.8×10^{-12} cm³/s whereas that for the unreactive one was smaller than the detection limit. The fraction of reactive component was estimated to be only 7%. The two component behavior was also observed for SiC_2^+. Within the reaction time up to 100 ms and O_2 gas pressure up to 2×10^{-5} Torr, the SiC_5^+ was observed to be unreactive.

The measured rate constants for the reactions of SiC_n^+ with O_2 together with the fractions of the reactive and less reactive components of SiC_2^+ and SiC_3^+ are

Table 2 Rate constants for the reaction of $SiC_n^+ + O_2$. Fractions of isomers are given in

n	1	2	3	4	5	6
Rate constants k (cm³/s)	~10^{-10}	1.8×10^{-10} (30%)	3.8×10^{-12} (7%)	8.6×10^{-11}	$< 10^{-13}$	1.1×10^{-10}
		3.3×10^{-12} (70%)	$< 10^{-13}$ (93%)			

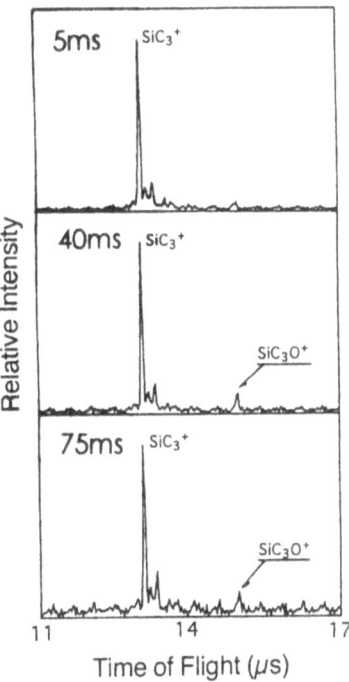

Fig. 6 TOF spectra for the reaction of $SiC_3^+ + O_2$.

summarized in Table 2. A strong odd-even alternation is observed in reactivity with O_2, which is in contrast to that observed for the carbon cluster ions. In addition, only the O atom adduct products, SiC_nO^+, have been observed.

One plausible explanation for the observation of the dual rate components is the presence of the SiC_n^+ geometric isomers. Parent[e] has pointed out that the small SiC_n^+ cluster ions most likely have linear structures with one terminal Si atom, based on the similarity between the reaction of SiC_n^+ and that of C_{n+1}^+ with acetylene. Thus, the reactive SiC_2^+, SiC_3^+ clusters are assigned to the linear isomers. In the present study, the reactivities of SiC_2^+ (higher value), SiC_4^+ and SiC_6^+ are almost the same as those of C_n^+. This result also indicates that these species could have linear structures terminating with one Si atom.

The observation of less reactive (or unreactive) isomers seems quite important since it implies the presence of a non-linear structure. The observation that the SiC_5^+ isomer did not react indicates that the larger cluster ions could terminate their reaction sites. According to a hypothesis proposed by Parent,[e] the reaction site of linear form SiC_n^+ should be a carbene, so the unreactive SiC_n^+ isomers are expected to have a structure without a carbene end, and most likely have a cyclic form.

It is interesting to compare the results for ScC_n^+ mentioned above and SiC_n^+. As in the SiC_n^+ case, an odd-even alternation in the rate constants has been observed for ScC_n^+. However, the size dependence was completely different. The cluster ions of odd n (ScC_3^+ and ScC_5^+) were much more reactive than those of even n. In contrast to the SiC_n^+ case, the Sc atom abstraction reaction occurred exclusively. Since the reaction likely occur at the scandium site, the addition of a Sc atom onto the carbon cluster does not inhibit the reaction. On the other hand, for SiC_n^+, the absence of an abstraction reaction implies that the Si atom forms a strong Si-C bond and also strengthens the C-C bond.

6.2 Si_2C^+ and $Si_2C_2^+$

As a serial study of the reaction of Si-C binary cluster ions, we have measured the rate constants and product ion species for the reactions of $Si_mC_n^+$ ($m = 1$ and 2, $n = 1$ and 2) with O_2 molecules.[9] Here, we present the results of Si_2C^+ and $Si_2C_2^+$, since those for SiC^+ and SiC_2^+ were mentioned above. Some striking features have been observed.

For the reactions of SiC^+, the product ions SiO^+ (70%), Si^+ (30%) and SiC^+ (very weak), which were produced by abstraction reaction of the parent ions, were observed, while no product ions by addition reaction was detected similar to the case of Si_3^+ previously reported by Creasy et al.[8] In addition, the rate constant determined in this work, 2.3×10^{-11} cm^3/s, is very close to that of Si_3^+ [8] but quite different from that of C_3^+ [b] By the theoretical work of Grev and Schaefer,[h] a bent structure was predicted with a Si-C-Si bond angle greater than 100° for the ground state of Si_2C. The present result suggests a similar geometric structure for Si_2C^+ ions.

In the case of $Si_2C_2^+$, surprisingly various product ion species were observed. TOF spectra obtained at different reaction times are shown in Fig. 7. The product ions and the branching ratios were SiC^+ (0.35), Si_2C^+ (0.35), $SiCO^+$ (0.1), SiC_2^+ (0.1), SiO^+ (0.05), Si_2O^+ (0.05), Si_2O^+ (0.05) and Si^+ (< 0.01). The attenuation curve consists of a single exponential and the rate constant was determined to be

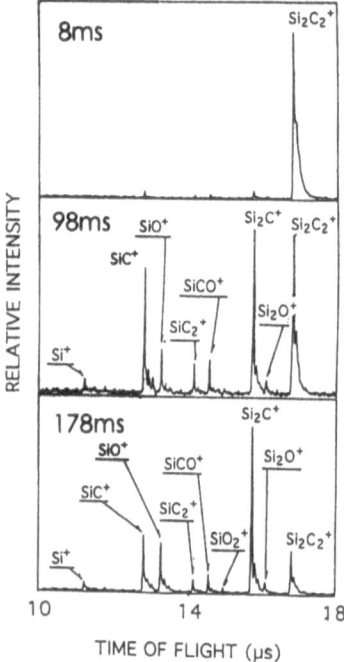

Fig. 7 TOF spectra for the reaction of $Si_2C_2^+ + O_2$.

1.8×10^{-10} cm^3/s.

It seems appropriate to compare the reactivity of $Si_2C_2^+$ with that of neat cluster ions Si_4^+ and C_4^+. For Si_4^+, it has been reported that the geometric structure is rhombic and Si_2O^+ and Si_2^+ are formed by the abstraction reaction with O_2.[g] On the other hand, for C_4^+ it is known that the cluster has a linear structure and C_4O^+ as well as C_2^+ are produced as mentioned above. This suggests that the structure of $Si_2C_2^+$ is similar to that of Si_4^+. The present result that all the product ions from $Si_2C_2^+$ were produced by the abstraction reaction has a strong similarity with the case of Si_4^+ having rhombic, rhomboidal or a more closely packed structure.

Calculations for Si_2C_2 have shown that the rhombic structure is the most stable and the linear (Si-C-C-Si) and the rhomboidal structures are close in energy.[i] The linear structure should not be favorable since the formation of Si_2C^+, observed to be one of main channels, requires that the C atom be abstracted following by recombination of Si and SiC. In addition, the formation of Si_2O^+ implies that the rhombic structure is also unfavorable since the two Si atoms in this structure are not combined directly. Therefore, the rhomboidal structure seems most likely. In this case, all the product ions observed can be formed.

From the similarity and disparity of the reactions of $Si_2C_2^+$ to those of Si_4^+ and C_4^+ and considering the structures theoretically predicted for neutral Si_2C_2 cluster, the rhomboidal structure is strongly suggested for the $Si_2C_2^+$ cluster ions studied in this work. On the other hand, from the fact that various kinds of product ion species have been observed, it seems natural to expect the existence of isomers. The attenuation curve of the parent ions is, however, obviously explained by a single exponential for a wide range of reaction time. Therefore, it is most likely that the $Si_2C_2^+$ cluster ions studied here consists of a single isomer with rhomboidal structure, although the presence of plural isomers with coincidentally the same reactivity is possible.

7. Conclusion

An rf ion trap has been successfully applied to the study of reactions of size-selected cluster ions with molecules in combination with a TOF mass spectrometer. We have demonstrated that it is useful to measure the rate constants of very slow reactions such as C_n^-- O_2 systems as well as various kinds of reactions. By using this new technique we have studied the reactions of positively and negatively charged carbon-metal and -silicon binary clusters with O_2 molecules. Rate constants and branching ratios of the product ions have been determined. The presence of geometric isomers has been suggested for some clusters from the structure observed in the attenuation curves of parent ions. The geometric structures of cluster ions studied have been discussed.

PUBLICATIONS

1. H. Shiromaru, M. Mizumachi, Y. Achiba, N. Yanase, N. Kobayashi and Y. Kaneko, *Physics and Chemistry of Fine Systems; From Clusters to Crystals*, **Vol. 1**. (Pe Jena et al., eds.), p.259, Kluwer Academic Publishers (1992).
2. T. M. Kojima, N. Kobayashi and Y. Kaneko, *Z. Phys.*, **D22**, 645 (1992).
3. T. M. Kojima, N. Kobayashi and Y. Kaneko, *Z. Phys.*, **D23**, 181 (1992).
4. H. Tanuma, M. Sakamoto and N. Kobayashi, *Surf. Rev. Lett.*, **3**, 205 (1996).
5. N. Watanabe, H. Shiromaru, N. Kurihara, Y. Achiba, N. Kobayashi, Y. Kaneko and J. Yoda,

Nucl. Instr. and Meth., **B69**, 385 (1992).

6. N. Watanabe, H. Shiromaru, Y. Negishi, Y. Achiba, N. Kobayashi and Y. Kaneko, *Z. Phys.*, **D26**, S252 (1993).

7. N. Watanabe, H. Shiromaru, Y. Negishi, Y. Achiba, N. Kobayashi and Y. Kaneko, *Chem. Phys. Lett.*, **207**, 493 (1993).

8. Y. Negishi, A. Kimura, N. Kobayashi, H. Shiromaru, Y. Achiba and N. Watanabe, *J. Chem. Phys.*, **103**, 9963 (1995).

9. Y. Negishi, N. Watanabe, H. Shiromaru, Y. Achiba and N. Kobayashi, *Surf. Rev. Lett.*, **3**, 661 (1996).

REFERENCES

a. W. Paul and H. Steenwedel, *Zeit. Naturforsh.*, **89**, 448 (1953).

b. S. W. McElvany, B. I. Dunlap and A. O'Keefe, *J. Chem. Phys.*, **86**, 715 (1987).

c. S. Suzuki, S. Kawata, H. Shiromaru, K. Yamauchi, K. Kikuchi, T. Kato and Y. Achiba, *J. Phys. Chem.*, **96**, 7159 (1992).

d. H. Shinohara, H. Sato, M. Ohkuchi, Y. Aundo, T. Komada, T. Shida, T. Kato and Y. Saito, *Nature*, **357**, 52 (1992).

e. D. C. Parent, *Int. J. Mass Spectrom. Ion Processes*, **116**, 257 (1992).

f. P. F. Greenwood, G. D. Willett and M. A. Wilson, *Org. Mass Spectrom.*, **28**, 831 (1993).

g. W. R. Creasy, A. O'Keefe and J. R. McDonald, *J. Phys. Chem.*, **91**, 2848 (1987).

h. R. S. Grev and H. F. Schaefer III, *J. Chem. Phys.*, **82**, 4126 (1985).

i. K. Lammertsma and O. F. Gruner, *J. Am. Chem. Soc.*, **110**, 5239 (1988).

31

Cluster Ions Emitted from a Liquid Metal Ion Source

Yahachi Saito[a], Kazuhiro Mihama[b] and Tamotsu Noda[c]

[a] Department of Electrical and Electronic Engineering, Mie University
1515 Kamihama-cho, Tsu 514-8507, Japan
[b] Department of Applied Electronics, Daido Institute of Technology
Minami-ku, Nagoya 457-0811, Japan
[c] Toyohashi Junior College
Ushikawa-cho, Toyohashi 440-0016, Japan

Purpose of the present study

Mass spectrometry has been one of the primary tools used to probe size dependent cluster properties, *e.g.*, size distribution, ionization energy, chemical reactivity. In the measurement of cluster abundances, local maxima or discontinuous variations at certain specific cluster sizes are frequently encountered, and these sizes are refereed to as "magic numbers." Clusters are normally ionized prior to detection. Depending on the ionization process more than two electrons can be removed, leaving the cluster multiply charged. It has been found that multiply charged clusters are observable only above a threshold size ("critical size"). A small multiply charged cluster will disintegrate into small fragments if the repulsive Coulomb forces between holes exceed the cohesive forces. We have so far studied cluster ions of monovalent (Li, Na), tetravalent elements (Si, Ge, Sn, Pb), and alloys (Li-Na, Li-Mg, Li-Al) by using a liquid metal ion source (LMIS). Since this source generates "born ionized" clusters, there is no need to ionize clusters for the following mass analysis. Therefore, in an interpretation of cluster abundances we are free from onerous problems due to size-dependent ionization cross-section and fragmentation pattern upon ionization. In the present study, we focused our attention on trivalent metals (Al, Ga, In), Bi and doubly-charged small Mg clusters. Collisional dissociation experiments were also performed for Bi in order to separate doubly charged even-numbered clusters Bi_{2n}^{2+} from Bi_n^+ and to study the stability of Bi clusters.

Contents of the present study

1. Liquid metal ion source and mass spectrometry

Figure 1 shows a schematic drawing of the ion source used in the present experiment. This was designed by Noda *et al.*[1] and has the following characteristics: (1) the source material is enclosed within a capped reservoir, (2) the reservoir and the needle tip are heated by electron bombardment, and (3) the needle can be moved back and forth from outside the vacuum. Because of the closed type reservoir, waste of source material through thermal evaporation is suppressed. Bake-out of the emitter tip by electron bombardment and the mechanical movement of the tip facilitate wetting of the source material over

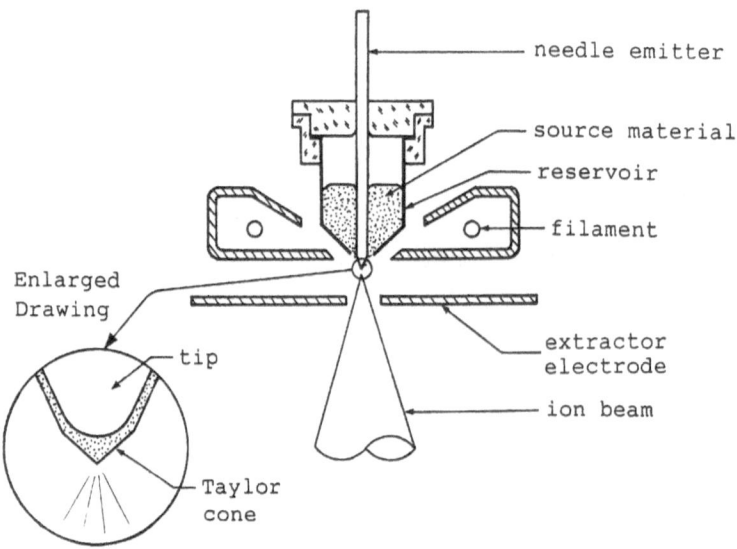

Fig. 1 Schematic drawing of a liquid metal ion source (LMIS) used.

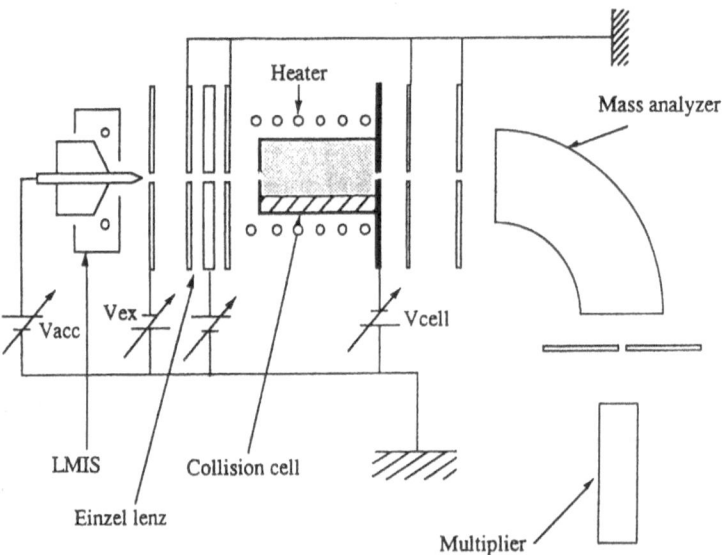

Fig. 2 Magnetic mass analyzer equipped with a LMIS and a collision cell.

the tip surface. A thin film of liquid metal covers the sharpened needle. Refractory metals, such as W and Fe, are employed as the needle emitter. However, the needle material to be employed depends on the liquid metal. There must be sufficient interaction between the needle and liquid metal for wetting to occur, but the solubility of the needle material in the melt must not be so high that substantial erosion of the needle occurs.

The liquid metal film is subjected to a high electric field held by the application of a positive high voltage, typically 4 to 10 kV, to the emitter relative to the extractor electrode (usually grounded). Under such conditions, the liquid surface is pulled outward due to electrostatic force. With increasing the strength of electric field near the tip, the liquid film protrudes outward and forms a cone (called the "Taylor cone").[a]

In practice, the electric fields near the tip of the protrusion are sufficiently large to cause emission of ions from the liquid phase. The mechanism of ion emission is not yet well solved, being a subject of considerable scientific debate. It is now believed that for total emission currents of 1 to 2 μA, the principal mechanism is field evaporation of ions from the end of the cone. At a current above a few μA, however, the emission mechanism is more chaotic and dynamic; the surface of the liquid no longer has a well-defined radius, but instead, pieces of the cone may be torn off as ionized clusters or droplets. According to Joyes and Van de Walle,[b] most of cluster ions observed in mass spectra would be stable fragments which remain after the disintegration process of highly charged and vibrationally excited droplets.

In the present experiments, the mass distribution of ions from the LMIS was measured by a single-focusing magnetic analyzer (magnetic radius 20 cm). For collisional dissociation experiments, a collision cell (3 cm in length and 1 cm in diameter) was inserted between the LMIS and the mass analyzer as shown in Fig. 2. All the ions emitted from the LMIS enter the cell with kinetic energy from 3 to 6 keV. The ions scattered in the forward direction from the cell was analyzed by the magnetic analyzer. Vapor of Mg and Pb was used as target gases.

2. Trivalent metals: Al, Ga, In

Figure 3 shows the mass spectrum of ions ejected from an aluminum LMIS. [2] Because of rapid dissolution of refractory metal needle supports, e.g., W and Ti, by molten Al, it is necessary to employ a nonmetallic needle. A carbon needle prepared from pencil lead was used as the emitter for Al. Carbon is fairly resistant to attack by molten Al, but Al compound clusters were always detected besides

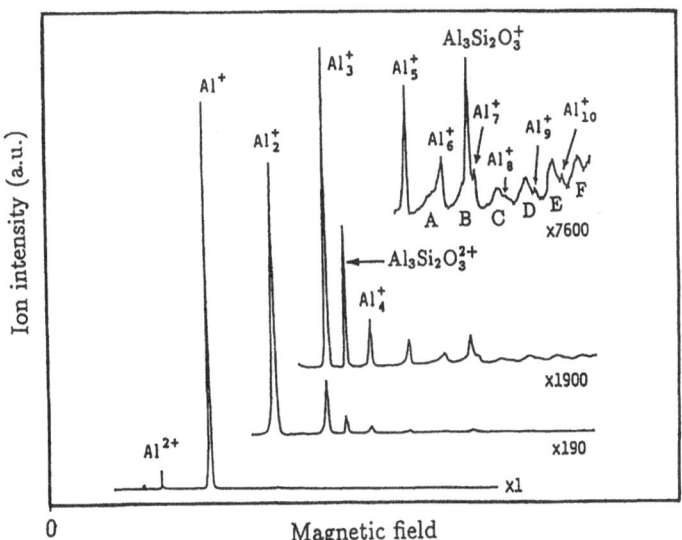

Fig. 3 Mass spectrum of Al clusters emitted from LMIS.
(From K.Sakaguchi, K.Mihama and Y.Saito, *J.Appl.Phys.*, **70**, 5051 (1991)).

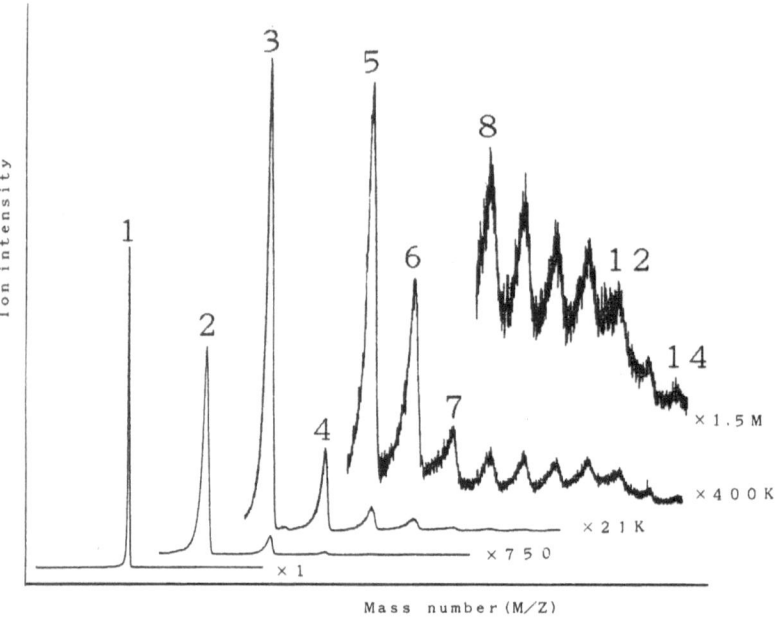

Fig. 4 Mass spectrum of In_n^+ clusters emitted from LMIS.

pure Al clusters. The intensity of $Al_3Si_2O_3^{2+}$ and $Al_3Si_2O_3^+$ was prominently high. Si and O were included in the pencil lead as impurities. Concerning pure Al clusters, singly charged Al_n^+ up to $n = 10$ were observed, but no multiply charged clusters were detected within the present mass-to-charge range. The intensity of Al_n^+ dropped after $n = 7$. This magic number can be explained by the electronic shell model[c]; 20 valence electrons in Al_7^+ just fill shells ($1s^2\ 1p^6\ 1d^{10}\ 2s^2$).

For Ga, singly charged clusters up to $n = 14$ were detected, and a local maximum at $n = 10$ was observed. In contrast to the present result, Barr[d] observed Ga_n^+ with n ranging up to 30 using a time-of-flight mass spectrometer, and found a few features in size distribution: a local maximum at $n = 7$, and a sudden drop in intensity at $n = 14$–16. These features in mass spectra cannot be explained simply by the shell model.

For In, sudden drops in ion intensity were round after $n = 6$ and 12 within a range of cluster observed up to $n = 14$ (Fig. 4), although the intensity decreases monotonically with increase of cluster size. Both sizes ($n = 6$ and 12) are smaller by one than those expected from the shell model for trivalent metals. The reason for this discrepancy is not yet understood.

3. Doubly charged Mg clusters (Mg_2^{2+} and Mg_3^{2+})

Among the ions ejected from an Li-Mg alloy LMIS, doubly charged diatomic and triatomic Mg clusters were observed.[3] Figure 5 shows a mass spectrum in a mass range around Mg^+. Peaks with half-integer mass-to-charge ratios 24.5 and 25.5 are observed, indicating the presence of doubly charged Mg dimers. The observed mass pattern between $m/z = 24$ and 26 cannot be explained by other doubly charged clusters, including heteronuclear clusters $Mg_xLi_y^{2+}$. Assuming the

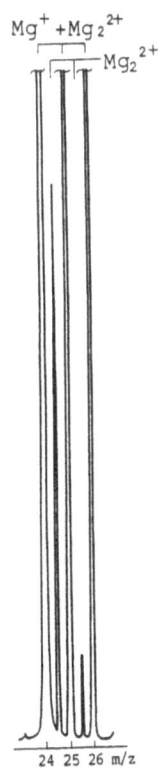

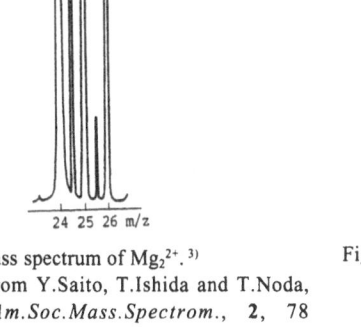

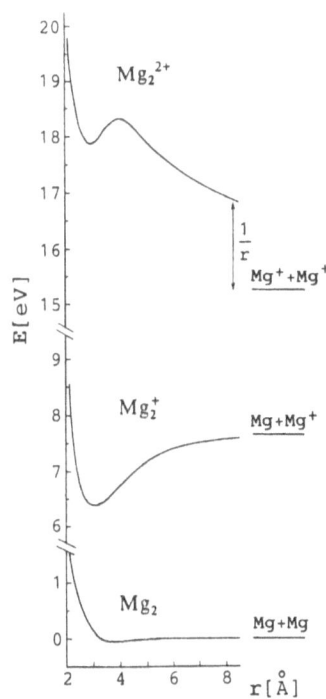

Fig. 5 Mass spectrum of Mg_2^{2+}.[3]
(From Y.Saito, T.Ishida and T.Noda,
J.Am.Soc.Mass.Spectrom., **2**, 78
(1991)).

Fig. 6 Potential energy curves of Mg_2, Mg_2^+ and
Mg_2^{2+}.

natural abundance of isotopes of Mg, we could reproduce well the mass pattern
by the mixture of Mg_2^+ and Mg^+ with intensity ratio $^{24}Mg_2^{2+}$ / $^{24}Mg^+$ = 0.06.
Therefore, it was concluded that Mg_2^{2+} was present in the ion beam.

Doubly charged Mg trimers Mg_3^{2+} were also observed. The observation of
Mg_2^{2+} and Mg_3^{2+} is surprising because the Coulomb repulsion between the two
holes (positive charges) should compel such small clusters to explode into two
singly charged fragments.

The free atoms of divalent elements have an s^2 closed-shell atomic
configuration. The atoms may be considered to keep essentially this ground-state
structure in small clusters. On the other hand, bulk Mg has a metallic character
that is due to the overlap between the filled s and empty p bands. Therefore, Mg
clusters smaller than a certain size have van der Waals bonding, and larger clusters
transform to metallic bonding. The dissociation energy D_0 of an Mg dimer in the
ground state is only 0.050 eV. The value of the binding energy per atom for the
bulk, E_B, is 1.53 eV.

The stability of doubly charged Mg dimer and trimer cannot be explained by
empirical models such as a charged liquid-drop and a chain-like cluster. The
metastability of the Mg_2^{2+} and Mg_3^{2+} is explained from the point of view of
molecular orbital theory. Whether the charges are on a single atom or two atoms,
the two s valence electrons of Mg_2^{2+} will find the s bonding molecular orbital and
leave the $(\sigma s)^*$ antibonding orbital empty. This strong bonding should create a

local minimum on the otherwise repulsive potential energy surface. Similar arguments apply to Mg_3^{2+}.

Using a model Hamiltonian including the polarization energy and the interactions between induced dipoles, Durand et al.[e] calculated potential energy surfaces of Mg_n^{z+} ($n \leq 5$; $z = 1, 2$) and showed that both Mg_2^{2+} and Mg_3^{2+} have local minima in the potential energy surfaces as shown in Fig. 6.

The existence of local minima in the potential energy surface of Mg_n^{2+} is not sufficient to ensure the observability of these clusters, because these minima are shallow and are significantly higher in energy than the potential energy of $Mg_p^+ + Mg_q^+$ ($p + q = n$) separated to an infinite distance from each other. Clusters might tunnel through the barrier, or, if excited to a state higher in energy than the barrier height, they will break into fragment ions. Another factor that prevents the observation of doubly charged clusters is the great difference between the geometry of the neutral Mg_n and that of Mg_n^{2+} in the local minimum. A vertical double ionization of the neutral species will not lead to the stable doubly charged cluster because it forms the cluster in a region outside the energy barrier.

Durand et al.[e] pointed out that the stable geometries of Mg_n^+ and Mg_n^{2+} are quite similar. This means that ionization from Mg_n^+ to Mg_n^{2+} would lead to a structure close to the bottom of the Mg_n^{2+} local minimum, providing an opportunity for this structure to have a significant lifetime. This theoretical finding indicates that the experimentally observed Mg_2^{2+} and Mg_3^{2+} are not produced by the simultaneous double ionization of free neutral Mg_2 and Mg_3, but rather are formed by ionization of Mg_2^+ and Mg_3^+, respectively. The Mg_2^+ and Mg_3^+ were presumably formed by field-evaporation (as ions) from the liquid surface. The successive ionization is known as postionization. This process successfully

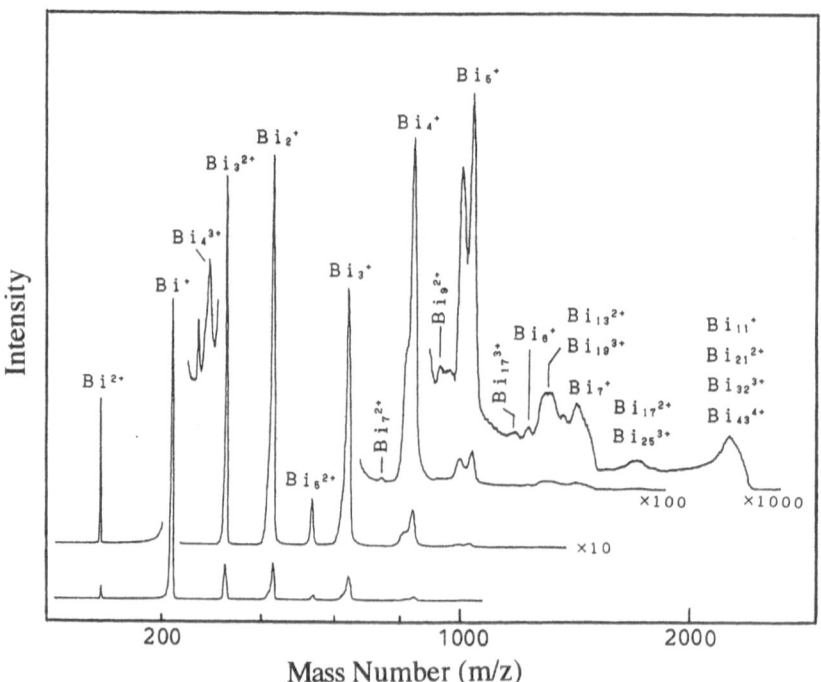

Fig. 7 Mass spectrum of Bi_n^{z+}.

explains the emission of doubly charged monomers from the LMIS for various metals. The present result suggests that postionization occurs for clusters as well as monomers.

4. Bi clusters

4.1 Abundance spectrum of Bi cluster ions emitted from LMIS

Figure 7 shows a mass spectrum of Bi cluster ions emitted from LMIS. Singly-charged clusters Bi_n^+ were observed up to $n = 15$, and multiply charged clusters such as Bi_3^{2+}, Bi_5^{2+} and Bi_4^{3+} were detected. Analyses of energy width and energy deficient of cluster ions suggested that the small clusters less than $n < \approx$ 4 were dominantly formed by field evaporation while the larger clusters of $n > \approx$ 8 were for the most part formed through disintegration of charged droplets. Multiply charges clusters might be formed by field-ionization of singly charged Bi_n^+.

The intensity distribution of singly charged Bi_n^+ shows a step and a local maximum at $n = 3$ and 7, respectively, suggesting that they are magic numbers. The stability of Bi_3^+ and Bi_7^+ can be explained by the shell model as follows. Each Bi atom has three electrons in the outermost 6p orbitals. The total number of valence electrons in a Bi_3^+ cluster becomes 8 ($= 3 \times 3 - 1$), being coincident with the shell closing number for the 1p shell. For Bi_7^+, the total number of valence electrons is 20, corresponding to the closure of the 2s shell. The features in the mass spectrum from the Bi LMIS reflect the stability of Bi_3^+ and Bi_7^+. The magic numbers found in the present study agree with those found for Bi_n^+ clusters produced by ion sputtering [f] and laser vaporization.[g]

4.2 Collisional dissociation of Bi cluster ions

Since Bi is monoisotopic, doubly-charged even-numbered clusters Bi_{2n}^{2+} cannot be separated from Bi_n^+ by mass spectrometry. Therefore, the mass peaks assigned as Bi_n^+ consist of Bi_n^+ and Bi_{2n}^{2+} (and also Bi_{3n}^{3+} etc., though the contribution may be small). In order to make clear what fraction the multiply-charged clusters contribute to the mass peaks, we carried out a collisional dissociation experiment of the Bi cluster beam. Bi clusters after collisions with gas molecules (Mg or Pb vapor) were analyzed by the magnetic sector analyzer. This experiment also provides us with information on the stability of clusters.

4.2.1 Separation of Bi_2^+ and Bi_4^{2+}

Daughter ions Bi_3^+ which were formed by the reaction, $Bi_4^{2+} \rightarrow Bi_3^+ + Bi^+$, were observed, revealing the presence of Bi_4^{2+} ions. The fraction of Bi_4^{2+} contributing to the mass peak assigned as Bi_2^+ was found to be 4%. This intensity of Bi_4^{2+} corresponds to more than 30% of singly charged tetramer Bi_4^+. The

Table 1 Intensity ratios of Bi_n^{2+} to Bi_n^+

cluster size, n	1	3	4	5
$I(Bi_n^{2+})/I(Bi_n^+)$	0.07	0.9	>0.3	2.5

intensity ratios of Bi_n^{2+} to Bi_n^+ for $n = 1$ to 5 (except for $n = 2$) are summarized in Table 1. It can be seen that the fraction of doubly-charged clusters increases with increase of size within the size range listed in the table.

4.2.2 Dissociation of Bi_n^+ ($n = 3, 4, 5$)

In the dissociation of Bi_3^+, two reactions forming Bi^+ and Bi_2^+ daughter ions were observed. The positive charge in the parent Bi_3^+ is expected to transfer to a fragment with the lower 1st ionization potential (IP_1). Although $IP_1(Bi)$ is lower than $IP_1(Bi_2)$, the reaction of $Bi_3^+ \rightarrow Bi_2^+ + Bi$ predominates.

In the dissociation of Bi_4^{2+} and Bi_5^{2+}, the reactions including the formation of Bi_3^+ daughter ions were predominantly observed irrespective of the target gases (Mg and Pb). This result reflects the high stability of Bi_3^+ ions. Similar results have been obtained by Ross and McElvancy,[f] who used Ar and He as the target gases.

4.2.3 Dissociation of Bi_n^{2+} ($n = 3, 5$)

In the dissociation of Bi_3^{2+}, Bi_2^+ was the major daughter ion, suggesting the following reaction process to be dominant:

$$Bi_3^{2+} + Mg \rightarrow Bi_3^+ + Mg^+ \rightarrow Bi + Bi_2^+ + Mg^+.$$

In the dissociation of Bi_5^{2+}, Bi_3^+ was the major daughter ion, suggesting the following two reactions to be dominant:

$$Bi_5^{2+} + Mg \rightarrow (1)\ Bi_2^+ + Bi_3^+ + Mg$$

and

$$\rightarrow (2)\ Bi_5^+ + Mg^+ = Bi_3^+ + Bi_2 + Mg^+.$$

PUBLICATIONS

1. T. Noda, T. Okutani and H. Tamura, *Ohyo-butsuri* (Applied Physics) **54**, 935 (1985) [in Japanese].
2. K. Sakaguchi, K. Mihama and Y. Saito, *J. Appl. Phys.*, **70**, 5049 (1991).
3. Y. Saito, T. Ishida and T. Noda, *J. Am. Soc. Mass Spectrom.*, **2**, 76 (1991).
4. Y. Saito and T. Noda, *Z. Phys. D*, **19**, 129 (1991).
5. Y. Saito, H. Ito and I. Katakuse, *Z. Phys. D*, **19**, 189 (1991).
6. A. Kajita, M. Kimura, S. Ohtani, H. Tawara and Y. Saito, *J. Phys. Soc. Jpn.*, **60**, 2996 (1991).
7. K. Hata, Y. Saito, A. Ohshita, M. Takeda, C. Morita and T. Noda, *Appl. Surface Sci.*, **76/77**, 36 (1994).
8. Y. Saito, *J. Aerosol Res. Jpn.*, **6**, 224 (1991) [in Japanese].

REFERENCE

a. G. Taylor, *Proc. Roy. Soc. A*, **280**, 383 (1964).
b. P. Joyes and J. Van de Walle, *J. Phys. B*, **16**, 3805 (1985).
c. W. D. Knight, K. Clemenger, W. A. de Heer, W. A. Saunders, M. Y. Chou and M. L. Cohen, *Phys. Rev. Lett.*, **52**, 2141 (1984).
d. D. L. Barr, *J. Vac. Sci. Technol. B*, **5**, 184 (1987).
e. G. Durand, J.-P. Daudey and J.-P. Malrieu, *J. Phys. (Paris)*, **47**, 1335 (1986).
f. M. M. Ross and S. W. McElvancy, *J. Chem. Phys.*, **89**, 4821 (1988).
g. M. E. Geusic and R. R. Freeman, *J. Chem. Phys.*, **89**, 223 (1988).

32

Neutron Irradiation of Carbon Nanoparticles

Atsuo Kasuya [a], Hideki Takahashi [a], Yahachi Saito [b], Toshiaki Mitsugashira [a], Tamaki Shibayama [a], Yoshinobu Shiokawa [a], Isamu Satoh [a], Michiko Fukushima [c] and Yuichiro Nishina [a]

[a] Institute for Materials Research, Tohoku University
   Sendai 980-8577, Japan
[b] Department of Electrical and Electronic Engineering, Mie University
   Tsu 514-8507, Japan
[c] Department of Basic Science, Ishinomaki Senshu University
   Ishinomaki 986-0031, Japan

Purpose of the present study

Carbon nanocapsules containing nanometer-size particles of molybdenum carbide were irradiated by neutrons of flux 10^{14} n/cm²·s for six days. The γ-ray spectrometry shows 10^{10} Bq/g of Mo-99 which is transformed from Mo-89 and decays into Tc-99. Our investigation by transmission electron microscope shows that the nanocapsule is rugged under neutron irradiation and serves as a container for radioactive materials and tracer elements. The contents of capsules on a nanometer scale may be activated and transformed into other nuclei.

Contents of the present study

1. Introduction

Carbon nanocapsules have the unique feature of encapsulating nanometer size materials in a cage composed of a few layers of graphitic sheet arranged in the form of polygon.[1] This feature may provide a variety of scientific and technological applications. As a container, the graphitic sheet is chemically inert, mechanically rugged and low in weight. In our previous work, for example. YC_2 was encapsulated and its electronic and magnetic properties analyzed.[2] It is paramagnetic at room temperature and becomes superconducting below 4 K.[a] Since it hydrolyzes easily in air,[b] one has difficulty in performing detailed long-term measurements, particularly in the form of nanoparticles. Our measurement shows high stability of the sample in air over a period of a year. Carbon nanocapsules can also be applied in the field of nuclear science and engineering. They may serve as a container for radioactive materials or isotope tracers. In order to investigate this possibility, a neutron irradiation experiment has been carried out on nanocapsules containing molybdenum.

2. Sample preparation

The sample of nanocapsules was prepared by a standard dc arc discharge between a negative graphite electrode and a molybdenum-filled positive graphite. The negative electrode has a diameter of 13 mm. The positive has a diameter of

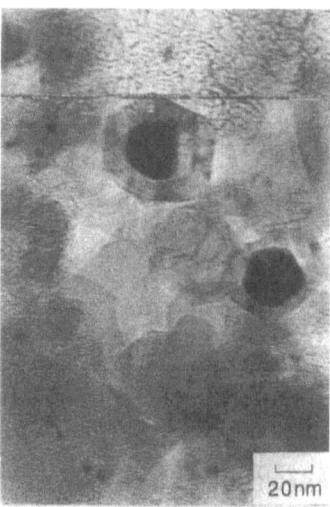

Fig. 1   Transmission electron microscope image of carbon nanocapsules and other deposits produced
by arc discharge of graphite electrodes with Mo filled in the positive electrode.

5 mm with a hole at the center of diameter 3.2 mm and depth 40 mm in which
powder of Mo mixed with graphite was loaded. The discharge took place for 5 min
under 600 Torr of He gas and 70 A of discharge current. Nanocapsules are formed
and deposited on the negative electrode.

Figure 1 shows a typical image of the deposit taken by our transmission
electron microscope. The image shows nanocapsules in the form of polygons
made of several layers of graphitic sheets, in addition to amorphous carbon and
molybdenum particles. The inside of the nanocapsules is filled with MoC_2 as
determined by electron diffraction and X-ray emission analysis. The size of
nanocapsules typically ranges between 20 and 100 nm.

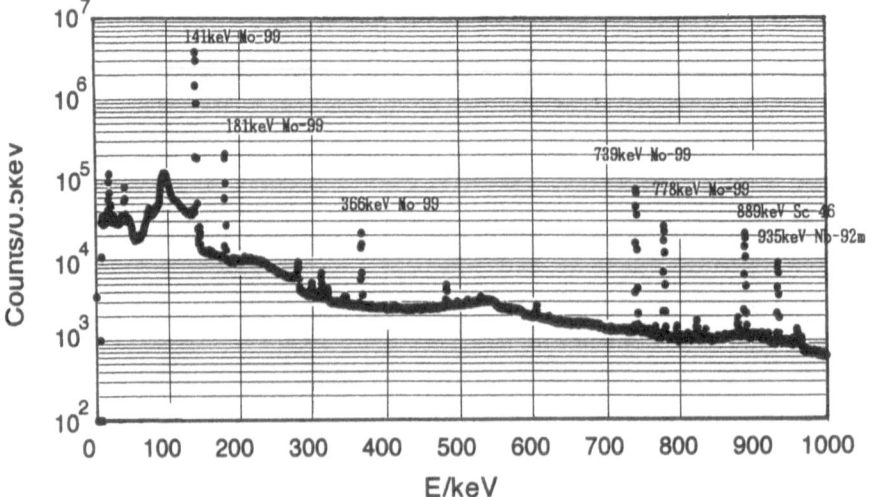

Fig. 2   γ-Ray emission spectrum of Mo encaged in nanocapsules irradiated by neutrons of flux
$10^{14}/cm^2 \cdot s$ for six days. The spectrum was measured 20 days after the neutron irradiation.

3. Experimental results

For neutron irradiation, 10 mg of the above sample was placed in a hydraulic rabbit for six days in the JMTR at the Japan Atomic Energy Research Institute. The neutron flux density was 10^{14} n/cm^2·s. Figure 2 is the γ-ray spectrum of the sample recorded 20 days after the irradiation. The spectrum shows peaks at 141, 181, 366, 739 and 778 KeV corresponding to γ emissions from Mo-99. The Mo powder loaded in the positive electrode contains isotopes in the natural abundance. Our Mo sample, therefore contains 24.1% of Mo-98 which transformed into Mo-99 by the neutron irradiation. The measured γ-ray spectrum in Fig. 2 shows the presence of 10^{10} Bq/g of Mo-99, which decays into Tc-99. The spectrum also shows small traces of impurity atoms. The intensity of Sc-46 indicates about 20 ppm of Ti.

Figure 3 shows images of nanocapsules after irradiation. There are several types of nanocapsules. Some capsules exhibit polygonal morphologies of graphitic layers typical of carbon nanocapsules. Most of these capsules contain less than 50% molybdenum carbide in their interior space. Most nanocapsules are nearly 100% filled with Mo before irradiation, as shown in Fig. 1. Those containing nearly 100% molybdenum carbide in Fig. 3 tend to show damage in graphitic layers after irradiation. Some are open or ruptured. These results indicate that damage of containers takes place mainly by nuclear reaction in the molybdenum carbide rather than direct reactions by neutron irradiation on the graphitic cage. Recoils associated the nuclear reaction induces the damage. Those containing less molybdenum carbide experience less nuclear reactions and hence, the graphitic cage is less damaged. Our investigation shows that the cage consisting of a few layers of graphite is rather durable against neutron irradiation and serves as a good container for radioactive materials. The contents of nanocapsules can be activated or transformed into other nuclei under moderate irradiation conditions.

Fig. 3   Transmission electron microscope image of carbon nanocapsules and other deposits after neutron irradiation for six days.

4. Summary

Carbon nanocapsules containing nanometer-size particles of molybdenum carbide were irradiated with neutrons to investigate their nuclear reactions. Our investigation shows feasibility of their use as containers for radioactive materials and tracers. It also shows that the contents of capsules are activated or transformed into other nuclei. Since some elements are more difficult to encapsulate than others, nuclear transformation may be a way to introduce such elements in the capsule.

Acknowledgment

The authors thank Messrs. Y. Ogawa and M. Narui for their support in the neutron irradiation experiment. This work was supported by the grant-in-aid for new programs from the Ministry of Education. Science and Culture.

PUBLICATIONS

1. Y. Saito *et al.*, *J. Phys. Chem. Solids*, **54**, 1849 (1993).
2. A. Kasuya *et al.*, *Surface Review and Letters* (1995) [in press].
3. A. Kasuya *et al.*, *J. Materials Science and Engineering A*, (1995) [in press].
4. A. Kasuya *et al.*, *Surface Review and Letters*, (1995) [in press].

REFERENCES

a. A. L. Giorgi *et al.*, *J. Less Comm. Met.*, 14, 247 (1968).
b. N. E. Topp, *Chemistry of the Rare-Earth Elements*, Elsevier, New York, (1965).

33

Size-dependent Characteristics of Single-wall Carbon Nanotubes

Atsuo Kasuya[a], Yahachi Saito[b], Yoshiro Sasaki[c], Michiko Fukushima[c], Toshiteru Maeda[c], Chuji Horie[c] and Yuichiro Nishina[a]

[a] Institute for Materials Research, Tohoku University
   Sendai 980-8577, Japan
[b] Department of Electrical and Electronic Engineering, Mie University
   Tsu 514-8507, Japan
[c] Department of Basic Science, Ishinomaki Senshu University
   Ishinomaki 986-0031, Japan

Purpose of the present study

Raman scattering by the phonon system is studied on nanotubes of diameter 1.1 nm. Spectra show multiple splittings in the Raman peak of optical phonon at 1580 cm^{-1} corresponding to E_{2g} mode of graphite. In addition, a new peak appears at 170 cm^{-1} in the region of acoustic phonon. These peaks all come from the extra boundary condition of a graphitic sheet rolled into a hollow cylinder. Our results show direct evidence for the size-dependent properties of nanotubes.

Contents of the present study

1. Introduction

Carbon nanotubes are characterized by their unique geometrical structure compared with other nanostructured materials.[a] It has the translational symmetry of graphitic network along the tube axis but has a cyclic symmetry along the circumference. As a result, the wavevector of both electrons and phonons takes continuous value in the direction corresponding to the tube axis in the Brillouin zone, but only a set of discrete values along the circumference. This restriction on the allowed wavevector in the tube plays the essential role in its basic properties that are strongly size dependent. In order to investigate such a fundamental aspect of nanotubes, the phonon system of both optical and acoustic branches are analyzed by Raman scattering measurements.

For this measurement, mono-sized and single-wall nanotubes are desirable. Single-wall nanotubes of diameter 1.1 nm are synthesized by a standard DC arc discharge of graphite electrodes in He atmosphere with the positive electrode filled with a mixture of Ni and Fe powders in equal weight.[b]

2. Raman scattering

Figure 1 shows our Raman spectra taken with Ar-ion laser at 488 nm for single-wall nanotubes (top), multi-wall nanotubes of mean diameter 10 nm (middle) and highly oriented pyrolytic graphite (bottom). The top spectrum of

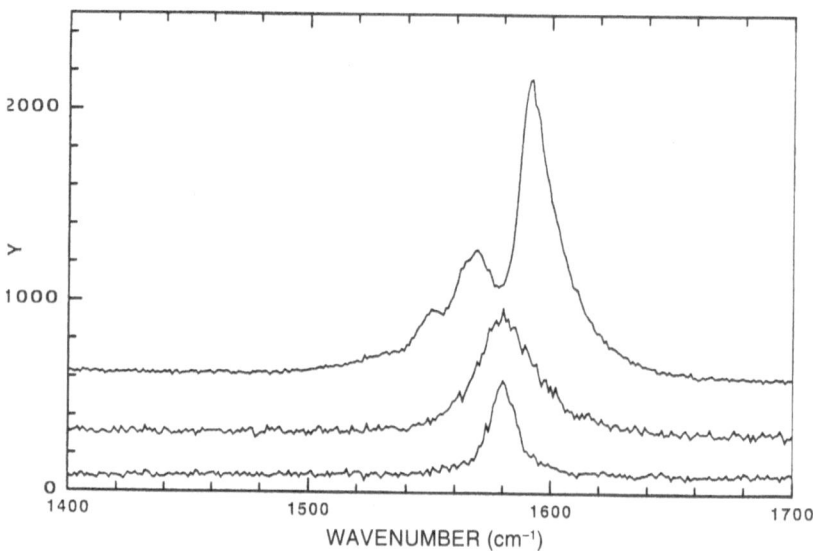

Fig. 1   Raman scattering spectra in the range from 1400 cm⁻¹ to 1700 cm⁻¹ of single-wall mono-sized
nanotubes (top), multi-wall tubes (middle) and graphite (bottom).

single-wall nanotubes shows multiple splitting of the peak at 1580 cm⁻¹. [C)] This
peak corresponds to the optical phonon of E_{2g} mode at $k = 0$ in graphite as
observed and shown in the bottom spectrum.

In graphite, only the optical phonon at the wavevector $k = 0$ is Raman active
because of the momentum conservation. In nanotubes, on the other hand, phonons
at a set of discrete wavevectors k_c also become possible by the cyclic boundary
condition. If the length of circumference is l, $k_c = 2$ n/l ($n = 1, 2, 3, ...$) in the
direction of circumference.

The dispersion curve of graphite shows that the optical phonon at $k = 0$ is split
into higher (longitudinal mode) and lower (transverse) energies for finite values
of k. Hence, as k increases, the two dispersion curves intersect at a series of
energies at $k = k_c$'s . Hence, the Raman peak splits into a series of sub-peaks on
both sides of 1580 cm⁻¹. The top spectrum of Fig. 1 shows broadening on the high
energy side and splitting on the low energy side. Since the dispersion curve is less
dispersed in the higher energy (longitudinal) mode compared with the lower
(transverse), only broadenings are observed on the high energy side.

The energy of acoustic phonon in graphite is zero at $k = 0$, but is finite at k
$= k_c$. Figure 2 shows our Raman spectrum of single wall nanotubes in the acoustic
energy range. The spectrum shows a peak at 170 cm⁻¹. The observed energy
value indicates that this peak is due to acoustic phonon at $k = k_c$. Based on the
dispersion curve of graphite, the peak value gives us the diameter of tube on the
order 1 nm as observed by transmission electron microscope.

3. Summary

Raman scattering from the single-wall mono-sized nanotube has been
measured. The observed splitting provides direct evidence that the nanotube

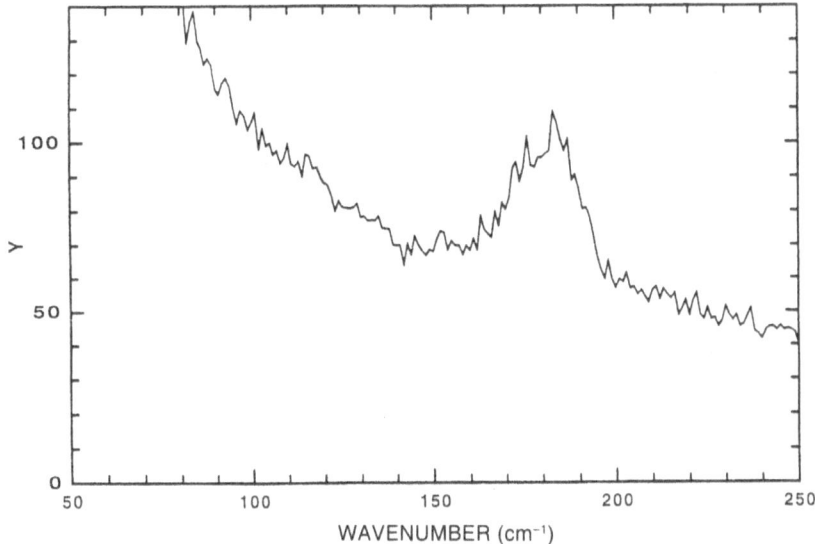

Fig. 2 Raman scattering spectrum of single-wall nanotubes in the spectral range of 50 to 250 cm⁻¹.

consists of a graphitic network sheet closed in a hallow cylindrical tube. This result shows that the diameter of the tube determines the size dependent characteristics of nanotubes.

PUBLICATION

1. A. Kasuya *et al.*, *J. Materials Science and Engineering A* (1995) [in press].

REFERENCES

a. N. Hamada, S. Sawada and A. Oshiyama, *Phys. Rev. Lett.*, **68**, 1579 (1992).
b. Y. Saito *et al.*, *Jpn. J. Appl. Phys.*, **33**, L526 (1994).
c. J. M. Hoden, *Chem. Phys. Lett.*, **220**, 186 (1994).

34

Laser-plasma Soft X-Ray Absorption Spectroscopy of Laser-ablated Silicon and Graphite Particles

Kouichi Murakami[a], Atsumi Miyashita[b] and Osamu Yoda[b]

[a] Institute of Materials Science, University of Tsukuba
   Tsukuba-shi, Ibaraki 305-0006, Japan
[b] Takasaki Research Establishment, Japan Atomic Energy Research Institute
   Takasaki-shi, Gunma 370-1207, Japan

Purpose of the present study

In order to study dynamics of formation of mesoscopic particles by measuring the time evolution of the electronic and geometrical structures of laser-ablated particles and their spatial distribution, we have developed time-resolved X-ray absorption spectroscopy using laser plasma. This method is demonstrated to have potential for pulsed-laser induced, fast-phase changes like laser annealing and laser ablation. In this work, laser-ablated particles of silicon and carbon are investigated by this method to clarify the dynamics of ablated particles. We show the results obtained for laser ablation of Si and C, the first experimental determination of Si-$L_{2,3}$ edges of Si atom and ions, and absorption lines suggesting formation of Si clusters and negative C ions.

Contents of the present study

1. Introduction

Laser ablation is a nonequilibrium potential technique for fabrication and synthesis of novel materials and introduction of novel impurity states. For example, C_{60} and C fullerenes[a] were for the first time synthesized by laser ablation as novel materials, and thin films consisting of multi-elements such as high Tc superconductors[b] have been successfully fabricated by the laser ablation and deposition method. We have also realized off-center nitrogen in silicon,[c] which is impossible by conventional thermal-quilibrium processes. The utility of laser ablation comes presumably from the advantages of laser plumes that include excited particles and those with high kinetic energy.[d]

To further develop the laser ablation method for synthesis and fabrication of novel materials with high quality, we need to exploit and develop new time-resolved and space-resolved measurements of geometrical and electronic structures of laser-ablated particles. In 1985, FOM Institute group exploited time-resolved and space-resolved soft X-ray spectroscopy using a pulsed soft X-ray source of laser-induced plasma.[e] We followed their work to develop an *in-situ* technique, and accomplished higher energy resolution and wider measurable-energy range[f] in order to investigate the dynamics of laser-ablated particles and the laser ablation/deposition process.

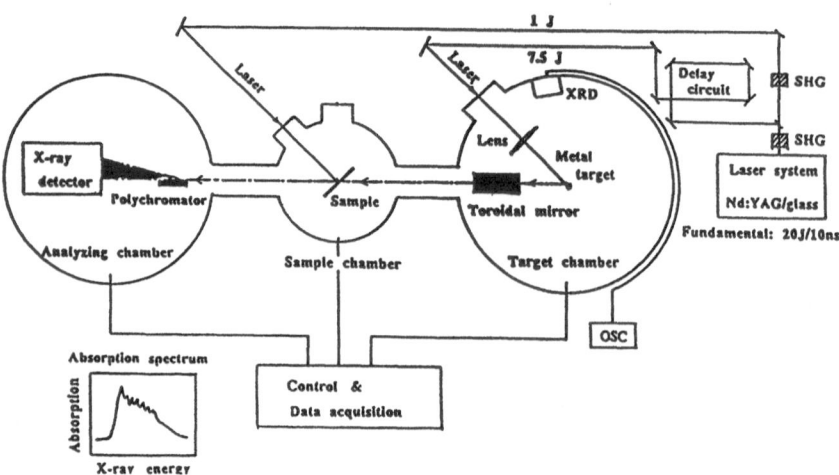

Fig. 1   Schematic diagram of the time/space-resolved soft X-ray absorption spectroscopy apparatus. (From K. Murakami *et al.*, *AIP. Conf.Proc.*, **288**, 376 (1994)).

In this paper, we demonstrate the usefulness of the time/space-resolved soft X-ray absorption spectroscopy for fast phase changes during pulsed laser irradiation, and report new results obtained for laser ablation of silicon (Si) and graphite (C) that are most ideal and simple materials.

2. Laser-plasma Soft X-ray Absorption Spectroscopy

Figure 1 shows a schematic diagram of the apparatus for the time/space-resolved X-ray absorption spectroscopy. The apparatus is composed of three parts: the laser system, the spectrometer and the data acquisition system. The most important spectrometer part is constituted by three vacuum chambers, *i.e.*, (i) the target chamber in which soft X-rays with a pulse width of 12 ns are generated from a Au target or a Ta target by a 12 ns pulsed laser with a wavelength of 532 nm as an X-ray source, and (ii) the sample chamber in which samples are installed and a high number of X-ray photons of approximately 1×10^{11} are collected at the sample position by a toroidal mirror, and (iii) the analyzing chamber in which the polychrometer and the micro-channel plate system are installed for dispersing and detecting the X-rays transmitted through a sample. The details have been in a previous paper.[8] Thus, the X-ray absorption measurement was performed by transmitting a broad continuum of soft X-rays through the created plume of Si or C, shown in Fig. 2, or Si foil. The X-ray probe beam was focused to a diameter of 0.1–0.2 mm for the space-resolved measurement and the distance d between the X-ray beam and the Si or C surfaces was varied for the measurement of the spatial distribution of ablated particles (see Fig. 2). The total measurement system enables us to record an absorption spectrum in a single shot with good statistics, because the number of X-ray photons included in one pulse is extremely high. Consequently, this time-resolved spectroscopy is very suitable to any single event occurring for a short time. The data were recorded in the energy range from 90 to 350 eV and the energy resolution was 0.7

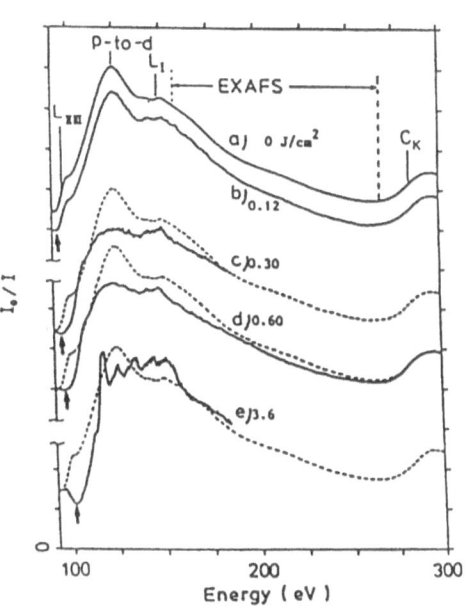

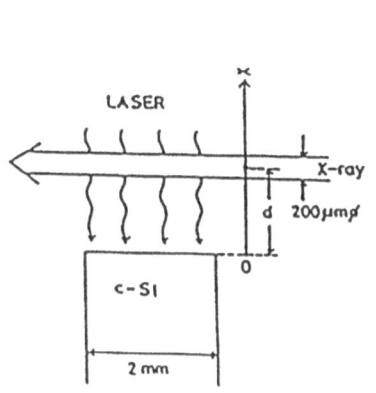

Fig. 2 Arrangement of samples with respect to the X-ray beam for measurement of spatial distribution of ablated particles.
(From K. Murakami *et al.*, *AIP. Conf. Proc.*, **288**, 377 (1994)).

Fig. 3 Typical X-ray absorption spectra of solid Si foil on C for various laser energy densities.
(From K. Murakami *et al.*, *Phys. Rew. Lett.*, **56**, 656 (1986)).

eV at around 100 eV. Variable delay times between the irradiation laser beam for annealing or ablation and the X-ray beam were controlled between 0 ns and 120 ns through the optical delay line shown in Fig. 1.

3. Laser ablation of silicon

Figure 3 shows previous results of typical X-ray absorption spectra of amorphous Si foil (600 Å thickness) on C (440 Å thickness), ranging from 90 to 300 eV at a delay time of 12 ns for various annealing energy densities.[h] The energy resolution was 4 eV, and was not good compared with that (0.7 eV) of our present apparatus. Below 0.17 J/cm^2 we observed no significant changes in spectra, while changes in the spectra appeared above this energy density. The changes are caused by melting of the Si foil. Above 1.0 J/cm^2 further drastic changes in the spectra are observed. This corresponds to the ablation threshold. For spectrum 3(a) without pulsed-laser irradiation, a clear edge and a broad peak are seen at 98 and 125 eV, respectively. The edge comes from $Si\text{-}L_{2,3}$ absorption that corresponds to excitation from the Si-2p core level to the bottom of the conduction band. The broad absorption is due to the "centripetal barrier" for p-to-d transition. The $Si\text{-}L_1$ edge is observed at about 150 eV: this edge corresponds to excitation from the Si-2s core level to the bottom of the conduction band. At 0.3 and 0.6 J/cm^2 the structure characteristics of the $Si\text{-}L_{2,3}$ absorption at 100 eV disappear as seen in Fig. 3(c) and (d). The observed spectra are much different from the original one of solid Si. There is a distinctly slower rise to the p-to-d maximum. If we define the point of maximum slope as an effective edge of the $L_{2,3}$

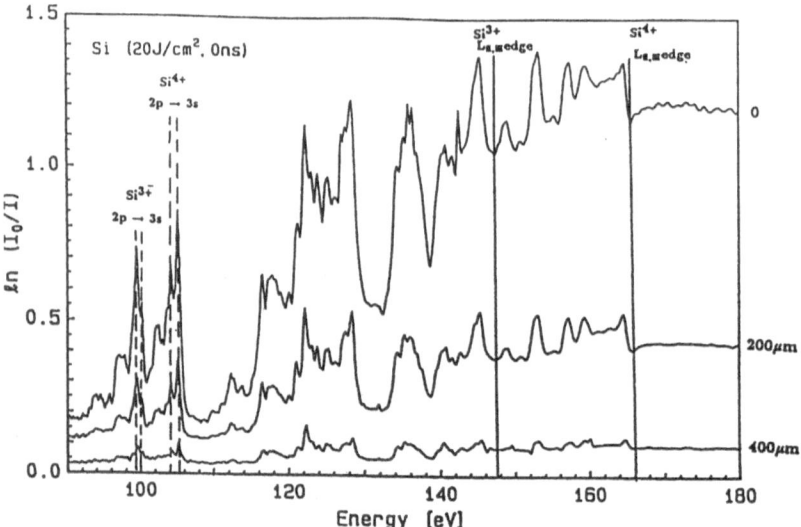

Fig. 4   Typical X-ray absorption spectra of laser-ablated Si particles measured at various distances
at a delay time of 0 ns (about 20 J/cm^2).
(From K. Murakami et al., AIP. Conf. Proc., **288**, 379 (1994)).

of the transient liquid Si, there is a clear edge shift of about 7 eV to higher energy. A decrease of the p-to-d maximum height is also seen in Fig. 3. This was the first observation of X-ray absorption of the laser-produced, short-lived transient liquid Si.[h] At an energy density of 3.6 J/cm^2, a complicated structure can be seen in spectrum (e) of Fig. 3. This is connected with the formation of a Si plasma, which is shown in detail below.

For bulk Si crystal, laser ablation and laser melting take place above a threshold energy of about 3 J/cm^2 and 0.4 J/cm^2, respectively. Figures 4 and 5 show the typical absorption spectra of ablated Si particles measured at each distance d at delay times of 0 ns and 120 ns, respectively, after the irradiation of an ablation laser. The laser energy density was about 20 J/cm^2. Similar experiments were previously performed with a poor energy resolution of 4 eV.[i] In this study the resolution became much better compared to the previous work, and from analysis of the spectrum of Fig. 4, it was found that Si ions like Si^{2+}, Si^{3+}, and Si^{4+} are produced at a delay time of 0 ns. With increasing probe distance from the Si surface, the absorption Rydberg peaks observed at 0 ns become weaker in amplitude. At a delay time of 120 ns, ablated particles reaching more distant points can be observed in Fig. 5. The absorption peaks are attributed to Si atoms, Si^+ and Si^{2+} ions at 0.2 and 0.6 mm, and only to Si^{2+} ions at 1.8 mm. If the Si ions have an average velocity of approximately 2×10^6 cm/s [d] and no recombination takes place, Si^{3+} and Si^{4+} ions should be detected at 120 ns. This suggests that Si^{3+} and Si^{4+} have recombination times or relaxation times less than 120 ns. It is noted that the Si atoms and Si^+ ions are still ejected from the Si surface 120 ns after the ablation-laser irradiation. In the spectra of Figs. 4 and 5, unidentified peaks are only the broad ones appearing at 122, 126, 132, and 141 eV that become more clearly observed at 120 ns. In the previous paper, [i] the broad absorption peak observed at 112 eV was tentatively assigned as Si clusters. However, this peak was

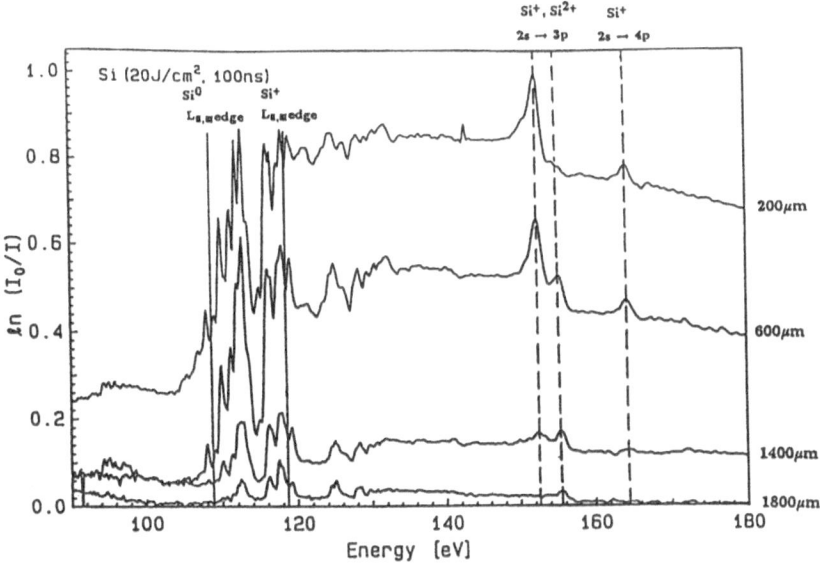

Fig. 5 Typical X-ray absorption spectra of laser-ablated Si particles measured at various distances
at a delay time of 120 ns (20 J/cm²).
(From K. Murakami *et al.*, *AIP. Conf. Proc.*, **288**, 380 (1994)).

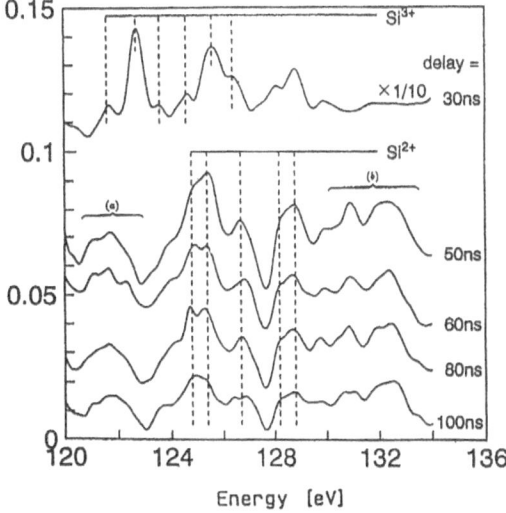

Fig. 6 X-ray absorption spectra of laser-ablated Si particles in vacuum observed at various delay times.

resolved into several sharp peaks in the present work and the peaks near 112 eV
are found to be assigned to Si⁺ ions. Si clusters are known to be formed by laser
ablation of Si,[j] so we believe the unidentified peaks are most likely related to Si
clusters. The absorption at energies higher than the $L_{2,3}$ edges of solid Si and Si
atom may be due to bond contraction of Si clusters or ionization of Si clusters.

The time/space-resolved measurements were performed also at lower energy
densities of about 10 and 5 J/cm². The unidentified absorption peaks were found
to be clearly observed at higher energy densities and longer delay times. This

Table 1 Laser-ablated Si particles at delay times of 0 to 120 ns

| valence electrons | | transitions (90–180 eV) | |
		Rydberg (from 2p)	(from 2s)
Si^0	$3s^2\,3p^2$	2p-3d, 4s, ...nd, ns	2s-3p, 4p
		$L_{2,3}$ edge (2p-continuum)	
Si^+	$3s^2\,3p^1$	2p-3d, 4s, ...nd, ns	2s-3p, 4p
		$L_{2,3}$ edge	
Si^{2+}	$3s^2$	2p-3d, 4s, ...nd, ns	2s-3p, 4p
		$L_{2,3}$ edge	
	$(Si^{2+})^*\ 3s^1\,3p^1$	2p-3s	
Si^{3+}	$3s^1$	2p-3s, 3d, 4s, ...nd, ns	2s-3p
		$L_{2,3}$ edge	
Si^{4+}		2p-3s, 3d, 4s, ...nd, ns	2s-3p
		$L_{2,3}$ edge	
liquid Si droplets		$L_{2,3}$ edge: 107 eV	at 120 ns delay
		(cf.solid Si: 100 eV)	

Table 2 Values of Si-$L_{2,3}$ edges obtained in this study and comparison with calculated values

| experimental values | | Hartree-Fock calculations |
		(atomic data and nuclear data Tables, Vol.14, No.3-4, 1974)
Si^0	109 ± 1 eV	114.9 eV
Si^+	120 ± 1 eV	125.4 eV
Si^{2+}	131 ± 1 eV	137.1 eV
Si^{3+}	148 ± 1 eV	152.0 eV
Si^{4+}	167 ± 1 eV	168.4 eV

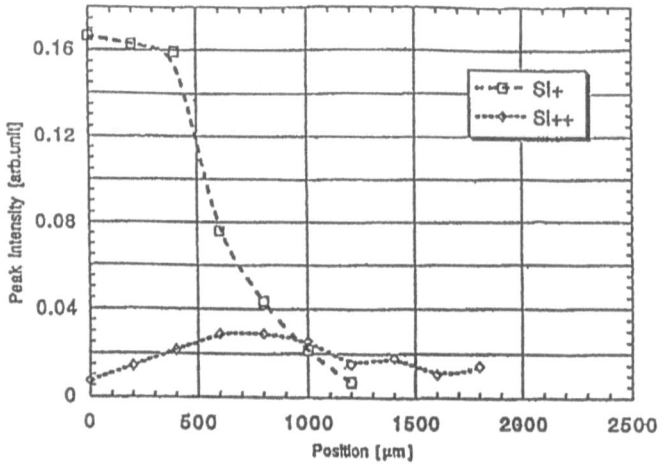

Fig. 7   Spatial distribution of laser-ablated Si particles measured by time-resolved X-ray absorption spectroscopy at a delay time of 120 ns. The laser energy density used was about 20 J/cm². (From K. Murakami *et al.*, *AIP. Conf. Proc.*, **288**, 382 (1994)).

suggests that, if they are attributed to Si clusters, high density Si particles are needed for formation through condensation of atoms and ions, and that another process of direct ejection of Si clusters from the Si surface takes place at later times.

Figure 6 shows time revolution of the unidentified lines near 122 and 132 eV from 30 to 100 ns. At 30 ns Si^{2+} and Si^{3+} are observed, while three lines and two broad lines are seen near 122 and 132 eV, respectively, during the period from 50 to 100 ns. It is likely that these come from ionized Si clusters such as Si_2 to Si_4.

The absorption lines and edges observed in this experiment are summarized in Table 1. Since Si has low vapor pressures at high temperatures and also a very high vaporization temperature, to our knowledge no accurate experimental values have been reported for the Si-$L_{2,3}$ edges of Si atoms and Si^{n+} ions ($n = 1, 2, 3$ and 4). The previous paper [i)] reported a value of 115 eV as the $L_{2,3}$ of Si atom. However, the value obtained in this work is 110 eV. Because we measured X-ray absorption spectra of Si atoms and ions with better energy resolution, we accurately determined their $L_{2,3}$ edges. The obtained values are summarized in Table 2, in comparison with theoretical ones.

The Rydberg lines of 2p to nd or to ns observed in Figs. 4 and 5 are very complicated while the lines of the transitions of 2s to np are clearly seen. Consequently, from the latter we determined the relative amount of ablated Si particles as functions of the distance d. A typical result is shown in Fig. 7. Ions with higher charges are found to reach a distant point and ions with lower charges are still ejected from the Si surface even at a delay time of 120 ns.

4. Laser ablation of graphite

For observation of ablated particles, a graphite rod (purity: 99.9999%) and pure C_{60} (Texas Fullerenes Co., purity: better than 99.9%) were used. In Fig. 8 we

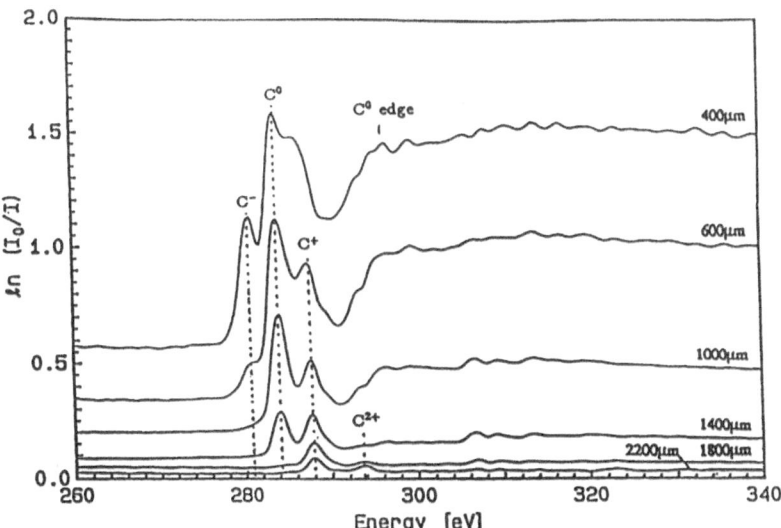

Fig. 8   Typical X-ray absorption spectra of laser-ablated C particles measured at various distances at a delay time of 120 ns (20 J/cm²).

show spatial dependence of the absorption spectrum of ablated C particles from the C rod in the energy range of C-K edge. The energy resolution in the energy range is 1.7 eV. The energy density was 20 J/cm^2 and the time delay was 120 ns. The spectra exhibit several peaks between 280 and 290 eV, and an intense broad absorption beginning at around 290 eV. As the probing distance increases, only two distinct peaks are seen at 284 and 288 eV and the absorption beyond 290 eV decreases. According to another report,[k] these two peaks are assigned to the transitions 1s to 2p of the neutral C atom and C$^+$ ion, respectively, and the peak at 294 eV seems to be due to C^{2+}. It should be noted that a peak is present at an energy of 281 eV, which is much lower than any values previously reported for C particles and C-based solid materials. Laser ablation of C produces negative C$^-$ ions and negative C clusters.[l] It is likely that the peak at 281 eV is assigned to the C$^-$ ion. On the other hand, we observed no negative ions at an energy range lower than 100 eV for laser ablation of Si. Laser ablation of C$_{60}$ pellet was found to yield a spectrum similar to that of graphite when the laser-energy densities used were higher than 0.1 J/cm^2.

5. Conclusions

We have demonstrated that the time/space-resolved X-ray absorption spectroscopy is useful for the study of fast phase changes induced by pulsed laser. The method will be combined with pulsed He gas jet to enhance formation of Si clusters Si clusters and C fullerenes and will be used to clarify the formation mechanism of microclusters on wider time scales, *i.e.* from nanoseconds to microseconds.

PUBLICATIONS

1. O. Yoda, A. Miyashita, K. Murakami, S. Aoki and N. Yamaguchi, Proc. SPIE vol.1503 *Excimer Lasers and Applications III*, pp.463–466 (Society of Photo- Optical Instru. Eng., 1991).
2. K. Murakami, *European Materials Research Society Monographs 4*, 125–140 (1992).
3. A. Miyashita, O. Yoda, K. Murakami, T. Ohyanagi, S. Aoki and N. Yamaguchi, Proc. on Laser Advanced Materials Processing, pp.1029–1034 (1992).
4. O. Yoda, A. Miyashita. T. Ohyanagi and K. Murakami, *JAERI*, **92-173**, pp.1–27 (1992).
5. O. Yoda, A. Miyashita, K. Murakami, T. Ohyanagi, S. Aoki and N. Yamaguchi, *Jpn. J. Appl. Phys.*, **32**, Suppl.32-2. 255–257 (1993).
6. K. Murakami, Proc. Intern. Workshop on Dynamic Response to Pulsed Heating, 29–40, Los Alamos (1993).
7. K. Murakami, T. Ohyanagi, K. Hara and K. Masuda, *Materials Science Forum*, **117/118**, 369–374 (1993).
8. K. Murakami, T. Ohyanagi, A. Miyashita and O. Yoda, Proc. Intern. Conf. on Laser Ablation, AIP Conf. Proc. vol.288, pp.375–384 (1994).
9. T. Ohyanagi, A. Miyashita, K. Murakami and O. Yoda :Jan. *J. Appl. Phys.*, **33**, 2586 (1994).
10. T. Ohyanagi, A. Miyashita, K. Murakami and O. Yoda, Proc. Intern. Conf. of TIARA, 881–890, Mito (1994).
11. A. Miyashita, T. Ohyanagi, O. Yoda and K. Murakami, Proc. Intern. Conf. of TIARA, 873–880, Mito (1994).
12. K. Murakami, T. Ohyanagi and K. Masuda, *Jpn. J. Appl. Phys.*, **33**, 4513 (1994).
13. T. Ohyanagi, A. Miyashita, K. Murakami and O. Yoda, *Surf. Rev. Lett.*, **3**, 187 (1996).
14. A. Miyashita, T. Ohyanagi, O. Yoda and K. Murakami, *Surf. Rev. Lett.*, **3**, 191 (1996).

REFERENCES

a. H. W. Kroto, J. R. Heath, S. C. O'Brien, R. F. Curl and R. E. Smalley, *Nature*, **318**, 162 (1985).
b. See for example, Laser Ablation of Electronic Materials—*Basic Mechanism and Applications*—, edited by E. Fogarassy and S. Lazare, European MRS Monographs, 4, North Holland (1992).
c. K. Murakami and K. Masuda, unpublished.

d. K. Murakami, *European Materials Research Society Monographs 4*, 125 (1992).

e. H. C. Genitsen, H. van Brug, F. Bijkerk and M. J. van der Wiel, *J. Appl. Phys.*, **59**, 2337 (1986).

f. A. Miyashita, O. Yoda, K. Murakami, T. Ohyanagi, S. Aoki and N. Yamaguchi, Proc. of 2nd Intern. Symp. on Laser Adv. Mater. Process. 1092 (1992).

g. O. Yoda, A. Miyashita, T. Ohyanagi and K. Murakami, *JAERI-M*, **92-173** (1992).

h. K. Murakami, H. C. Gerritsen, H. van Brug, F. Bijkerk, F. W. Saris and M. J. van der Wiel, *Phys. Rev. Lett.*, **56**, 655 (1986).

i. H. van Brug, K. Murakami, F. Bijkerk and M. J. van der Wiel, *J. Appl. Phys.*, **60**, 3438 (1986).

j. L. A. Bloomfield, R. R. Freeman and W. L. Brown, *Phys. Rev. Lett.*, **54**, 2246 (1985).

k. E. Jannitti, P. Nicolosi and G. Tondello, *Physica Scripta*, **41**, 458 (1990).

l. A. Kasuya and Y. Nishina, *Phys. Rev. Lett.*, **28**, 6571 (1983).

35

Optical Study of Free Clusters and Microcrystals by Time- and Space-resolved Spectroscopy Combined with Gas Evaporation Method

Shosuke Mochizuki and Hitoshi Nakata

Department of Physics, College of Humanities and Sciences, Nihon University
Setagaya-ku, Tokyo 156-0045, Japan

Abstract

We have developed a simple measurement method for optical spectra of free-clusters and microcrystals, by combining the time-resolved and space-resolved spectroscopies with the gas-evaporation method. The method is applied to the measurements of optical spectra of metal- and semiconductor vapors. The obtained spectra indicate the evolution of electronic states from atom to microcrystal and the quantum size effects of carriers and excitons.

1. Introduction

In recent years, considerable attention has been given to the theoretical and experimental investigations of clusters and microcrystals. The question of how the electronic and vibrational states change with increasing number of constituent atoms is of great interest in condensed matter physics. To answer such a question, it is best to produce a free cluster stream that contains size-selected clusters and to measure the optical properties as a function of cluster size. Many photoabsorption measurements of this type have been performed mainly on small clusters and reviewed by de Heer.[a] In the large-particle limit, optical spectra of microcrystal, of the order of ten nanometers, has been obtained by the gas evaporation method.[b] We have recently shown that, by setting appropriate conditions for the gas evaporation and by measuring time-resolved and space-resolved transmission spectra, the whole range of species size can be observed.

The present paper is a brief summary of our recent articles.[1–8]

2. Experimental method

It is well known that aggregates of various sizes between several angstroms and several microns can be produced easily by thermal evaporation under a low-pressure noble gas.[c,d] This method is called the "gas evaporation method". The structure of the vapor zone produced above an evaporation source by the gas evaporation method depends on the size and shape of the evaporation source, pressure and flow rate of the gas, the evaporation temperature, etc. If we elevate the temperature of the evaporation source gradually, we can observe optically that atoms evaporate first from the melted surface and then the atoms coalesce with surrounding atoms to produce dimers and clusters. However, the vapor zone

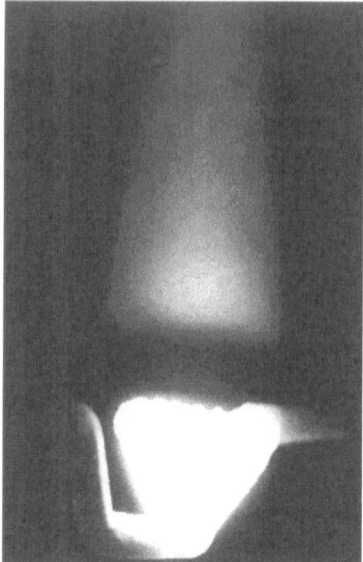

Fig. 1

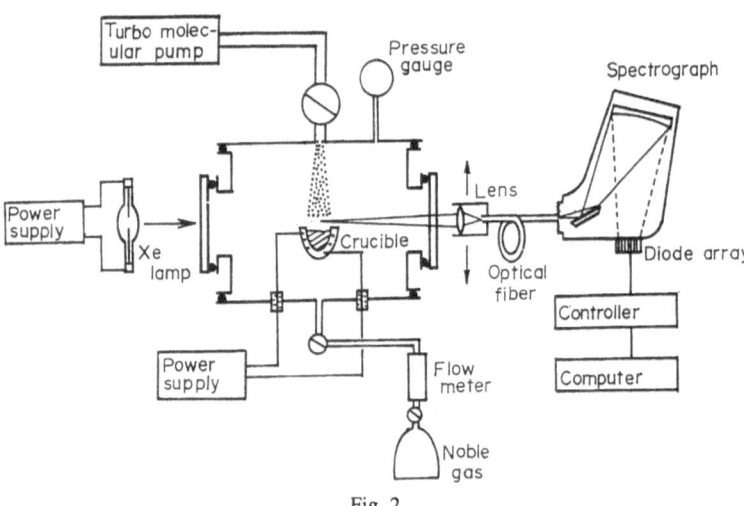

Fig. 2

reported in the literature was shaped like a candle flame around the evaporation source and consisted of inner, intermediate and outer zones. Such a zone structure is not suitable to optical measurement, since all zones on an optical axis contribute simultaneously to the optical transmission spectra. Thus, it was desirable to produce a well-separated vertical zone structure.

Recently, we have shown for the first time that, by setting appropriate conditions for gas evaporation of silver, two well-separated vertical zones have been produced above the evaporation source in a confined helium gas atmosphere, as shown in Figure 1.[1,2] Such a zone structure has been observed for many metallic elements in our laboratory. Thereafter, we measured the time evolution of the transmission spectra of vapor zones at various heights from the evaporation

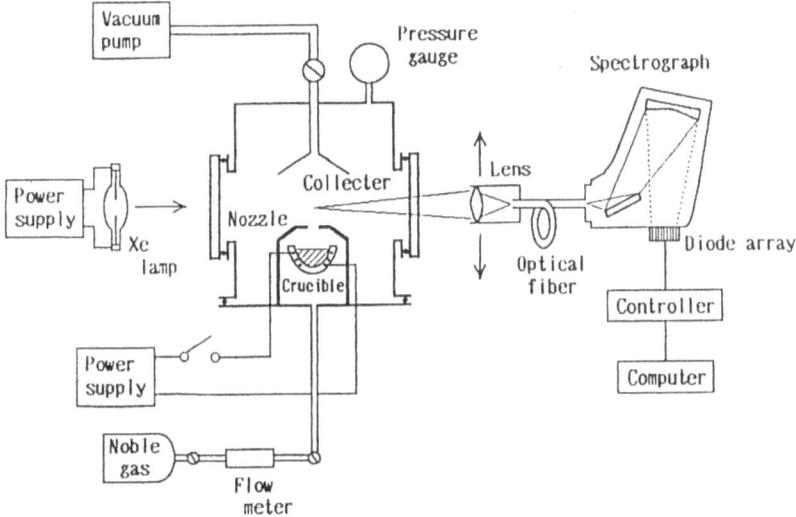

(a) free microcrystal measurement

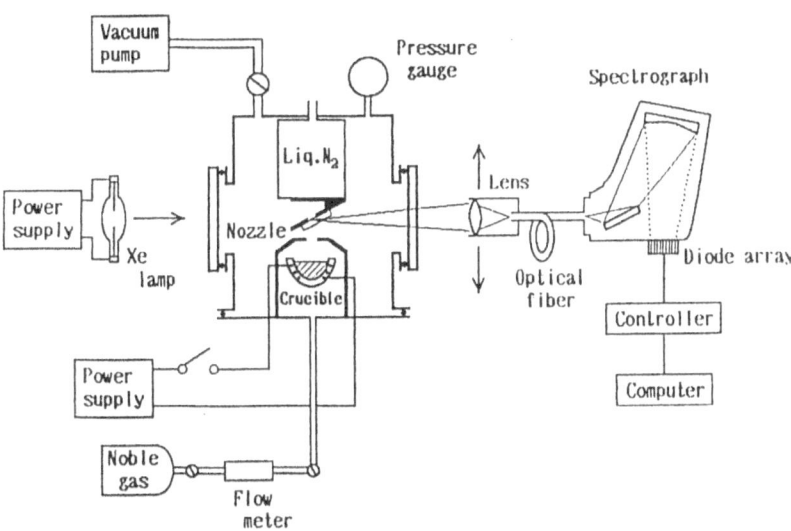

(b) quasi-free microcrystal measurement

Fig. 3

source by a multichannel optical analyzer system. We named the method "time-resolved and space-resolved spectroscopies of vapor phase." The block diagram for the measurements is shown in Figure 2. Using this apparatus, we have measured the absorption (extinction) spectra of vapor zones of alkali metals, [3-5] and noble metals at various positions as a function of time elapsed after the beginning of evaporation. Typical results are shown in section 3.

Very recently, we have improved the evaporation section of the apparatus to study the quantum-size effects of free semiconductor in the nanometer-size region. The block diagram for the measurement is shown in Figure 3(a). In order to produce well directional uniform-size microcrystal beam, the evaporation was carried in a following argon gas stream with a high velocity. A nozzle 6 mm in diameter is put on top of the evaporation chamber. Gas stream of high velocity was made by introducing argon gas continuously to the bottom of the evaporation chamber and by evacuating a beam chamber through the nozzle with a mechanical booster pump (exhaust speed of 800 l/min). Optical spectra of selected positions in the beam zone were recorded in a transmission configuration as a function of time elapsed after the beginning of the evaporation. Also, the experiments on supported (quasi-free) microcrystals were carried out by introducing a liquid nitrogen tank into the beam chamber described above. This is shown in Figure 3(b). A silica glass substrate was attached to the bottom of the liquid nitrogen tank and cooled. After the beginning of evaporation, the transmissivity spectra of the substrate were measured as a function of time using the same optical multichannel analyzer system as for the free microcrystals. The typical results for free and quasi-free microcrystals are shown in section 4.

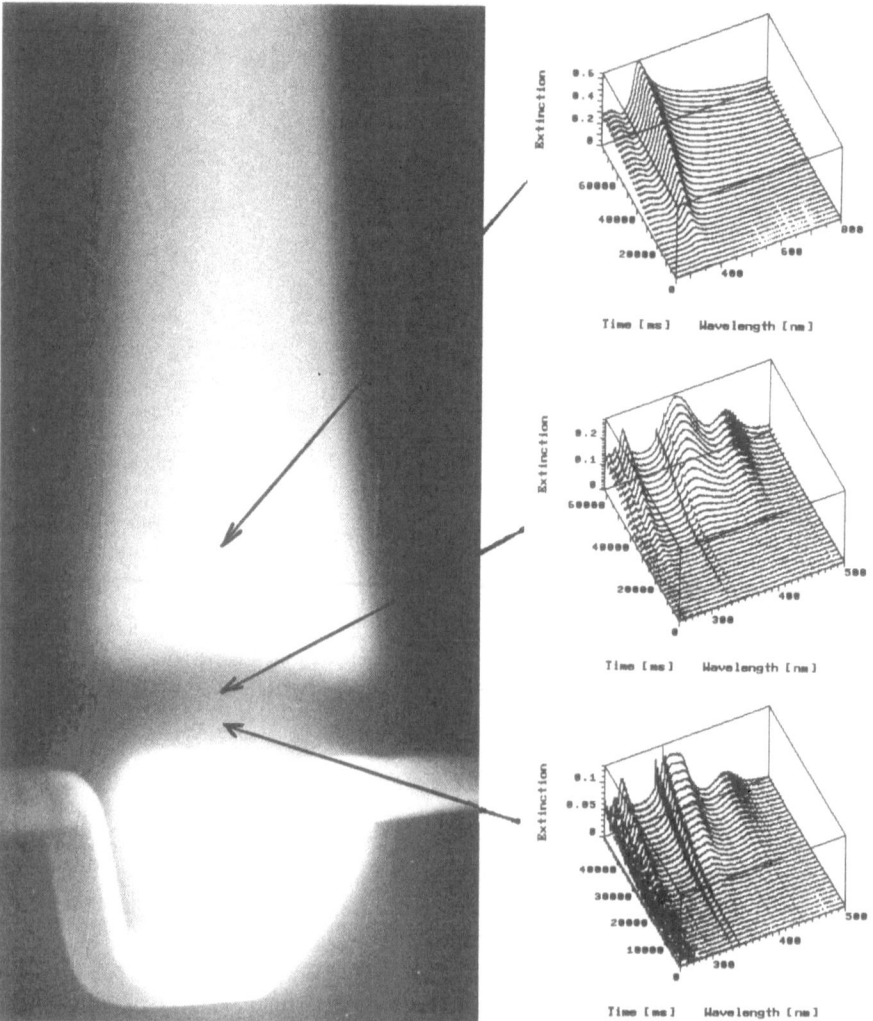

Fig. 4

3. Results: Metals

Figure 4 shows a typical view of silver vapor zones above a crucible and the time-resolved extinction spectra of each zone. [1,2] The evaporation was carried out in helium gas confined at 140 Torr. Two sharp lines at about 329 nm and 339 nm are due to the s-p transitions in atomic silver. Five bands at about 249 nm, about 255 nm, about 266 nm, about 276 nm and about 433 nm are due to the X-E, X-D, X-C, X-B and X-A transitions in silver dimers, and the broad band at about 355 nm is due to the surface plasmon absorption band in silver clusters. Time dependence of the extinction for each species in the vapor zones is plotted as a function of height from the crucible (source) in Figure 5. As seen in these figures, the spectral structure and the dependencies on time and on the distance from the source indicate that atoms evaporate first from the melted surface, and, thereafter, dimers and cluster grow by coalescence with atoms. Also, the size trend of the surface plasmon band is clearly observed.

We have observed similar results described above for copper, gold, sodium,[3] potassium[5] and rubidium.[4] Recently, many workers pay attention to the blue-shift nature of the surface-plasmon resonance frequency. From the frequency shift observed in sodium clusters[a] and potassium clusters,[e] it has been suggested that the observed blue shift arises mainly from the enhancement of electron spillout effect in small clusters. We have observed the nature of such blue shifts by time-resolved and space-resolved spectroscopies of vapor phase in sodium, potassium and rubidium. We show typical results obtained at the initial stage of gas evaporation of rubidium in Figure 6. The evaporation was carried out in a low-

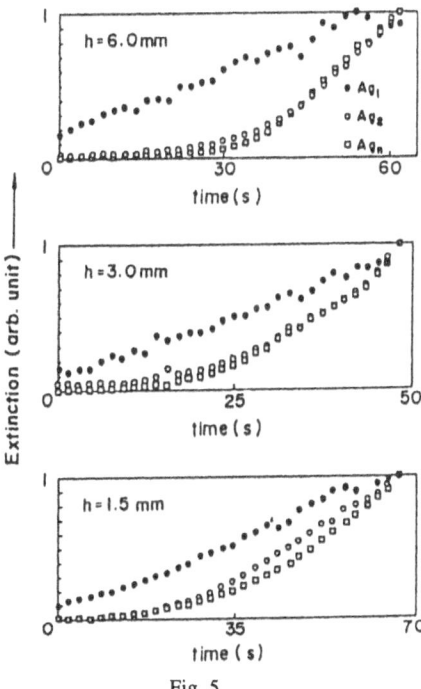

Fig. 5

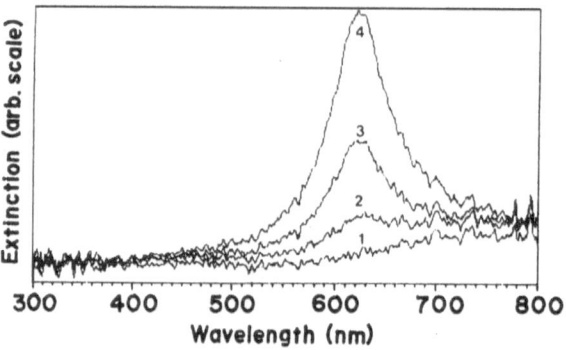

Fig. 6

pressure gas stream with a flow rate of 0.5 l/min at a pressure of 2 Torr. With progressing gas evaporation, after reaching maximum frequency, the blue-shifted surface plasmon frequency begins to go back to a longer wavelength (up to about 640 nm) and to increase in width. The nature of the red shift nature and broadening can be explained by calculating size dependence of the Mie scattering cross section.

4. Results: Semiconductors

Figure 7(a) shows a typical view of a CuBr microcrystal beam above the nozzle. This figure was taken at the final stage of the evaporation in an argon gas stream under illumination or a 150 W Xe lamp. The diameter or the beam was about 3 mm. In order to determine the size distribution of microcrystals in the beam, we collected microcrystals on the electron-microscopy grids covered with carbon-coated triacetylcellulose film at different evaporation stages. Their TEM images showed narrow distribution. During the evaporation, an average size increases from 4 nm to 22 nm. Figure 7(b) corresponds to the intermediate stage of the evaporation of CuBr.

Figures 8(a) and (b) are the time evolution of the exciton absorption bands ($Z_{1,2}$- and Z_3 bands) in free and quasi-free (collected) CuBr microcrystals, respectively. The beam temperature varied from 378 K to 393 K during the

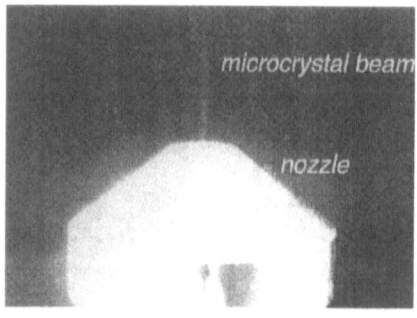

(a) CuBr microcrystal beam   (b) intermediate stage of CuBr evaporation

Fig. 7

353

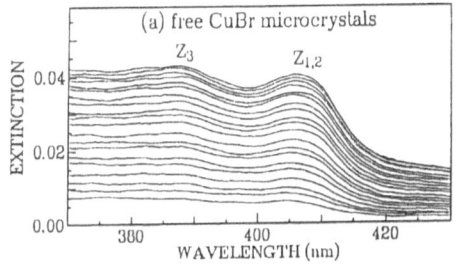

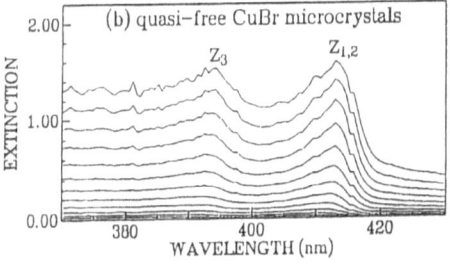

Fig. 8

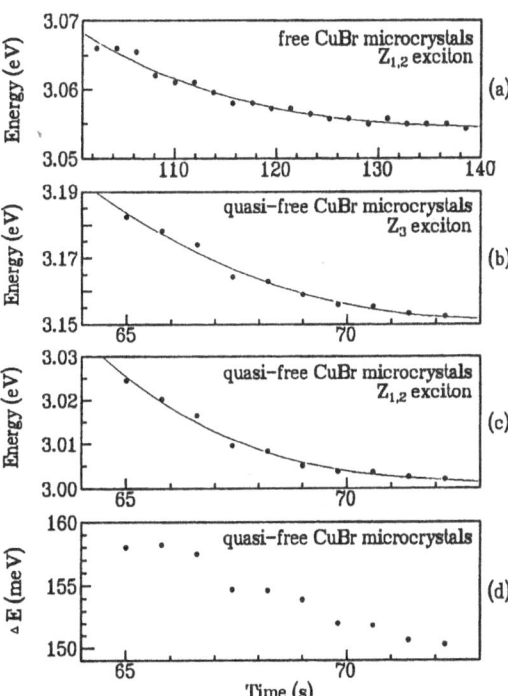

Fig. 9

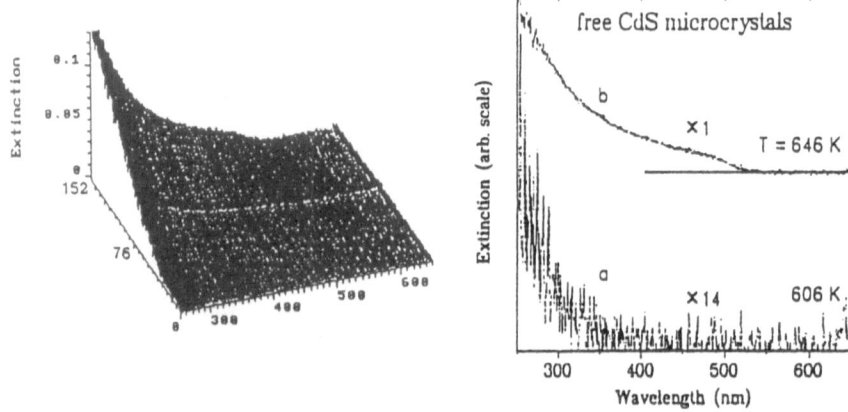

Fig. 10

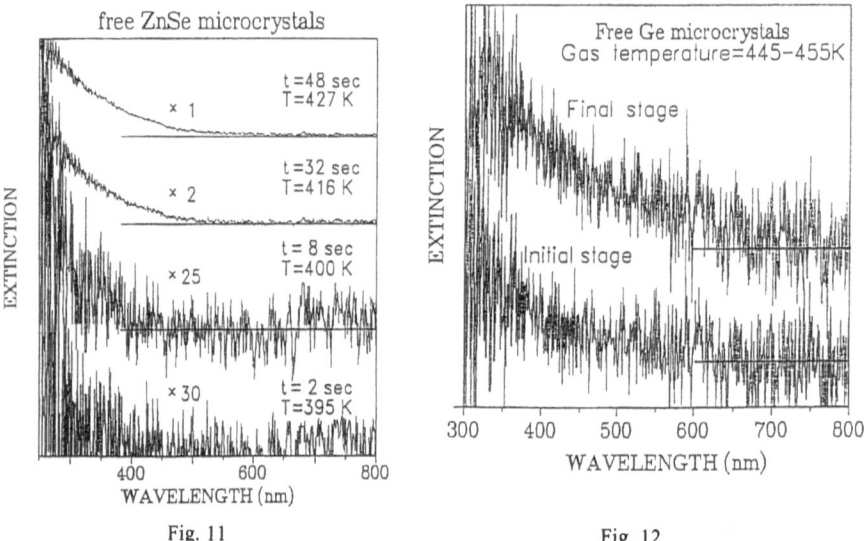

Fig. 11   Fig. 12

evaporation. The substrate temperature was 85 K for the quasi-free measurements. The time dependencies of the exciton energies are shown in Figures 9(a), (b), (c) and (d). As seen in these figures, each exciton energy decreases with increasing time (size). Approaching the steady state of the gas evaporation, the energy difference between the Z_3- and $Z_{1,2}$ excitons tends to smooth the energy difference (147 meV) observed in bulk CuBr. This indicates that the observed time evolution arises from exciton confinement in the CuBr microcrystals. Similar results are obtained for the exciton bands of CurI.[6]

Finally, we show the time evolution of the spectra for CdS,[7] ZnSe[8] and Ge[8] in Figures 10, 11 and 12, respectively. As seen in these figures, each absorption edge shifts toward lower energy with progressing evaporation. The structure of the spectra and their time dependencies indicate the quantum confinement effect of carriers.

PUBLICATIONS

1. S. Mochizuki, *Phys. Lett.*, **155A**, 510 (1991).
2. S. Mochizuki and R. Ruppin, J. Phys.: *Condens. Matters*, **5**, 135 (1993).
3. S. Mochizuki, *Phys. Lett.*, **A164**, 191 (1992).
4. S. Mochizuki, *Phys. Lett.*, **A176**, 382 (1993).
5. S. Mochizuki and R. Ruppin, *J. Phys.: Condens. Matter*, **6**, 7303 (1994).
6. S. Mochizuki, H. Nakata and R. Ruppin, *J. Phys.: Condens. Matter*, **6**, 1269 (1994).
7. S. Mochizuki and H. Nakata, *Phys., Lett.*, **A183**, 390 (1993).
8. S. Mochizuki, H. Nakata and M. Asanuma, Proc. ISSPIC7: *Surf Rev. and Letters* [in press].

REFERENCES

a. W. A. de Heer, *Rev. Mod. Phys.*, **65**, 611 (1993).
b. J. D. Eversole and H. P. Broida, *Phys. Rev.*, **B15**, 1644 (1977).
c. K. Kimoto, Y. Kamiya, M. Nonoyama and R. Ueta, *Jpn. J. Appl. Phys.*, **2**, 702 (1963).
d. C. G. Granqvist and R. A. Buhrman, *J. Appl. Phys.*, **47**, 2200 (1976).
e. C. Brechignac, Ph. Cahuzac, N. Kebaili, J. Leygnier, and A. Safati, *Phys. Rev. Lett.*, **68**, 3916 (1992).

36

Electronic Structures of Graphitic Carbon Cage Structures

Masaru Tsukada, Kazuto Akagi and Ryo Tamura

Department of Physics, Graduate School of Science, University of Tokyo
Bunkyo-ku, Tokyo 113-0033, Japan

Purpose of the present study

A variety of exotic caged nanostructures of graphitic carbon layers including fullerenes, carbon nanotubes, and onion-like nested forms have been found and their material properties are attracting much attention. In this work we aim to clarify some general properties of the electronic states common in such nanostructures of curved graphite surfaces. The problems relevant to these systems in general may be as follows: how are the band characters of the perfect 2D graphite reflected on the electronic structures of such nanostructures? And how do the geometrical and/or topological characteristics of the cage manifest themselves on the electronic state of the system? We study such fundamental issues of carbon cage structures using simple tight binding models.

Contents of the present study

1. Introduction

To clarify certain general properties of carbon cage structures in a conceptual rather than quantitative way, we focused on the problems listed below:

I) Nature of electronic states around disclination, *i.e.*, $N(\neq 6)$ membered ring and strange electronic states introduced by the fused disclinations,

II) Localized electronic states induced by the caps of the carbon nanotubes,

III) Electronic states of helically coiled carbon nanotubes and their relation to the phason line patterns.

Since subjects I and II are discussed in details elsewhere,[1-3] we focus in this paper on problem III.

Simple straight carbon nanotubules can be in a sense geometrically formed by rolling a complete graphitic sheet, but structures with non-zero curvature such as a cap or a helix need non 6-membered rings such as 7-membered or 5-membered rings in the graphitic sheets. The arrangement of atoms around the 5- or 7-membered rings is topologically equivalent to the one found in the structures obtained by removing or inserting a wedge of 60 degrees sector in a graphitic sheet. These defects, called disclination, introduce the points of positive or negative curvature. The family of the toroidal structures proposed by Ihara *et al.* have a common feature as the prototype, which was made from some C_{60} by

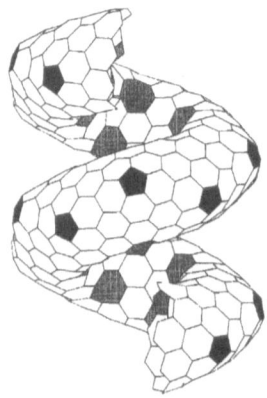

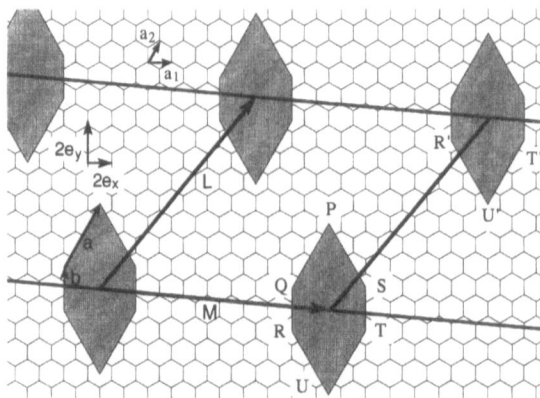

Fig. 1   The solid models of uniform tube helix C_{360}.

Fig. 2   The development maps of the uniform tube helix (type I) defined by the parameter set ($L_x = 13$, $L_y = 9$, $M_x = 19$, $M_y = -1$, $a_x = 3$, $b_x = 0$, $b_y = 1$).

substituting half of the 5-membered rings by 7-membered rings.[a–d] By changing the form of the centroid from a circle to a helix, a helically coiled tube is formed from a torus, and two types of helically coiled structures of nanometer scale using 5- and 7-ring pairs are proposed in ref. e.

The theoretically proposed helically coiled structures in ref. e were derived either from the tori proposed by Ihara and coworkers, [c] or those by Dunlap. [d] The torus proposed by Ihara *et al.* has the D_{5d} symmetry, and the chiral shift of the centroid along the C_5 axis produces a helix, which is called the helix, "C_{360}" hereafter. We call the helices obtained by the similar development maps (see Figs. 1 and 2) "uniform tube helices" hereafter, since they are based on uniform tubes and have only one type of chirality vector. Here the chirality vector makes the circumference of the tube. It specifies the electronic features in the case of straight carbon nanotubes. On the other hand, Dunlap proposed another family of tori [f] which have different geometrical features. In ref. e, a different kind of helix is also proposed using Dunlap's tori. These are called "non-uniform tube helices", since they are composed of sections of two different tubes and have two types of chirality vectors.

In a previous letter,[g] we intended to show that the helix C_{360} family has some remarkable features of band structures which are quite different from those found in straight tubules. In the present work, we attempt to reveal the systematic rule governing the band characteristics based on the method of the development map, and introduce the concept of "phason line." This concept makes the relation between structures and band characters visually clear, and the result implies that the electronic states of the nanostructures which have "classes" can be predicted by introducing such a new concept generally.

2. General construction of the uniform helix

The development maps belonging to the group of uniform tube helices have a two-dimensional periodical arrangement of the shaded zones defined by 8 parameters: $\vec{L}, \vec{M}, \vec{a}, \vec{b}$. The definition of these parameters are shown in Fig. 2. We need $\vec{a}$, $\vec{b}$ to determine the shape of the shaded zones. By cutting them out of tone graphic sheet and sticking side PQRU with PSTU, we get two 5-membered rings around P and U and two 7-membered rings around Q (or S) and R (or T). If there is no shaded zone on the development map, it becomes the straight carbon nanotube and $\vec{L}$ is the chirality vector: to form the 3D structure, the atoms joined by $\vec{L}$ are regarded as the same. Such a particular straight nanotube is called a "base tube" hereafter. $\vec{M}$ indicates the period along the 1D direction. A distorted shape of the hollow part is also possible, but to avoid complexity we discuss here only two types of arrangements classified in the previous work by Type I (Fig. 2) and Type II. These are restricted in that $\vec{b}$ or $\vec{a}$ is parallel to the sides of 6-membered rings. So, we need only 6 parameters $(\vec{L}, \vec{M}, a_x, b_y)$ or $(\vec{L}, \vec{M}, a_y, b_x)$ to describe the development maps of the helical tubes of type I and type II, respectively.

3. The relation of the parameters of development map and the topology of phason line

Now we will consider how to cover the development map with the Kekule structure. First, we put the typical Kekule structure on the graphitic sheet and obtain the model shown in Fig. 3. If there are disclinations such as 5- or 7-membered rings in the graphitic sheet, we find that the "•" positions must be placed on the centers of disclinations to form the Kekule structure. So we assign the "•" points to the disclination centers on the development map and try to expand the Kekule structures from each disclination center. If the positions of the neighboring disclinations are appropriate, the Kekule structures starting from different disclinations will coincide. In inappropriate cases, however, there is a border line on which the phase of the Kekule structures from the different disclinations changes (Fig. 4). This line has been named the "phason line" by Fujita. [h] What then is the appropriate condition of matching the Kekule structures without the phason line? The answer is that the x component of the relative vector between the two disclinations is a multiple of 3. Here the XY axes and the unit vectors are defined as shown in Fig. 3.

We then divide the development map of uniform tube helices by phason lines as noted above and pay attention to the x components of the vectors connecting the neighboring disclinations (Table 1). The number of such

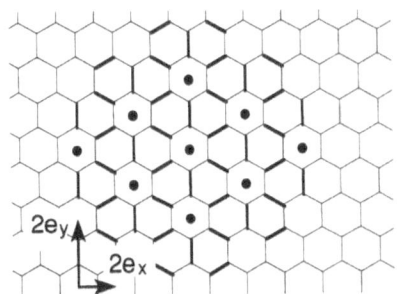

Fig. 3   The typical Kekule structure model on a graphitic sheet.

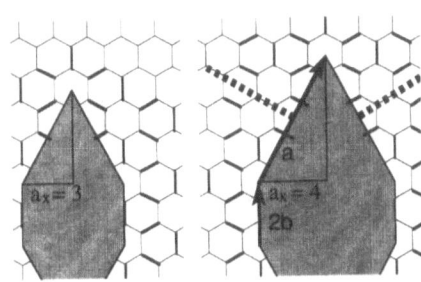

Fig. 4   A sample of phason lines; when a_x is not a multiple of 3, one phason line is formed between the 5-membered ring and the 7-membered ring.

Table 1 The x components of the relative vectors between the neighboring disclinations (see Fig. 5)

	general	type I	type II
t_{12}	a_x	a_x	$a_x(-3a_y)$
t_{23}	$2b_x$	0	$2b_x$
t_{24}	$M_x - \frac{1}{2}(a_x + 3a_y)$	$M_x - 2a_x$	M_x
t_{15}	M_x	M_x	M_x
t_{16}	$L_x - M_x - 2b_x - \frac{3}{2}(a_x - a_y)$	$L_x - M_x$	$L_x - M_x - 2b_x + 6a_y$
t_{17}	$L_x - 2b_x - \frac{3}{2}(a_x - a_y)$	L_x	$L_x - 2b_x + 6a_y$

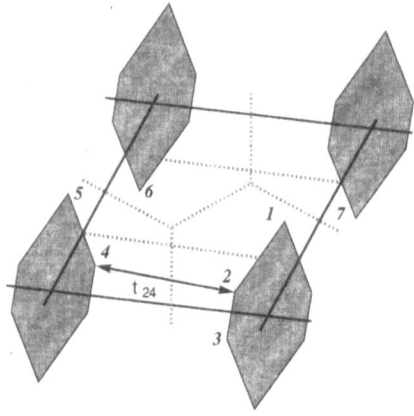

Fig. 5 The development map belonging to the group of uniform tube helix divided into some domains. Each domain includes a disclination.

components is 6 (Fig. 5). If they are not a multiple of 3, the corresponding border lines become the phason lines. Since the first column of the table is a general expression including not only type I or type II but also other cases, we write the expressions which consider the special cases of type I or II in the second and third columns for later discussion. Here, the limits of type I and II are realized by the condition, "$b_x = 0$, $a_x = a_y$" and "$a_x = -3a_y$," respectively. It must be pointed out that the x component of the chirality vector $\vec{L}$ itself is related to the existence of the phason line.

In this way, we find that the parameters which relate to the existence of phason lines are (L_x, M_x, a_x) for type I and (L_x, M_x, b_x) for type II. Fig. 6 shows the relation between these parameters and the pattern of phason lines for type I. Helices are covered with a single Kekule structure all over when all of the three parameters are multiples of 3, but phason lines appear in the other cases. More detailed classification of phason line patterns is necessary for the cases (4), (6) and (8) of Fig. 6. In the case of (4), two phason line patterns are generated according to the additional conditions satisfied by M_x, a_x; in case (4-1) mod $(M_x - 2a_x, 3) = 0$ and in case (4-2) mod $(M_x - 2a_x, 3) \neq 0$. Here mod $(n, 3)$ indicates the remainder of n divided by 3. In case (4-1) (case (4-2)), there is no phason line (there exists a phason line) enclosing the circumference of the helical tube. In the case of 6, two phason line patterns are also generated according to the additional conditions

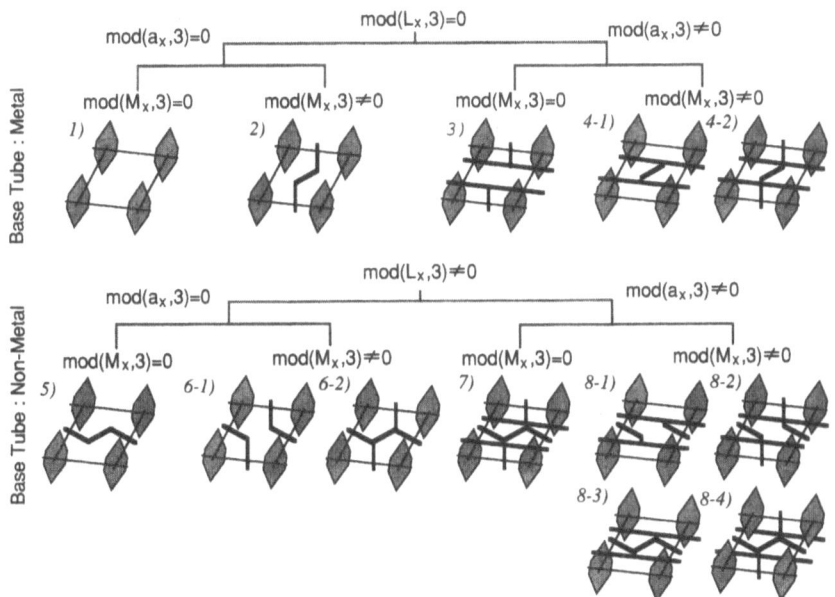

Fig. 6 The patterns of the phason lines on a uniform tube helix (type I).

Table 2 The four conditions which classify the phason line patterns (8-1)–(8-4) in Fig. 7

Figure	mod $(L_x - M_x, 3) = 0$	mod $(M_x - 2a_x, 3) = 0$
8-1	yes	yes
8-2	yes	no
8-3	no	yes
8-4	no	no

satisfied by L_x, M_x; (6-1) mod $(L_x - M_x, 3) = 0$ or (6-2) mod $(L_x - M_x, 3) \neq 0$. In the last case 8, four different phason line patterns (8-1)–(8-4) of Fig. 6 emerge corresponding to the four conditions shown in Table 2.

4. Tight-binding calculation

Since at least certain points on the surface of helices have a finite positive or negative curvature as a solid model indicates, there may be some effects of the σ bonding orbital. But such effects are not important for discussions of the qualitative feature of the π bands, as far as open-form systems are concerned. We have confirmed this by several calculations with and without σ bonding orbitals. [8] Each development map defines unit cells of the corresponding helical tube, and the 1D band structures can be calculated by varying the phase (ϕ) of the wave functions between neighboring cells. We use a single common hopping integral $-T$ ($T > 0$). The site energy is also taken commonly at each site.

A remarkable fact found by the band calculation is that the characteristics of the energy bands are governed by the phason line pattern. Namely, a family of helices with a common phason line pattern has the same feature of energy bands near the Fermi level. The systematics of the energy band feature is categorized in

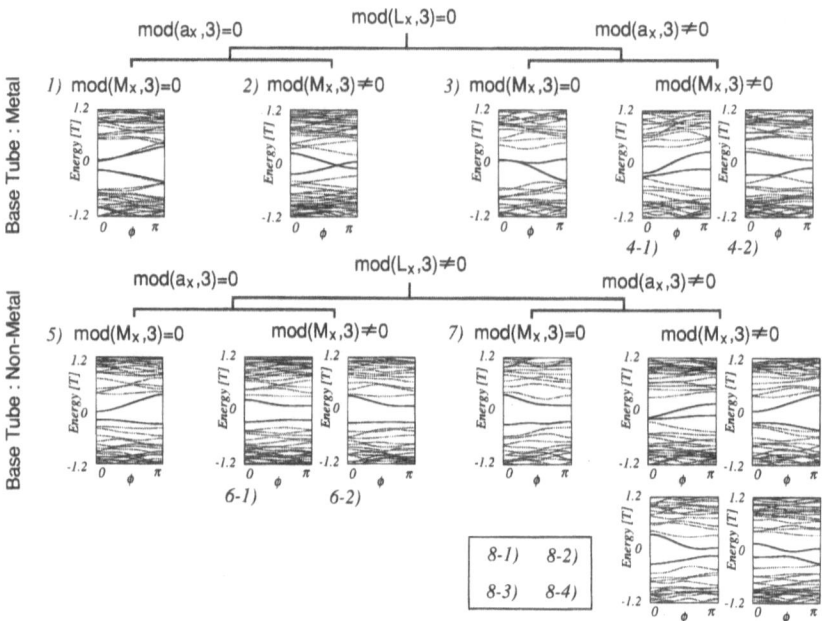

Fig. 7   The energy bands classified by parameters; uniform tube helix (type I). The unit of the energy is the absolute value T of the hopping integral between the nearest neighboring C $2p_z$ orbitals. The origin of energy is taken as the common site energy.

Fig. 7 in a way corresponding to that used for classification of the phason line pattern (Fig. 6). In this paper, we choose the band structures as the measure of the classification.

4.1 Uniform tube helix (type I)

This family includes C_{360}, which have the nearly degenerated partially flat bands of HOMO and LUMO as reported in previous works. [g) We calculated the band structures by varying 6 parameters (L_x, L_y, M_x, M_y, a_x, b_y), and found that only 3 parameters (L_x, M_x, a_x) determine their character. Moreover the remainder of each parameter divided by 3 is important. This classification matches exactly with that of the phason lines on uniform tube helices. The classification of the energy bands shown in Fig. 7 corresponds to the respective phason line patterns shown in Fig. 6. The other parameters (L_y, M_y, b_y) have no relation with the generation of phason lines.

If the chirality vector L is expressed as $\vec{L} = L_1\vec{a}_1 + L_2\vec{a}_2$ by the primitive vectors of graphite $\vec{a}_1$ and $\vec{a}_2$, the condition of mod $(L_x, 3) = 0$ is equivalent to mod $(|L_1 - L_2|, 3) = 0$. It is well known by the theory of straight nanotubes that the tube becomes metallic in the case of mod $(|L_1 - L_2|, 3) = 0$ and semiconductor in the other cases.[i-k)] Careful observation informs us whether or not L_x is a multiple of 3 has a dominant influence on the band structures; i.e. the HOMO-LUMO bands cannot touch each other irrespective of the other parameters when mod $(L_x, 3) \neq 0$. Therefore it is concluded that the uniform tube helix (Type I) cannot be metallic unless the corresponding base tube is metallic. However, it can be semiconducting, even if the corresponding base tube is metallic.

Some remarkable relations between the band structures and the pattern of phason lines on helices are found. Namely, i) when a complete Kekule structure is established all over the helix (in other words, there is no phason line), the band gap appears at Γ point ($\phi = 0$) even when the base tube has a metallic character, ii) helices become metallic when the phason line exist along the circumference around the tube and iii) the nearly degenerated HOMO and LUMO bands at the small ϕ region appear only when the three conditions mod $(L_x, 3) = 0$, mod $(a_x, 3) \neq 0$, mod $(M_x, 3) = 0$ are satisfied. The wave functions corresponding to these bands tend to be localized around the disclination centers. These features have been reported elsewhere.

As mentioned in the previous section, more detailed classification of the phason line patterns is necessary depending on the value of mod $(M_x - 2a_x, 3)$ and mod $(L_x - M_x, 3)$. They are shown in (4-1), (4-2), (6-1), (6-1), (8-1), (8-2), (8-3), (8-4) of Fig. 6a. In Fig. 7 the behaviors of the HOMO and the LUMO band are classified according to the same classification rule as for the phason line patterns. Therefore the feature of the energy bands of (8-1) in Fig. 7 corresponds to the general feature of the helices with phason line pattern (8-1) in Fig. 6, and so on. A remarkable fact we found is that we can make a one-by-one correspondence between all of the phason lines patterns and the features of the band structures. It is true that there are band structures quite similar to each other in Fig. 7, but the clear classification in the case of mod $(L_x, 3) = 0$ (*i.e.* the base tube is metallic) is important.

4.2 Uniform tube helix (type II)

In this case, the three parameters (L_x, M_x, b_x) play a dominant role in determining the band features. These parameters are the same as those which determine the patterns of the phason lines. Major relationships between the feature of the bands and the phason line patterns are similar to those of Type I. Namely we observe the following facts: i) the dominant classification is possible by whether L_x is a multiple of 3 or not. HOMO-LUMO bands do not touch together when L_x is not a multiple of 3, or when a phason lines run along the outer side parallel to the tube axis. ii) The band has a gap when the surface of the helix can be covered with a single Kekule structure, iii) the band becomes metallic (or narrow gap semiconductor) when the phason lines run around the tube. Since the ladder-like pattern of the phason line (Fig. 6, (3)) does not appear for type II, the nearly degenerated bands (Fig. 7, (3)) do not appear either. The figures for this family are omitted.

5. Summary

As discussed in the previous sections, the apparently complex band structures of helices can be classified on the parameter space of the development map. The domains of the parameter space for the band classification are the same as those which define the way of covering the helices with the Kekule structure. So we introduced the concept of the phason line and clarified the relation between its pattern and the band structures. Similar relations are also found in the group of non-uniform tube helices.[1] Discussion on the mechanism of the phason line in determining the band features remains to be made, but it is certain that such a clear rule as that found here, relating the topology of bond networks and band features, invokes many ideas towards its complete elucidation.

PUBLICATIONS

1. R. Tamura and M. Tsukada, *Phys. Rev.*, **B49**, 7697–7708 (1994).
2. M. Tsukada, K. Akagi, R. Tamura and S. Ihara, *Surface Review and Lett.* [in press].
3. R. Tamura and M. Tsukada, *Phys. Rev.*, **B52**, 6015–6026 (1995).
4. K. Akagi, R. Tamura, M. Tsukada, S. Itoh and S. Ihara, *Phys. Rev. Lett.*, **74**, 2307–2310 (1995).
5. K. Akagi, R. Tamura M. Tsukada, S. Itoh and S. Ihara, *Phys. Rev.*, **B** [in press].

REFERENCES

a. S. Itoh, S. Ihara and J. Kitakami, *Phys. Rev.*, **B 47**, 1703 (1993).
b. S. Itoh and S. Ihara, *Phys. Rev.*, **B 48**, 8323 (1993).
c. S. Ihara, S. Itoh and J. Kitakami, *ibid.*, 12908 (1993).
d. B. I. Dunlap, *Phys. Rev.*, **B 46**, 1933 (1992).
e. S. Ihara, S. Itoh and J. Kitakami, *Phys. Rev.*, **B 48**, 5643 (1993).
f. B. I. Dunlap, *Phys. Rev.*, **B 49** 5643 (1994).
g. K. Akagi, R. Tamura, M. Tsukada, S. Ihara and S. Itoh, *Phys. Rev. Lett.*, **74**, 2307 (1995).
h. M. Fujita, *Fullerene Sci. Technol.* (USA), 1, 365 (1993).
i. N. Hamada, S. Sawada and A. Oshiyama, *Phys. Rev. Lett.*, **68**, 1579 (1992).
j. G. Dresselhaus, M. S. Dresselhaus, and R. Saito, *Solid State Commun.*, **84**, 201 (1992).
k. J. W. Mintmire, B. I. Dunlap and C. T. White, *Phys. Rev. Lett.*, **68**, 631 (1992).
l. K. Akagi, R. Tamura M. Tsukada, S. Itoh and S. Ihara, *Phys. Rev.*, **B** [in press].

37

Ab initio Investigation of Metal-doped B_{12} Solids

Shigeki Gunji [a] and Hiroshi Kamimura [b]

[a] Institute of Physics, University of Tsukuba
Tsukuba-shi, Ibaraki 305-0006, Japan
[b] Institute of Physics, Graduate School of Science, Science University of Tokyo
Shinjuku-ku, Tokyo 162-0825, Japan

Purpose of the present study

All pristine boron (B) crystals have molecularly based exotic structures. Some of them have a marked similarity to the fcc solid C_{60}. The achievement of moderately high transition temperatures in the A_xC_{60} superconductors suggests that a superconducting phase may appear by metal doping into such semiconducting B solids. Especially for T_c, a high value is expected originating from the interaction of conduction electrons with high frequency phonons caused by B-B covalent bondings. The recent discovery of superconductivity at 23 K in a Y-Pd-B-C quaternary alloy [a] also illustrates the importance of exploring boron compounds as a candidate superconductor. In this report, a theoretical prediction is made on the stable geometry of new boron intercalation compounds. We have performed LDA Gaussian basis calculations of the total energy, heats of formation, lattice constant, electronic structure, density of states, and bulk modulus of the putative compounds, using *ab initio* pseudopotentials. The possibility of superconductivity is also discussed.

Contents of the present study

1. Introduction

The α-rhombohedral boron (α-rh B) has a structure of a slightly compressed fcc rhombohedral primitive cell along the body-diagonal (111) direction (the c-axis in the hexagonal description), and an icosahedral cluster occupies each vertex (cf. Fig. 1). This crystal is significantly analogous to the fcc phase of solid C_{60}, for both translational and rotational symmetry. Iwasa *et al.* have recently reported a two-dimensional C_{60} polymer [b] which is suggested to have a rhombohedral structure.[c] Its covalent-bonding-type cohesion and rhombohedral ordering of molecules indicate much closer resemblance between α-rh B and C_{60}-based solids. Only the 2D intermolecular bonding profile is different from the 3D fcc-like network [d] of icosahedral clusters in α-rh B.

Icosahedral cluster in α-rh B is an almost regular icosahedron B_{12} [e,f], where twelve B atoms are arranged at all the vertices of icosahedron (cf. Fig. 1). The intramolecular cohesion mechanism is a three-center covalent bonding type, where the highest amplitude of the total charge density distribution appears near

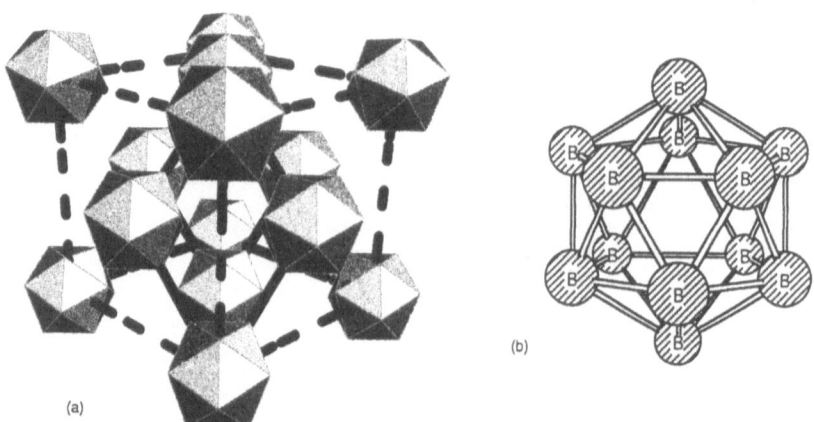

(b)

(a)

Fig. 1 Schematic views of (a) nearly fcc α-rh B crystal and (b) its icosahedral component, B_{12} cluster. The solid lines on (a) designate a rhombohedral primitive unit cell while the dotted lines designate a distorted fcc unit cell. Note that intra-B_{12} bondings in (b) are drawn for convenience and are different from the actual multi-center bondings.

the center of a triangular face of the icosahedron. Each B atom has an additional radial p_π bond pointing outward from B_{12} in the direction of every fivefold axis. These p_π orbitals are bonded together between the nearest-neighbor B_{12} and form two types of covalent crystals, α-rh and β-rh B.

The β-rh B also has a rhombohedral unit cell, but rather elongated along the (111) direction. In this crystalline phase, the corners and the medium points of the edges are all occupied by B_{12} icosahedra. In addition to these clusters, a single B atom is located at the center of the unit cell which corresponds to the octahedral site in fcc crystal, and two aggregates of three icosahedra are arranged symmetrically to it, along the major body-diagonal.[f] The β-rh B is a high temperature phase above the α-rh phase, and can be more easily prepared in experiments than α-rh B, because the melting point in B solids is as high as 2365 K,[g] whereas α-rh B is stable below 1370 K.[h] A metal doping to β-rh B is also of interest, but the crystalline complexity and already occupied octahedral sites are expected to give rise to a different situation as compared to the doped C_{60}. We have thus chosen the α-rh B as the parent material in spite of the experimental difficulty. An *ab initio* prediction of a new material requires considerable CPU time, but it allows greater opportunity to clarify the physical quantities of especially these experimentally hard-to-make materials. Working with another fcc system composed of icosahedra also leads to a better understanding of structural influence over the physical properties in A_xC_{60}.

In our previous papers,[1-4] we have pointed out the possible existence of α-rh B derivatives, Li_xB_{12} ($x = 1-3$). We simply calculated their electronic structures with use of density functional formalism within local density approximation (LDA), without considering any geometrical optitimizations of α-rh B. However, most electronic properties near the Fermi level are particularly sensitive to structural relaxation, and need to be more precisely evaluated, especially for a discussion on superconductivity. The aim of this report is to give accurate electronic structure studies on more extensive A_xB_{12} (A=Li, Ca; $x =1-4$) systems, and also to present an analysis of superconductivity in these wide ranges of putative compounds. A preliminary report has been given in a conference proceedings [4] only for Li_3B_{12}.

2. Models

The α-rh B is of interest as a parent material of new compounds since it has a number of interstices inside. The already known boron rich solids, boron carbide $(B_{12}C_3)$,[i] boron phosphide $(B_{12}P_2)$,[i,j] and boron are arsenide $(B_{12}As_2)$,[i,j] are regarded as α-rh B derivatives, where C, P, or As atoms are respectively inserted into these interstices of α-rh B. They show a striking feature in observed band gap energies,[j-l] which change widely from the infrared to ultraviolet region. However, no experimental studies on metal-doped α-rh B have been reported so far. Two fundamental questions have been raised in investigating metal-doped α-rh B theoretically. The first is at which position the metals should be doped, and the second is how the α-rh B unit cell is deformed in size and/or shape by doping. To respond to these questions, we assume the following four realistic sites. Then the stability is carefully analyzed allowing lattice relaxations: (a) An octahedral (O) site which corresponds to the body-center in the unit cell of α-rh B. (b) Two tetrahedral (T_d) sites per unit cell. (c) Both ends of an interstitial chain along the longest body-diagonal, that equal to P positions in $B_{12}P_2$ (called interstitial (int) sites). (d) The center of the hollow cluster B_{12}, which we have not dealt with in the previous papers.

Geometrical conditions of α-rh B are like those of pristine fcc C_{60}, but the intermolecular cohesion mechanism is quite different. In contrast to the intermolecular van der Waals cohesion in fullerite, intermolecular coupling in α-rh B is of a covalent bonding type due to transfer interactions between neighboring B_{12} icosahedra. It is therefore hard to dope relatively heavy alkali or alkaline-earth metals of large ionic radius into α-rh B. In order to aid construction of our models, we have calculated the site radius based on a simple rigid-sphere model, using available experimental information [i] on α-rh B and $B_{12}P_2$. The structural data of α-rh B and $B_{12}P_2$ are given in Ref. 2. The atomic radius of B atom required in our calculation has been estimated from the experimental B-B covalent bond length, which is approximately 0.88 Å on the average. The calculated results are summarized in Table 1, together with the ionic radius of alkali and alkaline-earth metal ions given by Pauling.[m] In our analysis, the IIA elements Be^{2+} and Mg^{2+} are not investigated in spite of the small values in ionic radius, because their chemical properties are quite different from those of the others.

Table 1 Estimated site radius by means of a simple rigid sphere model calculation. In this table, we also give the ionic radius of alkali and alkaline-earth metal ions. Li^+, Na^+ and Ca^{2+} ions are small enough and able to occupy all of the sites

Site	Radius (Å)	Element	Ionic radius (Å)	Element	Ionic radius (Å)
O	1.245	Li^+	0.60[a]		
T_d	1.167	Na^+	0.95[a]		
int	1.032	K^+	1.33[a]	Ca^{2+}	0.99[a]
B_{12}-center	0.826	Rb^+	1.48[a]	Sr^{2+}	1.13[a]

[a] Reference m.

The spheres of the four sites range over 0.826–1.245 Å in radius, following the order $O > T_d > int > B_{12}$-center. We have also found that even the most spacious O site can only accommodate relatively small ions, such as Li^+, Na^+, Ca^{2+} and Sr^{2+}. Thus the smallest alkali metal Li and the smallest alkaline-earth metal Ca have been chosen as dopants.

Depending on the positions and concentration, the realistic model systems, $LiB_{12}(O)$, $Li_2B_{12}(T_d)$, $Li_2B_{12}(int)$, and $Li@B_{12}$, are constructed for the Li-doping case. On each of the systems, a numerical work has been conducted as described in the next section. The $Li_2B_{12}(O)$ denotes the Li doping to every O site. In $Li_2B_{12}(T_d)$ or $Li_2B_{12}(int)$, two Li atoms per unit cell are doped into two T_d or interstitial sites respectively. The $Li@B_{12}$ represents the case in which Li atoms occupy the centers of all the B_{12} clusters. Based on total-energy analysis, we have also investigated other systems in the same way, corresponding to Li-dopings into more than one kind of the above-mentioned four sites at a time, including Li_3B_{12} — the same way of doping as for the superconducting composition A_3C_{60}. The Ca-doping cases, including $Li_{3-x}Ca_xB_{12}$, are also studied in order to search a new superconductor in the wide range of new materials.

3. Computation method

To make a theoretical prediction of a novel material, one must determine its stable structure. To make sure whether these compounds really exist or not, first-principles total energy analyses are performed by means of the *ab initio* pseudopotential [n] Gaussian basis calculation within the local-density-functional theory. [o,p] The details of the calculation are found in our previous paper [2] and also in Ref. 6. The exchange-correlation potential is included via the Perdew-Zunger parametrization [q] of the Ceperley-Alder functional. [r] In expanding Bloch functions, the 19-Gaussian basis set (spin degeneracies included) is employed per B atom (3 Gaussians for s-type orbitals, 2 for p-types, and 2 for d-types) and as many Gaussians per Li or Ca atom. Gaussian exponents used are as follows: (a) For B; (0.161, 0.455, and 1.77) as s-type Gaussians, (0.265 and 1.46) as p-types, and (0.140 and 0.240) as d-types. (b) For Li; (0.043, 0.145, and 0.374) as s-types, (0.072 and 0.828) as p-types, and (0.073 and 0.177) as d-types. (c) For Ca; (0.043, 0.363, and 5.21) as s-types, (0.042 and 0.274) as p-types, and (0.242 and 1.98) as d-types. These basis sets are enough to reproduce the already known band structure of each bulk crystal. High numerical precision is achieved by employing a large number of k points (up to 10 points) in 1/12 of the Brillouin zone.

As the basic structure is composed of B_{12} icosahedra, the energy-minimum is achieved with use of a pair of structural parameters, the length of the lattice vector a and the mean radius of the icosahedral B_{12} cluster r. The actual procedure for obtaining the optimized geometry is given in Ref. 6.

4. Results

4.1 Stability and energetics

First of all, the stability and energetics of these α-rh B derivatives are studied in detail by first-principles total energy calculations, taking into account changes in intra- and intermolecular distances. The total energy of isolated atom A (A=Li

Table 2 Calculated heats of formation ΔE of A_xB_{12}. $Li_2B_{12}(int)$, $Li@B_{12}$ and all of the Ca-doped A_xB_{12} are unstable, as the minus sign of ΔE shows. $Li_2B_{12}(T_d)$ is metastable. The others are stable enough and should be synthesized like K_xC_{60}

System	Site	ΔE eV/cell	eV/atom	System	Site	ΔE eV/cell	eV/atom
LiB_{12} (O)	O	2.38	0.183	CaB_{12}	O	−7.23	−0.556
Li_2B_{12} (T_d)	T_d	2.29	0.164	Ca_3B_{12}	$O + T_d$	−51.13	−3.409
Li_2B_{12} (int)	int	−2.70	−0.193	Li_2CaB_{12} (O)	O (Ca)+ T_d(Li)	−3.20	−0.213
$Li@B_{12}$	B_{12}-center	−2.86	−0.220				
				KC_{60}	T_d	8.5[a]	0.139
Li_3B_{12}	$O + T_d$	4.63	0.309	K_2C_{60}	T_d	17.7[a]	0.285
$Li@Li_3B_{12}$	$O + T_d + B_{12}$-center	2.92	0.182	K_3C_{60}	$O + T_d$	22.6[a]	0.359

[a] Reference s.

or Ca) has also been calculated for comparison, within the same framework of LDA, employing the same Gaussian exponents and *ab initio* pseudopotentials.

In Table 2, the calculated energy gains ΔE are given for a wide range of metal-doped B_{12} solids. The ΔE is defined by the following exothermic reactions, B_{12} (α-rh crystal) + xA (atom) $\leftrightarrow$ A_xB_{12} + ΔE. According to the analysis, Li-doped compounds have been demonstrated to give sufficient heats of formation ΔE, whereas all the Ca-doped compounds have negative ΔE, and it has been found that they, including Li_2CaB_{12}, are unstable against our simple rigid-sphere model estimate.

We have also clarified that Li atoms have a preference for O site rather than T_d site in Li_xB_{12}, while in the doped C_{60}, alkali atoms prefer T_d sites to O sites.[s] This result is consistent with our estimate by means of a simple rigid-sphere method. Lithium atoms, however, do not occupy interstitial site and B_{12}-center against our model calculation. The *ab initio* total-energy studies also demonstrate that $Li_2B_{12}(int)$ is unstable, although we have calculated both from the α-rh B and $B_{12}P_2$ structures. An intermediate structure, less deviated from α-rh structure than $B_{12}P_2$ in proportion to the ratio of Li:P in ionic radius, has also been used as initial data, but it remains unstable. We thus concluded that alkali metal atoms do not occupy interstitial sites, as in the doped C_{60}. Our LDA band structure calculation reveals the main reason why the $Li_2B_{12}(int)$ is unstable. According to the calculations, all stable and metastable Li_xB_{12} are "donor-type" metals, as seen in the next subsection, but the calculated energy band structure of unstable $Li_2B_{12}(int)$ instead shows an "acceptor-type" metallic behavior. a,b) Since some p_π orbitals point towards interstitial sites, Li 2s wave function in $Li_2B_{12}(int)$ overlaps substantially with these p_π orbitals, and as a result, two Li-B hybrid bands newly appear in the higher valence band region of α-rh B. Because they push up higher valence bands, E_F crosses the highest and second-highest valence bands, and hole-like carriers appear in this case, indicating that Li gets an extra electron and behaves as an anion in $Li_2B_{12}(int)$. This unusual behavior yields anomalously high total energy that leads to the instability of $Li_2B_{12}(int)$.

The total-energy calculations have also revealed that, as long as Li concentration remains in a lower stoichiometric level, the $LiB_{12}(O)$ is the most stable. The $Li_2B_{12}(T_d)$ also shows relatively large ΔE but it is a metastable compound, since in doping two Li atoms per B_{12}, one Li atom prefers to occupy the O site while the other is accommodated in the T_d site.

When Li concentration is further increased, the stoichiometric Li_3B_{12} appears. The Li_3B_{12} is the most stable compound among Li_xB_{12}, because of the largest ΔE of approximately 0.309 eV/atom. As seen in Table 2, it is a value comparable to that of the already obtained stable material, K_3C_{60},[s] and this comparison allows for the existence of Li_3B_{12}.

In the highest concentration, $Li@Li_3B_{12}$ is obtained. It corresponds to the maximum concentration, for interstitial sites are too close to O sites and the occupation of interstitial sites by additional Li atoms do not occur when other Li atoms already occupy the O sites.

The energy gain originated from the ion-ion electrostatic interactions for a single dopant Li is of the order of approximately 75 eV per cell and they increase linearly with increase of Li concentration. The calculated ΔE, however, does not show a linear dependency on concentration. According to our total-energy analysis, the Hartree energy is essentially responsible for the magnitude of ΔE, although the ion-ion electrostatic energy gains play an important role in stabilizing Li_xB_{12}.

4.2 Band structure

The calculated energy band structures of the stable and metastable Li_xB_{12} (x = 1 to 4) are presented in Fig. 2. The LDA band calculation indicates that the general features of the energy bands in Li_xB_{12} are mostly attributable to the band profile of α-rh B. The energy band structure of α-rh B is shown in Fig. 2(a). The undoped α-rh B is an indirect semiconductor with a gap energy of approximately 2 eV.[1] The calculated energy band structure of α-rh B within LDA compares quite well with experiments,[1] although the calculated band gap shows a smaller energy value of 1.53 eV. We have also compared it with other LDA calculations,[1] and have found reasonable agreement. The top of the valence band is located at Z point in the Brillouin zone shown in Fig. 2(f), and the bottom of the conduction band is at Γ point in the Brillouin zone. However, the highest occupied valence band is less dispersive along the Γ-Z direction, and one may regard α-rh B as an almost direct gap semiconductor. Although this highest valence band exhibits flat dispersion along the Γ-Z direction, it shows a dispersive feature along the Γ-L direction, suggesting a strong anisotropy near the Γ point. No localized electronic states appear as the ground states of α-rh B, although many experiments suggest the existence of intrinsic acceptor levels in β-rh B.[u]

Figure 2(b) show the calculated energy band structure of $LiB_{12}(O)$. Overall energy-band features of $LiB_{12}(O)$ are almost identical to those of α-rh B, but a Li-character band appears in a higher energy region above 20 eV beyond the range displayed in Fig. 2(b). Lithium 2s electrons are transferred from this band to the lowest conduction band, and in consequence, the Fermi level E_F crosses this lowest conduction band and $LiB_{12}(O)$ shows a "donor-type" metallic behavior. This band is nearly half-filled and all of the donated charges (one electron per B_{12}) are itinerant. Our charge density calculations also reveal that this band has a B(2s)-(2p) hybridized character. This B(2s)-(2p) conduction band extends in the narrow region between 10.2 and 11 eV, while in the pristine α-rh B, the corresponding band lies in the range between 9.5 and 11 eV. The same kind of narrowing of bandwidths occurs in several bands. It arises from the smaller orbital overlaps, due to both intra- and intermolecular relaxations. In addition to the

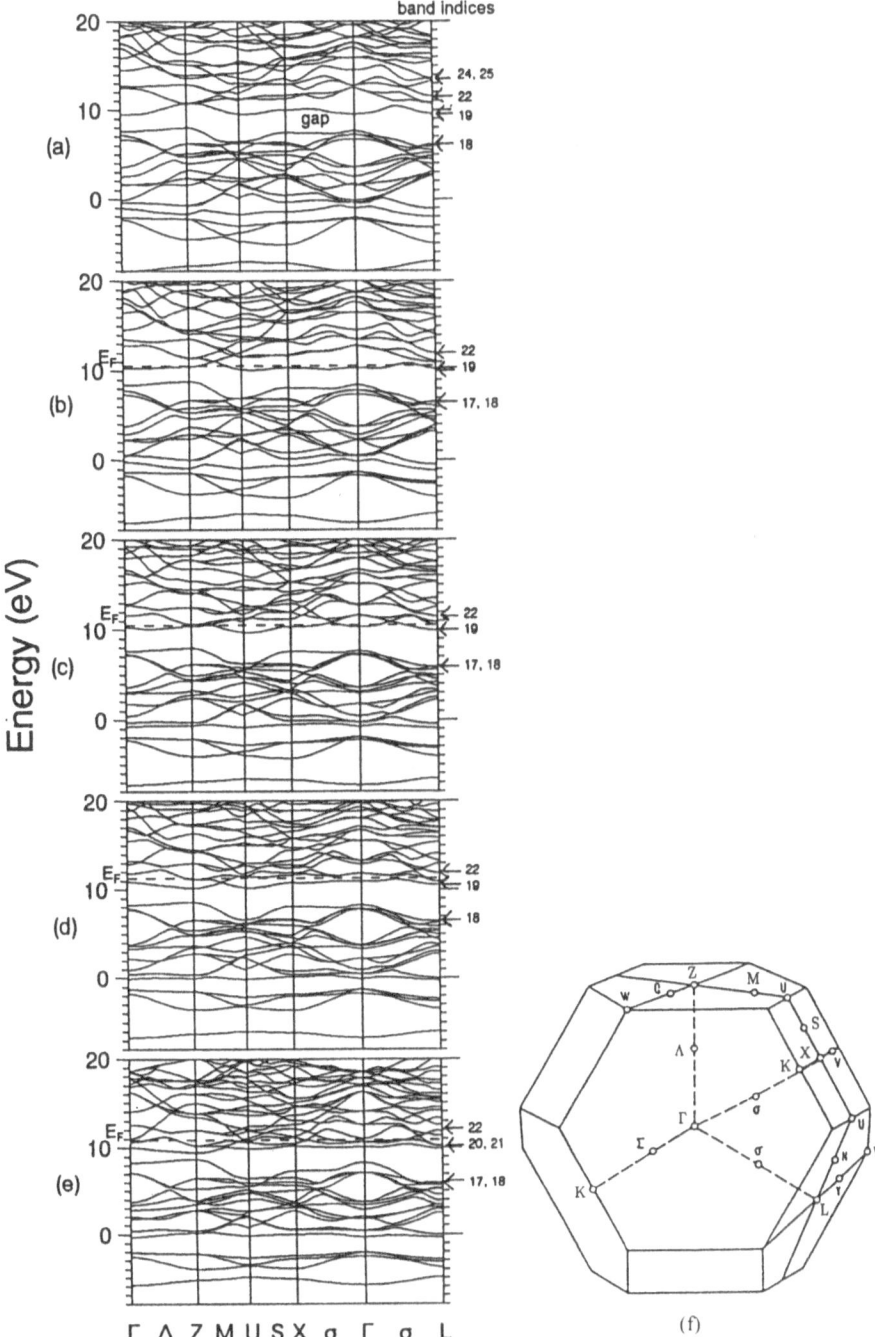

Fig. 2 Calculated energy band structure of (a) semiconducting α-rh B and its metallic derivatives; (b) true stable LiB$_{12}$(O), (c) the metastable Li$_2$B$_{12}$(T$_d$), (d) the most stable Li$_3$B$_{12}$, and (e) the stable Li@Li$_3$B$_{12}$. The band indices are numbered from the lowest band. (f) The Brillouin zone of α-rhmobohedral boron.

narrow bandwidth of the lowest conduction band (of order approximately 0.8 eV), most of its energy dispersion is almost flat in the neighborhood of E_F, and a very strong peak appears in the density of states at E_F, as seen in the next subsection. This will be very advantageous to the occurrence of superconductivity.

The calculated energy band structure of the metastable Li$_2$B$_{12}$(T_d) is shown in Fig. 2(c). Charge transfer similar to the previous case is brought about by doping. From the analysis of the charge density distribution at Γ point, we notice that the 22nd conduction band has some Li(2s) character, whose position is approximately 2 eV above E_F. This band originally has a B(2s)-character. The corresponding band in α-rh B is the 24th in Fig. 2(a), but has moved down through the hybridization with Li(2s) character in Li$_2$B$_{12}$(T_d). A charge density peak of this band appears just on the Li atom, but these electrons are transferred into lower two bands, the 19th B(2s)-(2p) hybridized band and the 20th B(2p) band with a little admixture of Li(2s) character. As the result of charge transfer, E_F crosses the lowest 19th and the second lowest 20th conduction bands, and Li$_2$B$_{12}$(T_d) also shows a metallic behavior. The lowest B(2s)-(2p) hybridized conduction band is almost full-filled, but it yields a few hole-like carriers around σ point. The second lowest B(2p) conduction band is little occupied, but a low concentration of electrons are found around the Z and L points. The calculated energy band structure has demonstrated that the metastable Li$_2$B$_{12}$(T_d) metallic system has two types of charged-carriers in the same order. A "band gap" of about 1.7 eV exists in Li$_2$B$_{12}$(T_d) below the two partially occupied conduction bands. The gap has changed its character from direct in α-rh B to indirect, where the "top" of valence bands in Li$_2$B$_{12}$(T_d) is at Z point and the "bottom" of conduction bands is located at U point. In the case of LiB$_{12}$(O), the almost direct gap of approximately 1.5 eV is open and not changed so much as compared to α-rh B.

The calculated energy band structures of the most stable Li$_3$B$_{12}$ are shown in Fig. 2(d), where Li atoms are doped into both O and T_d sites. Results similar to previous cases have been obtained. Associated with considerable hybridization with a Li(2s) character band, the 22nd band has moved further down into the energy region between 11.7 and 13.5 eV, whose position is about 1 eV above E_F. From the 22nd B(2s)-Li(2s) strongly hybridized band, Li 2s electrons are transferred into the lower 19th, 20th and 21st bands. Due to higher Li concentration, many more electrons than in previous cases are transferred, and the 19th band is filled completely. The E_F crosses the almost degenerate 20th and 21st conduction bands in Li$_3$B$_{12}$. Sufficient electrons are doped into these bands, and they occupy k states around the Z and L points. These bands originally have a B(2p) character in undoped α-rh B, but in Li$_3$B$_{12}$, the B(2p) character is mixed with the Li(2s) character in almost a one-to-one ratio. In this way, the "donor-type" metallic state is obtained as the ground state of Li$_3$B$_{12}$, and one can say that there are some similarities in the electronic structure as well as in the crystalline structure between the most stable Li$_3$B$_{12}$ and the superconducting A$_3$C$_{60}$.

A striking feature appears in the case of Li@Li$_3$B$_{12}$, where its energy band structure is shown in Fig. 2(e). The originally B(2s)-character band, whose energy range in α-rh B lies between 13 and 15 eV (the 24th band in α-rh B), is now considerably hybridized with Li(2s) character and appears in the region from 10.5 to 12 eV in Li@Li$_3$B$_{12}$. In this region, there already exist the two almost degenerate B(2p)-Li(2s) character bands, and these three bands interact with one another. As a result, the 20th band has dominant Li(2s) character along the U-X

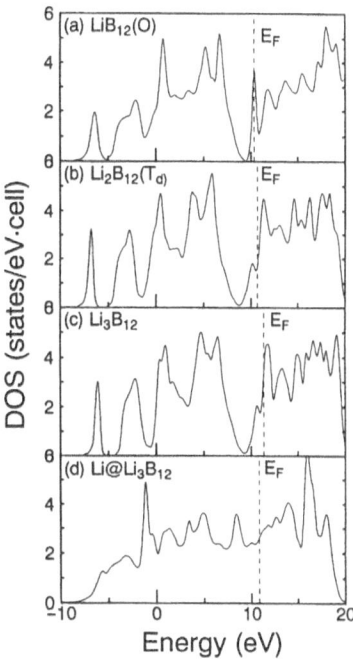

Fig. 3 Calculated density of states of the stable and metastable Li_xB_{12} (x = 1 to 4).

Table 3 DOS obtained at E_F of stable and metastable Li_xB_{12}. LDA calculation of $N(E_F)$ for superconducting K_3C_{60} is listed for comparison

System	$N(E_F)$ states/eV·cell	states/eV·atom
LiB_{12} (O)	3.39	0.26
Li_2B_{12} (T_d)	1.59	0.11
Li_3B_{12}	3.78	0.25
$Li@Li_3B_{12}$	2.55	0.16
K_3C_{60}	17.96[a]	0.28

[a] Reference 22. The calculated values of $N(E_F)$ are ambiguous, ranging from 12.0 to 25.0 states/eV·cell.[s, v] Here we have selected a moderate value for reference.

direction with a little admixture of B(2s) character, whereas it has the B(2p)-Li(2s) character for the other k points in the Brillouin zone. As for the 22nd band, the opposite occurs. Besides these changes, the 19th valence band that is fully occupied by the transferred Li 2s electrons is lowered, and the Fermi energy decreases in spite of the increase of Li concentration. The E_F crosses the 20th and 21st conduction bands. Since the 20th band has a Li(2s) character along the U-X direction, the electrons in this region remain around Li ions and do not transfer to other B-character bands.

4.3 Density of states

Knowledge of the density of states (DOS) $N(E)$ near the Fermi level is very important to understand the mechanism of superconductivity. In order to investigate the possibility of superconductivity in Li_xB_{12} metallic materials, $N(E)$ has been calculated and is delineated in Fig. 3 for pristine and doped α-rh B. The Fermi-level DOS, $N(E_F)$, is also given in Table 3, together with the calculated value of superconducting K_3C_{60} within LDA.[s,v] The tetrahedron method [23] is used in actual calculation of

$$N(E_F) = \Sigma_n \Sigma_{\vec{k}} \, \delta \, (E_F - \varepsilon_n \, (\vec{k})). \tag{1}$$

Since in $LiB_{12}(O)$, the lowest B(2s)-(2p) conduction band that crosses E_F is almost flat, this single narrow band yields a pronounced spike structure in $N(E)$ at E_F, which is expected to play an important role in superconductivity. Although a number of LDA calculations on $N(E_F)$ of the K_3C_{60} have been reported so far, the calculated $N(E_F)$ is not definite yet, ranging from 0.19 to 0.39 states/eV•atom.[s,v] The value of $N(E_F)$ per atom in $LiB_{12}(O)$ is comparable to the moderate $N(E_F)$ of K_3C_{60}, as seen in Table 3.

In $Li_2B_{12}(T_d)$, the energy band structure is similar to that of $LiB_{12}(O)$ and only minute differences in $N(E)$ are found. However, due to a large charge transfer, E_F lies at a slightly higher level away from the B(2s)-(2p) hybridized conduction band, and $N(E_F)$ decreases as compared with $LiB_{12}(O)$.

In Li_3B_{12}, the amount of charge transfer increases much more than in the previous two cases, and E_F, whose energy position is 11.4 eV, crosses two higher almost-degenerate B(2p)-Li(2s) hybridized conduction bands, and in consequence $N(E_F)$ becomes again very high. The most stable Li_3B_{12} is a match for, or even surpasses the superconducting K_3C_{60} in the obtained value of $N(E_F)$ per atom.

In $Li@Li_3B_{12}$, E_F instead takes a lower value of about 10.8 eV, contrary to the increase in Li concentration. The decrease in the Fermi energy is related to the lowering of bands, as mentioned in the previous subsection. The DOS structure of $Li@Li_3B_{12}$ is also changed drastically. The E_F crosses the almost degenerate 20th and 21st bands, but the obtained $N(E_F)$ is not so high, as compared to

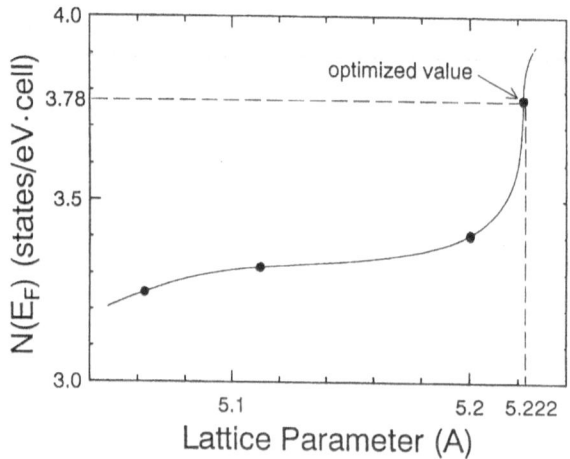

Fig. 4 Calculated $N(E_F)$ of Li_3B_{12} as a function of the lattice constant a. No simple linear relation exists in Li_xB_{12}, but obtained $N(E_F)$ (solid circles) shows monotonic increase in this region.

$LiB_{12}(O)$ or Li_3B_{12}.

Lithium doping yields changes in $N(E_F)$ according to the band occupation of transferred electrons, as mentioned above. With the exception of the metastable $Li_2B_{12}(T_d)$, Li_xB_{12} ($x = 1$ to 4) mostly show high density of states at E_F. Higher values of $N(E_F)$ are mainly due to the nature of the lowest and second-lowest conduction bands of the parent α-rh B. The $N(E_F)$ is also enhanced by the narrowing of the energy bandwidths associated with Li doping.

In A_xC_{60}, there exists an approximate linear relationship between $N(E_F)$ and the lattice constant, and the variation in T_c has been attributed to changes in the lattice constant. In Li_xB_{12}, no such simple linear relation exists, reflecting an intermolecular covalent bonding type cohesion, but the calculated $N(E_F)$ shows a monotonic increase, as a function of the lattice constant. Figure 4 illustrates the calculated $N(E_F)$ of Li_3B_{12} as a function of the lattice constant. It indicates that Li_3B_{12} would show a high value or T_c with the help of low pressure, affected by the rapid increase in $N(E_F)$.

4.4 Bulk modulus

The bulk modulus of the stable and metastable materials has also been calculated to obtain information about isotropic phonon mode.

The total energy surface E is assumed to have the following quadratic form, as a function of the length of the lattice vector a and the mean radius of the icosahedral B_{12} cluster r:

$$E(a, r) = E_{min} + k_a(a - a_{opt})^2 + k_r(r - r_{opt})^2 + k_{ar}(a - a_{opt})(r - r_{opt}) \quad (2)$$

Here the subscripts min and opt denote the corresponding physical quantities around the optimized geometry, and the coefficients k_a, k_{ar} and k_r are disposal parameters determined by the total-energy calculations. Since the calculated total energy shows a quadratic response to the deviation from a_{opt} or r_{opt}[3], Eq. (2) is a natural assumption.

Although the total energy obtained is a quantity for the definite volume (per unit cell volume V), the first order derivative with respect to r is well defined, for r is independent of the unit cell volume V. The condition of equilibrium at the E_{min} neighborhood is expressed as

$$\frac{\partial E(a,r)}{\partial r} = 2k_r(r - r_{opt}) + k_{ar}(a - a_{opt}) = 0. \quad (3)$$

Using Eq. (3), Eq. (2) is rewritten as

$$E(a) = E_{min} + \left(k_a - \frac{k_{ar}^2}{4k_r}\right)(a - a_{opt})^2. \quad (4)$$

On the other hand, the derivative $(\partial/\partial a)$ is related to $(\partial/\partial V)$ through a relation $V = \alpha a^3$ where $\alpha = 0.67850186$ for α-rh B and Li_xB_{12}, and thus we obtain

$$\frac{\partial^2}{\partial V^2} = -\frac{2}{9\alpha^2 a^5}\frac{\partial}{\partial a} + \frac{1}{9\alpha^2 a^4}\frac{\partial^2}{\partial a^2}. \quad (5)$$

Table 4   Optimized lattice constant a and calculated bulk modulus B around the optimized geometry of α-rh B and Li_xB_{12}. Other available LDA calculations of B for fcc C_{60} and K_3C_{60} are listed for comparison

System	a (Å)	B (GPa)	System	a (Å)	B (GPa)
LiB_{12} (O)	5.007	417	α-rh B	5.063[a]	233
Li_2B_{12} (T_d)	5.200	365			
Li_3B_{12}	5.222	325	C_{60}	9.64[b]	48[c]
$Li@Li_3B_{12}$	5.404	242	K_3C_{60}	9.44[b]	29[c]

[a] Present LDA calculation overestimates lattice constant a only by 1.1%, and obtained values of a and hence of bulk modulus B compare quite well with experiments.[t,x]
[b] Lattice constant in rhombohedral description, derived from Ref. s.
[c] Reference s.

From Eqs. (4) and (5), the bulk modulus B defined by

$$B = -V\frac{\partial P}{\partial V} = V\left(\frac{\partial^2 E}{\partial V^2}\right)_{E=E_{min}} \tag{6}$$

is calculated as

$$
\begin{aligned}
B &= -\frac{2}{9\alpha a_{opt}^2}\left(\frac{\partial E(a)}{\partial a}\right)_{a=a_{opt}} + \frac{1}{9\alpha a_{opt}}\left(\frac{\partial^2 E(a)}{\partial a^2}\right)_{a=a_{opt}} \\
&= \frac{2}{9\alpha a_{opt}}\left(k_a - \frac{k_{ar}^2}{4k_r}\right).
\end{aligned} \tag{7}
$$

The calculated bulk moduli are given in Table 4 for α-rh B and its stable and metastable derivatives, Li_xB_{12} ($x = 1$ to 4). The bulk modulus of α-rh B shows excellent agreement with an X-ray single-crystal study of α-rh B.[x]

As seen in Table 4, the bulk modulus initially increases by the doping, but then decreases monotonically with increasing Li concentration. The initial increase in the bulk modulus is attributed to additional contributions from the ion-ion electrostatic interaction associated with doping. The formation of metallic bondings, to which the interactions of conduction electrons with Li^+ ions mainly contribute, may also be responsible for it, but this is less important because the present total-energy calculation shows that the Ewald sum contribution due to electrostatic interactions is dominant, rather than the band energy contribution. The reason for the behavior of monotonic decrease is that dopant Li^+ ions weaken the intermolecular covalent bondings via the relaxations of inter-B_{12} clusters. As a result, the bulk moduli of $Li@Li_3B_{12}$ and Li_3B_{12} are comparable to that of undoped α-rh B, which is indicative of the same order of phonon frequencies.

X-ray diffraction studies on α-rh B and B_{12}-based crystals[i] have revealed that a B_{12} cluster is distorted from a regular icosahedron by the Jahn-Teller effect.[u,y] In addition, higher Debye temperature in α-rh B of approximately 1430 K is also derived[z] from specific heat capacity measurements on β-rh B and $B_{12}C_3$.[a] These studies suggest that there exists anomalously a strong electron-phonon interaction between valence electrons and some vibrational modes in α-rh B. Because the phonon frequencies are almost unaffected by doping, even after the doping, strong electron-phonon coupling is also expected, suggesting the discovery of new superconductors in Li_xB_{12} in the future.

5. Summary and conclusion

We have theoretically predicted the possible existence of new B_{12}-based materials, Li_xB_{12} with $x = 1$ to 4, on the basis of first-principle calculations within LDA. However, our total-energy calculations have also demonstrated that Ca_xB_{12} with $x = 1$ to 4 are all unstable. Among all the stable Li_xB_{12}, the Li_3B_{12} compound, with the same stoichiometry as the superconducting A_3C_{60}, is the most stable, and has also proved to be synthesizable because of its sufficiently large ΔE.

Our band structure calculations have revealed an essentially "donor-type" metallic behavior of Li_xB_{12}. From analysis of the energy band profile, we have noted that Li-doping produces changes that can be related to the dopant concentration, particularly in the bands near the Fermi level. These changes are advantageous to $LiB_{12}(O)$ and Li_3B_{12}, and these two compounds show high density of states at E_F, comparable to K_3C_{60} in the value per atom. As regards Li_3B_{12} and $Li@Li_3B_{12}$, we have also clarified that Li-doping does not cause any drastic changes in the bulk modulus. The observed Jahn-Teller effect and high Debye temperature in α-rh B are also expected in these compounds, indicative of anomalously strong electron-phonon coupling strength.

For many desired properties as a superconductor, Li_3B_{12} has both the high density states at E_F and strong electron-phonon coupling. A moderately high value of T_c is thus expected, mediated by the interaction of conduction electrons with high frequency phonons caused by B-B covalent bondings. In this context, the most stable Li_3B_{12} is a promising candidate for a new superconductor based on semiconductors.

Acknowledgments

This work was partly performed under the Project for Parallel Processing and Super Computing at Computer Centre, University of Tokyo. Financial support was provided by a Grant-in-Aid from the Ministry of Education, Science and Culture of Japan. One of the authors (S. G.) also acknowledges the Japan Society for the Promotion of Science (JSPS) for fellowships to graduate students.

PUBLICATIONS

1. S. Gunji, H. Kamimura and T. Nakayama, in *Proc. 21st Int. Conf. on Physics of Semiconductors, Beijing*, (Xie Xide, Kun Huang and L. L. Chang, eds.), p.1832, World Sci., Singapore (1993).
2. S. Gunji, H. Kamimura and T. Nakayama, *J. Phys. Soc. Jpn.*, **62**, 2408 (1993).
3. S. Gunji and H. Kamimura, *Jpn. J. Appl. Phys.*, Series **10**, 35 (1994).
4. S. Gunji and H. Kamimura, in *Proc. 22nd Int. Conf. on Physics of Semiconductors, Vancouver*, (D. J. Lockwood, ed.), p.2185, World Sci., Singapore (1995).
5. T. Hatakeyama and H. Kamimura, in *Surface Physics and Related Topics. Festschrift for Xie Xide*, (Fu-jia Yang, *et al.* eds.), p.166, World Sci., Singapore (1991); T. Hatakeyama and H. Kamimura, in *Proc. 20th Int. Conf. on Physics of Semiconductors*, (E. M. Anastassakis and J. D. Joannopoules, eds.), p.730, World Sci., Singapore (1991).
6. S. Gunji and H. Kamimura, *Phys. Rev. B*, **54**, 13665 (1996).

REFERENCES

a. R. J. Cava *et al.*, *Nature* (London), **367**, 6459 (1994).

b. Y. Iwasa *et al.*, *Science*, **264**, 5165 (1994).

c. G. Oszlanyi and L. Forro, *Solid State Comm.*, **93**, 4 (1995).

d. R. Naslain, A. Guette and P. Hagenmuller, *Less Common Met.*, **47**, 1 (1976).

e. J. L. Hoard, in *Boron* Vol.1, (J. A. Kohn *et al.* eds.), p.1, Plenum Press, New York (1960).

f. D. B. Sullenger and Ch. L. Kennard, *Sci. Am.*, **215**, 7 (1966).

g. C. E. Holcombe Jr. *et al.*, *High Temp. Sci.*, **5**, 349 (1973).

h. P. Runow, *J. Mater. Sci.*, **7**, 499 (1972).

i. D. R. Tallant, T. L. Aselage, A. N. Campbell and D. Emin, *Phys. Rev.*, **B40**, 5649 (1989); B. Morosin, A. W. Mullendore, D. Emin and G. A. Slack, in *Boron-Rich Solids* (AIP Conf. Proc. **140**), (D. Emin, T. L. Aselage, C. L. Beckel, I. A. Howard and C. Wood, eds.), p.70, AIP, New York (1986).

j. G. A. Slack, T. F. McNelly and IC. A. Taft, *J. Phys. Chem. Solids*, **44**, 1009 (1983); T. L. Chu and A. E. Hyslop, *J. Electrochem. Soc.*, **121**, 412 (1974).

k. H. Werheit, H. Binnenbrudc and A. Hausen, *Phys. Status Solidi*, B**47**, 153 (1971).

l. F. H. Horn, *J. Appl. Phys.*, **30**, 1611 (1959); E. P. Domashevskaya *et al.*, *Less Common Met.*, **47**, 189 (1976).

m. L. Pauling, *The Nature of the Chemical Bond*, 3rd ed., Cornell (1966).

n. D. R. Hamann, M. Schlüter and C. Chiang, *Phys. Rev. Lett.*, **43**, 1494 (1979); G. B.Bachelet, D. R. Hamann and M. Schlüter, *Phys. Rev.*, **B26**, 4199 (1982).

o. P. Hohenberg and W. Kohn, *Phys. Rev.*, **136**, B864 (1964).

p. W. Kohn and L. J. Sham, *Phys. Rev.*, **140**, A1133 (1965).

q. J. Perdew and A. Zunger, *Phys. Rev.*, **B23**, 5048 (1981).

r. D. M. Ceperley and B. J. Alder, *Phys. Rev. Lett.*, **45**, 566 (1980).

s. S. Saito and A. Oshiyama, *Phys. Rev.*, **B44**, 11536 (1991); A. Oshiyama, S. Saito, N. Hamada and Y. Miyamoto, *J. Phys. Chem. Solids*, **53**, 1457 (1992).

t. L. Kleinman, in *Boron-Rich Solids* (AIP Conf. Proc. **231**), (D. Emin, T. L. Aselage, A. C. Switendick, B. Morosin and C. L. Beckcel, eds.), p.13, AIP, New York (1991); S. Lee, D. M. Bylander and L. Kleinman, *Phys. Rev.*, **B42**, 1316 (1990).

u. H. Werheit, M. Laux and U. Kuhlmann, *Phys. Status Solidi*, B, **176**, 415 (1993); R. Franz and H. Werheit, *Europhys. Lett.*, **9**, 145 (1989).

v. M. Z. Huang, Y. N. Xu and W. Y. Ching, *Phys. Rev.*, **B46**, 6572 (1992), and related references therein.

w. J. Rath and A. J. Freeman, *Phys. Rev.*, **B11**, 2109 (1975).

x. R. J. Nelmes, J. S. Loveday, D. R. Allan, J. M. Besson, G. Hamel, P. Grima and S. Hull, *Phys. Rev.*, **B47**, 7668 (1993).

y. R. Franz and H. Werheit, in *Boron-Rich Solids* (AIP Conf. Proc. **231**), eds. D. Emin, T. L. Aselage, A. C. Switendick, B. Morosin and C. L. Beckel (AIP, New York, 1991) p.29, and related references therein.

z. G. A. Slack, D. W. Oliver and F. H. Horn, *Phys. Rev.*, **B4**, 1714 (1971).

a'. G. A. Slack, *Phys. Rev.*, **139**, A507 (1965).

38

Structure and Stability of Large Carbon Clusters

Yohji Achiba, Haruo Shiromaru, Tomonari Wakabayashi[a] and Shinzo Suzuki

Department of Chemistry, Tokyo Metropolitan University
Hachioji-shi, Tokyo 192-0397, Japan

present address:
[a] Department of Chemistry, Kyoto University
Sakyo-ku, Kyoto 606-8224, Japan

Purpose of the present study

The structures of large carbon clusters are described based on measurements of mass spectra and ^{13}C NMR in solution of HPLC (high performance liquid chromatography)-isolated and purified samples. The most interesting aspect deduced from the present systematic work on the large carbon clusters (higher fullerenes) up to C_{94} is that among 19 kinds of fullerenes different sizes and isomers, almost all commonly have at least one C_2 symmetry axis in their molecular frame. The spectral feature of UV/visible absorption obtained for the HPLC-isolated samples up to C_{116} fullerene gives, on the other hand, a strong indication that the numbers of co-existing isomers are extremely small, probably one or two isomers. Considering the huge numbers of possible isomer candidates (for example, over 5000 for the carbon clusters with size of $n = 116$), the experimental evidence may suggest the presence of very strong selectivity in the formation of stable large carbon clusters, which, in turn, is closely associated with the unknown growth process of a fullerene cage network. In order to understand the selectivity in more detail, we also performed the experiments to clarify the temperature dependence of isomer fractions.

Contents of the present study

1. Introduction

Soon after the discovery of large-scale preparation of C_{60} fullerene,[a] the presence of much larger all carbon molecules has been found and identified in carbon soot. From mass spectrometric characterizations, it has been suggested that these all carbon clusters possess similar molecular structures and properties to C_{60} and C_{70}. Actually, soon after Kratschmer and Huffman's findings, several kinds of stable higher fullerenes were isolated and characterized by ^{13}C NMR spectroscopy, revealing the presence of a polygonal structure with a five- and six-membered ring system.[b-e]

Studies of higher fullerenes, particularly those placing attention on the molecular structure are very important to understand not only the general properties of a novel molecular system consisting of a five- and six-membered ring network but also the unknown growth processes of fullerene cage structures.

Furthermore, on the basis of mass spectrometric studies, very large fullerenes up to over 1000 carbon atoms have been detected in carbon raw soot and solvent extract. However, the structural study of higher fullerenes is still very limited. C_{84} fullerene is the largest size molecule whose structure has been well identified. Among many intriguing aspects of the systematic research on the higher fullerenes, of particular interest is the general trends in the molecular structure of large fullerenes: is a spherical shape preferable? or tube-like structure? In the present work, we aim to understand better the general tendency appearing in the systematic structural studies on large fullerenes of up to C_{120}.

2. NMR structural identification of C_{86}, C_{88}, C_{90}, C_{92} and C_{94}

By means of high performance liquid chromatography (HPLC), previous recent experimental efforts have enabled us to is isolate and characterize the higher fullerenes up to C_{84}.[c-e] In the present work, we further extended the isolation and characterization task up to C_{120}. For isolation of these higher fullerenes, the previous HPLC method was modified by the use of a two-stage HPLC system (one is size exclusion type (polystyrene/CS_2) and the other one is adsorption (Cosmosil Buckyprep/toluene)).[g] As a result, we were safely able to obtain over 95% purified samples of the higher fullerenes up to C_{120}.

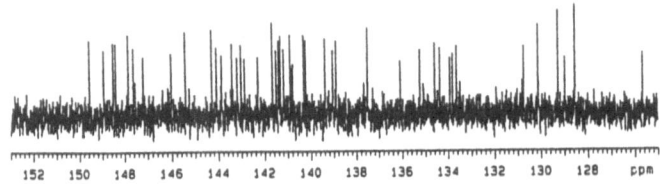

Fig. 1 ^{13}C NMR spectrum of C_{88} in CS_2 solution.

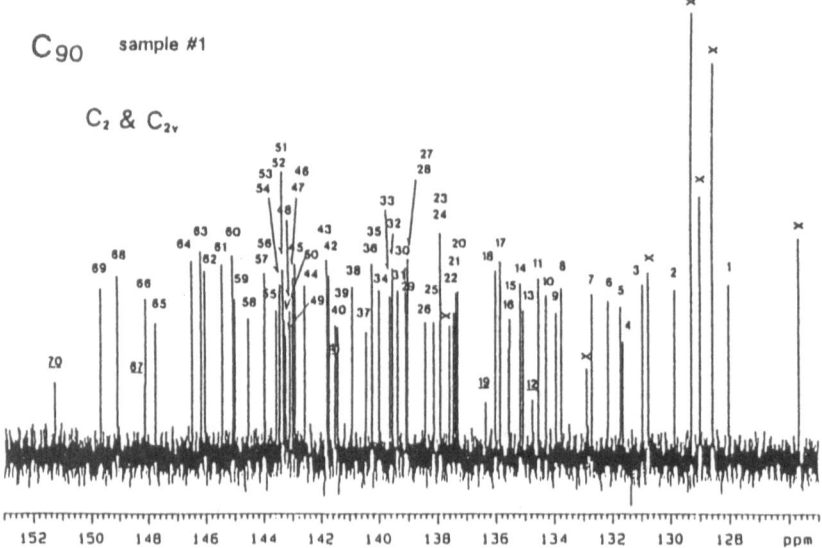

Fig. 2 ^{13}C NMR spectrum of the 1st HPLC fraction of C_{90} in CS_2 solution. The line numbers of 12, 19, 41, 67 and 70 are attributed to C_{2v}-C_{90} with a half intensity.

Table 1 Molecular symmetry of isolated fullerene isomers

Isomer	Symmetry
C_{60} (1)	I_h
C_{70} (1)	D_{5h}
C_{76} (2)	D_2
C_{78} (5)	C_{2v}, C_{2v}, D_3
C_{82} (9)	C_2
C_{84} (24)	D_2, D_2, D_3
C_{86} (29)	C_2, C_2
C_{88} (35)	C_2
C_{90} (46)	C_2, C_2, C_2, C_2, C_2, C_{2v}
C_{92} (86)	C_2, C_2, D_2, D_2
C_{94} (134)	C_2, C_2

The number of IPR-satisfying isomers is indicated in parentheses.

Since the fullerene molecule has been well characterized by its fragmentation patterns upon irradiation with high power laser, here in the present work also, all the HPLC purified samples were examined for fragmentation by a reflectron-type time of flight mass spectrometer. Under the low fluence of laser power, most of the samples showed only parent signals, but by increasing the laser power, the parent signal changed into one followed by C_2-loss fragments. Enhancements in the series of daughter ions were only stressed at C_{60} and C_{70}.

In order to clarify the structure of C_{86}–C_{94} fullerenes, ^{13}C NMR measurements were carried out in solution. The NMR spectra of C_{86} and C_{88} showed 43 and 44 distinct lines, respectively, strongly indicating the presence of fullerene cages with a C_2 symmetry as a major isomer. Figure 1 shows ^{13}C NMR spectrum of C_{88} in CS_2 solution. In addition to the major C_2 isomer of C_{88}, we have also found the presence of a minor isomer which has faster retention time in the liquid chromatography.

The major fraction of the C_{90} portion observed by the initial HPLC was further divided into three fractions after a recycle procedure with second-stage HPLC. ^{13}C NMR spectra were measured for each fraction, separately. The ^{13}C NMR spectra of these three fractions of C_{90} suggested that the first and third fractions consist of two isomers and the second one consists of a single isomer. As a conclusion, it was indicated that the C_{90} fullerene possesses at least 5 different isomers, namely, one C_{2v}, three C_2 and one C_1 symmetries. Here we note that the C_{2v}-C_{90} isomer is different from the one predicted by recent thermodynamic consideration.[h,i] According to the recent *ab initio* calculation on C_{90}, the most stable isomer has C_2 symmetry and the second one is C_{2v}. However, the C_{2v}-C_{90} recommended by the *ab initio* calculation should have 24 distinct lines among which 3 lines are a half in intensity. This is definitely not the present case, because, as can be seen in in Fig. 2, we can see 5 distinct lines with a half intensity. The most probable candidate for the observed C_{2v}-C_{90} isomer is the cage revealed by no. 18 in ref. 8. Table 1 is a summary of the molecular symmetries of all higher fullerenes examined so far.

3. Separation of large fullerenes up to C_{120}

In the course of the present work, we also aimed to isolate much higher

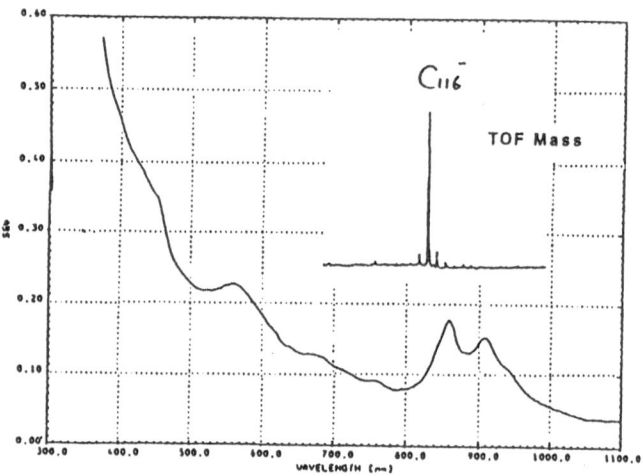

Fig. 3 UV/visible absorption spectrum of C_{116} in CS_2 solution. The insertion is a time-of-flight mass spectrum of the C_{116} sample after purification.

fullerenes up to C_{120}. Through the fullerene sizes from C_{100} to C_{120}, all even numbered carbon molecules were successfully separated. Unfortunately, however, since the amounts of isolated sample are quite small, we were only able to characterize these large fullerenes by means of UV/visible absorption in solution. A typical example of the HPLC-isolated sample with size of $n = 116$ is shown in Fig. 3. As can be indicated in the mass spectra inserted in Fig. 3, the purity of the C_{116} sample is is more than 90%.

Considering the fact that C_{120} fullerene has 10,774 distinct IPR-satisfying isomer candidates,[)] it is amazing that the UV/visible absorption feature of C_{116} is still very structure rich. This spectral characteristic may directly reflect the formation of very limited kinds of co-existing isomers of the same size in carbon soot. This, in turn, suggests the presence of a strong selectivity in the fullerene growth process. The driving force forming such a selectivity would not be simple thermodynamics, because, as shown in the case of C_{90}-C_{2v} isomer, there is no evidence that the thermodynamically more stable isomer is preferentially formed. Furthermore, considering the fact that almost all stable fullerenes examined so far have at least one C_2 symmetry component in their molecular frame, it is strongly suggested that very limited numbers of precursors with a C_2 symmetry would play an essential role in the selectivity of fullerene cage structure. One strong probable candidate for precursor may be something like a semispherical cap structure with six pentagons. This final implication leads to the suggestion of an open-end epitaxially growth model for the fullerene cage formation. In other words, large fullerenes may grow by adding hexagons to the semispherical precursors forming another semispherical network close by. Such a consideration is consistent with the present experimental evidence that C_{2v}-C_{90} structure is reproduced by adding 6 hexagons between the same cap structures as those of C_{2v}'-C_{78}. This situation is also rationalized by the structural relationship between C_2-C_{82} and C_2-C_{94}, in which the two cap structures are the same. Consequently, we tentatively propose the tube-like structure as the general trend in the structure of very large fullerenes.

4. Temperature dependence of fullerene isomer formation

The experimental setup for studying temperature dependence of fullerene cage formation is shown in Fig. 4. The apparatus used in the present work is essentially the same as that reported by Haufler et al.[k] Carbon vapor is produced by laser ablation of graphic (Nd-YAG pulsed laser, 1 J/pulse) under an atomsphere of Ar 300 Torr. The temperature of the reaction tube (25 mm in diameter) is controlled by a furnace.

In Fig. 5, the yields (absolute) of C_{60} are plotted as a function of inverse temperature keeping the pressure and laser intensity constant. We can clearly see a linear $1/T$ dependence of the yield in the temperature range between 600°C–900°C. By using the linear dependent region in the curve, we deduced the value of 1.0 + 0.2 eV as the Arrhenius activation energy of C_{60} production. However, for the temperature region at temperatures higher than 1000°C, the curve of the C_{60} yield tends to saturate, indicating the presence of a sort of dissociation channel in the cage-forming reaction.

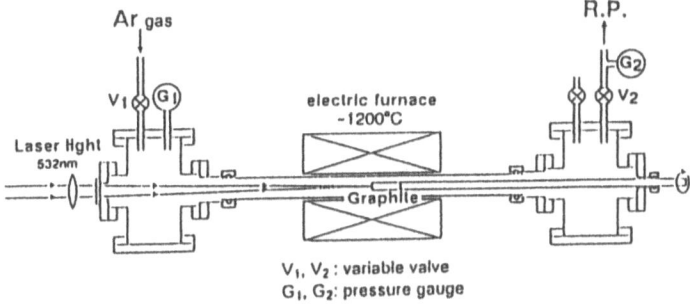

Fig. 4 A schematic drawing of the apparatus of the laser-furnace soot generator.

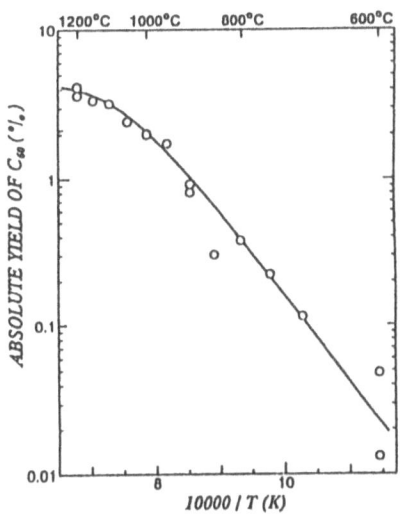

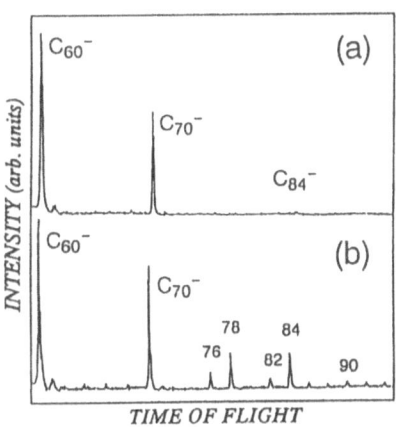

Fig. 5 The absolute yield of C_{60} as a function of $1/T$. The Arrhenius activation energy was obtained by the linear part of the curve.

Fig. 6 Laser desorption time-of-flight mass spectra of fullerenes produced at different temperature conditions (a) 900°C and (b) 1200°C.

Table 2   The Arrhenius activation energy of the 9 different fullerene isomers obtained by a laser-furnace method. The value of C_{84} is the one for the mixture of two isomers of D_2 and D_{2d}.

Isomer	Activation energy (eV)
C_{60} (I_h)	1.0 + 0.2
C_{70} (D_{5h})	1.0
C_{76} (D_2)	2.4
C_{78} ($C_{2v'}$)	2.9
C_{78} (C_{2v})	2.2
C_{78} (D_3)	2.6
C_{82} (C_2)	3.3
C_{84} (D_2 and D_{2d})	2.0

Figure 6 shows mass spectra of the soot materials after the toluene extraction obtained at different temperatures (a) 900°C and (b) 1200°C. The negative ions shown in the figures were directly produced by a time-of-flight laser desorption mass spectrometer. The most distinct feature suggested by the spectra is that the formation of higher fullerenes larger than C_{76} is negligibly small at 900°C, while the production of the higher fullerenes is significantly enhanced by increasing the furnace temperature up to 1200°C.

In order to further study the temperature dependence of the production of the higher fullerenes of different size and structure (isomer), we examined HPLC (liquid chromatography) task to quantify the fraction of isomers. As a result the Arrhenius activation energies for 9 different isomers were deduced as shown in Table 2. It is clear that generally, higher fullerenes larger than C_{76} have larger activaion energy in comparison with C_{60} and C_{70} and those particular isomers with C_{78} ($C_{2v'}$) and C_{82} (C_2) have significantly larger activation energies in comparison with other higher fullerenes. The high activation energy obtained for the formation of C_{2v}-C_{78} and C_2-C_{82} isomers is very consistent with our previous reports can the pressure-dependence of particular isomer formation in the are soot generator.[1]

PUBLICATIONS

1.   K. Mase, S. Mizuno, Y. Achiba and Y. Murata, *Surface Sci.*, **242**, 444 (1991).
2.   H. Shiromaru, T. Moriwaki, C. Kittaka and Y. Achiba, *Z. Phys. D*, **20**, 141 (1991).
3.   Y. Achiba, C. Kittaka, T. Moriwaki and H. Shiromaru, *Z. Phys. D*, **19**, 427 (1991).
4.   Y. Achiba and K. Kimura, *Chinese J. Phys.*, **29**, 223 (1991).
5.   S. Suzuki, T. Wakabayashi, H. Matsuura, H. Shiromaru, C. Kittaka and Y. Achiba, *Chem. Phys. Lett.*, **182**, 12 (1991).
6.   T. Kato, T. Kodama, T. Shida, T. Nakagawa, Y. Matsui, S. Suzuki, H. Shiromaru, K. Yamauchi and Y. Achiba, *Chem. Phys. Lett.*, **180**, 446 (1991).
7.   Y. Kajii, T. Nakagawa, S. Suzuki, Y. Achiba, K. Obi and K. Shibuya, *Chem. Phys. Lett.*, **181**, 100 (1991).
8.   Y. Achiba, T. Nakagawa, Y. Matsui, S. Suzuki, H. Shiromaru, K. Yamauchi, K. Nishiyama, M. Kainosho, H. Hoshi, Y. Maruyama and T. Mitani, *Chem. Lett.*, 1233 (1991).
9.   A. Hiraya, Y. Achiba, K. Kimura and E. C. Lim, *Chem. Phys. Lett.*, **185**. 303 (1991).
10.   Y. Luo, Y. D. Cheng, H. Agren, L. Maripuu, L. Ohlund, B. Carman, P. Ejeklint, W. Seibt, K. Z. Xing, Y. Achiba and K. Siegbahn, *J. Molec. Spectrosc.*, [in press].
11.   H. Hoshi, N. Nakamura, Y. Maruyama, T. Nakagawa, S. Suzuki. H. Shiromaru. and Y. Achiba, *Jpn. J. Appl. Phys.*, **30**, L1397 (1991).
12.   T. Kato, T. Kodama, M. Oyama, S. Okazaki and T. Shida, T. Nakagawa, Y. Matsui, S. Suzuki, H. Shiromaru, K. Yamauchi and Y. Achiba, *Chem. Phys. Lett.*, **186**, 35 (1991).
13.   A. A. Zakhidov, K. Imaeda, A. Ugawa, K. Yakushi, H. Inokuchi, Z. Iqubal, R. H. Baughman, B. L. Ramakrishna and Y. Achiba, *Physica C*, **185–189**, 411 (1991).

14. K. Kikuchi, N. Nakahara, M. Honda, S. Suzuki, K. Saito, H. Shiromaru, K. Yamauchi, I. Ikemoto, T. Kuramochi, S. Hino and Y. Achiba, *Chem. Lett.*, 1607 (1991).
15. K. Kikuchi, S. Suzuki, K. Saito, H. Shiromaru, I. Ikemoto, Y. Achiba, A. A. Zakhidov, A. Ugawa, K. Imaeda, H. Inokuchi and K. Yakushi, *Physica C*, **185–189**, 415 (1991).
16. T. Takahashi, T. Morikawa, S. Sato, H. Katayama-Yoshida, A. Yayuma, K. Seki, H. Fujimoto, S. Hino, S. Hasegawa, K. Kamiya, H. Inokuchi, K. Kikuchi, S. Suzuki, K. Ikemoto and Y. Achiba, *Physica C*, **185–189**, 417 (1991).
17. Y. Maruyama, T. Inabe, H. Ogata, H. Hoshi, N. Nakamura, Y. Mori, Y. Achiba, S. Suzuki, K. Kikuchi and I. Ikemoto, *Physica C*, **185–189**, 413 (1991).
18. T. Atake, T. Tanaka, H. Kawaji, K. Kikuchi, K. Saito, S. Suzuki. I. Ikemoto and Y. Achiba, *Physica C*, **185–189**, 427 (1991).
19. Y. Maruyama, T. Inabe, H. Ogata, Y. Achiba, S. Suzuki, K. Kikuchi and I. Ikemoto, *Chem. Lett.*, 1849 (1991).
20. K. Kikuchi, N. Nakahara, T. Wakabayashi, M. Honda, H. Matsumiya, T. Moriwaki, S. Suzuki, H. Shiromaru, K. Saito, K. Yamauchi, I. Ikemoto and Y. Achiba, *Chem. Phys. Lett.*, **188**, 177 (1992).
21. A. A. Zakhidov, A. Ugawa, K. Imaeda, K. Yakushi, H. Inokuchi, K. Kikuchi, I. Ikemoto, S. Suzuki and Y. Achiba, *Solid State Comm.*, **79**. 939 (1991).
22. Y. Fukuda, N. Sanada, M. Nagoshi, T. Takahashi, H. Katayama-Yoshida, K. Kikuchi, S. Suzuki, I. Ikemoto, Y. Achiba, Y. Syono and M. Tachiki, *Physica C*, **181**, 320 (1991).
23. N. A. Fortune, K. Murata, F. Iga, Y. Nishihara, K. Kikuchi, S. Suzuki, I. Ikemoto and Y. Achiba, *Physica C,* **185–189**, 425 (1991).
24. Y. Maniwa, K. Mizoguchi, K. Kume, K. Kikuchi, K. Saito, S. Suzuki, I. Ikemoto and Y. Achiba, *Solid State Commun.*, **80**, 609 (1991).
25. R. Danieli, V. Denisov, G. Ruani, R. Zamboni, C. Taliani, A. A. Zakhidov, A. Ugawa, K. Imaeda, K. Yakushi, H. Inokuchi, K. Kikuchi, I. Ikemoto, S. Suzuki and Y. Achiba, *Solid State Commn.*, **81**, 257 (1992).
26. S. Hino, K. Matsumoto, S. Hasegawa, K. Kamiya, H. Inokuchi, T. Morikawa, T. Takahashi, K. Seki, K. Kikuchi, S. Suzuki, I. Ikemoto and Y. Achiba, *Chem. Phys. Lett.*, **190**, 169 (1992).
27. I. Ikemoto, K. Kikuchi, K. Saito, S. Suzuki, Y. Achiba, A. Ugawa and K. Yakushi, *Jpn. J. Appl. Phys.*, Series **7**, 367 (1992).
28. H. Ogata, T. Inabe, H. Hoshi, Y. Maruyama, Y. Achiba, S. Suzuki, K. Kikuchi and I. Ikemoto, *Jpn. J. Appl. Phys.*, **31**, L166 (1992).
29. T. Wakabayashi and Y. Achiba, *Chem. Phys. Lett.*, **190**, 465 (1992).
30. T. Takahashi, T. Morikawa, S. Hasegawa, K. Kamiya, H. Fujimoto, S. Hino, K. Seki, H. Katayama-Yoshida, H. Inokuchi, K. Kikuchi, S. Suzuki, I. Ikemoto and Y. Achiba, *Physica C*, **190**, 205 (1992).
31. N. Watanabe, H. Shiromaru, N. Kurihara, Y. Achiba, N. Kobayashi, Y. Kaneko and J. Yoda, *Nuclear Instrumn. Methods in Phys. Research*, **B69**, 385 (1992).
32. T. Takahashi, S. Suzuki, T. Morikawa, H. Katayama-Yoshida, S. Hasegawa, H. Inokuchi, K. Seki, K. Kikuchi, S. Suzuki, K. Ikemoto and Y. Achiba, *Phys. Rev. Lett.*, **68**, 1232 (1992).
33. K. Kikuchi, N. Nakahara, T. Wakabayashi, S. Suzuki, H. Shiromaru, Y. Miyake, K. Saito, I. Ikemoto, M. Kainosho and Y. Achiba, *Nature*, **357**, 142 (1992).
34. A. A. Zakhidov, K. Imaeda, D. M. Petty, K. Yakushi, H. Inokuchi, K. Kikuchi, I. Ikemoto, S. Suzuki and Y. Achiba, *Phys. Lett.*, **A164**, 355 (1992).
35. K. Tanaka, A. A. Zakhidov, K. Yoshizawa, K. Okahara, T. Yamabe, K. Kikuchi, S. Suzuki, I. Ikemoto and Y. Achiba, *Solid State Commn.*, **85**, 69 (1993).
36. S. Suzuki, S. Kawata, H. Shiromaru, K. Yamauchi, K. Kikuchi, T. Kato and Y. Achiba, *J. Phys. Chem.*, **96**, 7159 (1992).
37. T. Moriwaki, H. Matsuura, K. Aihara, H. Shiromaru and Y. Achiba, *J. Phys. Chem.*, **96**, 9092 (1992).
38. Y. Maniwa, T. Shibata, K. Mizoguchi, K. Kume, K. Kikuchi, I. Ikemoto, S. Suzuki and Y. Achiba, *J. Phys. Soc. Jpn.*, **61**, 2212 (1992).
39. T. Arai, Y. Murakami, H. Suematsu, K. Kikuchi, Y. Achiba and I. Ikemoto, *J. Phys. Soc. Jpn.*, **61**, 1821 (1992).
40. S. Kawata, K. Yamauchi, S. Suzuki, K. Kikuchi. H. Shiromaru, M. Katada, K. Saito, I. Ikemoto and Y. Achiba, *Chem. Lett.*, 1659 (1992).
41. T. Wakabayashi, K. Kikuchi, H. Shiromaru, S. Suzuki and Y. Achiba, *Z. Phys.*, **D26**, 258 (1993).
42. T. Moriwaki, H. Shiromaru and Y. Achiba, *Z. Phys.*, **D26**, 320 (1993).
43. N. Watanabe, H. Shiromaru, Y. Negishi, Y. Achiba, N. Kobayashi and Y. Kaneko, *Z. Phys.*, **D26**, 252 (1993).
44. S. Suzuki, T. Wakabayashi, H. Matsuura, H. Shiromaru, C. Kittaka and Y. Achiba, *Z. Phys.*, **D26**, 317 (1993).
45. D. M. Poirier, J. H. Weaver, K. Kikuchi and Y. Achiba, *Z. Phys.*, **D26**, 79 (1993).

46. Y. Z. Li, J. C. Patrin, M. Chander, J. H. Weaver, K. Kikuchi and Y. Achiba, *Phys. Rev.*, **B47**, 10867 (1993).

47. T. Wakabayashi, H. Shiromaru, K. Kikuchi and Y. Achiba, *Chem. Phys. Lett.*, **201**, 470 (1993).

48. A. A. Zakhidov, I. I. Khairullin, P. K. Khabibullacv, V. Yu. Sokolov, K. Imaeda, K. Yakushi, H. Inokuchi and Y. Achiba, *Synthetic Metals*, **55–57**, 2967 (1993).

49. S. Hino, K. Matsumoto, S. Hasegawa, H. Inokuchi, T. Morikawa, T. Takahashi, K. Seki, K. Kikuchi, S. Suzuki, I. Ikemoto and Y. Achiba, *Chem. Phys. Lett.*, **197**, 38 (1992).

50. Y. Achiba and T. Wakabayashi, *Z. Phys.*, **D26**, 69 (1993).

51. S. Nagase, K. Kobayashi, T. Kato and Y. Achiba, *Chem. Phys. Lett.*, **201**, 475 (1993).

52. T. Atake, T. Tanaka, H. Kawaji, K. Kikuchi, K. Saito, S. Suzuki, Y. Achiba and I. Ikemoto, *Chem. Phys. Lett.*, **196**, 321 (1992).

53. S. Hino, K. Matsumoto, S. Hasegawa, K. Iwasaki, K. Yakushi, T. Morikawa, T. Takahashi, K. Seki, K. Kikuchi, S. Suzuki, I. Ikemoto and Y. Achiba, *Synthetic Metals*, **55–57**, 31 (1993).

54. Y. Maruyama, T. Inabe, H. Ogata, H. Hoshi, N. Nakamura, Y. Mori, Y. Achiba, S. Suzuki, K. Kikuchi and I. Ikemoto, *Mol. Cryt. Liq. Cryst. Sci. Technol.*, **A218**, 297 (1992).

55. Y. Maniwa, A. Ohi, K. Mizoguchi, K. Kume, K. Kikuchi, K. Saito, I. Ikemoto, S. Suzuki and Y. Achiba, *Synthetic Metals*, **55–57**, 3057 (1993).

56. Y. Maniwa, K. Kikuchi, T. Higono, A. Ohi, K. Kume, I. Ikemoto, S. Suzuki and Y. Achiba, *J. Phys. Soc. Jpn.*, **62**, 3822 (1993).

57. S. Shimomura, Y. Fujii, S. Nozawa, K. Kikuchi, Y. Achiba and I. Ikemoto, *Solid State Commun.*, **85**, 471 (1993).

58. H. Shiromaru, T. Moriwaki, H. Ikeda and Y. Achiba, *Z. Phys.*, **D26**, 216 (1993).

59. Y. Maniwa, K. Mizoguchi, K. Kume, K. Kikuchi, I. Ikemoto, S. Suzuki, Y. Achiba, K. Tanigaki, T. W. Ebbesen, S. Saito, J. Mizuki, J. S. Tsai and Y. Kubo, *Synthetic Metals*, **55–57**, 3063 (1993).

60. Y. Achiba, T. Wakabayashi, T. Moriwaki, S. Suzuki and H. Shiromaru, *Material Science and Engineering*, B19, 14 (1993).

61. M. Iwahashi, K. Kikuchi, Y. Achiba, I. Ikemoto, T. Araki, T. Mochida, S. Yokoi, A. Tanaka and K. Iriyama, *Langmuir*, **8**, 2980 (1992).

62. K. Imaeda, K. Yakushi, H. Inokuchi, K. Kikuchi, I. Ikemoto, S. Suzuki and Y. Achiba, *Solid State Commn.*, **84**, 1019 (1992).

63. K. Tanaka, A. A. Zakhidov, K. Yoshizawa, K. Okahara, T. Yamabe, K. Yakushi, K. Kikuchi, S. Suzuki, I. Ikemoto and Y. Achiba, *Phys. Rev.*, **B47**, 7554 (1993).

64. T. Inabe, H. Ogata, Y. Maruyama, Y. Achiba, S. Suzuki, K. Kikuchi and I. Ikemoto, *Phys. Rev. Lett.*, **69**, 3797 (1992).

65. K. Tanaka, A. A. Zakhidov, K. Yoshizawa, K. Okahara, T. Yamabe, K. Yakushi, K. Kikuchi, S. Suzuki, I. Ikemoto and Y. Achiba, Proceedings of the Adriatico Research Conference "Clusters and Fullerenes," Eds. V. Kumar, T. P. Martin and E. Tosatti, p.369 (World Scientific, 1993).

66. V. N. Denisov, A. A. Zakhidov, R. Danieli, G. Ruani, R. Zamboni, C. Taliani, K. Imaeda, K. Yakushi, H. Inokuchi and Y. Achiba, Proceeding of the Adriatico Research Conference "Clusters and Fullerenes," Eds. V. Kumar, T. P. Martin and E. Tosatti, p.435 (World Scientific, 1993).

67. Y. Murakami, T. Arai, H. Suematsu, K. Kikuchi, N. Nakahara, Y. Achiba and I. Ikemoto, *Fullerene Science and Technology*, **1**, 351 (1993).

68. F. Okino, H. Touhara, K. Seki, R. Mitsumoto, K. Shigemitsu and Y. Achiba, *Fullerene Science and Technology*, **1**, 425 (1993).

69. T. Arai, Y. Murakami, H. Suematsu, K. Kikuchi, Y. Achiba and I. Ikemoto, *Solid State Commun.*, **84**, 827 (1992).

70. A. Manivannan, H. Hoshi, L. A. Nagahara, Y. Mori, Y. Maruyama, K. Kikuchi, Y. Achiba and I. Ikemoto, *Jpn. J. Appl. Phys.*, **31**, 3680 (1992).

71. H. Shiromaru, M. Mizumachi, Y. Achiba, N. Yanase, N. Kobayashi and Y. Kaneko, *NATO ASI Ser.*, **C374**, 259 (1992).

72. T. Moriwaki, H. Shiromaru and Y. Achiba, *NATO ASI Scr.*, **C374**, 447 (1992).

73. K. Tanaka, A. A. Zakhidov, K. Yoshizawa, K. Okahara, T. Yamabe, K. Yakushi, K. Kikuchi, S. Suzuki, I. Ikemoto and Y. Achiba, *Phys. Lett.*, **A164**, 221 (1992).

74. Y. Achiba, T. Moriwaki and H. Shiromaru, *Proc. SPIE-Int. Sec. Opt. Eng.*, **1638**, 44 (1992).

75. A. A. Zakhidov, K. Yakushi, K. Imaeda, H. Inokuchi, K. Kikuchi, S. Suzuki, I. Ikemoto and Y. Achiba, *Mol. Cryst. Liq. Cryst. Sci. Technol.*, **A218**, 299 (1992).

76. Y. Achiba, K. Kikuchi, T. Wakabayashi, N. Nakahara and S. Suzuki, *World Sci. Adv. Ser. in Fullerenes*, **2**, 13 (1992).

77. T. Kitamoto, S. Sasaki, T. Atake, T. Tanaka, H. Kawaji, K. Kikuchi, K. Saito, S. Suzuki, Y. Achiba and I. Ikemoto, *Jpn. J. Appl. Phys.*, **32**, L424 (1993).

78. Y. Maniwa, A. Ohi, K. Mizoguchi, K. Kume, K. Kikuchi, K. Saito, I. Ikemoto, S. Suzuki and Y. Achiba, *J. Phys. Soc. Jpn.*, **62**, 1131 (1993).

79. N. Katayama, Y. Miyatake, Y. Ozaki, K. Kikuchi, Y. Achiba, I. Ikemoto and K. Iriyama,

Fullerene Sci. Technol., **1**, 329 (1993).

80. M. Yoshida, Y. Morinaga, M. Iyada, K. Kikuchi, I. Ikemoto and Y. Achiba, *Tetrahedron Lett.*, **34**, 7629 (1993).

81. Y. Maniwa, M. Nagasaka, A. Ohi, K. Kume, K. Kikuchi, K. Saito, I. Ikemoto, S. Suzuki and Y. Achiba, *Jpn. J. Appl. Phys.*, **33**. L173 (1993).

82. K. Kikuchi, S. Suzuki, Y. Nakao, N. Nakahara, T. Wakabayashi, H. Shiromaru, K. Saito, I. Ikemoto and Y. Achiba, *Chem. Phys. Lett.*, **216**, 67 (1993).

83. A. Ugawa, K. Yakushi, K. Kikuchi, S. Suzuki, Y. Achiba and I. Ikemoto, *Synth. Metals,* **56**, 2997 (1993).

84. K. Kikuchi, N. Nakahara, T. Wakabayashi, S. Suzuki, K. Saito, I. Ikemoto and Y. Achiba, *Synth. Metals*, **55–57**, 3208 (1993).

85. T. Kato, S. Suzuki, K. Kikuchi and Y. Achiba, *J. Phys. Chem.*, **97**, 13425 (1993).

86. T. Suzuki, Y. Maruyama, T. Kato, K. Kikuchi and Y. Achiba, *J. Am. Chem. Soc.*, **115**, 11006 (1993).

87. H. Kataura, N. Irie, N. Kobayashi, Y. Achiba, K. Kikuchi, T. Hanyu and S. Yamaguchi, *Jpn. J. Appl. Phys.*, **32**, L1667 (1993).

88. F. Negri, G. Orlandi, F. Zebetto, G. Ruani, A. A. Zakhidov, C. Taliani, K. Kikuchi and Y. Achiba, *Chem. Phys. Lett.*, **211**, 353 (1993).

89. E. Frankevich, Y. Maruyama, H. Ogata, Y. Achiba and K. Kikuchi, *Solid State Commun.*, **88**, 177 (1993).

90. T. Inabe, H. Pgata, Y. Maruyama, Y. Achiba, S. Suzuki, K. Kikuchi and I. Ikemoto. *Phys. Rev. Lett.*, **69**, 3797 (1993).

91. S. Hino, K. Matsumoto, S. Hasegawa, K. Iwasaki, K. Yakushi, T. Morikawa, T. Takahashi, K. Seki, K. Kikuchi, S. Suzuki, I. Ikemoto and Y. Achiba , *Phys. Rev.*, **B48**, 8418 (1993).

92. S. Hino, H. Takahashi, K. Iwasaki, K. Matsumoto, T. Miyazaki, S. Hasegawa, K. Kikuchi and Y. Achiba, *Phys. Rev. Lett.*, **25**, 4261 (1993).

93. T. Takahashi, T. Morikawa, H. Katayama-Yoshida, S. Hasegawa, H. Inokuchi, K. Seki, S. Hino, K. Kikuchi, S. Suzuki, I. Ikemoto and Y. Achiba, *Physica*, **B186–188**, 1068 (1993).

94. K. Maruyama, T. Tsuzuki, T. Ishiguro, H. Endo, K. Kikuchi, I. Ikemoto and Y. Achiba, *J. Phys. Soc. Jpn.*, **62**, 2889 (1993).

95. S. Morita, S. Kiyomatsu, M. Fukuda, A. A. Zakhidov, K. Yoshino, K. Kikuchi and Y. Achiba, *Jpn. J. Appl. Phys.*, **32**, L1173 (1993).

96. T. Kato, K. Kikuchi and Y. Achiba, *J. Phys. Chem.*, **97**, 10251 (1993).

97. K. Sugiura, T. Inabe, Y. Maruyama and Y. Achiba, *J. Phys. Soc. Jpn.*, **62**, 2757 (1993).

98. H. Suematsu, Y. Murakami, T. Arai, K. Kikuchi, Y. Achiba and I. Ikemoto, *Material Sci. Engineering*, **B19**, 141 (1993).

99. Y. Maruyama, T. Inabe, M. Ogata, Y. Achiba, K. Kikuchi, S. Suzuki and I. Ikemoto, *Material Sci. Engineering*, **B19**, 162 (1993).

100. D. M. Poirier, M. Knupfer, J. H. Weaver, W. Andrconi, K. Laasonen, M. Parrinello, D. S. Bethune, K. Kikuchi and Y. Achiba, *Phys. Rev.*, **B49**, 17403 (1994).

101. T. Tanaka, H. Kawaji, T. Atake, K. Kikuchi, K. Saito, S. Suzuki, Y. Achiba and I. Ikemoto, *Fullerene Sci. Technol.*, **2**, 121 (1994).

102. M. Tachibana, M. Michiyama, K. Kikuchi, Y. Achiba and K. Kojima, *Phys. Rev.*, **B49**, 14945 (1994).

103. Y. Saito, N. Fujimoto, K. Kikuchi and Y. Achiba, *Phys. Rev.*, **B49**, 14794 (1994).

104. T. Enoki, Y. Ohtsu, K. Suzuki, K. Imaeda, A. A. Zakhidov, K. Yakushi, K. Kikuchi, S. Suzuki and Y. Achiba, *Synth. Metals*, **64**, 329 (1994).

105. T. Wakabayashi, K. Kikuchi, S. Suzuki, H. Shiromaru and Y. Achiba, *J. Phys. Chem.*, **98**, 3090 (1994).

106. Y. Maniwa, T. Saito, A. Ohi, K. Mizoguchi, K. Kume, K. Kikuchi, I. Ikemoto, S. Suzuki, Y. Achiba, M. Kosaka, K. Tanigaki and T. W. Ebbesen, *J. Phys. Soc. Jpn.*, **63**. 1139 (1994).

107. S. Suzuki, H. Torisu, H. Kubota, T. Wakabayashi, H. Shiromaru and Y. Achiba, *Int. J. Mass Spectr. Ion Proc.*, **138**, 297 (1994).

108. K. Kikuchi, Y. Nakao, S. Suzuki, Y. Achiba, T. Suzuki and Y. Maruyama, *J. Am. Chem. Soc.*, **116**, 9367 (1994).

109. H. Suematsu, Y. Murakami, T. Arai, H. Kawata, Y. Fujii, N. Hamaya, O. Shimomura, K. Kikuchi, Y. Achiba and I. Ikemoto, *Mat. Res. Soc. Symp. Proc.*, **349**, 213 (1994).

110. M. Ishibashi, Y. Tomioka, Y. Taniguchi, S. Suzuki, T. Wakabayashi, Y. Kojima, K. Kikuchi and Y. Achiba, *Jpn. J. Appl. Phys.*, **33**. L1265 (1994).

111. K. Sueki, K. Kobayashi, K. Kikuchi, K. Tomura, Y. Achiba and H. Nakahara, *Fullerene Sci. Technol.*, **2**, 213 (1994).

112. K. Kobayashi, M. Kuwano, K. Sueki, K. Kikuchi, Y. Achiba, H. Nakahara, N. Kananishi, M. Watanabe and K. Tomura, *J. Radio Anal. Nucl. Chem.*, [in press].

113. K. Kikuchi, K. Kobayashi, K. Sueki, S. Suzuki, K. Tomura, M. Katada, H. Nakahara and Y. Achiba, *J. Am. Chem. Soc.*, **116**, 9775 (1994).

114. S. Suzuki, Y. Kojima, Y. Nakao, T. Wakabayashi, S. Kawata, K. Kikuchi and Y. Achiba, *Chem. Phys. Lett.*, **229**, 512 (1994).

115. T. Kodama, T. Kato, T. Morowaki, H. Shiromaru and Y. Achiba, *J. Phys. Chem.*, **98**. 10671 (1994).

116. S. Matsuura, T. Tsuzuki, T. Ishiguro, H. Endo, K. Kikuchi, Y. Achiba and I. Ikemoto, *J. Phys. Chem. Solids*, **55**, 853 (1994).

REFERENCES

a. W. Kratschmer, D. L. Lamb, K. Fostiropoulos and D. R. Huffman, *Nature*, **347**, 354 (1990).

b. R. Ettl, I. Chao, F. Diederich and R. L. Whetten, *Nature*, **353**, 149 (1991).

c. F. Diederich, R. L. Whetten, C. Thilgen, R. Ettl, I. Chao and M. M. Alvarez, *Science*, **254**, 1768 (1991).

d. K. Kikuchi, N. Nakahara, T. Wakabayashi, S. Suzuki, H. Shiromaru, Y. Miyake, K. Saito, I. Ikemoto, M. Kainosho and Y. Achiba, *Nature*, **357**, 142 (1992).

e. R. Taylor, G. J. Langley, T. J. S. Dennis, H. W. Kroto and D. R. M. Walton, *J. Chem. Soc. Chem. Commun.*, 1043 (1992).

f. K. Kikuchi and Y. Achiba, [to be published].

g. K. Kikuchi, S. Suzuki, Y. Nakao, N. Nakahara, T. Wakabayashi, H. Shiromaru, K. Saito, I. Ikemoto and Y. Achiba, *Chem. Phys. Lett.*, **216**, 6 (1993).

h. R. L. Murry and G. E. Scuseria, *J. Phys. Chem.*, **98**, 4212 (1994).

i. E. Osawa, private communication.

j. P. W. Fowler, S. J. Austin and D. E. Manolopoulos, in *Physics and Chemistry of the Fullerenes*, ed. by K. Prassides, Kluwer Academic Publishers 1994, p.41.

k. R. E. Haufler, Y. Chai, L. P. F. Chibante, J. Conceicao, C. Jin, L. S. Wang, S. Maruyama and R. E. Smalley, *Mater. Res. Symp. Proc.*, **206**, 627 (1991).

l. T. Wakabayashi, K. Kikuchi, S. Suzuki, H. Shiromaru and Y. Achiba, *J. Phys. Chem.*, **98**, 3090 (1994).

39

Carbon Clusters and Carbon Composite Particles

Yahachi Saito [a] and Tamotsu Noda [b]

[a] Department of Electrical and Electronic Engineering, Mie University
1515 Kamihama-cho, Tsu 514-8507, Japan
[b] Toyohashi Junior College
Ushikawa-cho, Toyohashi 440-0016, Japan

Purpose of the present study

C_{60} and other fullerenes, the third allotrope of carbon, have the potential of application to electronic devices, catalysis, organic materials, medicines, and so forth. Fullerenes are the first clusters that are obtained in macroscopic quantity as one-size clusters. C_{60} forms a regular lattice in a solid state and shows various interesting physical properties such as superconductivity and nonlinear optical response. The purposes of the present study concerning fullerenes are to clarify (1) production yield of fullerenes in arc-discharge method and (2) growth and structure of fullerene thin films. Carbon nanotubes, nanopolyhedra (nanoparticles), filled nanocapsules and so forth were discovered as an outgrowth of fullerene production with the arc-discharge. The nanocapsules, being air-tight, protect inner materials from oxidation and hydrolysis. This protective nature of nanocapsules promises the development of a new field of nanoscale material science. We also studied the growth and structures of (3) carbon nanotubes and (4) nanocapsules filled with rare-earth carbides and iron-group metals.

Contents of the present study

1. Production yield of fullerenes by arc method

Graphite rods were evaporated by DC arc discharge in an inert gas contained in a water-cooled stainless steel chamber (30 cm in diameter and 40 cm in height). The chamber was pumped by an oil diffusion pump to a base pressure of 10^{-5} Torr. After the chamber was isolated from the pump, He or Ar was introduced into the chamber up to a desired pressure ranging from 10 to 400 Torr. The anode was a graphite rod (purity 99.998%) with diameter of 13 mm and length of 50 mm, and the cathode was a graphite rod (99.998%) with diameter of 10 mm and length of 70–80 mm. These graphite electrodes were held vertically. The arc current was in range of 220–250 A, and voltage was about 20 V. The anode graphite was consumed by evaporation. The spacing of the arc gap was maintained within a few mm during the operation by advancing the cathode toward the anode.

Carbon soot deposited on the ceiling and side walls of the chamber was collected by gentle scratching with a brush. The raw soot obtained was subjected to Soxhlet extraction by refluxing toluene. The extraction was continued until the

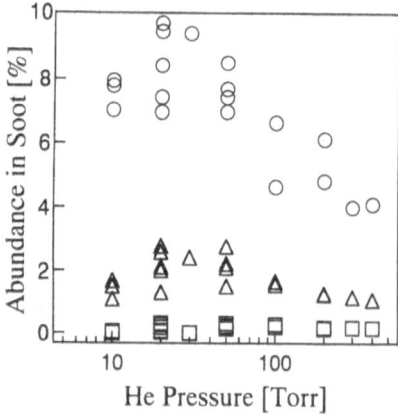

Fig. 1   Yield of C_{60}(○), C_{70}(△) and higher fullerenes (□) as a function of He pressure.[1]

brown color of the solution was completely faded. The solution was dried in vacuum, and a mixture of fullerenes was obtained. The ^{1}H NMR spectra of these fullerene mixtures suggested that only negligible amounts of hydrocarbon impurities were present. The mixture of fullerenes was fractionated by HPLC (high performance liquid chromatography) with an *n*-hexane mobile phase. The extracts were injected as a cyclohexane solution. The chromatogram was monitored by a UV detector ($\lambda = 350$ nm).

Figure 1 shows yields of toluene-soluble material plotted as a function of gas pressure for He. It is found that the fullerene yield strongly depends on the gas pressure. 1) Yields of fullerenes prepared in Ar gas showed a pressure dependence similar to that of He, although the absolute yield was lower by a factor of 2 or 3. The yield has a maximum value (about 13 wt% for He, 6 wt% for Ar) at 20 Torr. The yields of the three fractions, (1) C_{60}, (2) C_{70} and (3) the higher fullerenes (C_{76}, C_{78}, and C_{84}), also varied with the gas pressure. As is well known, C_{60} is the predominating fullerene in a toluene extract. Its yield in raw soot had a maximum at 20 Torr and decreased at higher pressures. C_{70} also showed a maximum of yield but at a slightly higher pressure (20–50 Torr). On the other hand, for the higher fullerenes whose yield was lower than 0.3%, it was difficult to draw systematic changes in their yield against the gas pressure. When the yield of the

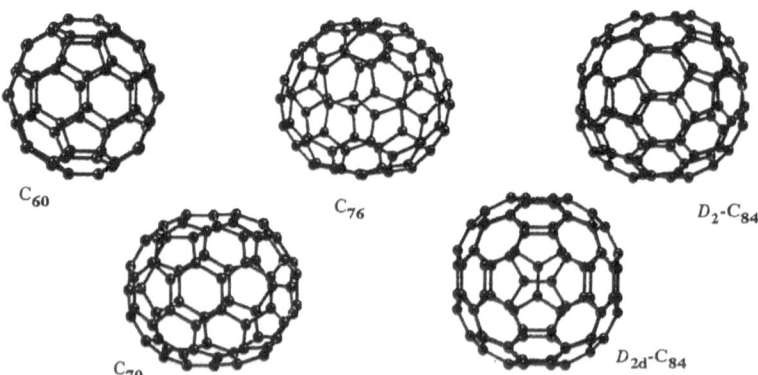

Fig. 2   Structures of C_{60}, C_{70}, C_{76}, and C_{84}.

higher fullerenes was plotted against that of C_{70}, a distinct correlation was revealed; the yield of the higher fullerenes seemed to change linearly with that of C_{70}.

2. Preparation and characterization of fullerene thin films

Thin films of C_{60}, C_{70}, C_{76}, and C_{84} were prepared on NaCl, mica and quartz substrates by vacuum evaporation.[2-5] The structure and morphology of the films were studied by electron microscopy. Separation and purification of fullerenes used for the film preparation were carried out by Prof. H. Shinohara (Nagoya Univ.), Prof. H. Nagashima (Toyohashi Univ. of Tech.), Prof. K. Kikuchi and Prof. Y. Achiba (Tokyo Metropolitan Univ.). Purities of C_{60}, C_{70} and C_{76} were higher than 99%, and that of C_{84}, more than 98%. The structures of these fullerenes are shown in Fig. 2.

Fullerene films grown on NaCl were polycrystalline with abundant planar defects such as twins and stacking faults. Fcc stacking was dominant but hcp stacking was observed locally. Even though the temperature of NaCl substrates was raised to 250°C, single crystalline films were not obtained in our experiment. Multiply-twinned particles (pentagonal bipyramids) were observed for C_{60}/NaCl.[4]

However, when mica was used as substrate, epitaxially grown single crystals with the fcc structure were obtained for all the fullerenes. The lattice constants observed for the fcc crystals are summarized in Table 1.

Table 1 Lattice parameters of the fcc-C_{60}, -C_{70}, -C_{76}, and -C_{84}

fullerene	lattice parameter (nm)	Ref.
C_{60}	1.43 ± 0.02	2
C_{70}	1.51 ± 0.02	2
C_{76}	1.53 ± 0.02	3
C_{84}	1.59 ± 0.02	5

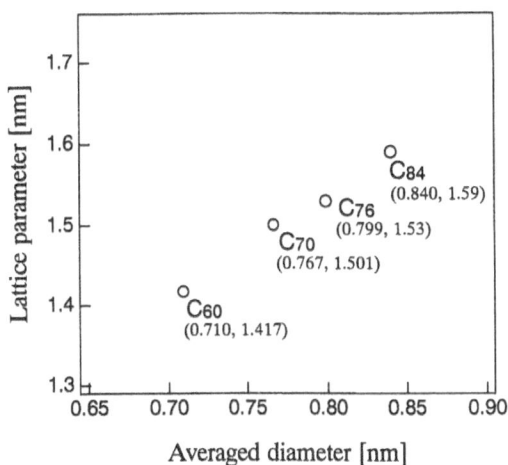

Fig. 3 Average diameters of fullerenes against the lattice parameters of fcc packing.[3]

It should be noted that the lattice parameters of the fcc-C_{60}, -C_{70}, -C_{76}, and -C_{84} plotted against the diameter of the fullerenes lie on a straight line as shown an Fig. 3.[3] The diameter of the fullerenes C_n used here is defined as follows: $d(C_n)$ = $0.71 \times (n/60)^{1/2}$ [nm], which is based on a diameter 0.71 nm for C_{60}. $d(C_{70})$ = 0.767 nm, $d(C_{76})$ = 0.799 nm, and $d(C_{84})$ = 0.840 nm are obtained. The diameter thus obtained is a kind of average diameter, being averaged over all the orientations of a fullerene whose shape deviates from a sphere. The linear relation between the lattice parameters and the average diameters seems to indicate that the orientation of fullerenes is disordered dynamically or statically, although the physical meaning of this linear relationship is not clear. The finding that even C_{76}, whose eccentricity is large, obeys the linear relation between the lattice parameters and the average diameters shows that the idea of the average diameter can be used conveniently to describe a linear dimension of fullerenes.

3. Growth of carbon nanopolyhedra and nanotubes

Nanopolyhedra (nanoparticles)[6] grow together with multi-wall nanotubes [a,b] in the inner core of a carbonaceous deposit formed on the top of the cathode. The size of nanopolyhedra falls in a range from a few to several tens of nanometers, being roughly the same as the outer diameters of multi-wall nanotubes. High-resolution TEM (transmission electron microscopy) observations reveal that polyhedral particles are made up of concentric graphitic sheets as shown in Fig. 4. The closed polyhedral morphology is brought about by well-developed graphitic layers which are flat except at the corners and edges of the polyhedra. When a pentagon is introduced into a graphene sheet, the sheet curves positively and the strain in the network structure is localized around the pentagon. The closed graphitic cages produced by the introduction of 12 pentagons will exhibit polyhedral shapes at the corners of which the pentagons are located. The overall shapes of the polyhedra depend on how the 12 pentagons are located.

The spacings between the layers (d_{002}) measured by selected area electron diffraction were in a range of 0.34 to 0.35 nm.[6] X-ray diffraction (XRD) of the cathode deposit, including nanopolyhedra and nanotubes, gave d_{002} = 0.344 nm,[7] being consistent with the result of electron diffraction. The interlayer spacing is wider by a few percent than that of the ideal graphite crystal (0.3354 nm). The wide interplanar spacing is characteristic of the turbostratic graphite.[c]

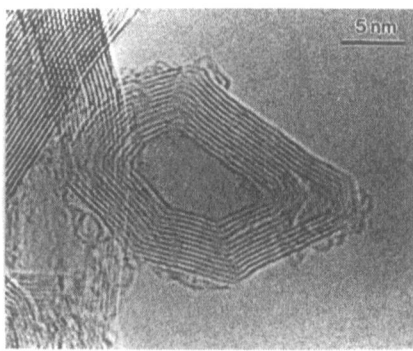

Fig. 4 TEM image of a nanopolyhedron.

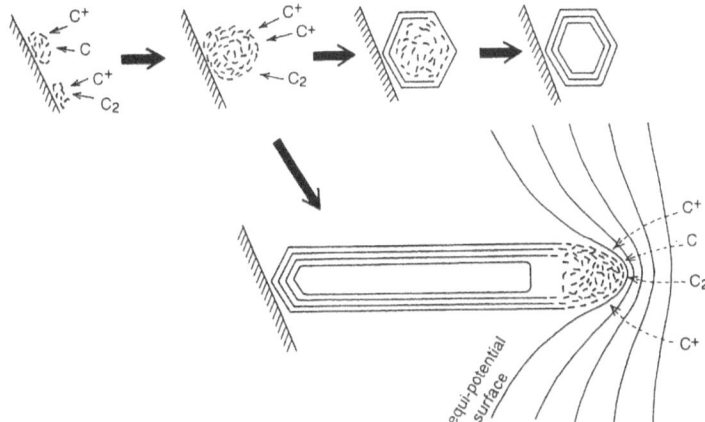

Fig. 5 Growth model of nanopolyhedra and nanotubes. [6]

Figure 5 illustrates a proposed growth process [6] of polyhedral nanopolyhedra and nanotubes. First, carbon neutrals (C and C_2) and ions (C^+) [8] are deposited, and then coagulate with each other to form small clusters on the surface of the cathode. Through an accretion of carbon atoms and coalescence between clusters, clusters grow into the particles of the size finally observed. The structure of the particles at this stage may be "quasi-liquid" or amorphous with high structural fluidity because of the high temperature ($\approx$ 3500 K) of the electrode and ion bombardment. Ion bombardment onto the electrode surface seems to important for the growth of nanoparticles as well as tubes. Near a cathode, the voltage drop of approximately 10 V occurs in a thin layer of 10^{-3} to 10^{-4} cm from the electrode surface. [d] Therefore, C^+ ions with an average kinetic energy of ~10 eV bombard the carbon particles and enhance the fluidity of particles.

The vapor deposition and ion bombardment onto quasi-liquid particles will continue until the particles are shadowed by the growth of tubes and other particles surrounding them, and then graphitization occurs. Since cooling occurs from the surface to the center of the particle, the graphitization initiates on the external surface of the particle and progresses toward its center. The internal layers grow, keeping their planes parallel to the external layer. The flat planes of the particle consist of nets of six-member rings, while five-member rings may be located at the corners of the polyhedra. The closed structure containing pentagonal rings diminishes dangling bonds and lowers the total energy of a particle. Since the density of highly graphitized carbon ($\approx$ 2.2 g/cm^3) is higher than that of amorphous carbon (1.3–1.5 g/cm^3), a pore will be left inevitably in the center of a particle after the graphitization. In fact, the corresponding cavities are observed in the centers of nanoparticles.

4. Synthesis and structures of filled nanocapsules

When a metal-packed graphite anode is evaporated by the arc method, nanocapsules filled with single-domain crystallites of metals are also synthesized. Nanocapsules containing LaC_2, which were discovered independently by us [9] and Ruoff et al., [e] were the first nanocapsules synthesized. The nanocapsules are made up of closed, concentric graphitic sheets, and their typical sizes were on the order of 10–100 nm. The outer graphitic shells protect the inner materials from oxidation and hydrolysis. Following the discovery of nanocapsules stuffed with rare-earth carbides, 3d-transition metals such as iron, cobalt and nickel were found to be wrapped in graphitic carbon by the arc evaporation of metal/carbon composites.

4.1 Rare earth carbides in nanocapsules

Figure 6 shows a typical TEM picture of carbon nanocapsules entrapping a single crystallite of YC_2. [10] The external shapes of nanocapsules for rare-earths were polyhedral. In the outer shell, (002) fringes of graphitic layers with 0.34 nm spacing are observed. The core YC_2 crystallite partially fills the inner space of the nanocapsule, leaving a small cavity inside.

Carbon nanocapsules stuffed with metal carbides were formed for most of the rare-earth metals, Sc, Y, La, Ce, Pr, Nd, Gd, Tb, Dy, Ho, Er, Tm and Lu.[11,12] The stuffed nanocapsules had common structural and morphological features: The outer shell, being made up of concentric multilayered graphitic sheets, is polyhedral in external shape, and the inner space was partially filled by a single-crystalline carbide. The carbides encapsulated were in the phase of RC_2 (R stands for rare earth elements) except for Sc, for which Sc_3C_4 was encapsulated.[13] It should be noted that the phases entrapped in nanocapsules were the metal carbides that have the highest content of carbon among the known carbides for the respective metals. This finding provides an important clue for elucidation of the growth mechanism of the filled nanocapsules.[10]

We have noted a good correlation between volatility of rare-earth metals and the encapsulation, namely, metals with low vapor pressure (non-volatile) were encapsulated (in the form of carbides) within polyhedral nanocapsules while those with high vapor pressure were not. [12] Interestingly, the same correlation of volatility with encapsulations is found for metallofullerenes. Moro et al.,[f] who studied the presence of metallo-fullerenes in arc-produced primary (crude) soot by laser desorption (LD) and thermal desorption (TD) mass spectrometry, also showed that the metals studied could be separated into two distinct groups; Sc, Y, La, Ce, Pr, Nd, Gd, Tb, Ho, Er, and Lu (group A), and Ca, Sr, Sm, Eu, and Yb (group B). Groups A and B in their study correspond exactly to the non-volatile

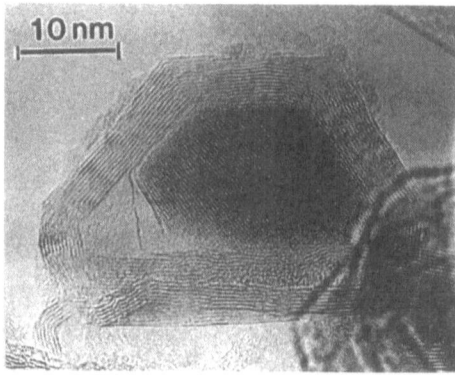

Fig. 6 TEM image of a nanocapsule staffed with YC_2. [10]
(From Y.Saito et al., Chem.Phys.Lett., **209**, 73 (1993)).

and volatile groups, respectively, found in the present study.

4.2 Iron, cobalt, and nickel wrapped with graphitic carbon

Nanocapsules of the iron-group metals (Fe, Co, Ni) showed structures and morphology different from those of rare-earth elements in the following points. First, most of core crystallites are in ordinary metallic phases, and minorities are carbides. The α (bcc)-Fe, [14,15] β (fcc)-Co [16,17] and fcc-Ni [18,19] are the major phases for the respective metals, and small amounts of γ (fcc)-Fe and α (hcp)-Co are also formed. Carbides formed for the three metals were of the cementite phases, viz., Fe_3C, Co_3C, and Ni_3C. The quantity of carbides formed depends on the affinity of metal toward carbon; iron forms the carbide most abundantly, nickel the least, and cobalt, intermediate between iron and nickel.

Second, the outer graphitic layers surround tightly the core crystallites without cavities for most of the particles, in contrast to the nanocapsules of rare-earth carbides for which the capsules are polyhedral and have cavities inside. The graphite layers wrapping iron (cobalt and nickel) particles bend to follow the curvature of the surface of a core crystallite. The graphitic sheets, for the most part, are stacked parallel to each other one by one, but defect-like contrasts suggesting dislocations have been observed, indicating that the outer carbon shells are made up of small domains of graphitic carbon stacked parallel to the surface of the core particle. The structure may be similar to that of graphitized carbon blacks, being composed of small segments of graphitic sheets stacked roughly parallel to the particle surface.

The protective nature of graphitic carbon against oxidation of the core crystallites was demonstrated by environmental test (80°C, 85% R. H., 7 days).[20] Even after this test, XRD profiles revealed that the encapsulated iron particles were not oxidized at all, while naked iron particles of similar size, ca. 50 nm, showed strong reflections due to oxides (rhombohedral-Fe_2O_3 (hematite) and cubic-Fe_2O_3 (maghemite)). Together with the excellent anti-oxidation property of the outer graphitic carbon, the lubricating nature of graphite makes the encapsulated particles prospective magnetic recording material.

PUBLICATIONS

1. Y. Saito, M. Inagaki, H. Shinohara, H. Nagashima, M. Ohkohchi and Y. Ando, *Chem. Phys. Lett.*, **200**, 643 (1992).
2. Y. Saito, T. Yoshikawa, Y. Ishikawa, H. Nagashima and H. Shinohara, *Mater. Sci. Eng.*, B, **19**, 18 (1993).
3. Y. Saito, N. Fujimoto, K. Kikuchi and Y. Achiba, *Phys. Rev. B*, **49**, 14794 (1994).
4. Y. Saito, Y. Ishikawa, A. Ohshita, H. Shinohara and H. Nagashima, *Phys. Rev. B*, **46**, 1846 (1992).
5. Y. Saito, T. Yoshikawa, N. Fujimoto and H. Shinohara, *Phys. Rev. B*, 48, 9182 (1993).
6. Y. Saito, T. Yoshikawa, M. Inagaki, M. Tomita and T. Hayashi, *Chem. Phys. Lett.*, **204**, 277 (1993).
7. Y. Saito, T. Yoshikawa, S. Bandow, M. Tomita and T. Hayashi, *Phys. Rev. B*, **48**, 1907 (1993).
8. Y. Saito and M. Inagaki, *Jpn. J. Appl. Phys.*, **32**, L954 (1993).
9. M. Tomita, Y. Saito and T. Hayashi, *Jpn. J. Appl. Phys.*, **32**, L280 (1993).
10. Y. Saito, T. Yoshikawa, M. Okuda, M. Ohkohchi, Y. Ando, A. Kasuya and Y. Nishina, *Chem. Phys. Lett.*, **209**, 72 (1993).
11. Y. Saito, T. Yoshikawa, M. Okuda, N. Fujimoto, K. Sumiyama, K. Suzuki, A. Kasuya and Y. Nishina, *J. Phys. Chem. Solids*, **54**, 1849 (1993).
12. Y. Saito, M. Okuda, T. Yoshikawa, A. Kasuya and Y. Nishina, *J. Phys. Chem.*, **98**, 6696 (1994).
13. Y. Saito, M. Okuda, T. Yoshikawa, S. Bandow, S. Yamamuro, K. Wakoh, K. Sumiyama and K.

    Suzuki, *Jpn. J. Appl. Phys.*, **33**, L186 (1994).
14. Y. Saito, T. Yoshikawa, M. Okuda, N. Fujimoto, S. Yamamuro, K. Wakoh, K. Sumiyama, K. Suzuki, A. Kasuya and Y. Nishina, *Chem. Phys. Lett.*, **212**, 379 (1993).
15. T. Hihara, H. Onodera, K. Sumiyama, K. Suzuki, A. Kasuya, Y. Nishina, Y. Saito, T. Yoshikawa and M. Okuda, *Jpn. J. Appl. Phys.*, **33**, L24 (1994).
16. Y. Saito, T. Yoshikawa, M. Okuda, N. Fujimoto, S. Yamamuro, K. Wakoh, K. Sumiyama, K. Suzuki, A. Kasuya and Y. Nishina, *J. Appl. Phys.*, **75**, 134 (1994).
17. Y. Saito and M. Masuda, *Jpn. J. Appl. Phys.*, **34**, 5594 (1995).
18. Y. Saito and T. Yoshikawa, *J. Cryst. Growth*, **134**, 154 (1993).
19. Y. Saito, M. Okuda, N. Fujimoto, T. Yoshikawa, M. Tomita and T. Hayashi, *Jpn. J. Appl. Phys.*, **33**, L526 (1994).
20. Y. Saito, *Recent Advances in the Chemistry and Physics of Fullerenes and Related Materials*, (K. M. Kadish and R. S. Ruoff, eds.) pp.1419–1432, Electrochemical Society, Pennington (NJ 1994).
21. S. Bandow and Y. Saito, *Jpn. J. Appl. Phys.*, **32**, L1677 (1993).
22. Y. Saito, H. Shinohara, M. Kato, H. Nagashima, M. Ohkohchi and Y. Ando, *Chem. Phys. Lett.*, **189**, 236 (1992).
23. H. Nagashima, A. Nakaoka, Y. Saito, M. Kato, T. Kawanishi and K. Itoh, *J. Chem. Soc. Chemical Comm. Issue 4*, 377 (1992).
24. Y. Saito, N. Suzuki, H. Shinohara, T. Hayashi and M. Tomita, *Ultramicroscopy*, **41**, 1 (1992).
25. S. Bandow, H. Oya, N. Akuzawa, H. Shinohara, H. Nagashima, A. Nakaoka, M. Ohkohchi, Y. Ando and Y. Saito, *Physics and Chemistry of Finite Systems: From Clusters to Crystals*, (P. Jena, S. N. Khanna and B. K. Rao, eds.), Vol. 2 (Nato-ASI Series C Vol. 374, Kluwer Academic, 1992) p.1311.
26. Y. Saito, N. Suzuki, M. Terauchi, R. Kuzuo, M. Tanaka, H. Shinohara, A. Ohshita, M. Ohkohchi and Y. Ando, *Physics and Chemistry of Finite Systems: From Clusters to Crystals*, (P. Jena, S. N. Khanna and B. K. Rao, eds.) Vol. 2 (Nato-ASI Series C Vol. 374, Kluwer Academic, 1992) p.1365.
27. H. Nagashima, A. Nakaoka, S. Tajima, Y. Saito and K. Itoh, *Chem. Lett.*, 1361 (1992).
28. J. Kawai, K. Maeda, M. Takami, Y. Muramatsu, T. Hayashi, M. Motoyama and Y. Saito, *J. Chem. Phys.*, **98**, 3650 (1993).
29. S. Bandow, H. Shinohara, Y. Saito, M. Ohkohchi and Y. Ando, *J. Phys. Chem.*, **97**. 6010 (1993).
30. M. Ohkohchi, Y. Ando, S. Bandow and Y. Saito, *Jpn. J. Appl. Phys.*, **32**, L1248 (1993).
31. H. Nagashima, H. Yamaguchi, Y. Kato, Y. Saito, M. Haga and K. Itoh, *Chem. Lett.*, 2153 (1993).
32. X. D. Wang, Q. Xue, T. Hashizume, H. Shinohara, Y. Saito, Y. Nishina and T. Sakurai, *Appl. Surface Sci.*, **76/77**, 334 (1994).
33. H. Kawamura, Y. Akahama, M. Kobayashi, H. Shinohara and Y. Saito, *J. Phys. Soc. Jpn.*, **63**, 2445 (1994).
34. R. Kuzuo, M. Terauchi, M. Tanaka and Y. Saito, *Jpn. J. Appl. Phys.*, **33**, L1316 (1994).
35. R. Kuzuo, M. Terauchi, M. Tanaka, Y. Saito and H. Shinohara, *Phys. Rev. B*, **49**, 5054 (1994).
36. Y. Saito, *Kagaku* (Chemistry), **47**, 409 (1992) [in Japanese].
37. Y. Saito, *J. Cryst. Soc. Jpn.*, **34**, 44 (1992) [in Japanese].
38. Y. Saito, *J. Cryst. Soc. Jpn.*, **34**, 197 (1992) [in Japanese].
39. Y. Saito, *J. Soc. Mater. Sci. Jpn.*, **41**, 1465 (1992) [in Japanese].
40. Y. Saito, The INTER Research Essay Mag. '93, p.136, UPU (1993).
41. Y. Saito, C_{60} fullerene chemistry (Special Issue of *Kagaku*), p.50 (1993) [in Japanese].
42. Y. Saito, *Kagaku Kohgyoh* (Chemical Industry) **44**, 390 (1993) [in Japanese].
43. Y. Saito, *Hyohmen* (Surface) **32**, 227 (1994) [in Japanese].
44. Y. Saito, *Electron-Microscopy* **29**, 74 (1994) [in Japanese].
45. Y. Saito, *Kikan* (Quarterly) *Fullerene*, Vol. 2, No. 4, 107 (1994) [in Japanese].

REFERENCES

a. S. Iijima, *Nature,* **354**, 56 (1991).
b. T. W. Ebbesen and P. M. Ajayan, *Nature,* **358**, 220 (1992).
c. M. S. Dresselhaus, G. Dresselhaus, K. Sugihara, I. L. Spain and H. A. Goldberg, *Graphite Fibers and Filaments*, p.42, Springer-Verlag, Berlin (1988).
d. W. Finkelnburg and S. M. Segal, *Phys. Rev.*, **83**, 582 (1951).
e. R. S. Ruoff, D. C. Lorents, B. Chan, R. Malhotra and S. Subramoney, *Science*, **259**, 346 (1993).
f. L. Moro, R. S. Ruoff, C. H. Becker, D. C. Lorents and R. Malhotra, *J. Phys. Chem.*, **97**, 6801 (1993).

40

ESR Studies on Endohedral Metallofullerenes

Hisanori Shinohara

Department of Chemistry, Nagoya University
Chikusa-ku, Nagoya 464-0814, Japan

Purpose of the present study

Endohedral metallofullerenes are novel forms of fullerene-related materials which encage metal atom(s) in various sizes of fullerene cages. In the present article, recent advances on the production, separation (isolation) and characterization of endohedral metallofullerenes are presented in an effort to clarify their structural and electronic properties. Endohedral metallofullerenes are produced by the DC arc-discharge method with metal/graphite composite rods as positive electrodes. The endohedral metallofullerenes, such as $Y@C_{82}$, $Sc@C_{82}$, $Sc_2@C_{84}$ and $Sc_3@C_{82}$, have been purified and isolated using the two-stage high performance liquid chromatography (HPLC). Using the purified metallofullerene sample, the electronic and structural features of the endohedral metallofullerenes are investigated by electron spin resonance (ESR).

Contents of the present study

1. Introduction

With the advent of the first macroscopic production and solvent extraction of $La@C_{82}$ by Smalley and coworkers,[a] fullerenes with metal atoms inside the carbon cage have attracted wide interest because of their novel structural and electronic properties.[b] The endohedral metallofullerenes produced so far can encage La,[a,c-f] Y,[g,h] Sc,[i-l] and most of the lanthanoid elements.[m,n] The metallofullerenes are normally produced by the so-called composite rod/arc-discharge method,[a,d,h] or, in some cases, by the high temperature laser-vaporization technique.[a,g]

The purification of the endohedral metallofullerenes was formerly difficult, mainly because the content of the metallofullerenes in the extract was very limited.[j] It took almost two years for metallofullerenes to be completely isolated by high performance liquid chromatography (HPLC)[o] after the first solvent-extraction of $La@C_{82}$ by the Rice group.[a] Success in purification was a great breakthrough for further progress in metallofullerene studies. We will first discuss production and sample purification procedures of the metallofullerenes then discuss the various spectroscopic characterizations of the metallofullerenes including electron spin resonance (ESR) and scanning tunneling microscopy

(STM).

ESR spectroscopy is a powerful method for studying the electronic structures of endohedral metallofullerenes,[c] since most of the mono- and tri-metallofullerenes studied so far have exhibited hyperfine structures (hfs) due to electron spin-nuclear couplings.[c,g-l] It will be shown from ESR hfs measurements that intrafullerene electron transfers from the encaged metal atom(s) to the carbon cages occur in the mono- and tri-metallofullerenes. The electron transfers within the metallofullerenes have been predicted by several theoretical calculations.[p-t]

2. Production and extraction of endohedral metallofullerenes

We prepared scandium-containing fullerenes by arc burning of a composite rod composed of Sc_2O_3 (99.9% purity; 2.5 g), graphite powder (99.995%; 4.3 g), and high-Strength pitch (5.0 g). The scandium / graphite composite rods were baked at 1000°C, 5 hours in vacuum (10^{-3} Torr) and then cured and carbonized at 1600°C for another 5 hours (10^{-5} Torr). These high-temperature heat treatments of the composite rods have been found to be crucial to the efficient production of endohedral metallofullerenes via the arc-discharge method. It has been shown [u] that metal carbides (MC_2)-enriched composite rods are good starting materials required for high yield synthesis of metallofullerenes.

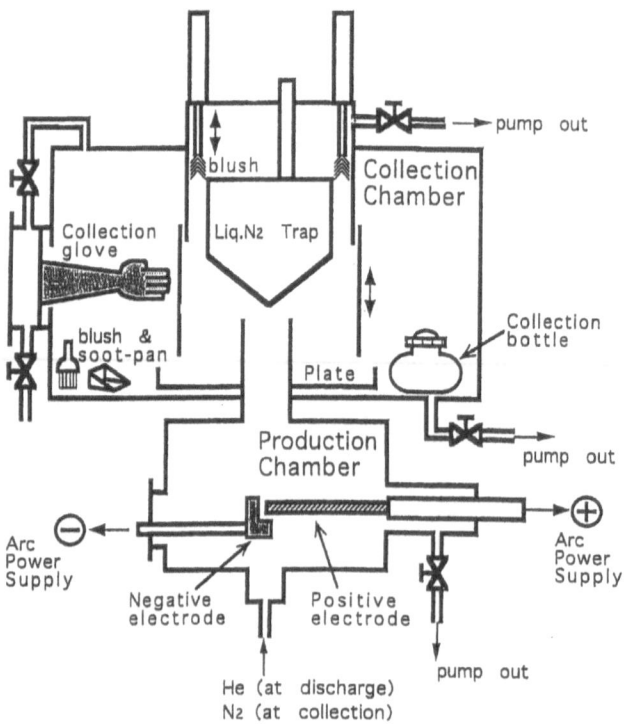

Fig. 1   DC arc-discharge apparatus with an anaerobic collection and sampling mechanism. The metallofullerene-containing soot produced is effectively trapped by the Liq. N_2 trap in the collection chamber. Typical arc discharge conditions: 40–100 Torr He flow, 200–500 A, and 25–30 V.

The mixture of Sc_2O_3, graphite powder and pitch was stuffed in graphite rods (20 mm diam. × 500 mm long) which were used as positive electrodes in the direct current (500 A) spark mode under 50–100 Torr He flow conditions (Fig. 1). The soot produced was collected under totally anaerobic conditions to avoid unnecessary degradation of the metallofullerenes produced during soot collection and handling (Fig. 1), because some of the metallofullerenes are air-sensitive.

The soot produced was soxhlet-extracted by carbon disulfide for 12 hours. We have found that metallofullerenes are further extracted from the residue by such solvents as pyridine and 1,2,4-trichlorobenzene. In the present study, the residue was further refluxed by pyridine for 6 hours. The metallofullerenes were found to be concentrated in this extracted fraction.

3. Separation and purification

The scandium fullerenes, such as $Sc@C_{82}$, $Sc_2@C_{84}$ and $Sc_3@C_{82}$, were separated and isolated from various hollow (C_{60}–C_{110}) fullerenes by, what we

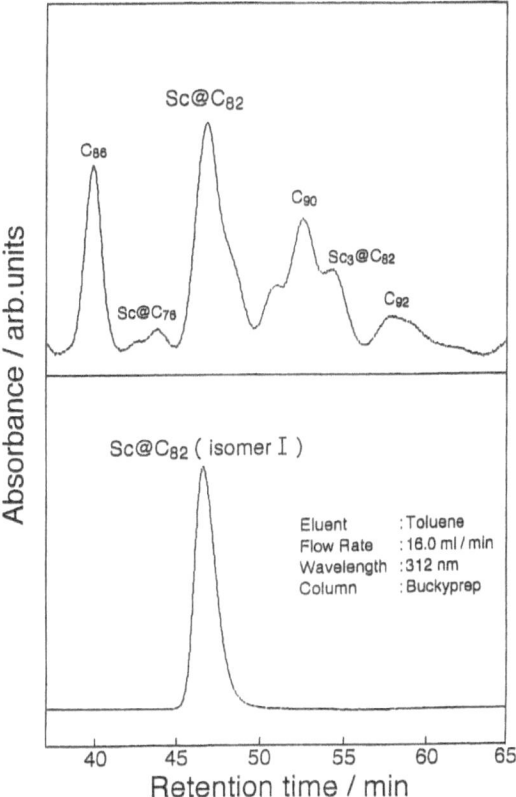

Fig. 2 (a) An overall high-performance liquid chromatography (HPLC) spectrum for the fraction which contains various scandium metallofullerenes in the first HPLC stage. (b) An isolated HPLC chromatogram for the $Sc@C_{82}$ (I) after the second HPLC stage. The experimental conditions are presented in the lower right of the figure.
(From H. Shinohara, *"Fullerenes" Recent.Advances in the Chemistry & Physics of Fullerenes and Related Materials, vol 2*, (K. Kadish *et al.*, eds.), **2**, p.338, Electrochemical Society (1995)).

Separation Scheme of Sc@C$_{82}$ (Isomer I)

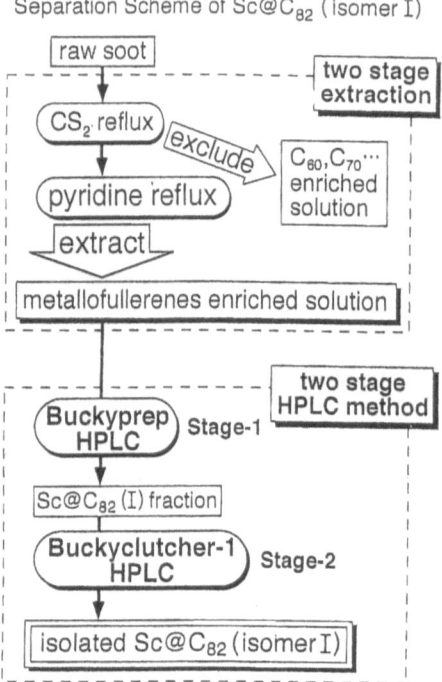

Fig. 3   Extraction and separation-isolation scheme of endohedral metallofullerenes. Two-stage
extraction includes CS$_2$ and pyridine refluxes. The pyridine extract of the residual soot after
CS$_2$ extraction is a metallofullerene-enriched extract. The separation and isolation is done by
the two-stage HPLC method.

call, the two-stage HPLC method.[o,v] The two-stage HPLC method uses two
complementary HPLC columns which have different types of fullerene adsorption
mechanisms. The two-stage HPLC method was first successfully applied to the
isolation of several di-scandium fullerenes including Sc$_2$@C$_{74}$, Sc$_2$@C$_{82}$ and
Sc$_2$@C$_{84}$ by the present group.[o] Kikuchi et al.[e] employed a similar method, but
in a somewhat different context, for the first isolation of La@C$_{82}$.

In the first HPLC stage, the toluene solution of the extracts was separated by
a preparative recycling HPLC system (Japan Analytical Industry LC-908-C60)
with a Trident-Tri-DNP column (Buckyclutcher I, 21 × 500 mm: Regis Chemical).
In the present study, the Buckyclutcher I column was used with a 100% toluene
eluent at a typical flow speed of 10 ml/min. In this HPLC process, the scandium
fullerene-containing fractions were separated from other fractions including C$_{60}$,
C$_{70}$ and higher fullerenes (C$_{76}$–C$_{110}$). The complete purification and isolation of
various scandium fullerenes were performed in the second HPLC stage by using
a Cosmosil Buckyprep Column (20 × 250 mm. Nacalai Tesque) with a 100%
toluene eluent. Fig. 2 shows the first and the second HPLC stages of the
purification of Sc@C$_{82}$. The overall extraction/separation scheme is shown in
Fig. 3.

Most of the mono-metallofullerenes, M@C$_{82}$, have at least two types of
structural isomers, I and II, which can separated by the present two-stage HPLC
technique.[f,w] The retention times of isomers I are shorter than those of isomers II
and, in general, the isomers I are much more stable than isomers II in air or in

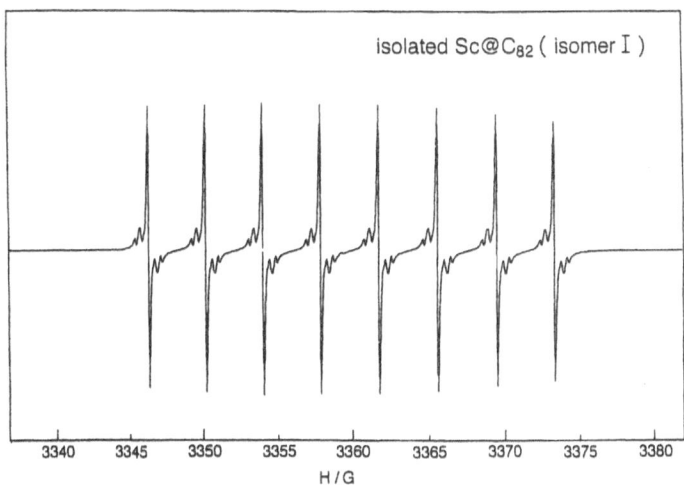

Fig. 4   X-band (9.435 GHz) ESR spectrum for the isolated Sc@C$_{82}$ (I) in carbon disulfide solution
at room temperature (g = 2.0002, A = 0.38 G, ΔH = 0.036 G).
(From H. Shinohara, *"Fullerenes" Recent.Advances in the Chemistry & Physics of Fullerenes
and Related Materials, vol 2*, (K. Kadish *et al.*, eds.), **2**, p.339, Electrochemical Society
(1995)).

various solvents. The isolation of various metallofullerenes was confirmed by
laser-desorption time-of-flight (LD-TOF) mass spectrometry.

4. Electron spin resonance of metallofullerenes

4.1 Sc@C$_{82}$ (isomer I)

Figure 4 exhibits an ESR spectrum of the isolated Sc@C$_{82}$ (isomer I) in a
degassed carbon disulfide solution at room temperature. The spectrum shows
equally spaced (0.38 G), eight narrow (0.036 G) ESR hyperfine splittings (hfs)
centered at g = 2.0002 owing to the the hyperfine coupling to the scandium
nucleus (I = 7/2).[i-l] The observed g value is similar to C$_{60}^-$ and C$_{70}^-$ radical anions,
and for La@$_{82}$ and Y@C$_{82}$ (*vide infra*), which range from 1.995 to 2.001,
indicating that the Sc@C$_{82}$ fullerene is in a doublet state. The ^{45}Sc hyperfine
constant (3.85 G) is very small compared with the reported coupling constants (60-
70 G) for Sc^{2+} impurities in a CaF$_2$ or SrF$_2$ lattice below 30 K.[x] Unusually small
hfc constants have been reported for La@C$_{82}$ (1.25 G)[c] and Y@C$_{82}$ (0.48 G).[g,h]
This indicates that the scandium atom is in the +2 or +3 oxidation state as reported
in La^{3+}@C$_{82}^{3-}$ and Y^{3+}@C$_{82}^{3-}$. Recent theoretical calculations suggest that the
electronic structure of Sc@C$_{82}$ is well represented by Sc^{2+}@C$_{82}^{2-}$.[s]

4.2 Sc$_3$@C$_{82}$

Figure 5 exhibits ESR spectrum of the isolated Sc$_3$@C$_{82}$ in a degassed CS$_2$
solution at 220 K. The spectrum shows perfectly symmetric, equally spaced (6.51
G), 22 narrow (0.770 G) ESR hyperfine splittings (hfs) centered at g = 1.9985,
which is a manifestation of the isotopic hyperfine coupling of three scandium
nuclei with I = 7/2 in the C$_{82}$ cage.[i,k] This hfs of Sc$_3$@C$_{82}$ was observed by the

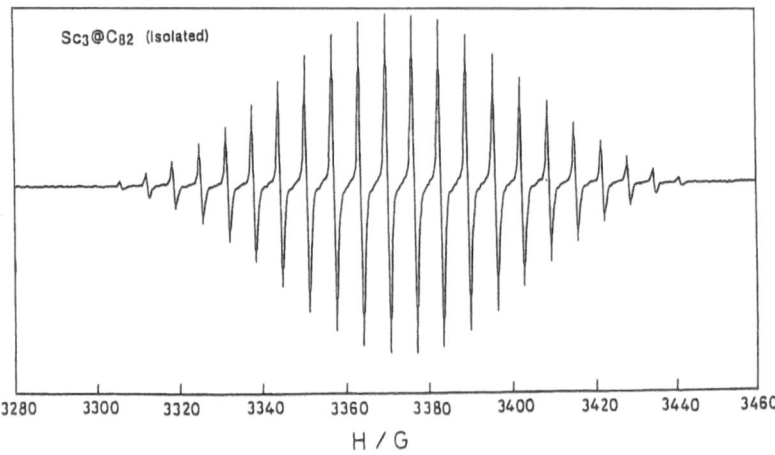

Fig. 5 X-band (9.4560 GHz) ESR spectrum for the isolated $Sc_3@C_{82}$ in CS_2 solution at 220 K, showing symmetric 22 hfs ($g = 1.9985$, $A = 6.51$, $\Delta H = 0.770$ G).

Nagoya[i] and IBM (Almaden)[k] groups. The presence of the perfectly symmetric 22 hfs lines suggests the geometrical equivalency of the three scandium atoms in the C_{82} cage.[i–k] The ^{13}C-NMR measurements[y] for the hollow C_{82} fullerene indicate the presence of at least three structural isomers: C_2, C_{2v} and C_{3v}. At present, no theoretical calculation has been reported for $Sc_3@C_{82}$. However, a recent *ab initio* calculation on $Sc_2@C_{84}$ (the D_{2d} isomer)[s] reveals that the encaged two scandium atoms are well separated from each other by 4.05 Å along the D_{2d} axis and that a substantial electron transfer from the scandium atoms to the C_{82} cage is taking place. According to the *ab initio* calculation, the formal net charge of the species can be described as $(Sc_2)^{4.4+}@C_{84}^{4.4-}$. A similar charge transfer has been already reported by Nagase and co-workers on $Sc@C_{82}$,[r] which leads to the electronic structure of $Sc^{2+}@C_{82}^{2-}$. Based on the appearance of the perfectly symmetric 22 hfs and the results of the theoretical calculations described above, three scandium atoms are separated from each other within the C_{82} cage so as to retain a three-fold axis as an entire $Sc_3@C_{82}$ molecule. This condition is only satisfied if the three scandium atoms (ions) are situated in a triangular position of the C_{3v} isomer of C_{82}.

The temperature dependence of the 22 hfs lines can provide further structural information on $Sc_3@C_{82}$. The ΔH value has a minimum at 220 K above which the hfs line width increases as temperature increases. A similar temperature dependence has been reported for $La@C_{82}$. Even at this temperature the hfs linewidth of $Sc_3@C_{82}$ is about 20 times as broad as that of $Sc@C_{82}$. From the temperature dependence we conclude that each hfs linewidth is homogeneously broadened due to dynamical averaging effects of the encaged scandium atoms.[z,a'] Figure 6 shows linewidths spectra for $Sc_3@C_{82}$ in CS_2 and liquid paraffin solutions at various temperature plotted against quantum number M_I of the Sc_3 nuclear magnetic moment. Kato *et al.*[b] interpreted the data by a dynamic Jahn-Teller distortion. At high temperatures, pseudorotation of the Sc metal among inequivalent environments becomes faster, and a motional narrowing mechanism could be expected as the total system is dynamically averaged to maintain a C_3 axis symmetry. In this limit, the ESR spectrum for $Sc_3@C_{82}$ should exhibit a symmetric 22-line pattern, each with a Lorentzian line shape. The intramolecular

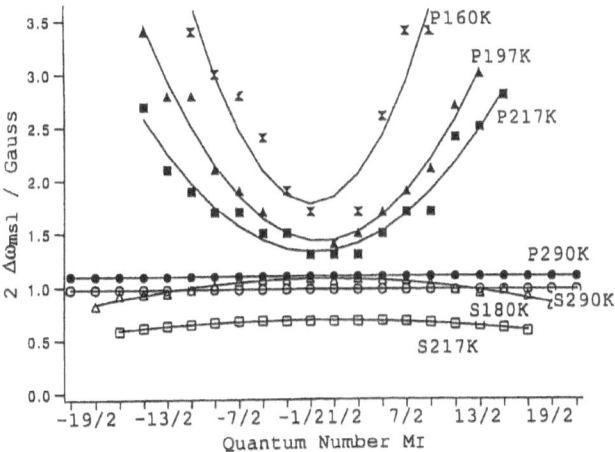

Fig. 6 Linewidth spectra for $Sc_3@C_{82}$ in CS_2 and liquid paraffin solutions at various temperatures plotted against quantum number M_I of the Sc_3 nuclear magnetic moment. P indicates the linewidth in liquid paraffin solution and S in CS_2 solution.
(From H. Shinohara *et al.*, *J. Phys. Chem.*, **99**, 857 (1995)).

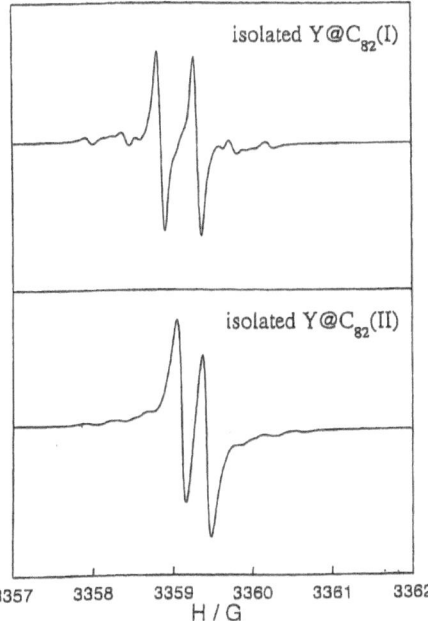

Fig. 7 X-band ESR spectra of the two isomers of $Y@C_{82}$ (I,II) in CS_2 at room temperature.
(From H. Shinohara, *"Fullerenes" Recent. Advances in the Chemistry & Physics of Fullerenes and Related Materials, vol 2*, (K. Kadish *et al.*, eds.), **2**, p.341, Electrochemical Society (1995)).

Table 1 Hyperfine parameters

Species	A/G	g value	ΔH/G
Sc@C$_{82}$ (isomer I)	0.38	2.0002	0.036
Sc$_3$@C$_{82}$	6.51	1.9985	0.770
Y@C$_{82}$ (isomer I)	0.48	2.0004	0.087
Y@C$_{82}$ (isomer II)	0.34	1.9999	0.12

dynamics is the inherent nature of the Sc trimer encapsulated in the C$_{82}$ cage.

4.3 Y@C$_{82}$ (isomers I, II)

There exist two structural isomers (I and II) for Y@C$_{82}$, which have been separated and isolated by the the two-stage HPLC method.[o] The yttrium metallofullerenes, Y@C$_{82}$ (isomers I and II), show distinct ESR hyperfine doublets due to the $I = 1/2$ yttrium nucleus. Figure 7 shows ESR spectra of Y@C$_{82}$ (I, II) in CS$_2$ solution at room temperature. The overall spectral pattern is similar, but the hyperfine splitting values are different with each other. Moreover, the appearance of the small satellite peaks due to ^{13}C, adjacent to the main doublets, are much less clear in Y@C$_{82}$ (II). Obviously, the electronic structure of the two isomers are

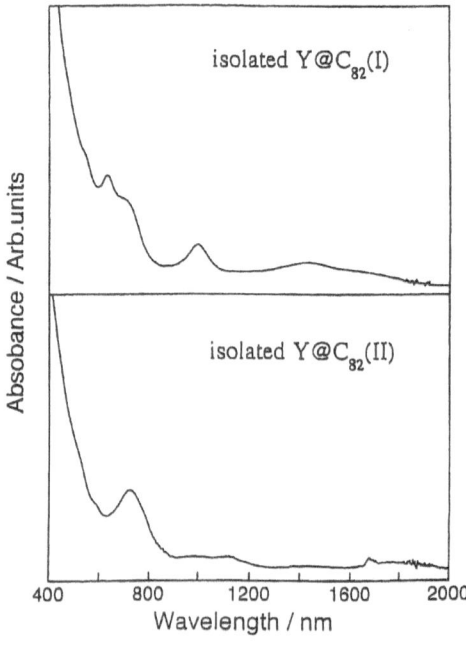

Fig. 8 UV-VIS absorption spectra for the isolated Y@C$_{82}$ (I and II) in carbon disulfide solution. (From H. Shinohara, *"Fullerenes" Recent.Advances in the Chemistry & Physics of Fullerenes and Related Materials, vol 2*, (K. Kadish *et al.*, eds.), **2**, p.342, Electrochemical Society (1995)).

different. The ESR parameters or $Y@C_{82}$ (I, II) together with that of $Sc@C_{82}$ (isomer I) are summarized in Table 1. The substantial difference in the electronic structures of the two $Y@C_{82}$ isomers is also evident from Fig. 8. Figure 8 shows UV-VIS absorption spectra of the two isomers of $Y @C_{82}$. The spectral features are quite different with each other. We believe that these differences stem from some structural differences which exist between $Y@C_{82}$ (I) and $Y@C_{82}$ (II). It has been found that isomer (II) is less stable in air and much more reactive towards various solvents than isomer (I).

It has been shown directly by a recent synchrotron X-ray diffraction study [b'] that $Y@C_{82}$ (I) has an "endohedral" structure in which the yttrium atom is encapsulated within the C_{82} cage and is strongly bound to the fullerene cage. However, at present, no structural analysis has been performed on $Y@C_{82}$ (II). Future X-ray structural analyses are definitely needed.

Acknowledgments

We express our thanks to M. Ohno and M. Inakuma, Nagoya University, for experimental help. The ESR study of $Sc_3@C_{82}$ was done in collaboration with Prof. T. Kato (Institute for Molecular Science). H.S. thanks the Japanese Ministry of Education, Science and Culture for Grants-in-Aid for Scientific Research on Priority Areas (No. 05233108) and General Scientific Research (No. 06403006).

PUBLICATIONS

1. A. Kasuya, W. Tanaka, R. Nishitani, Y. Saito, H. Shinohara and Y. Nishina, *Jpn. J. Appl. Phys. Suppl.*, **34**, Suppl. 34-1, 200 (1995).
2. T. Kato, S. Bandow, M. Inakuma and H. Shinohara, *J. Phys. Chem.*, **99**, 856 (1995).
3. T. Sakarai, X. D. Wang, T. Hashizume, V. Yurov, H. Shinohara and H. W. Pickering, *Appl. Surf. Sci.*, **87/88**, 405 (1995).
4. X. D. Wang, S. Yamazaki, T. Hashizume, H. Shinohara and T. Sakurai, *Phys. Rev.*, B, **52**, 2098 (1995).
5. M. Takata, B. Umeda, E. Nishibori, M. Sakata, Y. Saito, M. Ohno and H. Shinohara, *Nature*, **377**, 46 (1995).
6. M. Takata, B. Umeda, E. Nishibori, M. Sakata, Y. Saito, M. Ohno and H. Shinohara, (L. Nightingale: News and Views) *Nature*, **377**, 14 (1995).
7. H. Shinohara, M. Inakuma, M. Kishida, S. Yamazaki, T. Hashizume and T. Sakurai, *J. Phys. Chem.*, **99**, 13769 (1995).
8. H. Shinohara, M. Tansho, M. Inakuma, Y. Saito, H. Sato and N. Hayashi, *Surf. Sci. Lett.*, **3**, 799 (1996).
9. T. Takahashi, A. Ito, M. Inakuma and H. Shinohara, *Phys. Rev. B*, **52**, 13812 (1995).
10. T. Kimura, T. Sugai, H. Shinohara, T. Goto, K. Tohji and I. Matsuoka, *Chem. Phys. Lett.*, **246**, 571 (1995).
11. M. Inakuma, M. Ohno and H. Shinohara, in: F*ullerenes: Recent Advances in the Chemistry and Physics of Fullerenes and Related Materials*, **vol.2**, p.330 (1995).
12. H. Shinohara, M. Ohno, M. Kishida, S. Yamazaki, T. Hashizume and T. Sakurai, in: *Fullerenes: Recent Advances in the Chemistry and Physics of Fullerenes and Related Materials*, **vol.2**, p.763 (1995).
13. T. Sakurai, S. Yamazaki, Y. Hasegawa, Y. Ling, T. Hashizume and H. Shinohara, in: *Fullerenes: Recent Advances in the Chemistry and Physics of Fullerenes and Related Materials*, **vol.2**, p.709 (1995).
14. H. Shinohara, M. Ohno, M. Kishida, S. Yamazaki, T. Hashizume and T. Sakurai, *Structures and Dynamics of Clusters*: Proceedings of the XLIII Yamada Conference Tokyo Academic Press, 79 (1996).
15. H. Shinohara, M. Tanaka, M. Sakata, T. Hashizume and T. Sakurai, in: *Cluster-Assembled Solids*, (K. Sattler, ed.), Transtech Publication (1996).

REFERENCES

a. Y. Chai, T. Guo, C. Jin, R. E. Haufler, L. P. F. Chibante, J. Fure, L. Wang, J. M. Alford and R. E. Smalley, *J. Phys. Chem.*, **95**, 7564 (1991).

b. D. S. Bethune, R. D. Johnson, J. R. Salem, M. S. de Vries and C. S. Yannoni, *Nature*, **366**, 123 (1993).

c. R. D. Johnson, M. S. de Vries, J. Salem, D.S. Bethune & C. S. Yannoni, *Nature*, 355, 239 (1992).

d. M. M. Alvarez, E. G. Gillan, K. Holczer, R. B. Kaner, K. S. Min and R. L. Whetten, *J. Phys. Chem.*, **95**, 10561 (1991).

e. K. Kikuchi, S. Suzuki, Y. Nakao, N. Nakahara, T. Wakabayashi, H. Shiromaru, I. Saito, I. Ikemoto and Y. Achiba, *Chem. Phys. Lett.*, **216**, 67 (1993).

f. K. Yamamoto, H. Funasaka, T. Takahashi and T. Akasaka, *J. Phys. Chem.*, **98**, 2008 (1994).

g. J. H. Weaver, Y. Chai, G. H. Kroll, C. Jin, T. R. Ohno, R. E. Haufler, T. Guo, J. M. Alford, J. Conceicao, L. P. F. Chibante, A. Jain, G. Palmer and R. E. Smalley, *Chem. Phys. Lett.*, **190**, 460 (1992).

h. H. Shinohara, H. Sato, Y. Saito, M. Ohkohchi and Y. Ando, *J. Phys. Chem.*, **96**, 3571 (1992).

i. H. Shinohara, H. Sato, M. Ohchochi, Y. Ando, T. Kodama, T. Shida, T. Kato and Y. Saito, *Nature*, **357**, 52 (1992).

j. H. Shinohara, H. Yamaguchi, N. Hayashi, H. Sato, M. Inagaki, Y. Saito, S. Bandow, H. Kitagawa, T. Mitani and H. Inokuchi, *Mater. Sci. Eng.*, **B19**, 25 (1993).

k. C. S. Yannoni, M. Hoinkis, M. S. de Vries, D. S. Bethune, J. R. Salen, M. S. Crowder and R. D. Johnson, *Science*, **256**, 1191 (1992).

l. S. Suzuki, S. Kawata, H. Shiromaru, K. Yamauchi, K. Kikuchi, T. Kato and Y. Achiba, *J. Phys. Chem.*, **96**, 7159 (1992).

m. E. Gillan, C. Yeretzian, K. S. Min, M. M. Alvarez, R. L. Whetten and R. B. Kaner, *J. Phys. Chem.*, **96**, 6869 (1992).

n. L. Moro, R. S. Ruoff, C. H. Becker, D. C. Lorents and R. Malhotra, *J. Phys .Chem.*, **97**, 6801 (1993).

o. H. Shinohara, H. Yamaguchi, N. Hayashi, H. Sato, M. Ohkohchi, Y. Ando and Y. Saito, *J. Phys. Chem.*, **97**, 4259 (1993).

p. A. Rosen and B. Waestberg, *Z. Phys., D*, **12**, 387 (1989).

q. K. Laasonen, W. Andreoni and M. Parrinello, *Science*, **258**, 1916 (1992).

r. S. Nagase and K. Kobayashi, *Chem. Phys. Lett.*, **214**, 57 (1993).

s. S. Nagase and K. Kobayashi, *Chem. Phys. Lett.*, **231**, 319 (1994).

t. T. Guo, G. K. Odom and G. E. Scuseria, *J. Phys. Chem.*, **98**, 7745 (1994).

u. S. Bandow, H. Shinohara, Y. Saito, M. Ohkohchi and Y. Ando, *J. Phys. Chem.*, **97**, 6101 (1993).

v. H. Shinohara, M. Inakuma, N. Hayashi, H. Sato, Y. Saito, T. Kato and S. Bandow, *J. Phys. Chem.*, **98**, 8597 (1994).

w. M. Inakuma, M. Ohno and H. Shinohara, in: *Recent Advances in the Chemistry and Physics of Fullerenes Vol.II*, (R. Ruoff and K. Kadish, eds.), **vol.2**, p.330, The Electrochemical Society Inc. (1995).

x. U. T. Hoechli and T. L. Estle, *Phys. Rev. Lett.*, **18**, 128 (1967).

y. K. Kikuchi, N. Nakahara, T. Wakabayashi, S. Suzuki, H. Shiromaru, Y. Miyake, K. Saito, I. Ikemoto, M.'Kainosho and Y. Achiba, *Nature*, **357**, 142 (1992).

z. T. Kato, S. Bandow, M. Inakuma and H. Shinohara, *J. Phys. Chem.*, **99**, 856 (1995).

a'. P. H. M. van Loosdrecht, R. D. Johnson, D. S. Bethune, H. C. Dorn, P. Burbank and S. Stevenson, *Phys. Rev. Lett.*, **73**, 3415 (1994).

b'. M. Takada, B. Umeda, E. Nishibori, M. Sakata, Y. Saito, M. Ohno and H. Shinohara, *Nature*, **377**, 46 (1995).

41

Development of a Scanning Atom Probe
—A New Approach to the Study of Microclusters—

Osamu Nishikawa [a], Masahiro Kimoto [a], Masashi Iwatsuki [b], Susumu Aoki [b] and Yuuichi Ishikawa [c]

[a] Department of Materials Science and Engineering, Kanazawa Institute of Technology
     Nonoichi-cho, Ishikawa 921-8501, Japan
[b] Electron Optics Division, JEOL Ltd.
     Akishima-shi, Tokyo 196-0021, Japan
[c] Mechanical Engineering Research Laboratory, Hitachi Ltd.
     503 Kandatsu, Tsuchiura 300-0013, Japan

Purpose of the present study

One promising approach for clarifying the formation mechanism of microclusters is to mass analyze the field evaporated micro-fragments with an atom probe. However, conventional atom probes have serve intrinsic restrictions limiting their applicability. In order to remove these restrictions and extend the applicability of the atom probe, a new atom probe based on an innovative idea, a scanning atom probe (SAP), has been proposed. The purpose of the present study is to develop the SAP evaluating the performance and capability of the trial instrument. This article reports the principle and fundamental structure of the SAP and the results of preliminary experiments with the trial instrument.

Contents of the present study

1. Introduction

A unique phenomenon observed by the field ion microscope (FIM) [a] is the field-induced removal of surface atoms as positive ions, called field evaporation, breaking bonds between surface and substrate atoms. It has also been found that some materials are field evaporated as clusters while field evaporating semiconductive specimens such as Si, GaAs and GaP,[1] conducing polymers[2] and ceramics,[3] indicate the preferential break of weak bonds. This suggests that the identification of individual evaporated cluster species by the atom probe (A-P),[a,b,4] an instrument combining the FIM and the mass spectrometer, may provide basic information on the binding states, stability, and formation mechanism of clusters.

However, the trial application of the conventional A-P to the study of microclusters encountered a severe restriction originating in the basic structure of the A-P: the sample needs to be an extremely sharp tip and the analyzed area is an very small hemispherical tip apex. Furthermore, the fabrication of a sharp tip of semiconductors and polymers is very difficult and practically impossible to make a tip with a multilayer structure grown on a wafer at the tip end.

In order to overcome these difficulties a new A-P named the "scanning atom probe (SAP)" is proposed.[5] The specimen used for the SAP is not a single sharp tip but a flat surface with many spikes and cusps. A funnel-shaped micro-extraction electrode scans over the specimen and forms a minute field emission or field ion microscope when the electrode is positioned above one of the spikes or cusps. Then the high field generated by a bias voltage applied to the electrode or the specimen is confined to a small space between the electrode and the specific apex.

In the present study the feasibility of developing such as SAP was examined by computing the field distribution between a cusp apex and an extraction electrode for various tip lengths, cone angles of the spikes, and the position and configuration of the electrode. [5,6] The study was also extended to include a preliminary experiment modifying a conventional scanning tunneling microscope (STM),[c,7,8] and results supporting the feasibility of realizing the SAP were obtained.[9]

2. Scanning atom probe

Unique features of the SAP are a flat specimen and a micro-extraction electrode. A surface with many micro-spikes or cusps and a flat surface grooved or micro-photoetched in a checkerboard pattern forming many spikes or tips can be a specimen (Fig. 1). The assumed height of the spikes and the depth of the grooves are several to a few tens of microns and the spaces between the spikes and tips are tens to hundreds of microns. The funnel-shaped extraction electrode scans over the rugged surface and stands still above a particular apex (Fig. 2). Since the

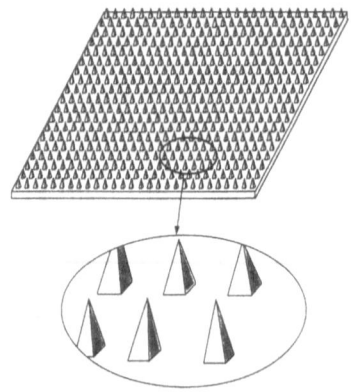

Fig. 1 Schematics of a grooved specimen.

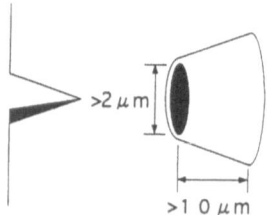

Fig. 2 Schematics of the extraction electrode and a tip formed by grooving.

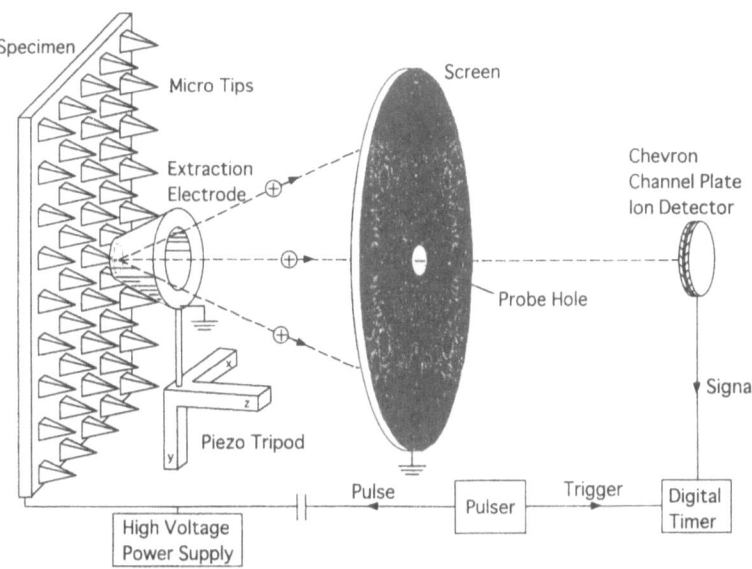

Fig. 3   Schematics of a scanning atom probe. One apex on the specimen, the exaction electrode and
the screen form an FEN and FIM. When a high pulse voltage is applied to the specimen, the
surface atoms at the apex are field evaporated and pass through the probe hole at the center
of the screen and are detected by the chevron.

diameter of the open hole at the end of the extraction electrode can be made as
small as a few to ten microns, the high field is well confined to a small space
between the tip and the electrode when a negative or positive bias voltage is
applied to the sample or the electrode. Then the combination of one apex of the
cusps or tips and the electrode enables us to extract field-emitted electrons, field-
ionized gas ions and field-evaporated surface atoms from the apex, and to project
a field emission or field ion image on a screen like a conventional field emission
microscope (FEM) and an FIM (Fig. 3).

3. Computation of field distribution

The confinement of the field depends on many factors setting the
configuration and arrangement of the apex and electrode. The most effective
factor is the apex-electrode gap; the narrower the gap, the smaller the high field
space and the lower the bias voltage. However, care must be taken to avoid field
emission from the open end of the electrode when the SAP is operated as an FIM
because the field strength at the electrode surface may be high enough to induce
a high-density emission current from the electrode and damage the apex. Thus, the
highest field strength on the electrode surface should be less than 0.55 V/Å: one
tenth the evaporation field of tungsten, 5.5 V/Å, which is known to be the highest
evaporation field for metals at 0 K.

Accordingly, the field distribution in the confined space was computed in
order to find the optimum apex and electrode configuration by varying the height
difference between the tip apex and the electrode d, tip radius of the apex r,
height of the cusp h, cone angle of the cusp θ, diameter of the open hole at the end
of the electrode ϕ, and cone angle and end curvatures of the electrode α_1, α_2 and

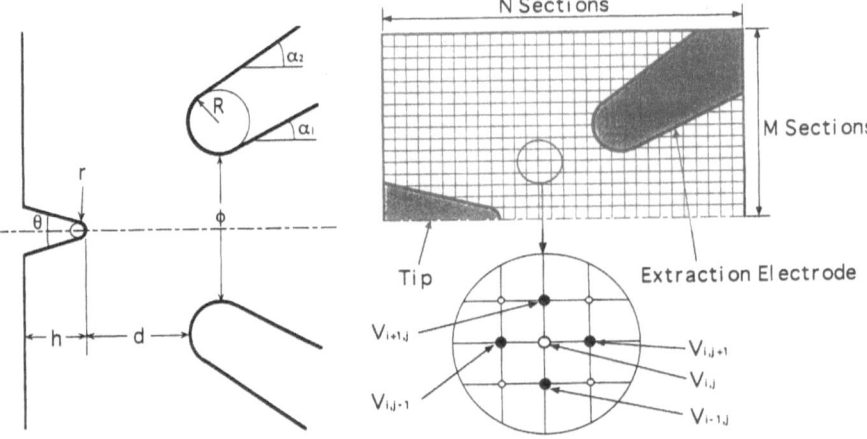

Fig. 4 Schematics indicating parameters to calculate the field distribution in the space between the electrode and the apex.

Fig. 5 Schematics of the divided space for computing the field distribution by difference equations.

R, respectively (Fig. 4). Since the apex-electrode assembly is symmetric around the cusp axis, cylindrical coordinates were employed to calculate the electric potential V. Then, Laplace's differential equation was transferred to difference equations and the r-z space was divided into $M \times N$ sections to compute the potentials at each i-jth crossing point $V_{i,j}$ (Fig. 5). In the present calculation M and N were 800 and 1600, respectively.

4. Field strength and field distribution

An intuitive estimate of the field strength suggests that the field strength at the apex, E_t, increases with decreasing r and θ, and increasing h. Similarly, the field strength at the end of the electrode E_e decreases with increasing ϕ, d and R. Accordingly, the ratio of the field strengths, $E_r = E_t/E_e$, should increase with decreasing r and θ, and with increasing h, ϕ, d and R. However, the confinement of the high field in a small space between the tip apex and the electrode requires small ϕ, d and R and large θ, which is favorable for fabricating the tips by grooving.

The bias voltage of the specimen V_t, E_e and E_r are calculated for $E_t = 5.5$ V/Å, $r = 1000$ Å, and for various θ, h, ϕ, d and R (Table 1). Variation ranges of V_t, E_e and E_r are rather narrow, from 6.8 to 12.2 kV, from 0.45 to 1.28 V/Å and from 4.3 to 12.3 V/Å, respectively, for the variation of h from 2.2 to 6.4 µm, θ from 20° to 90°, d from –0.5 to 1.0 µm, ϕ from 2.4 to 4.8 µm and R from 6000 to 9000Å. As presumed, V_t drops as the electrode approaches the apex, but E_e and E_r stay rather constant because E_t is constant. As θ increases, V_t and E_e increase significantly, lowering E_r. On the other hand, E_r increases sharply with h.

Field and equipotential lines and trajectories of He ions field-ionized above the tip surface are shown in Figs. 6 and 7. Although the high field regions are well confined to a narrow space around the tip apex and the field lines (left half of the figures) are smoothly curved, the ion trajectories (right half of the figures) are spread fairly straightly into the radial directions of the cusp apex, projecting an

Table 1 Variation of V_t, E_e and E_r with θ, h, ϕ, d, α_1, α_2 and R for $E_t = 5.5$ V/Å and $r = 1000$ Å

h (µm)	d (µm)	θ (deg)	ϕ(µm)	α_1 (deg)	α_2 (deg)	R (Å)	V_t (kV)	E_e (V/Å)	E_r
2.2	0	40	2.4	30	35	6000	7.5	0.93	8.1
2.2	1	20	3.4	30	35	6000	10.0	0.68	8.0
2.2	1	20	3.4	40	45	6000	8.1	0.85	9.5
3.2	0	20	2.4	30	35	6000	6.9	0.64	10.8
3.2	0	60	3.4	30	35	6000	9.0	0.68	8.1
3.2	0	90	3.4	30	35	6000	12.2	1.28	4.3
3.2	0	30	2.4	30	35	6000	7.3	0.57	9.7
3.2	−0.25	30	2.4	30	35	6000	7.0	0.58	9.5
3.2	−0.5	30	2.4	30	35	6000	6.8	0.59	9.3
3.2	1	30	2.4	30	35	9000	9.3	0.48	11.5
6.4	0	30	2.4	30	35	6000	7.2	0.45	12.2
6.4	0	30	4.8	30	35	6000	8.7	0.45	12.3

enlarged image of the hemispherical apex area on a screen like an ordinary FIM.

5. Trial scanning atom probe

In order to examine the feasibility of the SAP, a trial SAP was constructed modifying a low-temperature ultrahigh vacuum STM (Fig. 8). The STM tip mounted on a tube scanner is replaced by a specimen with multicusps. The STM specimen on the X-Y stage is also replaced by the extraction electrode. Since the other side of the electrode is an open space, the ions generated at the cusp apex by field ionization and field evaporation can fly to the screen or to the ion detector. The specimen can be cooled down to 30 K by liquid helium in a cold head through ten layers of vibration isolating silver foils of 10 mm wide and 0.1 mm thick which have the highest thermal conductivity at 20 K. The scanner and the sample are isolated thermally and electrically by a polychloro-trifluoro-ethylene (PCTFE) plate with low thermal conductivity (6×10^{-4} cal/s·deg: about one thousandth of stainless steel) and high resisitivity (1.2×10^{18} Ω·cm).

The sample approaches the electrode by an 8-mm slider and a motor-driven magnetic loader (up to 3 mm). At the final stage a PTZ scanner shifts the sample up to ± 0.6 µm. The scanner also shifts the sample in the X-Y direction up to ± 0.3 µm with an accuracy of 0.1 Å. The X-Y stage on which the extraction electrode is mounted also allows one to adjust the relative position between the electrode and a cusp on the sample up to 10 mm with an accuracy of 0.1 µm.

An open space behind the electrode is opened widely to allow field-emitted electrons and field-ionized ions to fly to the screen projecting an enlarged image. The maximum cone angle of the radially emitted electrons and ions is 36°, which is slightly narrower than a conventional FEM/FIM with a 75-mm-diameter screen 10 cm away from a tip.

For the preliminary experiment with this trial instrument, a 0.1-mm thick Mo disk with an open hole of 10 and 20 µm in diameter was employed as the extraction electrode and the scanning tip of the STM served as the single apex of a cusp on a flat specimen surface. The relative position of the cusp apex and the open hole of the electrode was observed by the SEM inserted into the vacuum chamber of the STM (Fig. 8). Initial approach of the tip apex to the center of the hole was conducted by a microscrew with a precision of better than 1 µm. The

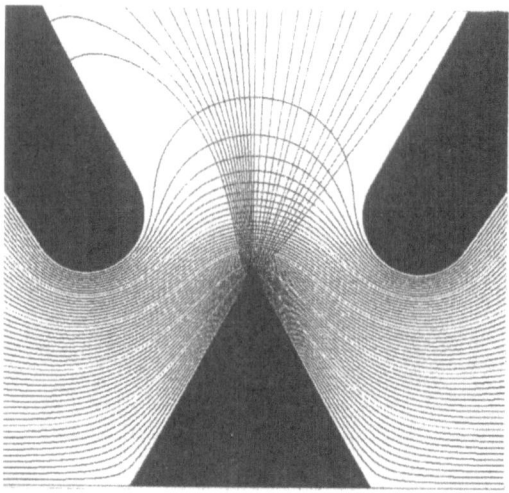

Fig. 6   Equipotential and field lines around a cusp and the electrode on the left half, and equipotential lines and trajectories of He ions emitted from the apex area covering 45° from the axis of the cusp on the right half. $r = 700$ Å, $h = 3.2$ μm, $\theta = 60°$, $d = 0$, $\phi = 3.4$ μm, $\alpha_1 = 30°$, $\alpha_2 = 35°$, and $R = 9000$ Å. E_t and E_e are 4.5 and 0.38 V/Å, respectively, at $V_t = 6.08$ kV resulting in $E_r = 11.8$. Voltage difference between two equipotential lines is $V_t/50 = 122$ V.

Fig. 7   Equipotential and field lines on the left half, and equipotential lines and trajectories of He ions on the right half. $r = 700$Å, $h = 3.2$ μm, $\theta = 90°$, $d = 0$, $\phi = 3.4$ μm, $\alpha_1 = 30°$, $\alpha_2 = 35°$, and $R = 9000$Å. E_t and E_e are 4.5 and 0.49 V/Å, respectively, at $V_t = 6.68$ kV resulting in $E_r = 9.2$. Voltage difference between two equipotential lines is $V_t/50 = 134$ V.

413

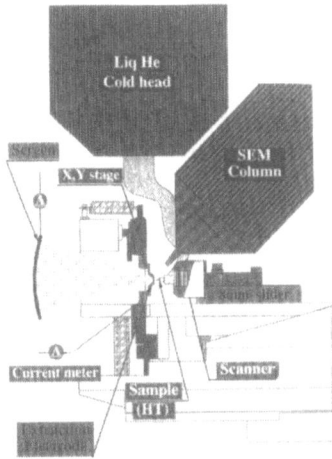

Fig. 8 Schematic figure of the trial SAP.

final approach was controlled by a piezo actuator as a conventional STM. When the tunneling current was detected, the tip was withdrawn a few microns and voltage high enough for field emission was applied to the tip. Then the tip scanned over the electrode surface searching the center of the open hole of the electrode. Although the SEM is very helpful for searching the electrode hole as shown in Fig. 9, no SEM image can be observed in the actual SAP operation because a flat specimen covers the viewing area of the SEM, an extremely narrow space between an apex of a cusp and the electrode. In order to overcome this difficulty, the variation in field emission current of scanning positions was examined.

6. Variation of field emission currents

When the tip was scanning about 5 μm above the electrode surface and crossed over an open hole 20 μm in diameter, field emission current from the tip at −110 V decreased from 300 pA to a few pA at the edge of the hole and increased to more than 20 pA at the other side of the edge (Fig.10). The large difference in the emission current at the two edges is due to the difference in the apex-surface distance. When the emission current was increased to nanoamperes, the emission current fluctuated wildly because the current density on the electrode surface 5 μm away from the tip apex is larger than 1 μA/cm^2 for the emission current of 1 pA assuming that the emitted current injects into a circle of 10 μm in

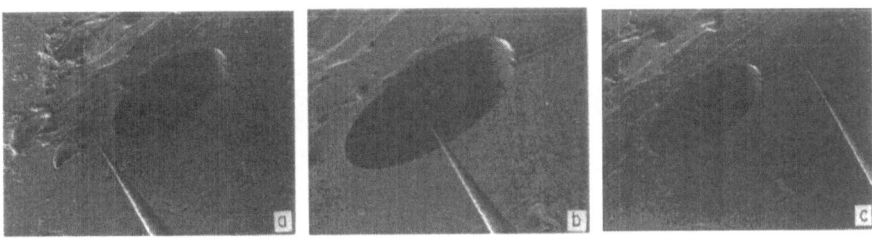

Fig. 9 Three SEM images of a scanning tip; (a), (b) and (c). The diameter of the open hole is 20 μm.

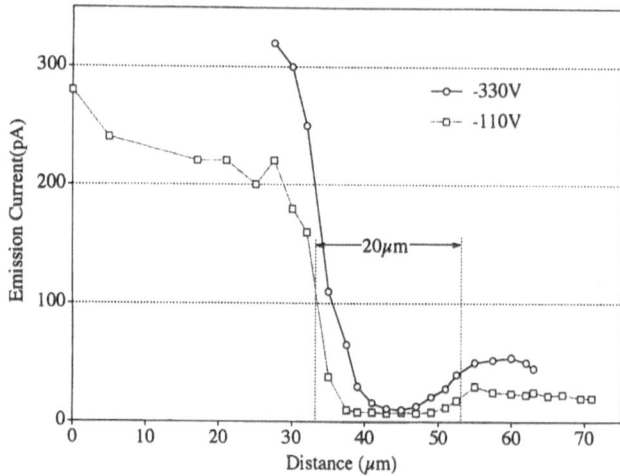

Fig. 10 Variation of field emission currents emitted from a pre-flashed sharp tip and a flashed tip crossing over a 20-μm diameter open hole. Tip voltages of the pre-flashed and flashed tips are −110 V and −330 V, respectively.

diameter. This current density is much larger than that of the ordinary incoming current density on the screen of an ordinary field emission microscope (FEM) and could be large enough to desorb the adsorbed gas, resulting in the current fluctuation.

The tip field emitting at −110 V flashed increasing its tip radius more than 20 times, nearly 2 μm, (Fig. 11). For an FEM with a single tip and a long tip-screen distance, the voltage required for field emission increases with tip radius. However, the observed rate of voltage increase is only 3 times, from −110 V to −330 V; this is less than one tenth the increase expected from the increase in the tip radius, possibly due to the generation of a sufficient high field in a narrow space between the electrode and the tip apex. It should be also noted that the field emission current from the flashed tip crossing over the open hole varies rather smoothly, and the scanning range emitting a small current corresponding to the diameter of the open hole appears narrower than the actual diameter. On the other hand, the current from the pre-flashed sharp tip varies fairly precisely following the profile of the hole and stays constantly low over the range corresponding to the hole, clearly indicating the size of the hole, (Fig. 10).

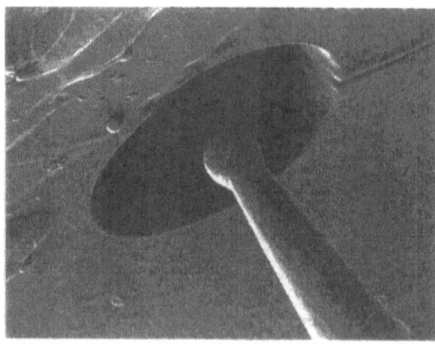

Fig. 11 SEM image of a flashed tip with tip radius of about 2 μm and the 20-μm diameter open hole.

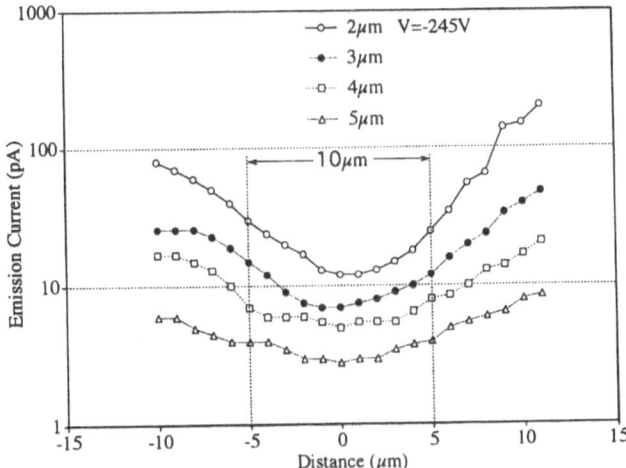

Fig. 12 Variation of field emission current emitted from a tip crossing over a 10-μm diameter open hole. The rate of current variation increases as the tip apex approaches to the electrode from 5 μm away to 2 μm. Tip voltage is −245 V.

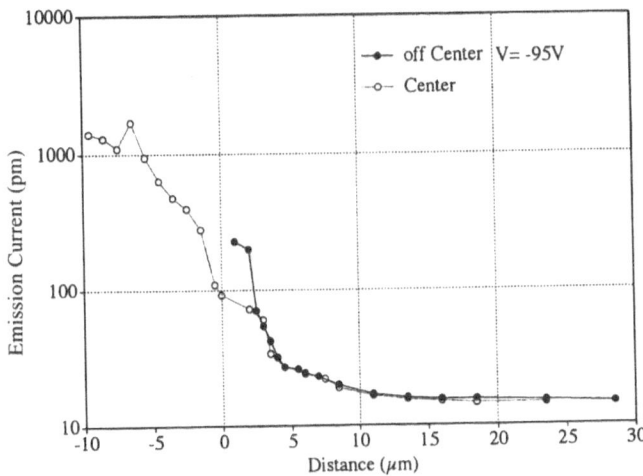

Fig. 13 Variation of field emission current with the distance between the tip apex and the electrode surface. Distance 0 corresponds to the position of the electrode surfaces. ○ represents the current from the tip passing the center of the hole and ● indicates the current from the tip approaching to a spot on the electrode surface outside the open hole. Tip voltage is −95 V.

Although the possibility of field emission from such a large tip was expected by the computer simulated field distribution,[5,6] the direct observation in the present experiment strongly supports the possibility of extending the applicable fields of the SAP to various specimen surfaces of metals, semiconductors, ceramics and organic materials with dull tips and cusps or even gently corrugated surfaces.

The variation of the emission current over an open hole 10 μm in diameter is less significant than that over a hole 20 μm in diameter because the variation in field strength over the smaller hole is less significant. Thus, it is desirable to bring the scanning tip closer to the electrode surface, that is, less than 5 μm, in order to enhance the variation of current (Fig. 12). The variation in field emission current with distance between the tip apex and the electrode surface was also measured (Fig. 13). The field emission current from the tip passing through the center of the open hole increases as the tip is inserted further into the hole but saturates when the inserted depth become nearly equal to the radius of the hole become the highest field strength is attained. Outside the hole the currents emitted from the tips passing the center of the hole and those departing from a spot on the electrode surface outside of the hole decrease nearly equally with distance: a fairly steep decrease up to a distance equal to about the diameter of the hole and a gentle decrease at greater distance.

7. Effects of field emission currents

Another subject to be examined is the applicability of the positive high voltage to a tip or cusp to field evaporate apex atoms. Since the high voltage is as high as ten times the negative voltage required for the field emission, it may generate a high field on the surface of the electrode and induce an extremely high current density of field emission from the electrode, resulting in serious damage to the tip apex. No change in the tip profile was observed by the SEM. The negative voltage required for a specified field emission current was measured before and after the application of the positive tip voltage. No voltage change was noticed.

8. Conclusion

The present study clearly indicates that the variation of the field emission current with the relative position between the center of the open hole of the electrode is large enough to bring the apex to the center of the hole implying that aiming at the center of the hole can proceed fairly easily by examining the variation of field emission currents even if the relative position is not inspected by the SEM while measuring the field emission current. The observed low voltage required for the field emission and the applicability of high positive field required for the field evaporation of apex atoms are also encouraging factors for the further development of the SAP.

Acknowledgements

This work is supported by a Grant-in-Aid for Scientific Research on Priority Areas, "Physical and Chemical Properties of Mesoscopic Materials" and "Tunneling Characteristic of Individual Surface Atoms," from the Ministry of Education, Science and Culture.

PUBLICATIONS

1. O. Nishikawa, E. Nomura, M. Yanagisawa and M. Nagai, *J. de Physique*, **47**, C2-30 (1986).
2. O. Nishikawa and H. Kato, *J. Chem. Phys.*, **85**. 6758 (1986).
3. O. Nishikawa and M. Nagai, *Phys. Rev. B*, **37**, 3658 (1988).
4. O. Nishikawa, K. Kurihara, M. Nachi, M. Konishi and M. Wada, *Rev. Sci. Instrum.*, **52**, 810 (1981).
5. O. Nishikawa and M. Kimoto, *Appl. Surf. Sci.*, **76/77**, 424 (1994).
6. O. Nishikawa, M. Kimoto, M. Iwatsuki and Y. Ishikawa, *J. Vac. Sci. Technol. B*, **13**, 599

(1995).

7. O. Nishikawa, M. Tomitori and F. Iwatsuki, *Mater. Sci. Eng. B*, **8**, 81 (1991).
8. O. Nishikawa, M. Tomitori and F. Iwatsuki, *Surf. Sci.*, **226**, 204 (1992).
9. O. Nishikawa, M. Iwatsuki, S. Aoki and Y. Ishikawa, *J. Vac. Sci. Technol. B*, **14**, 2110 (1996).

REFERENCES

a. E. W. Müller and T. T. Tsong, *Field Ion Microscopy, Principles and Applications*, Elsevier, New York (1969).
b. E. W. Müller and S. V. Krishnaswamy, *Rev. Sci. Instrum.*, **39**, 83 (1968).
c. G. Binnig, H. Rohrer, Ch. Gerber and E. Weibel, *Phys. Rev. Lett.*, **50**, 120 (1983).

42

STM/STS Study of Semiconductor Clusters

Masahiko Tomitori

School of Materials Science, Japan Advanced Institute of Science and Technology, Hokuriku Nomi-gun, Ishikawa 923-1212, Japan

Purpose of the present study

It is indispensable to use microscopies with atomic resolution in order to understand cluster characteristics such as morphology, the atomic arrangement, conformation to substrates sustaining clusters, electronic states and so on. Scanning tunneling microscopy/scanning tunneling spectroscopy (STM/STS), which has rapidly promoted surface science for the last decade, is one of the most promising candidates for this purpose. It utilizes the tunneling current passing between an atomically-sharpened tip and a sample surface, and is sensitive to variation in gap separation even as small as 0.01–0.1 Å. Thus we can now obtain the atom-resolved surface topography by scanning the tip over the sample surface, while maintaining a constant current. In addition to STM, STS can reveal the surface electronic states on an atomic scale by acquiring the tunneling current-bias voltage characteristics at each position while STM scanning. In this study, we report the morphology and the electronic states of Ge clusters grown on Si(001) substrates studied by STM/STS on an atomic scale.

Contents of the present study

1. Introduction

Heteroepitaxially grown films composed of various materials with different properties exhibit a diversity of morphology and electronic characteristics depending on substrates, growth temperature and other factors. Utilizing the diversity of the grown films, the realization of fast switching devices and optoelectronic devices which are superior to the present devices is promising. Recently, Si-based heterostructures have attracted much attention, because Si microfabrication technology, which is already well established, can be easily diverted to realize the Si-hetero devices. A Si-Ge heterostructure is a potential candidate: the crystal structure of Ge is the same as that of Si (diamond structure) and Ge is solid-soluble at any rate into Si. [a] However, since the lattices between Si and Ge are mismatched by 4.2%, the heterolayers contain the strain induced by the misfit so that the film growth mode of the heterostructure is complicated. It causes a problem of non-abruptness of the superlattice. In this study, STM images [b] are first presented for the morphology change of the Ge layers grown on Si(001) surfaces at various deposited amounts and growth temperatures. The STM images clearly show the Stranski-Krastanov growth mode of the Ge layers, where the layers grow layer-by-layer up to several monolayers, followed by 3-dimensional cluster formation at higher coverages. Then, from the viewpoint of the atomic reaction sites on the film growth, STS[c] is applied to Si(001) surfaces covered with a small amount of Ge to reveal the surface electronic states on the deposited atoms. The reactivity on the atomic sites is also discussed.

2. Experimental

The experiments were carried out in an ultra-high vacuum chamber with a lab-made STM/STS and a Ge deposition source.[1] The base pressure was in the middle of 10^{-11} Torr. Rectangular pieces cut from an n-type Si(001) wafer with a resistivity of 0.01 $\Omega \cdot$cm were used as substrates for Ge deposition. The substrate was chemically cleaned by the Shiraki method, then introduced into the chamber and outgassed at a heat stage in the chamber by resistively heating at 500°C--700°C. After flashing the substrate at about 1200°C for about 15 s, the substrate temperature was gradually cooled and held at 500°C, 400°C or 300°C, then Ge was deposited by heating Ge rods wound with a tungsten-rhenium filament. The deposition rate and the amount were monitored with a quartz thickness monitor. The typical deposition rate was 1–5 ML/min. After the deposition, the sample was immediately cooled to room temperature and transferred to the STM sample stage to observe the surface topography by the STM. The sample deposited by an amount of 5 ML at 400°C was annealed at 500°C for 5 min after the STM observation, and imaged again by the STM to examine the stability of the grown Ge clusters. The images were depicted at a tip voltage of 2.0 V and a tunneling current of 0.2–1.0 nA, corresponding to the filled states.

STS measurements were performed in a way similar to that proposed by Hamers et al.[c]: while scanning the tip over the surface and keeping the tunneling current constant, current-voltage curves were obtained at each pixel (64 × 64) by varying the bias voltage stepped by 128 points over several volts at the constant tip-sample separation by inactivating the STM feedback loop. For the STS, a sample was used with a Ge coverage of about 0.1 ML at a growth temperature of 300°C. A scanning tip was fabricated by electrochemically etching a [111]-oriented W single crystal wire, and cleaned by electron bombardment in the STM chamber.

3. Ge growth mode on Si(001)[2-7]

The deposited Ge atoms dimerize on the Si(001) substrate as the Si atoms on the substrate. The dimers in a line of the [110] or [1$\bar{1}$0] direction form a dimer row, and a bundle of rows on the terrace looks like hairlines in the STM images. The direction of the Ge dimer row on the Si(001) substrate rotates by 90° to that of the Si dimer row of the substrate. This is the same way the Si dimer rows on the upper terrace rotate by 90° to those on the lower terrace with a single-atomic height step on the Si(001) 2 × 1 surface.[d] Up to 1 ML coverage, the Ge overlayer grows extending the strength of the dimer rows to form a monolayer. In spite of the 4.2% lattice mismatch, there is no distinctive difference from Si homoepitaxial growth on Si(001) except strongly induced buckling of the dimers, exhibiting $c(4 \times 2)$ and

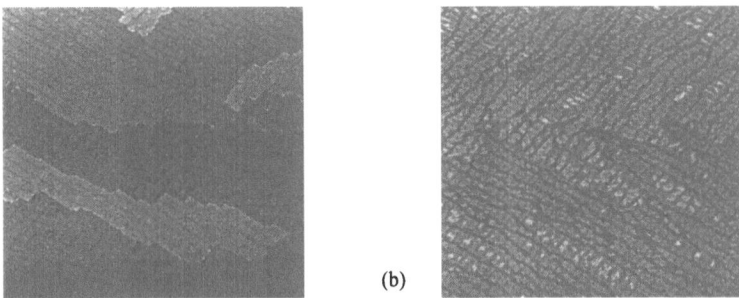

Fig. 1   STM images of the Ge overlayers grown layer-by-layer. The growth temperature was 500°C.
Scanning area: 960 Å × 910 Å. (a) Belt-like structure. The deposited amount of Ge was 2 ML.
(b) Patch-like structure. The deposited amount of Ge was 4 ML.
(From M. Tomitori *et al.*, *Appl. Surf. Sci.*, **76/77**, 323 (1994)).

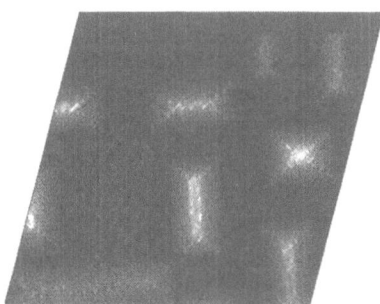

Fig. 2   STM image of hut clusters with the {015} facets of Ge on Si(001). The deposited amount of
Ge was 5 ML and the growth temperature was 500°C. The patch structures surrounded by
missing-dimer rows are seen between the hut clusters. Scanning area: 720 Å × 680 Å.
(From M. Tomitori *et al.*, *Appl. Surf. Sci.*, **76/77**, 323 (1994)).

$p(2 \times 2)$. At coverages higher than 1 ML, missing dimer rows are introduced,
cutting a chain of the dimers in a line by extracting a single dimer to minimize the
strain energy piled up along the dimer row. Then the missing dimers form a line
running perpendicular to the dimer row to minimize the strain energy through the
underneath layer. In STM images, the grown overlayer shows a belt-like structure
partitioned with the dark lines of the missing dimer row.

Figure 1(a) shows the Ge overlayer with belt structures at 2 ML at 500°C; the
step and the two-dimensional islands are seen partially. The dark lines of the
missing dimer rows run with spacings 8–11 times the lattice constant. At a higher
coverage of 4 ML, the overlayer exhibits patch structures surrounded by missing
dimer rows and trenches, which cross at a right angle owing to the 90° rotation of
the dimer rows on the single atomic overlayers, in Fig. 1(b).

Figure 2 shows the hut clusters[e] with four facets of the {015} planes grown
5 ML at 500°C. The plane indices were determined by measuring the slope angles
to the substrate and the orientation of the principal axes. The STM image exhibits
the regular zigzag patterns with a 2×1 reconstruction on the {015} and the
dimer rows with the 2×1 structure on the top (001) surface of the hut clusters.
The patch structures are also seen in flat spaces between the clusters. This change
in the growth mode from 2- to 3-dimensional can be attributed to the patch
structure induced by the misfit strain. In the 3 ML Ge overlayer, the top layer is

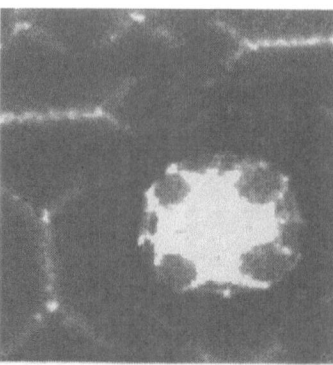

Fig. 3   STM image of a dome cluster grown by a deposition of 6.5 ML at 500°C. The cluster seems
symmetric and has several facet types. Scanning area: 500 Å × 500 Å.
(From M. Tomitori *et al.*, *Appl. Surf. Sci.*, **76/77**, 324 (1994)).

partitioned laterally by the surrounding crossing missing dimer rows. The Ge
atoms deposited on the layer cannot plug up the trenches on the boundary
consisting of the missing dimer rows: if the Ge atoms are forced to be bound with
the atoms on the boundary,the strain energy increases as discussed above. Thus
the growth of the deposited Ge atoms is restricted mainly within the patch region.
Consequently, the Ge overlayer does not grow laterally, but upward to cause 3D
cluster growth. For Ge-Si growth, the diffusion restriction due to the patch
structure plays an important role, leading the growth mode change from 2D to 3D.
This STM observation also shows the usefulness of exploring growth variation on
an atomic scale.

When the surface is almost completely covered with hut clusters by increasing
the amount of Ge, other types of clusters larger than hut clusters are found. As
shown in Fig. 3, a dome cluster on the Ge overlayer is grown with 6.5 ML at
500°C. It looks four-fold symmetric and the surface seems to be composed of the
(001), {015} and {113} facets. The hut clusters do not seem to grow lager because

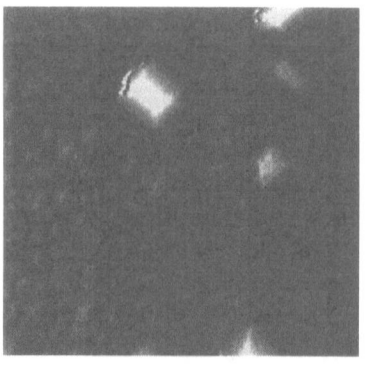

(a)   (b)

Fig. 4   STM images of the nuclei of macroscopic clusters. (a) The deposited amount of Ge was 5 ML
and the growth temperature was 400°C. The hut clusters which covered the whole surface are
found between the macroscopic clusters. Scanning area: 2400 Å × 2300 Å. (b) Enlarged
STM image of the cluster. The cluster has four facets; the indices are possibly {113}.
Scanning area: 720 Å × 680 Å.
(From M. Tomitori *et al.*, *Appl. Surf. Sci.*, **76/77**, 324 (1994)).

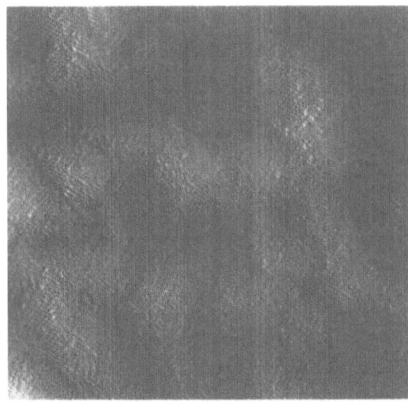

Fig. 5   STM image of the fused hut clusters grown with 8 ML Ge at 300°C. The dimer rows on the (001) surface can be observed on the top. Scanning area: 720 Å × 680 Å.
(From M. Tomitori *et al.*, *Appl. Surf. Sci.*, **76/77**, 325 (1994)).

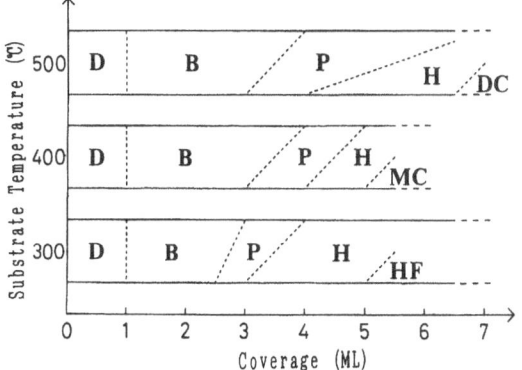

D :Dimer Chain
B :Belt Structure
P :Patch Structure
H :Hut Cluster
HF:Hut Cluster Fusion
MC:Macroscopic Cluster
DC:Dome Cluster

Fig. 6   Phase diagram of the Ge growth mode on the Si(001) substrate at 300°C, 400°C and 500°C. It is constructed from the STM images obtained in this study. The dashed lines roughly indicate the coexisting range of two phases at both sides of the lines.
(From M. Tomitori *et al.*, *Appl. Surf. Sci.*, **76/77**, 326 (1994)).

the contact of the clusters on the base plane impedes further growth. Thus, the growth mode changes from the hut cluster to the dome cluster formation. On the other hand, at a growth temperature of 400°C, the macroscopic clusters with four facets of the {113} planes grow at coverages higher than 5 ML, in Fig. 4. The symmetrical axis of the macroscopic cluster rotates 45° with respect to that of the hut cluster. Several hut clusters coalesce with the neighbors rotating the boundaries by 45°. These are possibly the nuclei of the macroscopic clusters.

At growth temperature of 300°C with a high coverage of 8 ML, the hut clusters coalesce with each other and fuse to form a rugged surface after the whole surface is covered with the clusters, as shown in Fig. 5. The regular zigzag patterns on the {015} facets and the dimer rows on the top still remain. There are no other facets even at thick coverage. At the low temperature of 300°C, the {015} facets in the strained region close to the Ge-Si interface are stable compared with the {113} facets or the macroscopic clusters. Since the {015} consists of two-dimer rows on the small (001) terraces and single atomic height steps, the strain

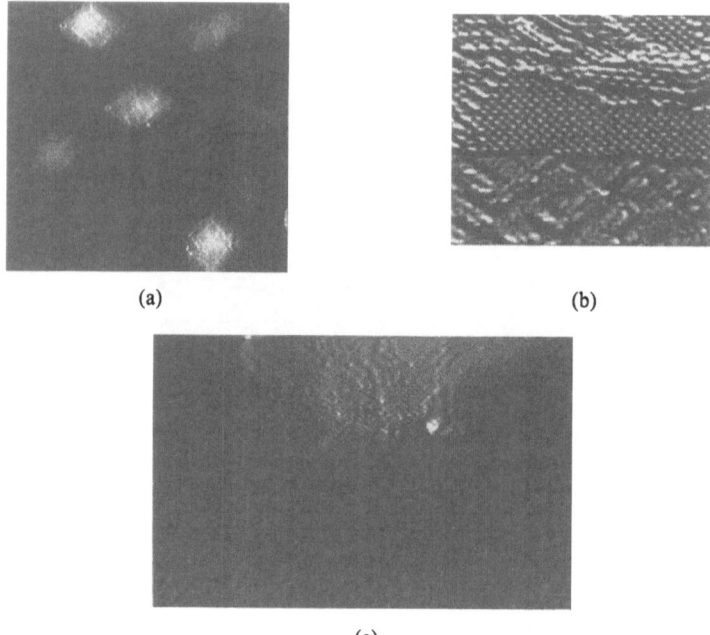

(a)   (b)

(c)

Fig. 7   (a) STM image after the annealing of the sample shown in Fig. 4 at 500°C for 5 min. Other
         types of macroscopic clusters with complicated facets grew. The hut clusters disappeared and
         the patch structure appeared again between the macroscopic clusters. Scanning area: 2400 Å
         × 2300 Å. (b) Image of the foot of the cluster. The zigzag pattern on the {015} surface are
         clearly recognized. Scanning area: 240 Å × 200 Å. (c) Enlarged image of a half of the cluster.
         Scanning area: 710 Å × 450 Å.
         (From M. Tomitori *et al.*, *Appl. Surf. Sci.*, **76/77**, 327 (1994)).

accumulated along the dimer row is easily released at the step edges. At higher
temperatures the {113} facets of the macroscopic clusters are energetically
favorable. The strain may partially relax inside the clusters except on the cluster
facets, because the facets in the top region of large and high macroscopic clusters
are far away from the strained interface between the Ge overlayer and the Si
substrate.

The detailed growth mode with the transition from 2D- to 3D-island
formation depends on the growth temperature. Figure 6 shows the phase diagram
of the Ge growth mode on Si(001) substrates; the diagram summarizes the STM
observations in this study. The morphology of the Ge clusters changes drastically
after the formation of hut clusters. The critical thickness of Ge from 2D- to 3D-
dimensional growth, that is, from the patch structure to hut clusters, is more than
3 ML, and slightly increases with temperature. Si and Ge atoms near the interface
may intermix at 500°C, and the lattice mismatch can be reduced slightly to
increase the critical thickness.

For the 5 ML Ge overlayer grown at 400°C, annealing effects for 5 min at
500°C were also examined. As shown in Fig. 7, the macroscopic clusters disappear
and new types of complicated clusters grow. These clusters have the {015} facets
on the foot of the clusters, the {113} facets on the breast, and the 2 × 1
reconstructed (001) surface with many steps on the top. The hut clusters disappear
by annealing as if the macroscopic clusters absorb the hut clusters, and are then

transformed into complicated clusters. After the hut clusters' disappearance, the patch structures with missing dimer rows reappear between the complicated clusters. The overlayer with the patch structures, whose thickness is perhaps 3 ML, may be strongly bound to the substrate, while the hut clusters may be loosely bound to the substrate and are easily transformed by the annealing. The reappearance of the {015} facets on the foot of the clusters after the annealing indicates the stability of the {015} facets in the region close to the interface. The growth mode of Ge on the (015) substrate is interesting where the critical thickness from 2D to 3D growth increases dramatically.[8]

4. Surface electronic states of Ge on Si(001) [8]

STS was applied to a Si(001) surface deposited with a small amount of Ge atoms at 300°C, forming dimers and short dimer rows. The deposited atoms diffuse along the Si dimer rows of the substrate and are easily adsorbed at the ends of the growing Ge dimer row, resulting in asymmetrical 2D growth. Thus the electronic states concerning the binding near the end of the row should play an important role in the growth mechanism. At the ends of dimer rows in the STM images, notable features are found in that some Ge atoms exist on the extension lines of Ge dimer rows but are detached from the ends of the rows with a gap of one surface lattice constant, appearing "two-eyed" (Fig. 8). The model of the atomic arrangement is also shown. Moreover, a "three-eyed" structure was also found in the image, coupling with one more atom apart from the two-eyed dimer by a gap. Normally, in STM images of filled states, the middle of a symmetric dimer appears bright, and the buckled-up atom of an asymmetric dimer appears brighter than the buckled-down atom. Accordingly, the observed two-eyed dimer, which is isolated a little from the dimer chain of the row, is probably different from the normal dimer due to the different atomic configuration and electronic states. In the empty state images, the two-eyed dimer looks like a 1×1 feature

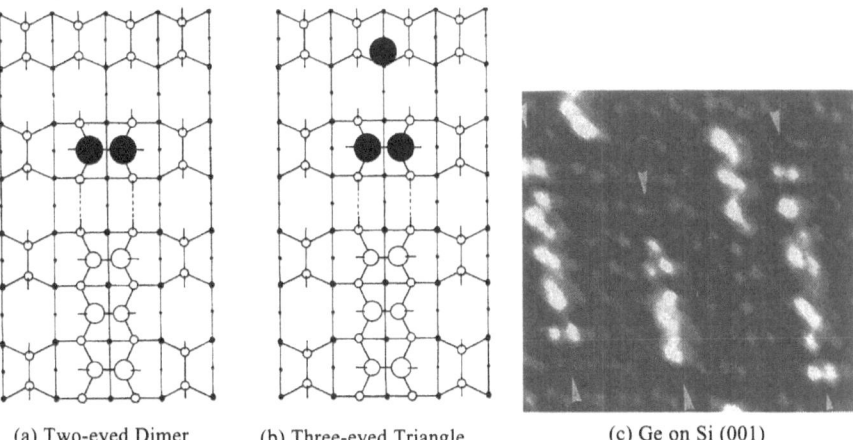

| (a) Two-eyed Dimer | (b) Three-eyed Triangle | (c) Ge on Si (001) |

Fig. 8   Schematic model of the atomic arrangements of (a) the two-eyed dimer and (b) the three-eyed triangle.
(c) STM image of the two- and three-eyed arrangements for Ge deposition on Si(001). Sample bias voltage: –2.0 V.
(From M. Tomitori *et al.*, *J. Vac. Sci. Technol. B*, **12**, 2023 (1994)).

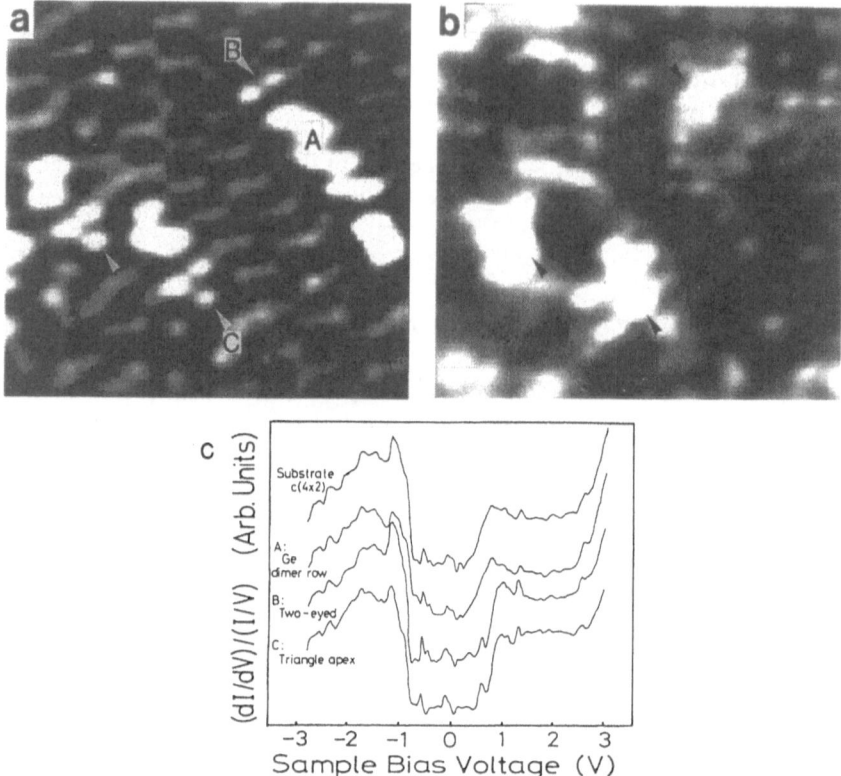

Fig. 9 Simultaneously obtained images of the topography and the normalized differential
conductance (dI/dV)/(I/V) of a small amount of Ge on Si(001), (a) and (b), respectively,
and STS spectra at several positions (c). Tunneling current: 0.5 nA, sample bias voltage: −2
V for the topography. The conductance image corresponds to the density of the empty state
from 1 to 2.5 eV.

similar to the normal dimer, where the bright spots correspond to the dangling
bond sites. The atoms of the two-eyed dimer could be configured symmetric,
estimated from the similar images in empty and filled states.

The STS spectra with the topography in Fig. 9 show that the energy gap
over the two-eyed dimer and the apex atom in the three-eyed become wider than
that of the normal dimers mainly on the empty state side. This means that the
characteristics of the dangling bonds of the two-eyed dimer and the apex atom
may change from the s and p hybridized orbital to the s-like and the p-like orbitals,
which have a wider energy gap. Then the dangling bond sites can be brighter in
the filled state image due to an increase in the density of the s-like filled state. The
peaks at the two-eyed and the apex atom near −1 eV are higher than at the normal
dimers. Furthermore, the conductance image (dI/dV)/(I/V) from 1 to 2.5 eV in the
empty states exhibits higher density states over the two-eyed and the apex atoms.
These results suggest that, compared with the normal dimers, the atoms on the
sites become stable with lowering electronic energy, where the migrating
deposited atoms tend to be captured to grow the islands anisotropically on Si(001).

5. Summary

The growth mode of Ge overlayers on Si(001) was observed on an atomic scale by STM. The change in mode from 2D to 3D, the so-called Stranski-Krastanov mode, was presented in detail, varying the amount of Ge deposition and the growth temperature. It is suggested that the strain between Ge and Si causing the patch structure plays an important role in the growth mode change. The stability of the facets {015} near the strained interface was discussed from the annealing experiment of the Ge clusters on Si(001). Moreover, the initial growth of Ge dimers was investigated on an atomic scale from the standpoint of surface electronic states using STS.

PUBLICATIONS

1. O. Nishikawa, M. Tomitori and F. Iwawaki, *Mater. Sci. Eng.*, **B8**, 81 (1991).
2. F. Iwawaki, M. Tomitori and O. Nishikawa, *Surf. Sci. Lett.*, **253**, L411 (1991).
3. F. Iwawaki, M. Tomitori and O. Nishikawa, *Ultramicroscopy*, **42-44**, 902 (1992).
4. F. Iwawaki, M. Tomitori, H. Kato and O. Nishikawa, *Ultramicroscopy*, **42-44**, 859 (1992).
5. F. Iwawaki, M. Tomitori and O. Nishikawa, *Surf. Sci.*, **266**, 285 (1992).
6. M. Tomitori, K. Watanabe, M. Kobayashi and O. Nishikawa, *Appl. Surf. Sci.*, **76/77**, 322 (1994).
7. M. Tomitori, K. Watanabe, M. Kobayashi, F. Iwawaki and O. Nishikawa, *Surf. Sci.*, **301**, 214 (1994).
8. M. Tomitori, K. Watanabe, M. Kobayashi and O. Nishikawa, *J. Vac. Sci. & Technol.*, **B12**, 2022 (1994).

REFERENCES

a. E. Kasper and H. Jorke, *J. Vac. Sci. Technol. A*, **10**, 1927 (1992).
b. G. Binnig, H. Rohrer, Ch. Gerber and E. Weibel, *Phys. Rev. Lett.*, **50**, 120 (1982).
c. R. J. Hamers, R. M. Tromp and J. E. Demuth, *Phys. Rev. Lett.*, **56**, 1972 (1986).
d. R. M. Tromp, R. J. Hamers and J. E. Demuth, *Phys. Rev. Lett.*, **55**, 133 (1985).
e. Y.-W. Mo, D. E. Savage, B. S. Swartzentruber and M. G. Lagally, *Phys. Rev. Lett.*, **65**, 1020 (1990).

43

Structure and Physical Properties of Atomic Clusters in MgO Crystals

Nobuo Tanaka [a], Tokushi Kizuka [a,b] and Kazuhiro Mihama [c]

[a] Department of Applied Physics, School of Engineering, Nagoya University
Chikusa-ku, Nagoya 464-8603, Japan
[b] Research Center for Advanced Waste and Emission Management, Nagoya University
Chikusa-ku, Nagoya 464-8603, Japan
[c] Department of Applied Electronics, Daido Institute of Technology
Minami-ku, Nagoya 457-0811, Japan

Purpose of the present study

In the present study, atomic structure of metal and semiconductor clusters less than 5 nm size in MgO single crystalline films prepared by a UHV co-deposition method were analyzed by high-resolution electron microscopy (HREM) and nm-sized electron diffraction. The physical properties of the cluster-assembled samples were measured for magnetic, electric transport, optical absorption and catalytic properties. Object of the study is to clarify the relationship between the atomic structure and physical properties.

Contents of the present study

1. Structure and physical properties of metal-MgO single crystalline composite films [1-14]
   1.1 Growth features and structure of Fe-MgO composite films
   1.2 Magnetic properties of Fe-MgO composite films
   1.3 Infrared optical absorption of Fe, Cr and Au-MgO composite films
   1.4 Structure of CaF_2-MgO composite films
   1.5 Growth features of Ni-Al alloy clusters in MgO films

2. Structure and physical properties of metals and semiconductors-CaF_2, MoS_2 [12,15,16] composite films
   2.1 Preparation and optical absorption of CdSe-CaF_2 composite films
   2.2 Preparation of Au-MoS_2 composite films

3. Metal and fullerene clusters on MgO single crystalline films [17-30]
   3.1 Dynamic behavior of Au clusters on MgO films
   3.2 Dynamic behavior of W clusters on MgO films
   3.3 Observation of C_{60} crystals and clusters on MgO films
   3.4 Observation of $Gd@C_{82}$ clusters on MgO films
   3.5 Observation of Au-Cu alloy clusters on MgO films

In this paper, the contents of 1 and 2 above are summarized.

Results of the present study

1. Structure and physical properties of metal-MgO single crystalline composite films

We have prepared metal clusters of 1–2 nm size embedded in MgO films as new type cluster assembled samples, by co-deposition of metal and MgO on heated cleaved-surfaces of NaCl. The samples were first analyzed in atomic level by using high-resolution electron microscopy, then various types of physical properties were measured.[1–8] This study is related to that of "cermet films,"[a] and the characteristic is the use of single crystalline samples. Also, the fact that the clusters are studied without exposure in air (anaerobic conditions) is related to studies of carbon nanocapsules.[b] During the study, we studied magnetic and optical properties of Fe and Cr-MgO composite films.

1.1 Growth features and structure of Fe-MgO composite films [4–8]

Figure 1 shows an electron micrograph and the diffraction pattern of a Fe-MgO composite film. The diffraction pattern shows that Fe grows with bcc structure in two kinds of epitaxial orientations $(1)(011)[100]_{Fe}//(001)[100]_{MgO}$, $(2)(001)[1\bar{1}0]_{Fe}//(001)[100]_{MgO}$. No trace of iron oxide is detected. In the micrograph, Fe clusters are identified by moiré fringes of 0.45 nm spacing. The clusters with moiré fringes as A are with the orientation (1), and the other, B, that of (2).

After annealing the sample at 660°C, migration of Fe atoms occurs and structure transformation from α-Fe into γ-Fe was found. Further annealing introduces coalescence of the transformed clusters and finally by annealing at 1000°C for a few minutes, we have obtained a spinel, $FeMg_2O_4$.

The γ-Fe sample prepared by annealing is not used for the study of magnetic properties of γ-Fe clusters embedded in MgO crystals, because α-Fe clusters with the orientation (2) remain. After many trials in the preparation of the Fe-MgO

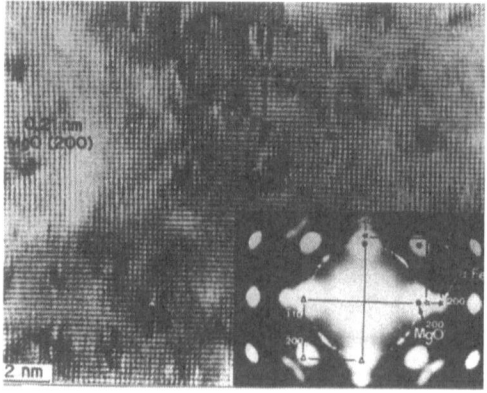

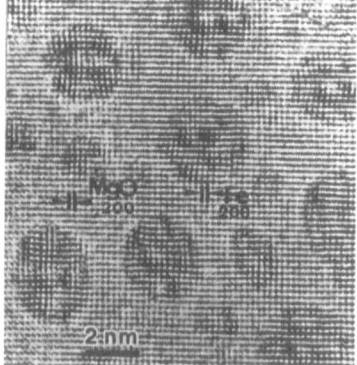

Fig. 1 HREM image and diffraction pattern of α-Fe-MgO composite films.
(From N. Tanaka et al., Mater. Trans. JIM., **31**, 588 (1990)).

Fig. 2 HREM image of γ-Fe-MgO composite films.

composite films, we have successfully obtained only γ-Fe clusters embedded in MgO films.[6] The growth conditions discovered here are (1) the deposition thickness of Fe is less than 2 nm, and (2) the ratio of deposition rates of Fe and MgO is around unity.

Figure 2 shows an HREM image of the γ-Fe clusters in MgO films. The lattice fringe inside the crystals is (200) lattice fringe of fcc-Fe ($a = 0.36$ nm). An interesting finding is the strong bending of the lattice fringe at the periphery region. The bending is due to accommodation of lattice misfit of 13% between γ-Fe and MgO. The bending structure was studied in detail using nm-area electron diffraction.

1.2 Magnetic properties of Fe-MgO composite films [5-9]

Figure 3 shows thermomagnetic curves of α-Fe-MgO and γ-Fe-MgO composite films. The curves have the following nature. (1) In lower temperature regions, they have two branches as zero-field cooling (ZFC) and field cooling (FC)

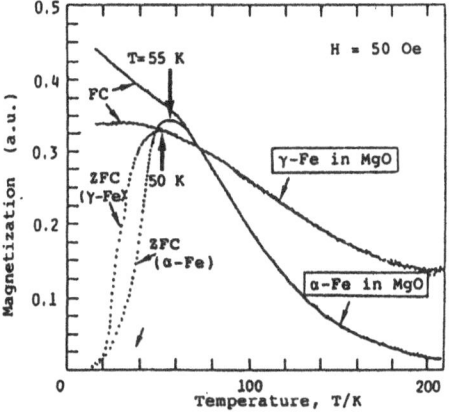

Fig. 3 Thermomagnetic curves of α-Fe-MgO and γ-Fe-MgO composite films.
(From N. Tanaka *et al.*, *Mater. Trans. JIM.*, **31**, 588 (1990)).

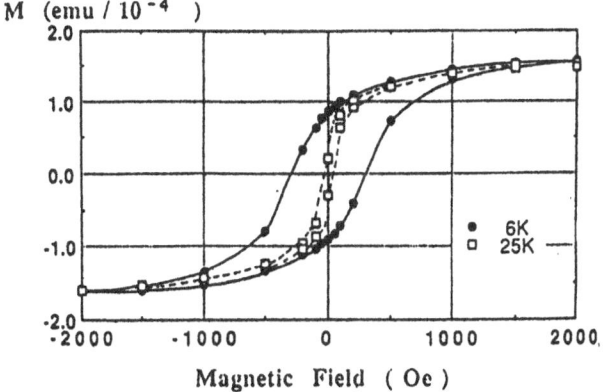

Fig. 4 Magnetization curves of γ-Fe-MgO composite films.
(From N. Tanaka *et al.*, *Acta. Met. Mater.*, **40**, S275 (1992)).

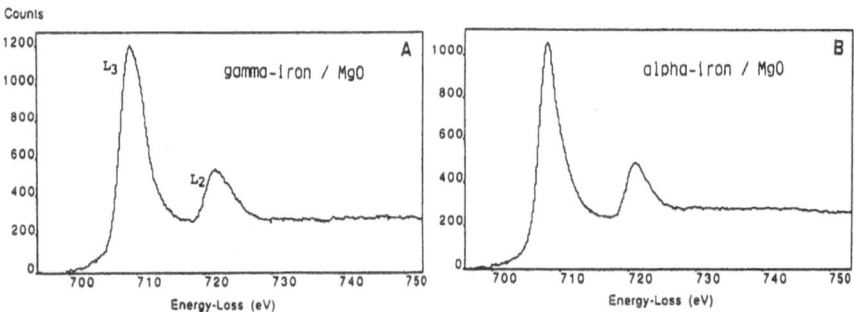

Fig. 5 L-absorption edge in EELS of γ-Fe-MgO and α-Fe-MgO composite films.
(From H. Kurata and N. Tanaka, *Microsc. Microanal. Microst.*, **2**, 183 (1991)).

curves. In the ZFC curve, we have a broad maximum around 30–100 K. (2) Increasing the deposition duration, the temperature giving the maximum moves to higher temperatures. (3) With increasing external magnetic field, the temperature moves to lower temperatures. These features are similar to those of so-called spin glass in diluted magnetic alloys.[c] However, the present sample is composed of distinct nm-sized crystallites in an MgO matrix, so it is not spin glass with randomly arranged and correlated spins. The data show that the γ-Fe crystallites in MgO show a ferromagnetic nature similar to that of α-Fe-MgO composite films.

Figure 4 shows magnetization curves of the γ-Fe-MgO composite film measured at 6 and 25 K, showing characteristic hysterisis curves with a coercive force of about 400 Oe at 6 K.

The coercive force is larger than that of bulk α-Fe, 300 Oe, extrapolated by using the power law from RT to lower temperatures. The magnetic moment per atom was measured to be 1.8 μ_B by determining the absolute amount of iron in the composite films by atomic absorption spectroscopy. This value is a little smaller than that recently obtained for fcc-Fe/Cu double layers by Matsui *et al.*[d] It has also been recently established from band calculations of iron that the magnetic transition from α-Fe into γ-Fe occur around a lattice constant of 0.36 nm. The appearance of the ferromagnetism may be due to the lattice expansion of the periphery region of the crystallites, as shown in Fig. 2.[5-7]

The ferromagnetic state was confirmed by measuring the white-line ratio in electron energy loss spectroscopy as shown in Fig. 5.[10] The ratio of the L_3/L_2 absorption edge is related to the occupancy of 3d state of electrons.[e] In the present measurement, the ratio of γ-Fe and α-Fe is 3.2 and 3.5, respectively, showing ferromagnetic (high-spin) states.

1.3 Infrared optical absorption of Fe, Cr and Au-MgO composite films

In the present composite films, clusters of 1–3 nm size are embedded in MgO single crystalline films. The effect of the clusters on the lattice vibration of the matrix is related to a problem of the so-called local mode. In the present experiment, measurements were performed for Fe and Cr clusters, which have a strong interaction with the MgO matrix, as well as Au clusters for reference.

Figure 6 shows absorption spectra of vacuum-deposited MgO films of varying

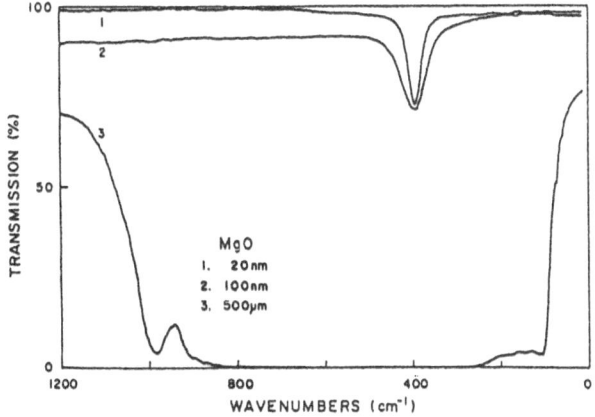

Fig. 6 Infrared absorption spectra of MgO films.
(From M. Wada and N. Tanaka, *Jpn. J. Appl. Phys.*, **29**, L1497 (1990)).

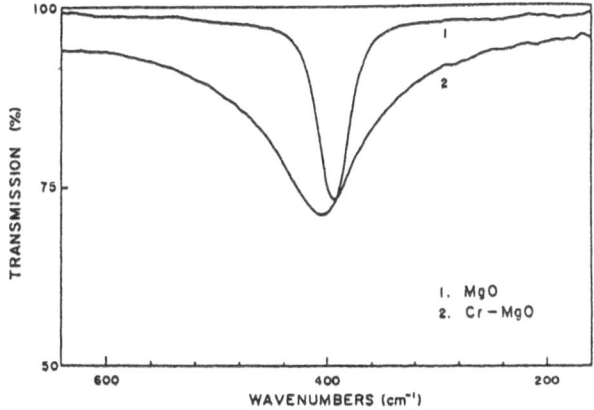

Fig. 7 Absorption spectra of Cr-MgO and MgO films.
(From M. Wada and N. Tanaka, *Jpn. J. Appl. Phys.*, **29**, L1497 (1990)).

thickness. The spectra of (1) and (2) are obtained from the present composite films, and that of (3), from a thinned bulk crystal of MgO. The peak at 400 cm^{-1} is due to the degenerated TO phonon mode of the structure of O^5_n-Fm3m. All the peaks agree with those measured previously in the bulk crystal. In the spectrum of (2), the peak is broader than that of (1), which may be due to the effect of anharmonic lattice vibrations. When the film thickness of vacuum-deposited films is larger and atomic clusters are embedded inside the single crystalline MgO matrix, the single crystalline nature becomes worse and the translational symmetry may be lost. The normal mode is then gradually disturbed, which may make the peak-width broader. Fig. 7 shows a comparison of peak width of Cr-MgO composite films and MgO films. Peak in (2) is much broader than that of only MgO. This suggests that Cr atom clusters interact with MgO matrix. The peak position shifts slightly in the Cr-MgO composite films. The origin of the shift is not yet clarified. The density of the clusters is not high, 20–30 volume%, below the percolation threshold. It is anticipated that smaller atom clusters which are not detected in the present electron microscopy may affect the lattice vibration of MgO.

1.4 Structure of CaF_2-MgO single crystalline composite films [12)

CaF_2 has a fluorite structure with a lattice constant of $a = 0.54$ nm. Recent studies focus mainly on the overlayers epitaxially grown on silicon or germanium crystals. However, little study has been performed with regard to the structure and physical properties of fine crystallites and clusters, in particular, those embedded in other crystals. In the present study, we prepared composite films of CaF_2 and MgO.

The present CaF_2-MgO composite films were prepared on NaCl substrate at 150°C or 400°C by co-deposition of the materials. The CaF_2 and MgO were evaporated from a conventional tungsten boat and an electron beam heating gun, respectively. The average deposition thicknesses are 10 nm and 40 nm, respectively. Fig. 8 shows the diffraction pattern of the sample, showing that the material grows with <111> fiber orientation can the MgO(001) substrate at 180°C. Fig. 9 shows bright and dark-field images of the sample prepared at 300°C. The diffraction pattern shows that the CaF_2 crystallites have three kinds of epitaxial orientations of [111], [001] and [112]. Among these, the [111] one is dominant. From the electron micrograph, the size of CaF_2 crystallites is 6 nm. Detailed

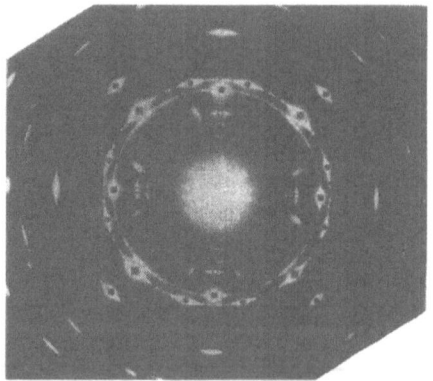

Fig. 8 Diffraction pattern of CaF_2-MgO composite films.
(From K. Mihama, S. Iwama and N. Tanaka, *Z. Phys. D*, **26**, S243 (1993)).

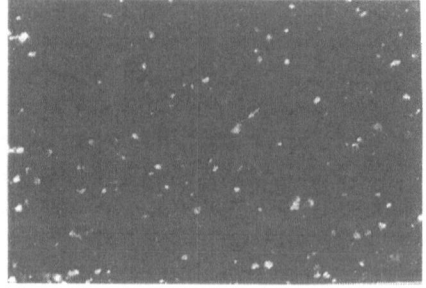

Fig. 9 Bright and dark field images of CaF_2-MgO composite films.
(From K. Mihama, S. Iwama and N. Tanaka, *Z. Phys. D*, **26**, S243 (1993)).

analysis of the diffraction intensity suggests the deficiency of fluorine atoms in CaF_2 crystals during the preparation by electron beam heating. We checked the annealing effect of the sample and observed no change in structure in the sample annealed at 300°C for a few hours.

1.5 Growth of Ni-Al alloy clusters in MgO single crystalline films [13,14]

Binary alloys have a disordered state in which constituent atoms occupy atomic sites in relation to the composition ratio above a given temperature, and an ordered state with a regular arrangement of atoms at the atomic sites below the same temperature. The phase-transition is second order and is studied theoretically as a problem of statistical mechanics. The present problem is the dependence of the transition temperature on the size of crystallites. In the present experiment, we have prepared Ni-Al clusters inside MgO crystalline films and studied the structure in relation to the order-disorder transition.

The samples were prepared on heated NaCl(001) substrates by vacuum codeposition of various kinds of Ni-Al mixtures and MgO.

First, we studied the growth feature of Ni-Al alloy clusters on MgO(001) films. Electron diffraction patterns show that the mixture of Ni-Al (7 : 3) gives a Ni_3Al phase with L_{12} structure and that of 6 : 4, a NiAl phase with B2 structure

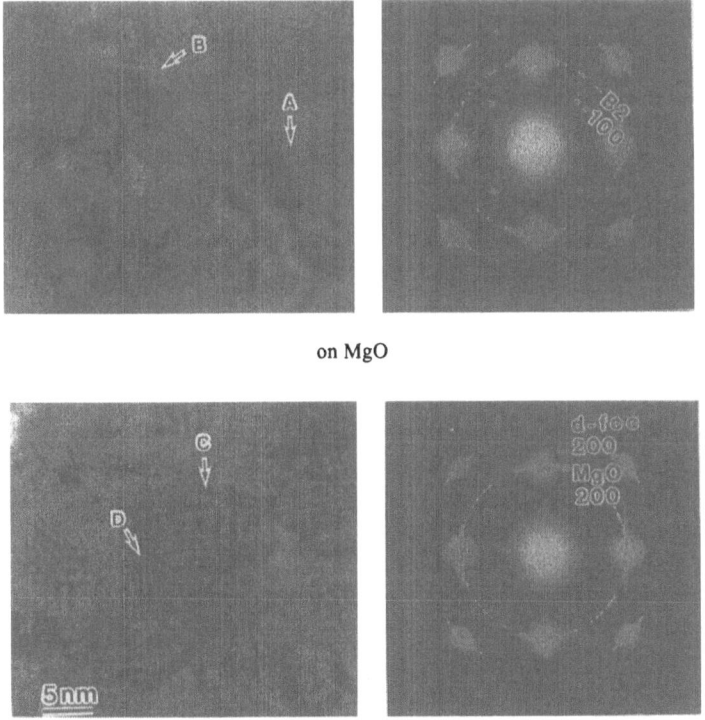

on MgO

in MgO

Fig. 10 HREM images and diffraction patterns of Ni-Al (7 : 3) on/in MgO.
(From T. Kizuka, N. Mitarai and N. Tanaka, *J. Mater. Sci.*, **29**, 5599 (1994)).

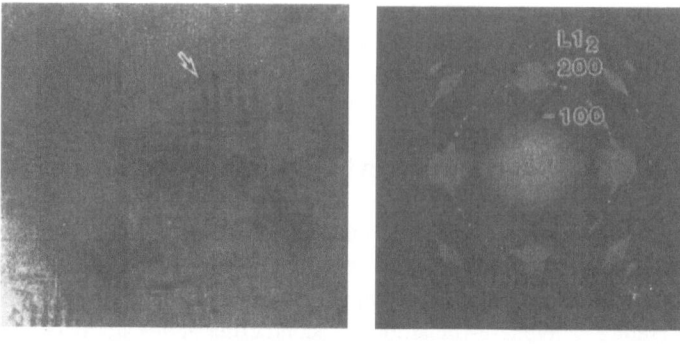

on MgO

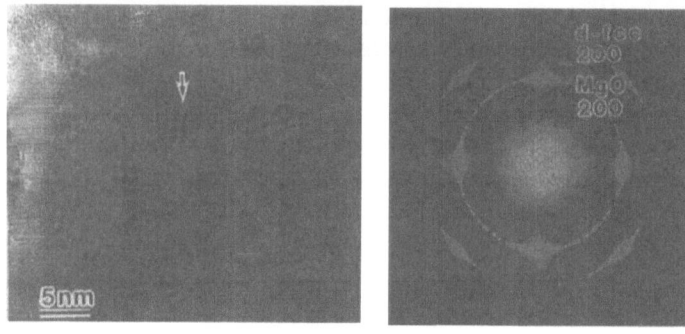

in MgO

Fig. 11 HREM images and diffraction patterns of Ni-Al (6 : 4) on/in MgO.
(From T. Kizuka, N. Mitarai and N. Tanaka, *J. Mater. Sci.*, **29**, 5599 (1994)).

both in single phases. The size of the clusters is around 1–3 nm as determined from the electron micrograph. The orientation relationships are:

$$[100](001)_{Ni_3Al}//[100](001)_{MgO}$$
$$[100](001)_{NiAl}//[100](001)_{MgO}.$$

For composite films prepared by co-deposition of the mixture and MgO, the mixtures of 7 : 3 and 6 : 4 give similarly a disordered Ni-Al alloy with fcc structure. Figs. 10 and 11 show micrographs of the deposition films and the composite films obtained from the 7 : 3 and 6 : 4 mixtures. In both cases, the size of the alloy clusters was reduced to 3 mn. During the study, we also prepared Au-Cu alloy clusters in MgO, and the size was also less than 3 nm. This result suggests that critical thickness from the ordered state to the disordered state is 4–5 nm. In composite films, the MgO matrix is composed of thin lamellar crystals with epitaxial orientation to each other, so the vertical size of the embedded metal clusters is strongly suppressed, so the ordered state cannot exist.

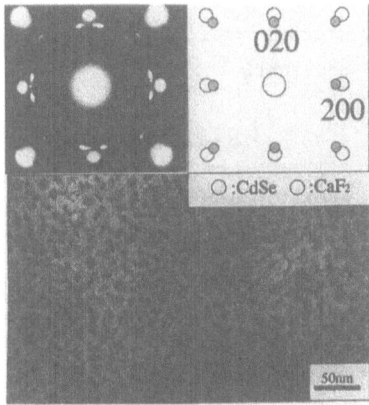

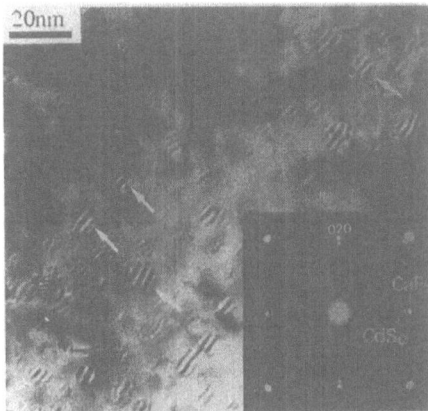

Fig. 12 Electron micrograph and diffraction pattern of as-grown CdSe-CaF$_2$ composite films. (From N. Tanaka *et al.*, *Z. Phys. D*, **26**, S225 (1993)).

Fig. 13 Electron micrograph and diffraction pattern CdSe-CaF$_2$ composite films annealed.

2. Structure and physical properties of composite films of semiconductors and ceramics

2.1 Preparation and optical absorption of CdSe-CaF$_2$ composite films [15,16]

Semiconductor clusters embedded in glass or insulator crystals show a nonlinear absorption due to the confinement effect of excitons inside the clusters. In the present study, we prepared Si-MgO, Ge-MgO, Ge-CaF$_2$, CdSe-CaF$_2$, PbS-PbTe and PbTe-MgO composite films. In this section, we describe the structure and optical absorption properties of CdSe-CaF$_2$ composite films.

The samples were prepared by UHV co-deposition of CdSe and CaF$_2$ on NaCl(001) substrates at 300°C. The CaF$_2$ was evaporated from an electron heating gun and the CdSe, from a conventional tungsten boat. The deposition thickness of CdSe and CaF$_2$ was measured to be 5–10 nm and 30 nm, respectively, using a quartz monitor. The structure of the samples was analyzed with a high-resolution electron microscope (JEM-2010). The optical absorption measurement was performed with a transmission optical spectrometer (Hitachi: U-3200) in a temperature range from −180°C to 400°C.

Figure 12 shows an electron micrograph and the diffraction pattern of as-grown CdSe-CaF$_2$ composite film. The diffraction pattern shows that CaF$_2$ is a single crystal with [001] orientation and that CdSe has a ZnS structure. The electron micrograph shows a maze pattern composed of CdSe clusters of 2–5 nm extending in the <110> direction. The origin is due to void lattices prepared through the desorption of F atoms by electron bombardment during the evaporation. The CdSe clusters are aligned along the void lattices. Fig. 13 shows an electron micrograph and diffraction pattern of the same sample after annealing at 500°C for 1 hour. The maze pattern disappears and the clusters grow up to 5 nm size. Also strain inside as-grown CdSe clusters was released by the heat treatment.

Figure 14 is an optical absorption spectrum of as-grown CdSe clusters embedded in a CaF$_2$ single crystalline film. The absorption edge is located at

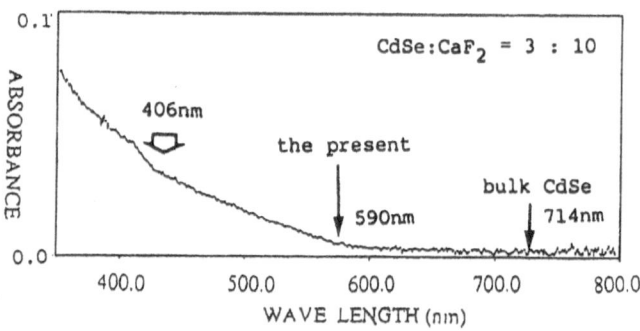

Fig. 14 Optical absorption spectrum of as-grown CdSe-CaF₂ composite films.
(From N. Tanaka *et al.*, *Z. Phys. D*, **26**, S225 (1993)).

590 nm (E = 2.10 eV), which is smaller than 714 nm (E = 1.73 eV) of the bulk crystal. That is a "blue shift". The shift is due to modification of the band by size reduction. In these semiconductor clusters, absorption is caused by the confinement effect of excitons. [f] The Bohr radius is a representative parameter to determine the phenomenon. That of CdSe is 4.5 nm. The average size of the present CdSe clusters is around 3 nm, so we do not have confinement of excitons, but the independent confinement of an electron and a hole. According to the theory of exciton confinement, many absorption peaks should be observed with a constant energy separation. The present sample, however, shows a broad peak as shown in Fig. 14. This is because the present CdSe clusters have a size distribution due to the preparation of vacuum deposition. This is in contrast with CdSe clusters embedded in glass, which have a round shape of unique size. However, these kinds of samples have a disadvantage in that they cannot be analyzed at the atomic level by HREM.

In the present study, CdSe clusters of relatively controlled size were prepared by vacuum codeposition of CdSe and CaF₂. The characteristic is its single crystalline nature for clusters and matrix. Nonlinear absorption spectra were successfully obtained, but the peaks are not distinct and rather broad, because the effective size distribution is large.

2.2 Preparation of Au-MoS₂ single crystalline composite films [12]

As described above, we studied the structure and physical properties of metal and semiconductor clusters embedded in cubic crystals of MgO and CaF₂. On the other hand, layered composites such as the present MoS₂ have a strong anisotropy along the c-axis, which produce so-called intercalation compounds. The interaction along the c-axis is not so strong due to the Van der Waals force, which gives us the possibility to insert various kinds of materials, that is, intercalations. However, the intercalation process may be limited for alkali metals. In the present study, we have tried to prepare composite films of noble metals and MoS₂ by vacuum deposition.

Au-MoS₂ composite films were prepared by RF magnetron co-sputtering (200 W, Ar-20 Pa). The target is 100 mmϕ MoS₂ with 0.3 mmϕ Au-wire. Control of the quantity of gold is performed by changing number of Au-wires on the

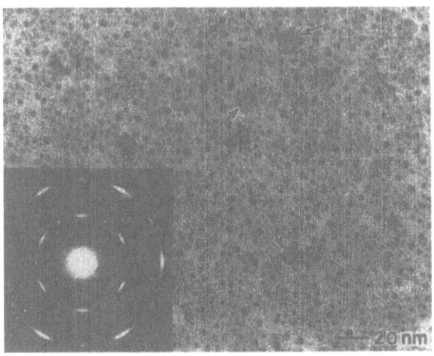 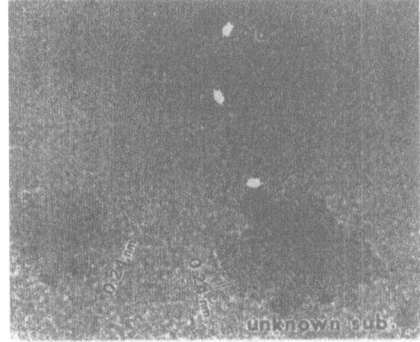

Fig. 15 Electron micrograph and diffraction   Fig. 16 HREM image of Au-MoS$_2$ composite
pattern of Au-MoS$_2$ composite films.   films.

target. The total thickness of the composite films is around 10 nm. The substrate
is (00.1) cleaved surfaces of mica and MoS$_2$.

Figure 15 shows an electron micrograph and diffraction pattern of Au-MoS$_2$
composite film prepared on a mica substrate at 300°C. The MoS$_2$ grows in a
fiber-orientation, with a lattice constant of 0.262 nm and 5% shrinkage. The
shrinkage may be due to the deficiency of sulfide atoms in the sputtering of
MoS$_2$. On the other hand, gold clusters are identified with an epitaxial orientation
against the MoS$_2$. In the electron micrograph, the size of the gold clusters is 1–2
nm. The clusters were not changed by annealing at 300°C for 6 hours. Fig. 16
shows high-resolution electron micrograph of the sample prepared on MoS$_2$(00.1)
surfaces. In this case, gold clusters and MoS$_2$ matrix films have the epitaxial
orientation with the substrate. The size of the gold clusters is around 1 nm and a
gold-sulfide compound is also identified inside the matrix. The composite films
do not show any image contrast indicating existence of strain, so the gold cluster
is located at the interlayers of flakes of thin MoS$_2$ crystals prepared by sputtering.

Summary

In the present study, we have used MgO single crystalline films as matrix
films for supporting metal clusters, and prepared metal-MgO single crystalline
composite films. The characteristics of the composite films and the studies are as
follows:

(1) The clusters embedded in MgO films are well epitaxially oriented.
(2) The size distribution of the clusters is rather sharp, around 1.5–3 nm.
(3) Oxidation of metal clusters is prevented due to embedding of the clusters in
    MgO matrix —a kind of capsule—.
(4) The measurement of physical properties is performed with samples analyzed
    in atomic level by using high resolution electron microscopy.

The results obtained by structure analysis are as follows:

(1) We have developed a new technique for identifying clusters composed of a few

atoms by using the off-Bragg diffraction condition (off-Bragg HREM method). [3]

(2) We have found the growth condition of γ-Fe clusters embedded in MgO single crystalline films. [6]

(3) We succeeded in preparing CdSe-CaF$_2$ samples in which CdSe quantum dots are embedded epitaxially in matrix film. [15,16]

In the study of physical properties of composite films, we have measured electric, magnetic, optical and catalytic properties of various kinds of composite films analyzed in advance by HREM. High-sensitivity apparatus such as SQUID were intensively used for the purpose.

The conclusion obtained from the present study is that reliable identification of clusters of less than 1 nm and detailed analysis of interfaces between metals and MgO are essential for a full understanding the metal-MgO composite films. For the latter study, analysis using nm-sized electron beam developed by the present authors is very effective. [31] The MgO vacuum-deposited film is a good supporting film for observation of atomic level objects such as clusters composed of a few atoms and carbon clusters. [17-30] We have also observed C$_{60}$ single molecules and Gd@C$_{82}$ metal fullerene molecules adsorbed on MgO film. The surface diffusion process was also studied using 1/60 time resolved HREM technique. [23,24] These advanced techniques in HREM can be fully applicable to other studies in materials science, and we can expect further advances in these research fields in the near future.

Finally, the present authors would acknowledge profs. A. Nakamura and H. Shinohara for kind collaboration in the present study.

PUBLICATIONS

1. M. Nagao, N. Tanaka and K. Mihama, *Jpn. J. Appl. Phys.*, **25**, L215 (1986).
2. N. Tanaka, M. Nagao and K. Mihama, *Ultramicrosc.*, **25**, 241 (1988).
3. N. Tanaka, K. Kimoto and K. Mihama, *Ultramicrosc.*, **39**, 395 (1991).
4. M. Nagao, N. Tanaka and K. Mihama, *Jpn. J. Appl. Phys.*, **25**, L614 (1986).
5. N. Tanaka, K. Kimoto and F. Yoshizaki and K. Mihama, *Material Trans JIM*, **31**, 588 (1990).
6. F. Yoshizaki, N. Tanaka and K. Mihama, *J. Electron Microsc.*, **39**, 459 (1990).
7. N. Tanaka, F. Yoshizaki, K. Katuda and K. Mihama, *Acta Metall. Mater.*, **40**, S275 (1992).
8. F. Yoshizaki, N. Tanaka and K. Mihama, *Z. Phys. D*, **19**, 259 (1991).
9. S. Matsuo, T. Matuura, I. Nishida and N. Tanaka, *Jpn. J. Appl. Phys.*, **33**, 3907 (1994).
10. H. Kurata and N. Tanaka, *Microsc. Microanal. Microst.*, **2**, 183 (1991).
11. M. Wada and N. Tanaka, *Jpn. J. Appl. Phys.*, **29**, L1497 (1990).
12. K. Mihama, S. Iwama and N. Tanaka, *Z. Phys. D*, **26**, S243 (1993).
13. T. Kizuka, N. Mitarai and N. Tanaka, *J. Mater. Sci.*, **29**, 5599 (1994).
14. T. Kizuka and N. Tanaka, Proc. 3rd Jap. International SAMPE Sym., p.1165 (1993).
15. N. Tanaka, T. Ishikawa and K. Mihama, *Z. Phys.*, D, **26**, S225 (1993).
16. N. Tanaka, T. Ishikawa, M. Sugita and K. Mihama, Proc. 10th European Cong. on E. M., 671 (1992).
17. T. Kizuka, T. Kachi and N. Tanaka, *Z. Phys. D*, **26**, S58 (1993).
18. T. Kizuka and N. Tanaka, Proc. 13th Int. Cong. on E. M., Vol.2 p.411 (1994).
19. T. Kizuka and N. Tanaka, *Surf. Rev. and Lett.*, **3**, 1187 (1996).
20. T. Kizuka and N. Tanaka, *Philo. Mag. Lett.*, **69**, 135 (1994).
21. T. Kizuka and N. Tanaka, Proc. 13th Int. Cong. on E. M., Vol.2 p.509 (1994).
22. T. Kizuka and N. Tanaka, *ibid.* Vol.2 p.369.
23. N. Tanaka, T. Kitagawa and T. Kizuka, *Mater. Sci. Eng.* B, **19**, 53 (1993).
24. N. Tanaka, H. Kimata and T. Kizuka, *Surf. Rev. and Lett.*, **3**, 955 (1996).
25. N. Tanaka, T. Kitagawa, T. Kachi and T. Kizuka: *Ultramicrosc.*, **52**, 533 (1993).
26. K. Mihama and N. Tanaka, Proc. 13th Int. Cong. on E. M., Vol.2 p.389.
27. T. Kizuka and N. Tanaka, *Philo Mag. Lett.*, **69**, 135 (1994).

28. T. Kizuka and N. Tanaka, Proc. 13th Int. Cong. on E. M., Vol.2 p.1303 (1994).
29. T. Kizuka and N. Tanaka, *J. Cryst. Growth*, **131**, 439 (1993).
30. T. Kizuka and N. Tanaka, *Surf. Rev. and Lett.*, **3**, 1181 (1996).
31. N. Tanaka and K. Mihama, *Ultramicrosc.*, **26**, 37 (1989).

REFERENCES

a. J. C. C. Fan, *Thin Solid Films*, **54**, 139 (1978).
b. M. Tomita *et al.*, *Jpn. J. Appl. Phys.*, **32**, L280 (1993).
c. V. Cannela *et al.*, *Phys. Rev.*, **B6**, 4220 (1972).
d. S. Mitani *et al.*, *J. Mag. Mag. Mater.*, **126**, 76 (1993).
e. T. I. Morrison *et al.*, *Phys. Rev.*, **B32**, 3107 (1985).
f. A. Nakamura *et al.*, *Phys. Rev.*, **B40**, 8585 (1989).

44

Vibrational Modes and Structure of C_{60} on Si(111) 7×7 and Graphite Surfaces —Study by HREELS-STM

Shozo Suto

Department of Physics, Graduate School of Science, Tohoku University
Aoba-ku, Sendai 980-8578, Japan

Purpose of the present study

The first goal of this study is to construct a new system to investigate the vibration modes and electronic excitations of nanostructure materials at surfaces with atomic resolution. We have set up a system combining high-resolution electron-energy-loss spectroscopy (HREELS) and scanning tunneling microscopy (STM), HREELS-STM. Our second goal is to apply the HREELS-STM as a powerful tool to real nanostructures. We investigated vibrational modes of C_{60} films adsorbed on Si(111) 7×7 and graphite surfaces using HREELS-STM, and found that the energy loss spectra are strongly structure dependent. In the case of monolayer film, the differences in energies and in oscillator strengths from a thick C_{60} film (*i.e.* bulk C_{60}) are discussed in terms of the charge transfer from silicon dangling bonds to C_{60} molecules.

Contents of the present study

1. Introduction

The buckminsterfullerene, C_{60}, and its relatives exhibit the third allotropic form of carbon in addition to diamond and graphite structures. Extensive experimental and theoretical studies have been carried out to investigate the chemical and physical properties of this new material.[a] But little is known about the properties of monolayer (ML) C_{60} films on metals and semiconductors. [1,b,c] It is important to understand the film growth and the interaction between C_{60} molecules and surfaces to develop new material functions of C_{60} molecules. The large molecular ionization potential of C_{60} molecules and the deep lowest unoccupied molecular orbital of the T_{1u} level show the possibility of its forming an acceptor-type compound.[a] The clean Si(111) surface reconstructs to be a 7×7 structure and has 19 dangling bonds in the unit cell.[d] Since such dangling bonds are chemically active, some charge transfer interaction is expected between the silicon surface and the C_{60} molecule. High-resolution electron-energy-loss spectroscopy (HREELS) is very sensitive for measuring the vibrational modes of surface adsorbates and it is the only method for measuring the vibrational modes of C_{60} monolayer film adsorbed on the surface.[e] Scanning tunneling microscopy (STM) has atomic resolution. We intend to extract information from atomic scale structures using measurements combining macroscopic beam by HREELS and microscopic tunneling current by STM.

The icosahedral structure of the C$_{60}$ molecule has the symmetry of the I$_h$ point group. The C$_{60}$ molecule has four infrared-active intramolecular vibration modes with T$_{1u}$ symmetry and ten Raman-active modes with two Ag modes and eight Hg modes.[f] The vibrational modes for bulk C$_{60}$ are measured by infrared absorption spectroscopy at $v_1 = 65.2$, $v_2 = 71.4$, $v_3 = 146.6$, $v_4 = 177.2$ meV[g] and at $v_1 = 66$, $v_3 = 147$ and $v_4 = 178$ meV by HREELS.[h,i] The v_2 mode is scarcely observed at 72 meV by HREELS due to the low resolution of 10 meV compared to infrared absorption spectroscopy of less than 0.1 meV resolution.

With interest in the mechanism of superconductor of K$_3$C$_{60}$, energy shifts to lower frequencies of v_1 and v_4 modes are observed by infrared absorption spectroscopy by increasing the alkaline doping rate.[j,k] Theoretically, the linear shifts of the two peaks are predicted in the weak electron-molecular-vibration coupling calculation by Rice and Choi.[l] Recently, Modesti *et al.* have measured the energy shift of the v_1 and v_4 modes of C$_{60}$ adsorbed on the gold surface and estimated the charge state of C$_{60}$ monolayer film with the help of theoretical and experimental results.[b]

2. HREELS-STM system

We have constructed a UHV system which consists of two analysis chambers and a sample evaporation chamber, as shown in Fig. 1. The first analysis chamber is equipped with HREELS, a rear-view LEED and Auger electron spectrometer. The second analysis chamber is equipped with STM. Samples are transferred among these three chambers under UHV condition. The HREELS is made up of two 127° cylindrical electrostatic deflectors. One is used as the monochromator and the other as the analyzer. The incident energy of the electron beam was 4.8

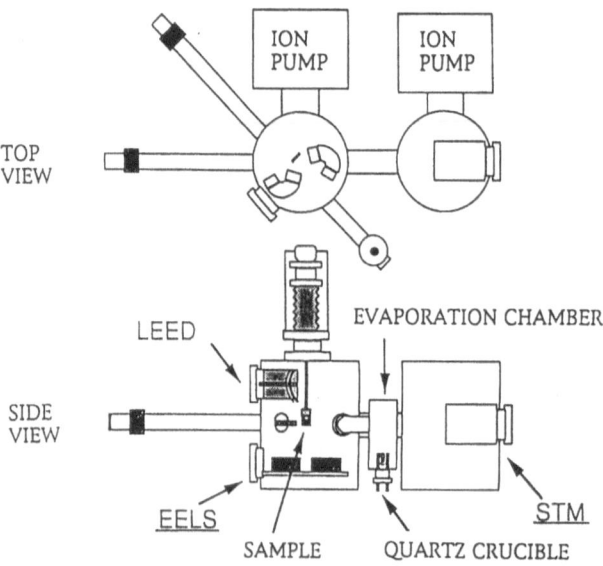

Fig. 1 Schematic drawing of the HREELS-STM system.

eV and the half width of the elastically scattered peak was about 15 meV for the clean Si(111) surface and 10 to 17 meV after the deposition of C_{60}. The spectral resolution was within ± 2 meV. The incident and the scattering angles are 65° from the surface normal (specular condition). The base pressure in the analysis chambers was 1.4×10^{-8} Pa and the evaporation chamber 1.4×10^{-7} Pa.

The C_{60} powder was prepared and purified carefully by the following procedure. First, the C_{60} powder was chromatographically separated from carbon soot. Second, the C_{60} was rinsed in tetrahydrofuran (THF) with ultrasonic cleaner in order to eliminate hydrocarbons and other impurities. Finally, C_{60} was distilled in vacuum. Without the second step, we observed a strong loss peak at 355 meV, which is due to the stretch vibration of C-H molecule. [2] The C_{60} powder was loaded in a quartz crucible and heated with tungsten wire in the evaporation chamber. The sample was carefully outgassed below 300°C for over 24 hours prior to evaporation. The thickness of the C_{60} layer was monitored by a quartz crystal oscillator. A thickness of 10 Å is estimated to be 1 ML of the C_{60} film by STM. The deposition rate was approximately 1 monolayer/min.

The Si(111) 7×7 surface was prepared chemically followed by the procedure reported by Shiraki.[m] After annealing at 850°C, the crystal surface shows sharp 7×7 spots as observed in LEED. For the STM measurements, the annealing temperature was raised to 1200°C to observe a clear 7×7 structure. The graphite surface was cleaved in air and heated in UHV at 400°C. The cleanliness of the surface was verified by Auger electron spectroscopy and by the absence of any loss peaks due to vibration modes of adsorbed molecules on Si(111) surface.

3. Structure dependent energy loss spectra

Figure 2 shows the STM image of C_{60} film of 50 Å thickness grown on the Si(111) 7×7 surface with the scale of 70×70 nm². Many small islands of C_{60} molecules are shown in the image, and all the C_{60} islands are ordered. The black part in the background is the valley of the film. The area of each small island is

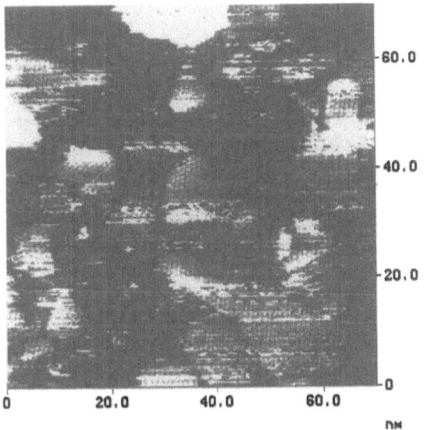

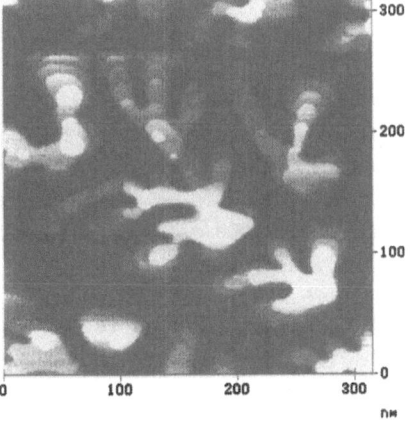

Fig. 2   STM image of C_{60} film on Si(111) 7×7 surface with thickness of 50 Å. (From S. Suto *et al.*, *Surf. Rev. Lett.*, **3**, 928 (1996)).

Fig. 3   STM images of C_{60} film on graphite surface with thickness of 50 Å. (From S. Suto *et al.*, *Surf. Rev. Lett.*, **3**, 929 (1996)).

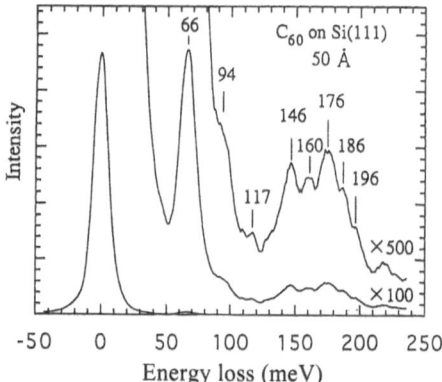

Fig. 4 Electron-energy-loss spectrum of C_{60} film on Si(111)7 × 7 with thickness of 50 Å. (From S. Suto *et al.*, *Surf. Rev. Lett.*, **3**, 929 (1996)).

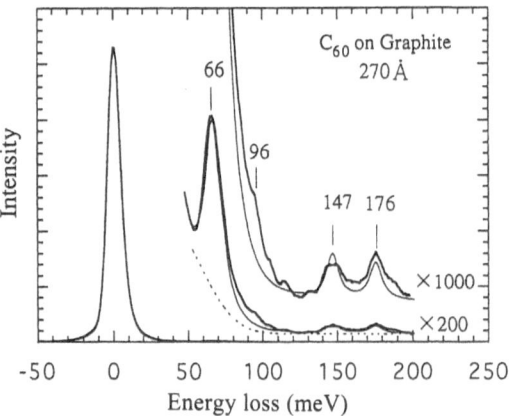

Fig. 5 Electron-energy-loss spectrum of C_{60} film on graphite with thickness of 270 Å. The thick lines show the energy-loss spectrum and the thin lines are theoretical calculation with dielectric function. The dotted line is the background.
(From S. Suto *et al.*, *Surf. Rev. Lett.*, **3**, 929 (1996)).

less than 10 × 10 nm² and the islands are made up of well-ordered C_{60} molecules. Since the tip bias of STM is 2.1 eV, the image is scanned over the wave function 2.1 eV higher than the Fermi level. Although the morphology should be dependent on the deposition rate, we use the evaporation rate of 10 Å/min. The C_{60} molecules are not ordered but form a rather smooth monolayer film on Si(111) 7 × 7 surface with the evaporation rate.[3]

Figure 3 shows the STM image of C_{60} film grown on a graphite surface with the scale of 320 × 320 nm². Large islands of C_{60} molecules are shown and the islands grow dendrically. Each island is made up of well-ordered terraces. Every part of the C_{60} islands is well ordered and the area is more than 20 × 100 nm². The area is ten times larger than that on Si(111) 7 × 7 surface.

Figure 4 shows the electron-energy-loss spectra of C_{60} molecules adsorbed on the Si(111) 7 × 7 surface with a thickness of 50 Å. Strong peaks appear at 66, 146, 160, and 176 meV, and shoulders at 94, 117, 186, and 196 meV. The peak energies and intensities are the same as the results reported in ref. 3. With the

Table 1 Peak positions (in meV) measured by HREELS, infrared absorption spectroscopy and Raman scattering. The dominant peaks are underlined. The last column is the mode assignment

Si(111)	Our results Graphite	HREELS Si(100)	Raman Thick Film	IRAS Thick Film	Assignment Ordered Film
			34		Hg
	43				T$_{2u}$
		53	54		Hg
			62		Ag
66	66	66		65	T$_{1u}$
		72		71	T$_{1u}$
			88		
94	96	94	96		Hg
117		119			
		133	136		Hg
146	147	147		147	T$_{1u}$
160		161	155		Hg
176	176	178	177	177	T$_{1u}$
186			182		Ag
196		192	195		Hg

results, 66, 146, and 176 meV peaks are assigned to the vibrational modes of dipole-active T$_{1u}$ symmetry. On the other hand, 94, 186, and 196 meV peaks are attributed to the Raman-active Hg, Ag, Hg modes, respectively.

Figure 5 shows the electron-energy-loss spectra of C$_{60}$ molecules adsorbed on the graphite surface with a thickness of 270 Å. The thick lines are the experimental data and the dotted line is the background due to the interband transition of graphite. The thin lines are the theoretical curve with classical dielectric function according to the procedure by Gensterblum et al. [i] First, we assumed the Lorentzian line shape for the dielectric function $\varepsilon(\omega)$ with the data obtained by infrared absorption spectra. Then the loss function, i.e. Im$(-1/(\varepsilon(\omega)+1))$, was calculated. We used the same constants listed in ref. i. The T$_{1u}$ modes are observed at 66, 147 and 176 meV, and Hg mode at 96 meV. Thin lines fit the experimental data quite well without the 96 meV peak, and the result is consistent with the dipole scattering theory. The spectra are the same in energy and intensity with the thickness from 10 to 400 Å except the background. The background at 50 Å is twice higher than that at 270 Å. All the energies of peaks and shoulders are summarized in Table 1 together with the results by HREELS on Si(100) [h] and those for thick C$_{60}$ films by Raman scattering [g] and by infrared absorption spectroscopy.[g] The absence of energy shifts indicates that the C$_{60}$ molecules adsorb on the graphite surface mainly by van der Waals interaction.

4. Estimation of charge transfer at monolayer film

Figure 6 shows the STM image of C$_{60}$ molecules deposited on the Si(111) 7 × 7 surface at a coverage of 1.1 ML. The C$_{60}$ molecules do not order perfectly but no islands of C$_{60}$ molecules are observed. If the deposition rate is higher than our rate, C$_{60}$ molecules make island film. The several bigger white balls are the second layer. The black part in the background is silicon surface. In the right bottom of the image, C$_{60}$ molecules grown at the edge of the silicon step are seen as a straight line. The C$_{60}$ molecules are not ordered but form a rather smooth

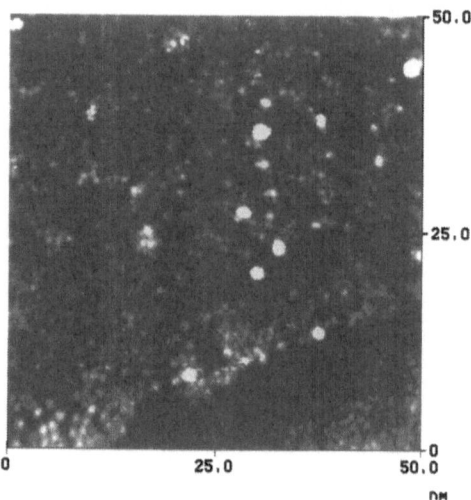

Fig. 6   STM images of C_{60} molecules on the Si(111) 7 × 7 surface. The coverage is 1.1 monolayer.
Sample bias is 4.0 V, tunneling current 500 pA.
(From S. Suto *et al.*, *Jpn. J. Appl. Phys. (Lett.) Part 2.*, **33**, L1490 (1994)).

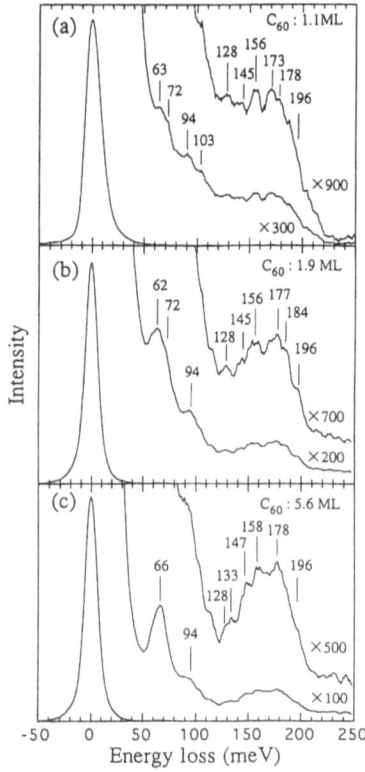

Fig. 7   Electron-energy-loss spectra of C_{60} molecules on the Si(111) 7 × 7 with coverage of 1.1 ML
(a), 1.9 ML (b) and 5.6 ML (c).
(From S. Suto *et al.*, *Jpn. J. Appl. Phys. (Lett.) Part 2.*, **33**, L1491 (1994)).

Table 2 Peak positions (in meV) and relative oscillator strengths obtained by the convolution with three
Gaussian curves at 1.1, 1.9, 5.6 ML together with the results obtained by infrared absorption
spectroscopy (IRAS) and HREELS for the thick film

1.1 ML		1.9 ML		5.6 ML		IRAS		HREELS	
peak	strength	peak	strength	peak	strength	peak	strength	peak	strength
63	1.0	64	1.0	66	1.0	65.2	1.00	65.5	1.00
72	1.3	72	0.7	72	0.3	71.4	0.29	71.7	0.27
94	1.3	94	0.5	94	0.3	—	—	93.7	0.14

monolayer film. The images are well reproduced with the same evaporation rate. This microscopic information cannot be obtained by other methods.

Figure 7 shows the electron-energy-loss spectra of C_{60} molecules adsorbed on the Si(111) 7×7 surface. The coverages are 1.1 ML (a), 1.9 ML (b) and 5.6 ML (c). At a coverage of 1.1 ML, peaks appear at 63, 72, 94, 103, 128, 145, 156, 173, 178 and 196 meV. With increasing coverage up to 1.9 ML, a strong 62 meV peak is observed and the 173 meV peak at 1.1 ML shifts to 177 meV. At 5.6 ML, the peak at 72 meV disappears and the profiles are the same as those of the thick C_{60} film (i.e. bulk C_{60}). The Gaussian curves well reproduce the experimental data and the peak energies and intensities (i.e. oscillator strengths),[3] which are shown in Table 2. At 1.1 ML, it is noticed that the intensity of the 72 meV peak is stronger than that of the 63 meV peak. At 1.9 ML, the 62 meV peak shifts to 64 meV after deconvolution. At 5.6 ML, the peak energies of 66 and 72 meV and their intensity ratio are the same as those of the thick C_{60} film.

In the film of 5.6 ML, the peaks at 66, 147, 178 meV are very similar in energy to the absorption spectrum of the thick films shown in Table 1. The electron scattering intensity of 66 meV peak shows strong dependence on the scattering angle. We, therefore, assign them as four T_{1u} dipole-active modes of v_1 = 65.2, v_2 = 71.4, v_3 = 146.6, v_4 = 177.2 meV measured by infrared absorption spectroscopy. [g] In the film of 1.1 ML, v_1 and v_4 peaks are shifted towards lower energy at 63 and 173 meV. The peaks at 94 and 196 meV at 1.1 ML correspond to the Raman active mode of 97 and 196 meV, respectively, observed by the impact scattering mechanism.[e] Since the peak at 156 meV at 1.1 ML shows up even for the thick film of 200 ML on Si(111) and Si(100), and does not appear for the film of the same thickness on graphite, the C_{60} molecule grows ordered film on the layered semiconductor of graphite.[4] The peak should be assigned to Raman active mode of 156 meV peak observed due to the surface disorder of C_{60} film.[4,i]

We have explained the energy shift and the change in oscillator strength of T_{1u} modes by the charge transfer from silicon dangling bonds to C_{60} molecules. Theoretical calculation by Rice and Choi[1] shows that the energies of v_1 and v_4 modes shift linearly towards low energy by increasing the number of electrons transferred from alkaline-metal to T_{1u} level of C_{60} molecule. The energies of v_2 and v_3 do not shift. This calculation agrees with our results. The energy shifts are plotted in a straight line in Fig. 8.

Recently, Pitcher et al.[j] and Martin et al.[k] measured the infrared absorption spectra of alkaline-doped thick C_{60} films. The observed energies of potassium-doped C_{60} are plotted in Fig. 8 with open circles. The energy shift of the v_4 mode in the bulk k_xC_{60} ($x = 0, 3, 4, 6$) is nearly linear with respect to the charge transfer of -1.8 meV/electron. The energies of v_2 and v_3 modes do not shift with charge

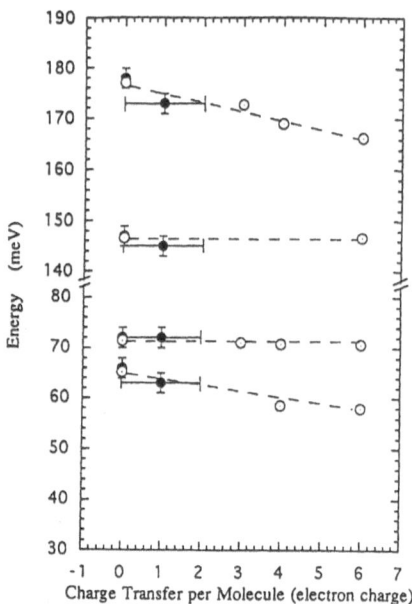

Fig. 8   Energies of the four T$_{1u}$ vibrational modes of C$_{60}$ molecules adsorbed on Si(111) 7 × 7 surface. The filled circles are our measurements and the open circles are measurements by Martin et al. [k] The straight lines are the theoretical calculation by Rice and Choi. [l] (From S. Suto et al., Jpn. J. Appl. Phys. (Lett.) Part 2., 33, L1492 (1994)).

transfer. The ν_1 mode disappears in K$_3$C$_{60}$ and shows up in K$_4$C$_{60}$ and in K$_6$C$_{60}$. First, the ratio of the oscillator strength ($\nu_1/\nu_2 \cong 0.8$) suggests that the transferred charge is more than 0 electron but not more than 3 electrons. Second, the energies of the ν_1 to ν_4 modes in the present experiment are plotted in Fig. 8 with filled circles. The values are in good agreement with the calculated shift. The resolution of our experiment, however, is not good enough, and the amount of the charge transfer is estimated to be 1 ± 1 electron(s) with Fig. 8. Although there should be some contribution from the distortion of the C$_{60}$ molecule due to the bonding to the dangling bond, the energy shift and the change in oscillator strength are consistent with the results of the charge transfer. The 7 × 7 structure has 19 dangling bonds in the unit cell. Approximately seven C$_{60}$ molecules adsorb on the 7 × 7 unit cell at 1 monolayer. Our analysis indicates that one third of the electron charge transfers from the dangling bond to a C$_{60}$ molecule.

5. Summary

We have measured the vibration modes and the structures of C$_{60}$ adsorbed on Si(111) 7 × 7 and graphite surfaces using HREELS-STM. The energy loss spectra have been found to be strongly structure dependent. In the case of the monolayer film, the energy shifts of the ν_1 and ν_4 modes and change in oscillator strength of the ν_2 mode are observed. We have interpreted the results as being due to the charge transfer from dangling bonds to C$_{60}$ molecule. The amount of the charge transfer is estimated to be 1 ± 1 electron(s) per C$_{60}$ molecule. The initial stage of C$_{60}$ film growth and reaction on Si(111) 7 × 7 and graphite surfaces have been studied by HREELS-STM. [5,6,7]

PUBLICATIONS

1. S. Suto, A. Kasuya, O. Ikeno, N. Horiguchi, Y. Achiba, T. Goto and Y. Nishina, *J. Electron Spectrosc. Relat. Phenom.*, **64/65**, 877 (1993).
2. S. Suto, A. Kasuya, O. Ikeno, N. Horiguchi, A. Waro, T. Goto and Y. Nishina, *Sci. Rep. RITU*, **A39**, 47 (1993).
3. S. Suto, A. Kasuya, O. Ikeno, C.-W. Hu, A. Waro, R. Nishitani, T. Goto and Y. Nishina, *Jpn. J. Appl. Phys.*, **33**, L1489 (1994).
4. S. Suto, A. Kasuya, C.-H. Hu, A. Wawro, T. Goto and Y. Nishina, *Surf. Rev. and Lett.*, **3**, 927 (1996).
5. C. W. Hu, A. Kasuya, S. Suto, A. Wawro and Y. Nishina, *Surf. Rev. and Lett.*, **3**, 933 (1996).
6. S. Suto, A. Kasuya, C.-H. Hu, A. Wawro, K. Sakamoto, T. Wakita, T. Goto and Y. Nishina, *Materials Science and Engineering A*, **217/218**, 34 (1996).
7. S. Suto, A. Kasuya, C.-H. Hu, A. Wawro, K. Sakamoto, T. Goto and Y. Nishina, *Thin Solid Films*, **281–282**, 602 (1996).

REFERENCES

a. See for example, *The Fullerenes*, (H. W. Kroto *et al.* eds.), Pergamon, Oxford (1993).
b. S. Modesti, S. Cerasari and P. Rudolf, *Phys. Rev. Lett.*, **71**, 2469 (1993).
c. A. Sellidj and B. E. Koel, *J. Phys. Chem.*, **97**, 10076 (1993).
d. K. Takayanagi, Y. Tanishiro, S. Takahashi and M. Takahashi, *Surf. Sci.*, **164**, 367 (1985).
e. H. Ibach and D. L. Mills, *Electron Energy Loss Spectroscopy*, Academic, New York (1982).
f. R. E. Stanton and M. D. Newton, *J. Phys. Chem.*, **92**, 2141 (1988).
g. D. S. Bethune *et al.*, *Chem. Phys. Lett.*, **179**, 181 (1991).
h. A. A. Lucas *et al.*, *Phys. Rev.*, **B45**, 13694 (1992).
i. G. Gensterblum *et. al.*, *J. Phys. Chem. Solids*, **53**, 1427 (1992); *Appl. Phys.*, **A56**, 175 (1993).
j. T. Pitcher, M. Matus and H. Kuzmany, *Solid State Commun.*, **86**, 221 (1993).
k. M. C. Martin, D. Koller and L. Mihaly, *Phys. Rev.*, **B47**, 14607 (1993).
l. M. J. Rice and H.-Y. Choi, *Phys. Rev.*, **B45**, 10173 (1992).
m. A. Ishizuka and Y. Shiraki, *J. Electrochem. Soc.*, **133**, 666 (1986).

Index